T0399786

Heavy Metals in Plants
Physiological to Molecular Approach

Editors:

Jitendra Kumar
Assistant Professor of Environmental Sciences
Dr. Shakuntla Misra National Rehabilitation University, Lucknow, India

Shweta Gaur
DD Pant Interdisciplinary Research Laboratory
Department of Botany
University of Allahabad, Prayagraj, India

Prabhat Kumar Srivastava
Assistant Professor, Department of Botany
KS Saket PG College, Ayodhya, India

Rohit Kumar Mishra
Research Assistant, Ranjan Plant Physiology and Biochemistry Laboratory
Department of Botany
University of Allahabad, Prayagraj, India

Sheo Mohan Prasad
Professor of Botany and In-charge of Ranjan
Plant Physiology and Biochemistry Laboratory
University of Allahabad, Prayagraj, India

Devendra Kumar Chauhan
Professor of Botany and In-charge of DD Pant Interdisciplinary Research Laboratory
University of Allahabad, Prayagraj, India

CRC Press
Taylor & Francis Group
Boca Raton London New York

CRC Press is an imprint of the
Taylor & Francis Group, an **informa** business
A SCIENCE PUBLISHERS BOOK

First edition published 2022
by CRC Press
6000 Broken Sound Parkway NW, Suite 300, Boca Raton, FL 33487-2742

and by CRC Press
4 Park Square, Milton Park, Abingdon, Oxon, OX14 4RN

© 2022 Taylor & Francis Group, LLC

CRC Press is an imprint of Taylor & Francis Group, LLC

Library of Congress Cataloging-in-Publication Data (applied for)

ISBN: 978-0-367-62739-3 (hbk)
ISBN: 978-0-367-62740-9 (pbk)
ISBN: 978-1-003-11057-6 (ebk)

DOI: 10.1201/9781003110576

Typeset in Times New Roman
by Radiant Productions

Preface

Heavy metals have been used since the beginning of human civilization. Some metals in the form of macronutrients and micronutrients are essential for plant growth, while some other metals that have no proven biological role in living beings are toxic even in trace amounts. Heavy metal insertion into ecosystems has occurred due to rapid industrialization and anthropogenic activities viz. vehicular traffic and other petroleum combustions, power plants, refineries, industries and rampant use of metals in fertilizers. Heavy metal contamination of the earth exaggerates due to unscientific and unsystematic measures of disposal. Heavy metals are hazardous to ecosystems due to their toxic nature, long-term persistence and bioaccumulation. The trophic transfer and consequent magnification of heavy metals in aquatic and terrestrial food webs are hazardous for living beings. Heavy metal contamination has now become a global environmental issue. The unplanned use of these metals is rapidly paving the way for heavy metal-build-up in the environment and is consequently are potentially toxic and affect both plants and animals.

Heavy metals have adversely affected the food and vegetable crops and other plants constraining survival, biomass production and yield. Heavy metals generate reactive oxygen and nitrogen species like superoxide anion, hydroxyl radical, singlet oxygen, peroxynitrite and nitrogen dioxide, etc., that induce oxidative stress, consequently, different biomolecules like DNA, proteins, lipids and bio-membranes are damaged. Heavy metal accretion in the soil is a major issue for agricultural yield due to their detrimental impact on food safety and market potential and phytotoxicity induced in crops. The present book extensively covers the prevalence of heavy metals in our ecosystem, the sources of heavy metals in the ecosystem, their toxicity in plants, particularly in crop plants, toxicity perception in plants at physiological and cellular levels, heavy metal induced oxidative stress and antioxidant defence system, heavy metal detoxification and sequestration in plants, amelioration of heavy metal toxicity by natural and synthetic hormones. Amelioration by mineral nutrition, i.e., ionomic studies, characterization of genes, i.e., genomics and molecular aspects of metal tolerance and hyperaccumulation, stress-inducible proteins and their roles under heavy metal stress (proteomics) have also been covered. Along with these plants with the capability to grow on high levels of heavy metals concentration (metallophytes) and various phytoremediation techniques like hyperaccumulators have been dealt.

This book is the result of a concerted effort of many scholars working in different parts of India along with all the six editors. The compilation of various studies in the form of an edited book enriches the existing knowledge about metal pollution and opens newer avenues to be exercised. The students and scholars would find many studies, researches, reviews of literature, views, opinions in one book. All the editors thankfully acknowledge their contributions. All the editors gratefully acknowledge the CRC Press, Taylor & Francis Group, Florida, The USA, which made possible the book in the present form. We hope that this book will remain relevant for upcoming many years for the students of stress physiology, environmental sciences, agronomy, life sciences and crop sciences at the university level.

Jitendra Kumar
Shweta Gaur
Prabhat Kumar Srivastava
Rohit Kumar Mishra
Sheo Mohan Prasad
Devendra Kumar Chauhan

Contents

1

Heavy Metals in our Ecosystem

Richa Upadhyay

ABSTRACT

Heavy metals pollution has become a global environmental issue as it has contaminated every sphere of the earth. Due to the increase in population with rapid growth in industrialization and urbanization metal contamination has increased at an alarming rate. Metal pollution is caused by both natural and human activities. Metals are not biodegradable, they are persistent in the environment, and tend to accumulate in the organism due to biomagnification. The metals contaminate all ecosystems, affect organisms residing in them, and eventually deteriorate human health through the successive transfer of metals in the entire food chain. Since plants can absorb heavy metals from both soil and water, metals hamper crop productivity and reduce its nutritional value. Thus, metal affects the health of human beings both directly and indirectly. This chapter explains the different sources of heavy metal, their forms and distribution in soil, the cycling of metals in the ecosystem, its trophic transfer through the food chain, its effect on the different ecosystems, crop productivity, and human health. This chapter comprehensively deals with the impact of heavy metals on the ecosystem in different contexts.

1. Introduction

Heavy metal contamination has now become a global environmental issue. These metals can enter the ecosystem through multiple routes. Because of the tremendous increase in the use of metals in several fields viz; industry, agriculture, domestic, and technology, metals have disturbed several ecosystems and hampered the ecological balance (Masindi and Muedi, 2018). Heavy metals are largely present in soil and aquatic ecosystems. Some of the metals called micronutrients (Mn, Co, Cr, Cu, Mo, Se, Fe, and Zn) are essential for plant growth at very low concentrations

Department of Botany, Mihir Bhoj Postgraduate College, Dadri, G.B. Nagar-203207, U.P., India
Email: ru4004@gmail.com

while macronutrients (Na, Mg, Ca, P, and S) are required in a large amount. While, some metals (Ni, As, Pb, Cd, and Hg) are toxic even in a trace amount. High metal concentration in the environment is hazardous to ecosystems due to its toxic nature, long-term persistence, and bioaccumulation (Conceiçao et al., 2012). The successive transfer of the heavy metals in the aquatic and terrestrial food web is hazardous for the residing organism and human health. The metals generate reactive oxygen species like superoxide anion, hydroxyl radical, singlet oxygen, etc., that induce oxidative stress by damaging different biomolecules like DNA, protein, lipids and, bio-membranes, etc. This stress induces inflammation that leads to the development of different cardiovascular, neurodegenerative, and other chronic diseases. In this chapter, there is a comprehensive description of the effects of different heavy metals on various ecosystems, crop productivity, and human health.

2. Sources of metal contamination

Metals enter into the ecosystem via natural as well as man-made activities (Fig. 1). Natural phenomena, i.e., weathering of rocks and volcanic eruptions are the major cause of metal pollution (Masindi and Muedi, 2018; Shallari et al., 1998). Environmental contamination also occurs through several anthropogenic sources viz; petroleum combustion, power plants, refineries, industries, and use of metals in fertilizers (Arruti et al., 2010). Heavy metals reach the soil through the use of phosphate fertilizers. In the process of extracting phosphate fertilizers from

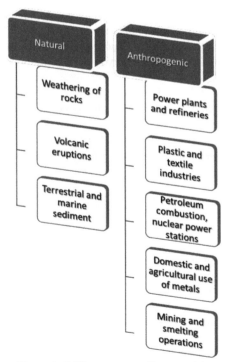

Figure 1: Different sources of heavy metals.

phosphate rock, known as acidulation, different heavy metals are produced as a by-product (Mortvedt, 1996). The application of fertilizers in agricultural soils not only contaminates the soil but also leaches into groundwater and eventually contaminates it (Dissanayake and Chandrajith, 2009). Vehicular traffic is one of the key contributors of metal pollution among different man-made sources (Ferretti et al., 1995). This is the reason behind the high metal contamination of the soil and plants situated beside the road in urban and metropolitan areas.

3. Geochemical aspects of the distribution and forms of heavy metals in soils

The heavy metal presence and distribution in the soil depend on several parameters such as the composition of parent rock and its weathering, physiochemical and biological characteristics of soil, and climatic conditions (Arunakumara et al., 2013). For example, soil receiving fertilizers and Cu fungicide are rich in metal contents compared to virgin soils (Semu and Singh, 1996). In urban areas, soils possess a high concentration of Pb (Mackay et al., 2013). Metals such as Pb, Ni, Cu, Cd, Zn, and Cr are the major soil pollutant (Hinojosa et al., 2004).

Basic knowledge of the heavy metal forms and distribution in the soil is a prerequisite for soil management practices as well as minimizing metal's effect on the ecosystem. Heavy metals may occur in the soil in four different forms— (a) dissolved, (b) exchangeable, (c) as a structural component of the soil lattices, and (d) insoluble precipitates. The dissolved and exchangeable forms are utilized first by the plants while the rest two forms may be utilized later. Different Physico-chemical properties of soil viz; organic matter content, pH, cation exchange capacity (CEC), quality and quantity of clay particles, and its redox potential determine the soil's heavy metal retaining and mobilizing ability. The availability of metals increases with an increase in these parameters and when the metal concentration reaches above the threshold, it becomes toxic. The presence of any element in the soil depends on its equilibrium between the solid and solution state of the soil, which is influenced by soil pH (Lindsay, 1979). The metal mobility increases with an increase in pH and attains a peak under mildly alkaline conditions. However, the mobility of different elements varies under different pH conditions (Fuller, 1977). Heavy metals remain in a transferrable form with the organic matter complexes and the metal's affinity for organic matter varies (Stevenson, 1982). For example, Cu remains in an unavailable form with complexes, while Cd occurs in the most available form (Kirkham, 1977). The cation exchange capacity (CEC) of the soil is determined by the organic matter content and clay particles. The soils with high clay content have higher CEC that can retain more heavy metals, e.g., montmorillonite. Thus, it is imperative to determine the forms of heavy metals (available and unavailable) in the soil to minimize metal pollution.

4. Cycling of heavy metals in ecosystems

Since metals exist in different forms in the environment and are toxic to organisms, therefore it is important to understand the biogeochemical cycle of the element.

The metal cycle involves different biological, geological, and chemical processes simultaneously. The cycle of every heavy metal involves diverse processes, e.g., transport and mobilization of elements between different living and non-living components, the bioconcentration, and biotransformation of elements by living organisms. The transport of elements from soil or water to the environment takes place through the process of volatilization very readily. However, the atmospheric residence of elements in the environment depends on many other factors. Every element has a different residence time in the environment, e.g., mercury and lead have higher residence time than others. Bioconcentration is a process that leads to the accumulation of an element in a large amount in an organism than in the environment. While, in the biotransformation, the substance is transformed into different forms by the action of microorganisms. Unlike organic pollutants, heavy metals are non-biodegradable. They may transform or become persistent contaminants in the environment that ultimately accumulate in the soil and sediments (Ahmed et al., 2018). The metals that have been accumulated in soils and sediments may enter different ecosystems. After being released from different sources, metals contaminate different components of the environment, i.e., hydrosphere, lithosphere, biosphere, and sediments. The metal contamination of aquatic and terrestrial ecosystems is deleterious to human health. Figure 2 illustrates the general biogeochemical cycle of heavy metals.

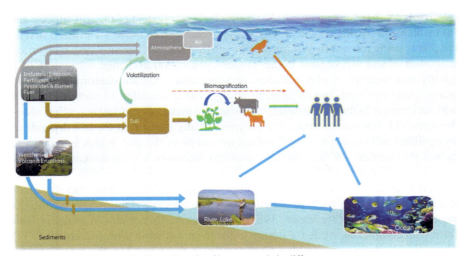

Figure 2: General cycle of heavy metals in different ecosystem.

5. Metals in the soil-plant continuum (SPC)

The contamination of plants with metals may occur through its interaction with soil, water, and air, but the plant interaction with the soil surface is the main cause of metal contamination. There is a highly significant correlation between heavy metal concentration in soil and crops (Khan et al., 2015). Metal transmission from soil to plant is an integral component of the food chain. The metals are first absorbed by the plants from the contaminated soil and then transferred to herbivores through trophic transfer. The contamination of cereals and vegetables is the main source of contamination in human beings. The texture of the soil determines the transfer of heavy metal from soil-to-plant. For example, plants growing on sandy soil contain a higher amount of metal than those growing on loamy soil (Treder and Cieslinski, 2005). Since metals are highly mobile in sandy soil as compared to loamy soil, therefore former shows a higher rate of bioaccumulation. Several indices describe the transfer of metals from soil to plant, e.g., bioconcentration factor (BCF), transfer factor (TF), and pollution load index (PLI). The BCF value is the best parameter for determining soil to plant transfer of metals. Various studies have reported the BCF values of different crops in the following order leafy vegetables > tuberous crops > horticultural and fruit crops (Chang et al., 2014; Yang et al., 2018).

Bioconcentration Factor = Concentration in crop or vegetable/concentration in soil

6. Metal availability and accumulation in plants

Metal bioavailability to plants is affected by metal-organic matter complex, their distance from the source, and types of metals. The presence of organo-arsenic compounds reduces the availability of metals (Juhasz et al., 2006). The metal bioavailability to plants is higher in high-traffic areas (Galal and Shehata, 2015), while, soil treated with compost reduces the availability of metals (Smith, 2009). Liu et al. (2005) reported that metal contaminated soils have the highest bioavailability of Cd, while the lowest bioavailability for As.

The metal chelators, e.g., organic acids, phenolics, and siderophores, released by plants and microbes enhance the bioavailability of metals (Pilon-Smits, 2005). These chelators cause the release of metal cations from soil and in turn make it more bio-available to plants. Volkering et al. (1998) reported that bacteria releases biosurfactants, e.g., rhamnolipids that convert hydrophobic substances into hydrophilic. Plant extracts also release some lipophilic compounds that increase the solubility of heavy metals (Siciliano and Germida, 1998). Heavy metal bioavailability varies for different plant organs, for example, it is highest for roots as compared to other plant organs (Goni et al., 2014). Thus, metal accumulation in plants depends upon different factors viz; metal concentrations, bioavailability, the efficiency of metal absorption, season and species of plant (Khan et al., 2008; Yang et al., 2009).

Based on heavy metal accumulation, there are three categories of plants- accumulator, hyperaccumulator, and excluders. The hyperaccumulators or metallophytes absorb and sustain a high concentration of metals from the soil by three adaptations—

(1) high efficiency of heavy metal absorption,
(2) high root-to-shoot translocation ratio, and
(3) detoxification of heavy metal.

Generally, leafy vegetables are more efficient in absorbing heavy metals than nonleafy vegetables (Yu et al., 2006).

7. Metal contamination at different trophic levels

Metals present in the environment are assimilated by various living organisms. As a result, metals reach every organism of entire food chains. While the concentration of metals varies at successive trophic levels, i.e., it may increase or decrease. Both biotic and abiotic component leads to the transmission of metals to the body of an organism. The rate of metal assimilation in an organism depends on its rate of absorption and elimination from the body. Metal retention in an organism depends on the type of metal and strategies adopted for heavy metal regulation and detoxification. For example, methyl mercury can accumulate in large amounts in the food chain due to biomagnification and its lipophilic nature. Heavy metals have varying half-lives indifferent organisms. Different food sources viz; cereals, vegetables, and fishes lead to the accumulation of toxic metals in human beings. Metal contamination of water bodies and agricultural fields causes bioaccumulation of these metals in residing organisms and crops, respectively. Figure 2 illustrates the transfer of heavy metals from aquatic and terrestrial ecosystems to humans. Biomagnification of metals is reported in several food chains,in which, organisms of higher trophic levels are at greater risk. Thus, it is imperative to regularly monitor the food chains of human beings for bioaccumulation and biomagnification. Furthermore, to avoid metal pollution in the food chain, untreated waste waters should not be drained into water bodies and agricultural lands.

8. Effect of metals on aquatic ecosystem

Contamination of the aquatic ecosystem by metals is a serious environmental problem and it is harmful to aquatic organisms and human beings (Rezania et al., 2016). All aquatic ecosystems viz; freshwater and marine are susceptible to metal pollution (Ali et al., 2019). Since metals are consistently present in the environment and toxic in very trace amount, it affects aquatic ecosystem very severely. The metals are resistant to microbial degradation and remain present in the aquatic ecosystem permanently (Bruins et al., 2000). The majority of metals are non-biodegradable, and once introduced in the ecosystem, they accumulate in the aquatic environment, sediments, and are consumed by aquatic organisms. After absorption of heavy metals by aquatic organisms its concentration increases in the food chain, so the top consumers especially human beings are at great risk of metal contamination (Baby et al., 2010; Saha et al., 2013). It disturbs the ecological balance and species diversity of the aquatic ecosystem (Farombi et al., 2007). Heavy metals generally, enter the aquatic environment through different natural and human activities. The toxic metals such as Cu, Cr, Cd, Pb, Hg, Pb, Zn, and Ni are present in large amounts in the aquatic environment (Hashem et al., 2017). In the aquatic environment, sediments act as both sources and sinks of metals. Continuous deposition of metals in sediment leads to the contamination of groundwater (Sanyal et al., 2015). The metals deposited in riverine sediments are deleterious to benthic organisms (Decena et al., 2018). The accumulation of metals in the body of aquatic organisms leads to several diseases. The high concentration of metals hinders the development of aquatic organisms viz; phytoplankton, zooplankton, and fishes (Atici et al., 2010; Bere et al., 2012). Among all aquatic organisms, fishes are most susceptible to heavy metal pollution

(Olaifa et al., 2004). The most common toxic pollutants for fishes are Pd, Cu Fe, Cd, Zn, and Mn, which either act directly or synergistically. Fishes and molluscs are most commonly used to evaluate the metal toxicity of aquatic ecosystems because they often accumulate large amounts of certain metals (Farkas et al., 2002). Metals affect the different physiological processes and enzymatic metabolism of aquatic organisms like oxygen consumption, reproductive processes, molluscs development, and byssus formation. Heavy metals induce deformities in the fish population, impacting their survival, growth rates, and nutritional value (Sfakianakis et al., 2015). Metals act as neurotoxins in fishes and interrupt their chemical communication (Baatrup, 1991). The metals cause gill necrosis and the degeneration of the liver in the fishes and crustaceans (Brraich and Kaur, 2015; Sevcikova et al., 2016). The metals inhibit the respiratory enzymes of crustaceans and cause death. After chronic exposure metals inhibited several antioxidant enzymes in the fishes and crustaceans (Mishra and Mohanty, 2008; Jiang et al., 2016). The metals inhibit the organism's growth and development by inhibiting enzymes involved in the synthesis of protein and cell division. Heavy metal exposure causes several histopathological damages in the exposed organism by damaging protein, lipid, and DNA through ROS generation (Das et al., 2019). Metals such as As, Cu, Cd,Zn, Ni, Pb, Zr, and Ti showed varying toxicity to *Cyprinus carpio*, which could lead to great loss for fisheries (Gheorghe et al., 2017). Moreover, the authors also found that the metals involved in the metabolic processes showed more bioaccumulation than the toxic ones.

9. Effect of metals on the soil

The metals present in the soil impact every component of the biosphere (Briffa et al., 2020). High metal concentration disturbs the microbial community of the soil and reduces their fertility. Metals alter various properties of the soil, e.g., color, porosity, pH, and natural chemistry, and also contaminate the water (Muchuweti et al., 2006; Gupta et al., 2012). Metal toxicity not only affects plant productivity but also changes various properties of the microbial population, i.e., number, size, composition, and activity (Yao et al., 2003). Change in the activity of the microbial population brings about change in the enzymatic activities of the microbes (Shun-hong et al., 2009). Thus, metals exert their toxic effects by modifying the number and activities of soil microorganisms (Xie et al., 2016). Metal exposure for a long duration increases the resistance of microbes, which are further helpful in the remediation of polluted systems (Mora et al., 2005). Thus, it is imperative to assess the properties of soil microbes exposed to metals for a longer duration. It has been reported that heavy metal exposure causes a change in species diversity and biomass of the soil microbes (Chen et al., 2010). Different metals affect the toxicity of the enzymes indifferent ways (Karaca et al., 2010). For example, Cd exerts more toxicity than Pb to enzymes because the latter possess high mobility and less affinity to the soil. Similarly, the β-glucosidase enzyme is more inhibited by Cu than cellulase. Pb inhibits the activities of different antioxidant enzymes viz; acid phosphatase, catalase, invertase and, urease, etc. Arsenic inhibits phosphatase and sulfatase but could not affect urease. Similarly, Cd inhibits the alkaline phosphatase, arylsulfatase, protease, and urease but it has no significant effect on the invertase. The degree of inhibition depends on the sensitivity of the enzymes to heavy metals.

10. Effect of heavy metals on crop productivity

The plants absorb metals from the soil that are either present in soluble form or dissolved by root exudates (Blaylock and Huang, 2000). Some of the metals are essential for growth of the plant while others are non-essential. Although essential elements are required for the growth, an excessive amount of these elements also become toxic to plants. The inherent ability of the plants to absorb essential metals also allows them to absorb nonessential ones (Djingova and Kuleff, 2000). All metals whether essential or non-essential induce common toxic systems on the plant such as chlorosis, inhibition of growth and photosynthesis, lower biomass, senescence, and death. Plants growing in metal-contaminated areas generally accumulate more heavy metals, which in turn, contaminate the entire food chain. As metals cannot be destroyed, when concentrations of these metals in the plant increase above the threshold, they adversely affect crop productivity.

Heavy metals interact with plants in two ways. In one way heavy metals exert a negative influence on the plant while in another way plants develop several resistant mechanisms against heavy metal damage. The heavy metal effect on the plant depends on several factors, such as environmental conditions, pH, media composition, fertilizer application, and plant species (Cheng, 2003). Plants develop several resistant mechanisms to protect themselves against metal toxicity such as combining metals with proteins, expression of the detoxifying enzymes and, nucleic acid.

Metals induce toxicities in plants by different mechanisms. For example, (i) metals competition with the nutrient cation (ii) interaction of metals with different functional groups of proteins, (iii) replacement of essential cations from specific binding sites of the enzymes, and (iv) generation of reactive oxygen species (ROS) (DalCorso et al., 2013). The metals affect crop productivity both directly and indirectly by inhibiting various cellular, molecular, biochemical, and physiological activities of the plants. Metals reduce plant productivity by inhibiting different physiological processes such as seed germination, photosynthetic efficiency, enzyme activity, plant growth, and yield. The different direct and indirect toxic effect of the plants affecting plant productivity is summarized in Fig. 3.

Since roots of the plants are the first that come in contact with heavy metals, therefore, several studies have been done to assess the impact of metals on roots. The metals suppress root growth by decreasing the mitotic activity in several plant species (Pena et al., 2012; Thounaojam et al., 2012). The metals affect the plant by changing the level of growth hormones, growth parameters, and yield (Silva, 2012; Yuan et al., 2013; Anjum et al., 2014). Therefore, it is evident from these findings that metals inhibit root growth which in turn hamper water and mineral absorption, and their transport to the aboveground parts. The ill-developed root ultimately inhibits shoot growth and biomass accumulation.

Metal absorption causes stunting of plant organs, and induces chlorosis in younger parts mainly leaves that may further out spread to the older ones (Srivastava et al., 2012). Metal toxicity causes enzyme and protein denaturation, induces hormonal imbalance, DNA and protein synthesis (Silva, 2012; Wani et al.,

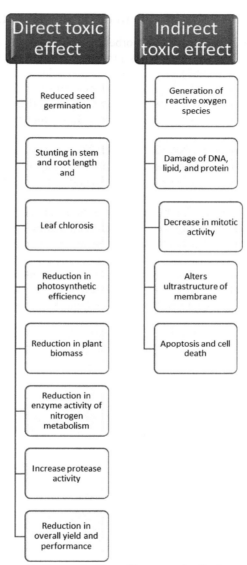

Figure 3: Direct and indirect toxic effects of heavy metals affecting crop productivity.

2012). Cd, Cu, Mn, Zn, and Ni are the main inhibitors of photosynthesis. It alters the ultrastructure of membranes and as a result hinders several plant processes (He et al., 2012; Ali et al., 2013a,b). Metals retard nitrogen metabolism, enzyme activity, growth, and development in plants (Lea and Miflin, 2004). It affects seeds by reducing their germination, seedling growth, dry weight, nutrient loss, and altered sugar and protein metabolisms (Pourrut et al., 2011). Besides individual effects, the antagonistic and synergistic interaction between heavy metals is also toxic to plant.

11. Threat to human health due to heavy metal contamination

The metals present in soil are first absorbed by the crops and then transferred to different food chain organisms. Metals produce toxic effects if not metabolized properly (Sobha et al., 2007). They accumulate in the soft tissues and chronic exposure results in a long-lasting effect. Metals affect different cell organelles, e.g., mitochondria, endoplasmic reticulum, lysosome, cell membrane, and nuclei. Metal damages various enzymes involved in cellular metabolism, detoxification, and repair of DNA damage (Wang and Shi, 2001). Beyersmann and Hartwig (2008) reported that metal induces DNA damage and changes conformations of the protein that eventually leads to cell cycle modification and apoptosis. The toxicity and carcinogenic properties of the several metals depend on their ability to generate ROS that damages membranes and biomolecules (Tchounwou et al., 2001; Tchounwou et al., 2004a,b; Yedjou and Tchounwou, 2006; Sutton and Tchounwou, 2007; Patlolla et al., 2009). The mechanism of toxicity of different heavy metals on human health is summarized in Fig. 4. The five metals namely arsenic, cadmium, chromium, lead, and mercury are most hazardous to human health even at a low concentration.

It has been reported that heavy metals exposure induces several neurological problems, e.g., insomnia, depression, irritation, and gastric troubles (Agnihotri and Kesari, 2019). Metal exposure retards development, kidney damage, and death on

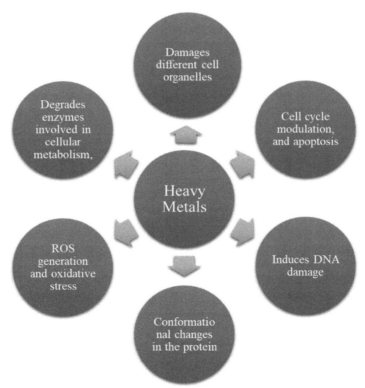

Figure 4: Mechanism of heavy metal toxicity on human health.

extreme exposure (Lentini et al., 2017). Methyl mercury absorption leads to severe health implications after intake of aquatic barriers. It can damage the lungs and kidneys as well. Baba et al. (2013) reported that cadmium accumulation in human bodies causes kidney malfunction and a severe bone disease called Itai-Itai. High cadmium concentration leads to carcinoma and stones in the gallbladder (Shukla et al., 1998; Rai and Tripathi, 2007). Lead (Pb) is a well-known physiological and neurological toxin. Acute Pb exposure leads to malfunctioning of the liver, brain, kidney, reproductive, nervous, and cardiovascular system (Odum, 2000; Assi et al., 2016). While chronic exposure leads to anemia, fatigue, anoxia, gastrointestinal trouble, and urinary tract infection. Lead generally affects children and retards brain development.

Arsenic is a well-known carcinogen that affects several organs. Acute toxicity symptoms are nausea, vomiting, and diarrhea while chronic toxicity leads to multiorgan damage (Ratnaike, 2003). It has been reported to inhibit approximately 200 enzymes involved in cellular metabolism. Cadmium induces toxicity in the pancreas, endocrine, cardiovascular, immune, and reproductive systems (Buha et al., 2017, 2018; Mezynska and Brzoska, 2018).

12. Conclusion and future perspectives

Heavy metals are the ubiquitous environmental pollutants that exist in different ecosystems. Metals are hazardous due to their persistent occurrence, toxic nature, and bio-accumulative properties. The successive transfer of the metals through food chains impacts the life of aquatic organisms, plants, wildlife, and human beings. Thus, it is imperative to regularly assess the concentration of metals in different ecosystems. Moreover, there is a need to minimize metal pollution in different ecosystems viz; aquatic and terrestrial to protect the crop productivity, resident biota, and their consumers. There must be effective wastewater treatment before its discharge into water bodies and public awareness about metal toxicity.

References

Agnihotri SK, and Kesari KK. 2019. Mechanistic effect of heavy metals in neurological disorder and brain cancer. *In*: Kesari K (eds.). *Networking of Mutagens in Environmental Toxicology. Environmental Science and Engineering*. Springer, Cham. https://doi.org/10.1007/978-3-319-96511-6_2.

El-Kady AA, and Abdel-Wahhab MA. 2018. Occurrence of trace metals in foodstuffs and their health impact. *Trend Food Sci Tech* 75: 36–45.

Ali B, Qian P, Jin R, Ali S, Khan M, Aziz R, Tian T, and Zhou W. 2013a. Physiological and ultra-structural changes in *Brassica napus* seedlings induced by cadmium stress. *Biol Plant* 58: 131–138.

Ali B, Wang B, Ali S, Ghani MA, Hayat MT, Yang C, Xu L, and Zhou WJ. 2013b. 5-Amino levulinic acid ameliorates the growth, photosynthetic gas exchange capacity, and ultrastructural changes under cadmium stress in *Brassica napus* L. *J Plant Growth Regul* 32: 604–614.

Ali H, Khan E, and Ikram Ilahi I. 2019. Environmental chemistry and ecotoxicology of hazardous heavy metals: environmental persistence, toxicity, and bioaccumulation. *Hind J Chem* 2019: 1–14.

Anjum NA, Gill SS, Gill R, Hasanuzzaman M, Duarte AC, Pereira E, Ahmad I, Tuteja R, and Tuteja N. 2014. Metal/metalloid stress tolerance in plants: role of ascorbate, its redox couple, and associated enzymes. *Protoplasma* 251: 1265–1283.

Arruti A, Fernández-Olmo I, and Irabien A. 2010. Evaluation of the contribution of local sources to trace metals levels in urban PM2.5 and PM10 in the Cantabria region (Northern Spain). *J Environ Monit* 12: 1451–1458.

Arunakumara KKIU, Walpola BC, and Yoon MH. 2013. Current status of heavy metal contamination in Asia's rice lands. *Review Environ Sci Technol* 12: 355–377.

Assi MA, Hezmee MNM, Haron AW, Sabri MYM, and Rajion MA. 2016. The detrimental effects of lead on human and animal health. *VetWorld* 9: 660–671.

Atici T, Obali O, Altindag A, Ahiska S, and Aydin D. 2010. The accumulation of heavy metals (Cd, Pb, Hg, Cr) and their state in phytoplanktonic algae and zooplanktonic organisms in Beysehir Lake and Mogan Lake, Turkey. *Afr J Biotechnol* 9: 475–487.

Baatrup E. 1991. Structural and functional effects of heavy metals on the nervous system, including sense organs, of fish. *Comp Biochem Physiol Part C: Comp Pharmacol* 100: 253–257.

Baba H, Tsuneyama K, Yazaki M et al. 2013. The liver in itai-itai disease (chronic cadmium poisoning): pathological features and metallothionein expression. *Mod Pathol* 26: 1228–1234.

Baby J, Raj SJ, Biby ET, Sankarganesh P, Jeevitha MV, Ajisha Su, and Rajan SS. 2010. Toxic effect of heavy metals on aquatic environment. *Int J Biol Chem Sci* 4: 939–952.

Bere T, Chia MA, and Tundisi JG. 2012. Effects of Cr III and Pb on the bioaccumulation and toxicity of Cd in tropical periphyton communities: implications of pulsed metal exposures. *Environ Pollut* 163: 184–191.

Beyersmann D, and Hartwig A. 2008. Carcinogenic metal compounds: recent insight into molecular and cellular mechanisms. *Archiv Toxicol* 82: 493–512.

Blaylock MJ, and Huang JW. 2000. Phytoextraction of metals. pp. 53–70. *In*: Raskin I, and Ensley BD (eds.). *Phytoremediation of Toxic Metals: Using Plants to clean up the Environment*. Wiley, New York, USA.

Briffa J, Sinagra E, and Blundell R. 2020. Heavy metal pollution in the environment and their toxicological effects on humans. *Heliyon* 6: e04691. doi: 10.1016/j.heliyon.2020.e04691.

Brraich OS, and Kaur M. 2015. Ultrastructural changes in the gills of a cyprinid fish, *Labeo rohita* (Hamilton, 1822) through scanning electron microscopy after exposure to lead nitrate (Teleostei: Cyprinidae). *Iran J Ichthyol* 2: 270–279.

Bruins MR, Kapil S, and Oehme FW. 2000. Microbial resistance to metals in the environment. *Ecotoxicol Environ Saf* 45: 198–207.

Buha A, Matovic V, Antonijevic B, Bulat Z, Curcic M, Renieri EA, Tsatsakis AM, Schweitzer A, and Wallace D. 2018. Overview of cadmium thyroid disrupting effects and mechanisms. *Int J Mol Sci* 19: 1501.

Buha A, Wallace D, Matovic V, Schweitzer A, Oluic B, Micic D, and Djordjevic V. 2017. Cadmium exposure as a putative risk factor for the development of pancreatic cancer: Three different lines of evidence. *Bio Med Res Int* 2017: 1981837.

Chang CY, Yu HY, Chen JJ, Li FB, and Zhang HH. 2014. Accumulation of heavy metals in leaf vegetables from agricultural soils and associated potential health risks in the pearl river delta, South China. *Environ Monit Assess* 186: 1547–1560.

Chen GQ, Chen Y, Zeng GM, Zhang JC, Chen YN, Wang L, and Zhang WJ. 2010. Speciation of cadmium and changes in bacterial communities in red soil following application of cadmium-polluted compost. *Environ Eng Sci* 27: 1019–1026.

Cheng S. 2003. Effects of Heavy metals on plants and resistance mechanisms. *Environ Sci Pollu Res* 10: 256–264.

Conceiçao Vieira M, Torronteras R, Córdoba F, and Canalejo A. 2012. Acute toxicity of manganese in goldfish *Carassius auratus* is associated with oxidative stress and organ specific antioxidant responses. *Ecotoxicol Environ Saf* 78: 212–217.

Dal Corso G, Fasani E, and Furini A. 2013. Recent advances in the analysis of metal hyper accumulation and hyper tolerance in plants using proteomics. *Front Plant Sci* 4: 280.

Das KK, Honnutagi R, Mullur L, Reddy RC, Das S, Dewan Syed AbdulMajid DSA, and Biradar MS. 2019. Heavy metals and low-oxygen microenvironment-its impact on liver metabolism and dietary supplementation. pp. 315–332. *In*: Watson RR, and Preedy VR (eds.). *Dietary Interventions in Liver Disease*.

Decena SC, Arguilles M, and Robel L. 2018. Assessing heavy metal contamination in surface sediments in an urban river in the Philippines. *Polish J Environ Stud* 27: 1983–1995.

Dissanayake CB, and Chandrajith R. 2009. Phosphate mineral fertilizers, trace metals and human health. *J Nat Science F Sri Lanka* 37: 153–165.

Djingova R, and Kuleff I. 2000. Instrumental techniques for trace analysis. pp. 137–185. *In*: Vernet JP (ed.). *Trace Elements: Their Distribution and Effects in the Environment*, Elsevier, London, UK.

Farkas A, Salanki J, and Specziar A. 2002. Relation between growth and the heavy metal concentration in organs of bream *Abramis brama* L. populating lake Balaton. *Arc Environ Contam Toxicol* 43: 236–243.

Farombi EO, Adelowo OA, and Ajimoko YR. 2007. Biomarkers of oxidative stress and heavy metal levels as indicators of environmental pollution in African Cat fish (*Clarias ariepinus*) from Nigeriaogun river. *Int J Environ Res Public Health* 4: 158–165.

Ferretti M, Cenni E, Bussotti F, and Batistonin P. 1995. Vehicle induced lead and foodstuffs and their health impact. *Trends Food Sci Technol* 75: 36–45.

Fuller WH. 1977. *Movement of selected metals, asbestos and cyanide in soil: Application to waste disposal problem*. EPA600/2-77-020. Solid and hazardous waste research division, U.S. environmental protection agency, Cincinnati, OH.

Galal TM, and Shehata HS. 2015. Bioaccumulation and translocation of heavy metals by *Plantago major* L. grown in contaminated soils under the effect of traffic pollution. *Ecol Indi* 48: 244–251.

Gheorghe S, Stoica C, Vasile GG, Nita-Lazar M, Stanescu E, and Lucaciu IE. 2017. *Metals toxic effects in aquatic ecosystems: modulators of water quality*. Intechopen.

Goni MA, Ahmad JU, Halim MA, Mottalib MA, and Chowdhury DA. 2014. Uptake and translocation of metals in different parts of crop plants irrigated with contaminated water from DEPZ area of Bangladesh. *Bull Environ Contam Toxicol* 92: 726–32.

Gupta N, Khan DK, and Santra SC. 2012. Heavy metal accumulation in vegetables grown in a long-term wastewater-irrigated agricultural land of tropical India. *Environ Monit Assess* 184: 6673–6682.

Hashem MA, Nur-A-Tomal MS, Mondal NR, and Rahman MA. 2017. Hair burning and liming in tanneries is a source of pollution by arsenic, lead, zinc, manganese and iron. *Environ Chem Lett* 15: 501–506.

He H, Zhan J, He L, and Gu M. 2012. Nitric oxide signalling in aluminium stress in plants. *Protoplasma* 249: 483–492.

Hinojosa MB, Carreira JA, Ruiz RG, and Dick RP. 2004. Soil moisture pre-treatment effects on enzyme activities as indicators of heavy metal contaminated and reclaimed soils. *Soil Biol Biochem* 36: 1559–1568.

Jiang H, Kong X, Wang S, and Guo H. 2016. Effect of copper on growth, digestive and antioxidant enzyme activities of juvenile Qihe Crucian Carp, *Carassius carassius*, during exposure and recovery. *Bull Environ Contam Toxicol* 96: 333–340.

Juhasz AL, Smith E, Weber J, Rees M, Rofe A, Kuchel T, Sansom L, and Naidu R. 2006. *In vivo* assessment of arsenic bioavailability in rice and its significance for human health risk assessment. *Environ Health Perspec* 114: 1826–1831.

Karaca A, Cetin SC, Turgay OC, and Kizilkaya R. 2010. Effects of heavy metals on soil enzyme activities. *In*: Sherameti I, and Varma A (eds.). *Soil Heavy Metals, Soil Biology*. Heidelberg 19: 237–265.

Khan A, Khan S, Khan MA, Qamar Z, and Waqas M. 2015. The uptake and bioaccumulation of heavy metals by food plants, their effects on plants nutrients, and associated health risk: a review. *Environ Sci Pollut Res* 22: 13772–13799.

Khan S, Aijun L, Zhang S, Hu Q, and Zhu YG. 2008. Accumulation of polycyclic aromatic hydrocarbons and heavy metals in lettuce grown in the soils contaminated with long-term wastewater irrigation. *J Hazard Mater* 152: 506–515.

Kirkham MB. 1977. Trace elements sludge on land: Effect on plants, soils, and ground water. pp. 209–247. *In*: Laehr RC (ed.) *Land as a Waste Management Alternative. Ann Arbor Sci Pub, Ann Arbor*, MI.

Lea PJ, and Miflin BJ. 2004. Glutamate synthase and the synthesis of glutamate in plants. *Plant Physiol Biochem* 41: 555–64.

Lentini P, Zanoli L, Granata A, Signorelli SS, Castellino P and Dell'Aquila R. 2017. Kidney and heavy metals—The role of environmental exposure (Review). *Mol Med Rep* 15: 3413–3419.

Lindsay WL. 1977. *Chemical equilibria in soils*. John Wiley & Sons, New York, NY.

Liu H, Probst A, and Liao B. 2005. Metal contamination of soils and crops affected by the Chenzhou lead/zinc mine spill (Hunan, China). *Sci Total Environ* 339: 153–166.

Mackay AK, Taylor MP, Munksgaard NC, Hudson-Edwards KA, and Burn-Nunes L. 2013. Identification of environmental lead sources and pathways in a mining and smelting town: mount Isa, Australia. *Environ Poll* 180: 304–311.
Masindi V, and Muedi KL. 2018. Environmental contamination by heavy metals. *Heavy Metals* 10: 115–32.
Mezynska M, and Brzoska MM. 2018. Environmental exposure to cadmium-A risk for health of the general population in industrialized countries and preventive strategies. *Environ Sci Pollut Res* 25, 3211–3232.
Mishra AK, and Mohanty B. 2008. Acute toxicity impacts of hexavalent chromium on behaviour and histopathology of gill, kidney and liver of the freshwater fish, *Channa punctatus* (Bloch). *Environ Toxicol Pharmacol* 26: 136–141.
Mora AP, Calvo JJO, Cabrera F and Madejon E. 2005. Changes in enzyme activities and microbial biomass after "*in situ*" remediation of a heavy metal-contaminated soil. *Appl Soil Ecol* 28: 125–137.
Mortvedt JJ. 1996. Heavy metal contaminants in inorganic and organic fertilizers. *Fert Res* 43: 55–61.
Muchuweti M, Birkett JW, Chinyanga E, Zvauya R, Scrimshaw MD, and Lester JN. 2006. Heavy metal content of vegetables irrigated with mixtures of wastewater and sewage sludge in Zimbabwe: implications for human health. *Agric Ecosyst Environ* 112: 41–48.
Odum HT. 2000. Background of published studies on lead and wetland. *In*: Odum HT (ed.). *Heavy Metals in the Environment Using Wetlands for their Removal*. Lewis Publishers, New York, USA, pp. 32.
Olaifa FG, Olaifa AK, and Onwude TE. 2004. Lethal and sublethal effects of copper to the African Cat fish (*Clarias gariepnus*). *Afr J Biomed Res* 7: 65–70.
Patlolla A, Barnes C, Yedjou C, Velma V, and Tchounwou PB. 2009. Oxidative stress, DNA damage and antioxidant enzyme activity induced by hexavalent chromium in Sprague Dawley rats. *Environ Toxicol* 24: 66–73.
Pena LB, Barcia RA, Azpilicueta CE, Méndez AA, and Gallego SM. 2012. Oxidative post translational modifications of proteins related to cell cycle are involved in cadmium toxicity in wheat seedlings. *Plant Sci* 196: 1–7.
Pilon-Smits EAH. 2005. Phytoremediation. *Annu Rev Plant Biol* 56: 15–39.
Pourrut B, Shahid M, Dumat C, Winterton P, and Pinelli E. 2011. Lead uptake, toxicity, and detoxification in plants. *Review Environ Conta Toxicol* 213: 113–136.
Rai PK, and Tripathi BD. 2007. Heavy metals in industrial wastewater, soil and vegetables in Lohta village, India. *Toxicol Environ Chem* 1–11.
Ratnaike RN. 2003. Acute and chronic arsenic toxicity. *Postgrad Med J* 79: 391–396.
Rezania S, Taib SM, Md Din MF, Dahalan FA, and Kamyab H. 2016. Comprehensive review on phytotechnology: heavy metals removal by diverse aquatic plants species from wastewater. *J Hazard Mat* 318: 587–599.
Saha N, and Zaman MR. 2013. Evaluation of possible health risks of heavy metals by consumption of foodstuffs available in the central market of Rajshahi City, Bangladesh. *Environ Monit Assess* 185: 3867–3878.
Sanyal T, Kaviraj A, and Saha S. 2015. Deposition of chromium in aquatic ecosystem from effluents of handloom textile industries in Ranaghat-Fulia region of West Bengal, India. *J Advan Res* 6: 995–1002.
Semu E, and Singh BR. 1996. Accumulation of heavy metals in soils and plants after long-term use of fertilizers and fungicides in Tanzania. *Fert Res* 44: 241–248.
Sevcikova M, Modra H, Blahov J, Dobsikova R, Plhalova L, Zitka O, Hynek D, Kizek R, Skoric M, and Svobodova Z. 2016. Biochemical, haematological and oxidative stress responses of common carp (*Cyprinus carpio* L.) after sub-chronic exposure to copper. *Veterinarni Medicina* 61: 35–50.
Sfakianakis DG, Renieri E, Kentouri M, and Tsatsakis AM. 2015. Effect of heavy metals on fish larvae deformities: a review. *Environ Res* 137: 246–255.
Shallari S, Schwartz C, Hasko A, and Morel JL. 1998. Heavy metals in soils and plants of serpentine and industrial sites of Albania. *Sci Total Environ* 19209: 133–142.
Shukla VK, Prakash A, Tripathi BD, and Reddy DCS. 1998. Biliary heavy metal concentration in carcinoma of the gall bladder: case-control study. *BMJ* 317: 1288–1289.

Shun-hong H, Bing P, Zhi-hui Y, Li-yuan C, and Li-cheng Z. 2009. Chromium accumulation, microorganism population and enzyme activities in soils around chromium-containing slag heap of steel alloy factory. *Trans Nonfer Metal Soc China* 19: 241–248.

Siciliano SD, and Germida JJ. 1998. Mechanisms of phytoremediation: biochemical and ecological interactions between plants and bacteria. *Environ Rev* 6: 65–79.

Silva S. 2012. Aluminium toxicity targets in plants. *J Bot* 2012: 219462.

Smith SR. 2009. A critical review of the bioavailability and impacts of heavy metals in municipal solid waste composts compared to sewage sludge. *Environ Int* 35: 142–156.

Sobha K, Poornima A, Harini P, and Veeraiah K. 2007. A study on biochemical changes in the fresh water fish, catla catla (hamilton) exposed to the heavy metal toxicant cadmium chloride. *Kathm Uni J Sci Engin and Technol* 1: 1–11.

Srivastava G, Kumar S, Dubey G, Mishra V, and Prasad SM. 2012. Nickel and ultraviolet-B stresses induce differential growth and photosynthetic responses in *Pisum sativum* L. seedlings. *Biol Trace Elem Res* 149: 86–96.

Stevenson FJ. 1982. Humus chemistry: genesis, composition reactions. John Wiley and Sons, New York, NY.

Sutton DJ and Tchounwou PB. 2007. Mercury induces the externalization of phosphatidylserine in human proximal tubule (HK-2) cells. *Int J Environ Res Public Health* 4: 138–144.

Tchounwou PB, Ishaque AB, and Schneider J. 2001. Cytotoxicity and transcriptional activation of stress genes in human liver carcinoma cells (HepG2) exposed to cadmium chloride. *Mole Cell Biochem* 222: 21–28.

Tchounwou PB, Centeno JA, and Patlolla AK. 2004a. Arsenic toxicity, mutagenesis and carcinogenesis—a health risk assessment and management approach. *Mol Cell Biochem* 255: 47–55.

Tchounwou PB, Yedjou CG, Foxx D, Ishaque A, and Shen E. 2004b. Lead induced cytotoxicity and transcriptional activation of stress genes in human liver carcinoma cells (HepG2). *Mole Cell Biochem* 255: 161–170.

Thounaojam TC, Panda P, Mazumdar P, Kumar D, Sharma GD, Sahoo L, and Sanjib P. 2012. Excess copper induced oxidative stress and response of antioxidants in rice. *Plant Physiol Biochem* 53: 33–39.

Treder W, and Cieslinski G. 2005. Effect of silicon application on cadmium uptake and distribution in strawberry plants grown on contaminated soils. *J Plant Nutr* 28: 917–929.

Volkering F, Breure AM, and Rulkens WH. 1998. Microbial aspects of surfactant use for biological soil remediation. *Bioremediation* 8: 401–417.

Wang S, and Shi X. 2001. Molecular mechanisms of metal toxicity and carcinogenesis. *Mol Cell Biochem* 222: 3–9.

Wani PA, Khan MS, and Zaidi A. 2012. Toxic effects of heavy metal on germination and physiological processes of plants. pp. 45–66. *In*: Zaidi A, Wani PA, and Khan MS (eds.). *Toxicity of Heavy Metal To Legumes And Bioremediation*. Springer-Verlag Wien.

Xie Y, Fan J, Zhu W1, Amombo E, Lou Y, Chen L, and Fu J. 2016. Effect of heavy metals pollution on soil microbial diversity and bermudagrass genetic variation. *Front Plant Sci* 7: 755.

Yang P, Mao R, Shao H, and Gao Y. 2009. The spatial variability of heavy metal distribution in the suburban farmland of Taihang piedmont plain, China. *C R Biol* 332: 558–566.

Yang Y, Chang AC, Wang M, Chen W, and Peng C. 2018. Assessing cadmium exposure risks of vegetables with plant uptake factor and soil property. *Environ Pollut* 238: 263–269.

Yao H, Xu J, and Huang C. 2003. Substrate utilization pattern, biomass and activity of microbial communities in a sequence of heavy metal polluted paddy soils. *Geoderma* 115: 139–148.

Yedjou CG, and Tchounwou PB. 2006. Oxidative stress in human leukemia cells (HL-60), human liver carcinoma cells (HepG2) and human Jerkat-T cells exposed to arsenic trioxide. *Metal Ions Biol Med* 9: 298–303.

Yu L, Yan-bin W, Xin G, Yi-bing S, and Gang G. 2006. Risk assessment of heavy metals in soils and vegetables around non-ferrous metals mining and smelting sites, Baiyin, China. *J Environ Sci* 18: 1124–1134.

Yuan HM, Xu HH, Liu WC, and Lu YT. 2013. Copper regulates primary root elongation through PIN1-mediated auxin redistribution. *Plant Cell Physiol* 54: 766–778.

2

Heavy Metal Contamination in Plants
An Overview

Rashmi Mukherjee,[1] Soumi Datta,[2] Dwaipayan Sinha,[3],*
Arun Kumar Maurya[4] and Sambhunath Roy[5]

ABSTRACT

Heavy metals (HMs) are those metals whose density is more than 5 grams per cc. They are present in the earth's crust as ores and minerals. They have been in use by humans since the dawn of civilization. However, the unplanned use of these metals has resulted in contamination of the earth over the past few centuries. They are potentially toxic and affect both plants and animals. Contamination by HMs has adversely affected the plants and has negatively affected the crops by reducing their productivity. This is largely due to rapid industrialization, which is forcing people to make rampant use of these heavy metals without taking scientific and systematic measures for disposal. This negligence is rapidly paving the way for HM build-up in the environment and is consequently up taken by the plants. Conventional methods of decontamination are adopted but it attracts a higher price and is not sustainable. Thus, scientists have resorted to green methodologies to decontaminate excess HMs from the environment. Thus, the process of phytoremediation is adopted for the 'clean up' process. This chapter is an attempt to overview HMs and pollution/problems caused

[1] Department of Botany, Raja N.L. Khan Women's College (Autonomous), Gope Palace, Midnapore-721102; West Bengal, India.
[2] Department of Pharmacognosy & Phytochemistry, School of Pharmaceutical Education & Research, Jamia Hamdard, New Delhi-110062, India.
[3] Department of Botany, Government General Degree College, Mohanpur, Paschim Medinipur, West Bengal-721436, India.
[4] Department of Botany, Multanimal Modi College, Modinagar, Ghaziabad, Uttar Pradesh-201204, India.
[5] Department of Geography, Raja N.L. Khan Women's College (Autonomous), Gope Palace, Midnapore-721102; West Bengal, India.
* Corresponding author: dwaipayansinha@hotmail.com
ORCID iD: https://orcid.org/0000-0001-7870-8998.

by them. Efforts are made to make elaborate documentation of HM contamination, its accumulation within plants, and the use of plants for phytoremediation purposes.

1. Introduction to heavy metals

The twentieth century has seen a stupendous increase in the demand for land resources from human activities due to the fast progress of the social economy, industrialization, urbanization, and agricultural revolution (Cheng et al., 2015). As a result, harmful alterations of the soil characteristics take place including soil erosion, desertification, increased salinity, pollution, and decreased soil fertility (Vaverková et al., 2019). Contamination of natural soils occurs through the gathering of heavy metals (HM) and metalloids from industrial discharge, mining and smelting, non-treatment and careless discarding of metal wastes, manufacturing and applicating of gasoline and paints containing lead (Pb), incessant use of chemical fertilizers/insecticides/pesticides, wastewater from sewage or irrigation, fossil fuel or petrochemicals combustion remains, as well as atmospheric deposition (Awa and Hadibarata, 2020). One of the most trending areas of environmental research is how to address HM pollution affecting all ecosystems (Ali et al., 2019).

Defining HMs has always been a debatable issue. Csuros and Csuros (2002), defined an HM as metals with a density ≥ 5 g/cc. Later in the same year Duffus (2002) used the term HMs for the group of metals and semimetals causing contamination and are toxic to the living system. Currently, HMs are considered to be those whose atomic number is > 20 and with a density ≥ 5 g/cc (Ali and Khan, 2018). HMs, an ill-defined group that generally occur at polluted sites, are categorized by their lengthy existence, high density, and high toxicity in the environment resulting in various health issues in living organisms at very minute concentrations of < 2 µg (Selvi et al., 2019). In some cases, the term "trace elements" is also used for HMs due to their presence in trace concentrations ($<$ less than 10 ppm) in the environment (Hossain et al., 2021). The bioavailability of HMs is predisposed via physical factors including temperature, sequestration, adsorption, pH, and phase association as well as by chemical factors namely, solubility, the kinetics of complexation, and octanol/water partition coefficients (Iwona et al., 2018, Tchounwou et al., 2012).

Eight HMs—mercury (Hg), cadmium (Cd), lead (Pb), arsenic (As), chromium (Cr), zinc (Zn), copper (Cu), and nickel (Ni)—are categorized as the maximum prevalent highly poisonous HMs as per the United States Environmental Protection Agency (USEPA). Along with these 8, silver (Ag), gold (Au), palladium (Pd), bismuth (Bi), selenium (Se), and tin (Sn) are also highly toxic. Cobolt (Co) and iron (Fe) are classified as HMs which function as macro-nutrient elements in living systems. On the contrary, Cu, Ni, Cr, Manganese (Mn), and molybdenum (Mo) are considered micronutrient elements. The HMs platinum (Pt), Ag, Au, Pd, and ruthenium (Ru) are precious metals. Moreover, the HMs uranium (Ur), thorium (Th), radium (Ra), cerium (Ce), and praseodymium (Pr) function as radionuclides. Table 1 gives a brief overview of the properties of the 08 elements which are considered highly toxic to the environment.

Several studies were performed to evaluate Pb toxicity on plants, animals, and humans. Pb absorption in the food chain has become a significant health danger for

Table 1: Physico-chemical properties of 8 heavy metals.

Heavy metals	Position in periodic table [Group (period)]	Atomic number	Atomic mass	Density (g/cm^3)	Melting point (°C)	Boiling point (°C)
Lead (Pb)	IV (6)	82	207.2	11.4	327.4	1725
Chromium (Cr)	VIB (4)	24	52	7.19	1875	2665
Arsenic (As)	VA (4)	33	75	5.72	817	613
Zinc (Zn)	IIB (4)	30	65.4	7.14	419.5	906
Cadmium (Cd)	XII (5)	48	112.4	8.65	320.9	765
Cu (Cu)	IB (4)	29	63.5	8.96	1083	2595
Mercury (Hg)	XII (6)	80	200.6	13.6	13.6	357
Nickel (Ni)	X (4)	28	58.69	8.908	1453	2732

living organisms at all trophic levels (Kumar et al., 2020). After reaching the soil, Pb enters roots ($\geq$ 95%) where it may accumulate or, in some cases, be translocated to the shoot portion (Rai et al., 2019). Extensive research has been performed with *Allium sativum* (Jiang and Liu, 2010), *Avicennia marina* (Yan et al., 2010), *Pisum sativum* (Małecka et al., 2008), *Lathyrus sativus* (Brunet et al., 2009), *Nicotiana tabacum* (Gichner et al., 2008), and *Zea mays* (Metanet et al., 2019) to understand the mechanism of Pb transport, toxicity response, and bioaccumulation. The Agency for Toxic Substances and Disease Registry (CERCLA Priority List, 2017) ranked Cr as the 17th among the most hazardous substances. As per the International Agency for Research on Cancer, Cr has also been categorized as the number one carcinogen (IARC monographs, 1987). Hence, research investigations are ongoing to provide better insights for a precise knowledge of its source, transport, and bioaccumulation, and exhaustive exploration of the soil-plant systems (Sharma et al., 2020a). By combining with oxygen, Cr (VI) forms chromate (CrO_4^{2-}) or dichromate ($Cr_2O_7^{2-}$) (Mou et al., 2021). Both of these forms are highly lethal to living organisms especially plants. Cr (VI) goes inside the cells and ultimately gets reduced to Cr (III), thiol radicals, and hydroxyl radicals which bring about oxidative damage to proteins, deoxyribonucleic acid (DNA), and membrane lipids (Stambulska et al., 2018). Cr toxicity is associated with the reactive oxygen species (ROS) generation resulting in plant oxidative stress (OS), and influencing the plant growth by damaging their essential metabolic processes (Tchounwou et al., 2012).

Pre-dominantly occurrings as arsenite As(III) and arsenate As(V), As(III) is more poisonous because of its greater mobility and solubility (Rasul et al. 2002). As(V) is transformed to As(III) in plant root cell and is transported to other plant parts as As(III) (Garg and Singla 2011). Interestingly, As(III) also converts to As(V) inside the plant cell under various physiological conditions. Both ions obstruct numerous cellular metabolisms by different methods.

Zn is present in the soil but its contamination is escalating abnormally, due to anthropogenic activities (Nkwunonwo et al., 2020). In soil, Zn occurs as insoluble complexes (90% of soil Zn). It can also exist in adsorbed and exchangeable form. Though, an additional portion exists in a water-soluble form that is easily accessible to plants (Gupta et al., 2016). Higher accumulation of Zn in soil results in excess uptake by the plants which is deleterious to their physiological system. It can also negatively influence the microbial activity or the activity of earthworms, thereby inhibiting organic matter degradation (Wuana and Okieimen, 2011).

Cd is another important HM that is not associated with any important biological function. It occurs as the divalent Cd(II) ion (Rafati et al., 2017). Cd retards the mineral uptake, transport, and use by plants by interfering with the availability of soil minerals or killing beneficial microbes present in soil (Goss et al., 2013). It also decreases the absorption and transportation of nitrate through inhibition of shoot nitrate reductase (NR) activity (Mao et al., 2014). The third most utilized metal in the world is Cu. Cu strongly complexes to the organic material in the soil suggesting that a minor portion of Cu will be available in the dissolved state as Cu(II). The solubility of Cu increases considerably at pH 5.5 (Martínez and Motto, 2000). Cu is particularly significant in plants for the production of seeds, resistance to diseases, and water balance (Anderson et al., 2018). Excess Cu levels in soil damage plant roots, a significant decline in the stele diameter as well as alteration to its cell structure (Minkina et al., 2020).

Hg chiefly occurs in 3 states as element (Hg), inorganic [Hg(II) and Hg(I)], and organic forms [e.g., methylmercury (MeHg), ethylmercury (EtHg), dimethyl mercury $(Me)_2Hg$, and phenylmercury (PhHg)] (Hu et al., 2013). The toxicity of organomercury compounds is tenfold as compared to the inorganic forms (Tangahu et al., 2011). Being an ever-existing pollutant bioaccumulating in fish, animals, and human beings, both organic and inorganic forms of Hg are highly toxic (Chang et al., 2009). Hg deposited on leaf surfaces is absorbed by plants (Hanson et al., 1995). Hg gets accumulated in plants as Hg(0), Hg(II), and organic Hg. Generally, hydrophytes have the greater quantity of mercurial organic forms as compared to terrestrial plants (Rea et al., 2001, Li et al., 2017).

Ni occurs as divalent ionic form, Ni(II), and precipitates under neutral-slightly alkaline conditions as nickel (ii) hydroxide $Ni(OH)_2$. $Ni(OH)_2$ easily dissolves in acid solutions and produces Ni(III); whereas Hydroxy(oxo)nickel $(HNiO_2)$ is formed in very alkaline conditions which is readily water-soluble (Deng et al., 2018). It is usually present in the soil at a very minute quantity and is indispensable. Ni becomes harmful if its level exceeds the maximum tolerable limits and causes several types of cancers in human beings and animals (Singh et al., 2011). Plants absorb Ni mainly as Ni^{2+} (Dinu et al., 2021).

2. Distribution of heavy metals

HMs area global threat at present (Alengebawy et al., 2021). With the advent of the rapid development of society, HM contamination has progressively become very prevalent worldwide (Jaishankar et al., 2014). The worldwide distribution of the major HMs is shown in Fig. 1.

There is a presence of Zn, As, Fe, Cd, and Cu in North America with Zn and Cd being the predominant HMs (Aponte et al., 2020). There are a large proportion of As-affected aquifers in South Africa along with the presence of Cu, Cd, and Fe in the soil (Abedin and Shaw, 2013). Although in low percentage, Zn and Ni are also present (Coakley et al., 2019). The continent of Africa has a widespread distribution of

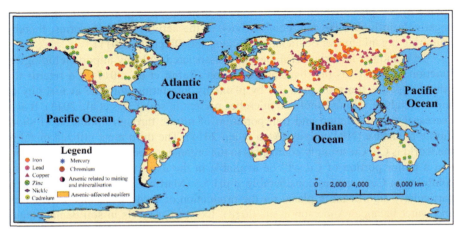

Figure 1: Worldwide distribution of major heavy metals.

Fe, Cu, and Zn in its soil whereas Australia has a predominantly higher percentage of Zn and Fe followed by Ni and Cd. European countries have a very dense distribution of Cd, Pb, Zn, Cu, and Fe. Cd is very prevalent in the soil of Asian countries followed by Fe, As, and Zn (Shankar et al., 2014). Among all the continents of the world, Africa has the least As contamination (Kumar et al., 2019).

Anthropogenic activities namely, industrial, agrochemicals, paint, cosmetics, oil-processing activities, smelting, stained glass, batteries, lead-glazed dishes, mining, vinyl mini-blinds, firearms with lead bullets, refining, informal recycling of Pb, jewellery, plumbing materials and alloys, toys, ceramics, pottery, water from old pipes, and radiators for cars and trucks are the chief causes of environmental Pb contamination (Rai et al., 2019; Kumar et al., 2020). Cr is found naturally in rocky soils and volcanic dust (10 and 50 mg kg^{-1}) (Sharma et al., 2020a). Its hexavalent form is used in numerous industrial activities, e.g., electroplating, leather processing, dyeing of textiles, steel production, and tannery leading to effluent discharge contaminated with the metal which finally increases Cr in the environment (Ertani et al., 2017). This can also result in Cr accumulation of upto 350 mg kg^{-1} in agricultural land (Joutey et al., 2015). Several anthropogenic activities like mining, smelting, fossil fuel combustion, semiconductors production, fertilizers, glass production, and chemotherapeutic drug of cancer release As in to the environment (Field and Withers, 2012). Natural activities (volcanic eruption weathering of rock and microbial colonization) contribute to the bulk of As contamination in the atmosphere (Kumar et al., 2019). Zn is found in soil, its concentrations are rising abnormally due to anthropogenic sources (Jan et al., 2015). Industries release their untreated waste waters into rivers, etc., and as a result, Zn-polluted sludge is repeatedly being accumulated on riverbanks (Gupta et al., 2016). Natural phenomena (dust storms, volcanic activities, sea salt spray, weathering, erosion, and wildfire) are the main reasons for elevated environmental Cd concentrations (Kubier et al., 2019). Non-polluted soil solutions contain ~ 0.04–0.32 mM Cd while those containing 0.32–1 mM Cd concentration are considered to be as polluted to a moderate level (Charfeddine et al., 2019). The various anthropogenic sources of Cd are Ni-Cd batteries, power stations,

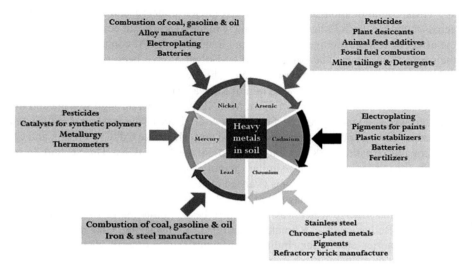

Figure 2: Heavy Metals—their anthropogenic sources in soil.

coatings and platings, stabilizers for plastics, heating systems, fossil fuel combustion, phosphate fertilizer urban traffic, waste incineration, and cement production (Rydh and Svärd, 2003). Depending upon geology, Cu concentrations vary between 2 and 50 mg Cu/kg in soil (Oorts et al., 2013). Soil properties regulate Cu toxicity and control its availability in the soil through several physicochemical processes (Olaniran et al., 2013). Various agrochemical and petrochemical based industries are responsible for Hg contamination (Resaee et al., 2005). Other sources of Hg include bleach, acids (hydrochloric acid and sulfuric acid), medical instruments, dental fillings, latex paint, Hg vapour lamps, pesticides, pharmaceuticals, laboratory chemicals, inks, wiring devices and switches, lubrication oils, and textiles (Tangahu et al., 2011). Combustion of fossil fuels, metal plating, mining, and electroplating is the common source of Ni pollution in the environment (Genchi et al., 2020). A large proportion of all environmental Ni compounds adsorbs to sediment or soil particles resulting in immobilization (Wuana and Okieimen, 2011). The various anthropogenic sources of HMs in soil are depicted in Fig. 2.

3. Heavy metal contamination in plants: A global scenario

The scientific and industrial developments are not only advantageous but also disadvantageous to the society as it generates soil contamination by HMs, which are mainly produced as industrial effluents. The concentration of the HMs is industry-specific and directly related to environmental pollution, and thus, has become a global issue due to its unfavourable environmental effects, extensive occurrence, and toxicity, both in acute and chronic forms (Ali et al., 2019). Stable metal elements or metals whose density is higher than 5 grams per cc, like lead, copper, nickel, cadmium, zinc, mercury and chromium, etc., cannot be degraded or destroyed, are non-biodegradable, and non-thermo-degradable; and therefore get deposited in the soil.

HMs contamination in plants is a worldwide concern and many researchers across the globe, both India and abroad have made a significant contribution towards the effect of sewage storage, drainage, and irrigation on HM contaminated land and their effect on plants. The soils of Turkey-Bulgaria have been reported to have a reduction in HMs accumulation in flora established in particular regions (Aydinalp and Marinova, 2003). The fertile terrain of Konya, Iran is contaminated by enhanced absorption of HMs, such as Cd, Cu, Cr, Mn, Ni, Pb, and Zn through sewage and the plant growth is adversely affected (Mustafa et al., 2006). The waste water irrigated vegetables from Varanasi were contaminated with HMs like Cd, Cr-, Pb, Zn, and Cu (Mishra and Tirpathi, 2008); the emitted particulates like Cd, Cu, Hg, Se, and Zn by Mount. Etna volcano, Italy has similar quantities of these HMs that are released in the Mediterranean area by humans (Varrica et al., 2000); increased levels of Zn and Pb were found at mine drainage area of Boles law, Poland (Grobelak and Napora, 2015); and the sewage water was reported to be the main source of contamination for soils of Bellandur lake, Karnataka (India) (Lokeshwari and Chandrappa, 2006). Leather tanneries at Kanpur (India) are a significant source of HMs pollution in nearby soils, as effluents discharged are often used for watering crops (Gupta et al., 2007); while in Delhi, India, the Keshopur effluent irrigation scheme resulted in accumulation of diethylenetriamine pentaacetate extractable Zn, Cu, Fe, Ni, and Pb in fields irrigated by sewage (Rattan et al., 2005). The highly populated Asiatic countries and some impoverished African nations have reported heavy metal contamination in crops irrigated by wastewater and sludge patterns, with deleterious ecological consequences. In India, lengthy-term wastewater irrigation has resulted in the accumulation of HMs in food crops making them health hazards (Rattan et al., 2005; Rai and Tripathi, 2008; Garg et al., 2014; Saha et al., 2015; Chabukdhara et al., 2016). China, being the world's most populous country, has placed a strong emphasis on ecotoxicological, environmental, and food safety issues to accommodate the expansion of industrial lands. The country also relies on wastewater for irrigation in certain pockets similar to those of other countries of Asian and African continents (Zhang et al., 2015; Junhe et al., 2017). Particulate matters (PMs) and modern intensive agricultural practices form a major source of HMs in Europe, America, and Oceania and thus the geographical pattern of HMs source is different as compared to Africa and Asia.

In the United States shooting activities form the largest contributor of lead (more than 60,000 tons annually) to the environment (Wani et al., 2015). Cadmium contamination occurs due to sludge, fertilizer, and manures, and also through the natural geogenic process and its effect vary from one country to another (Liu et al., 2016). As a result, developing a reliable strategy and stable treatment systems is required to address the issue of rising wastewater generated by home and industrial activities.

4. Heavy metal contamination in plants

The rapid growth of population and industries has resulted in environmental pollution. The pollutants from different industrial sources as well as unplanned anthropogenic uses contaminate air water and land. This HMs contamination is posing a serious risk to the plants in general and more specifically to the crops on which the human civilization is dependent for food. HMs are toxic to the plants, induce chlorosis, retard growth, negatively affect biomass and reduce productivity (Onakpa et al., 2018). Consumption plants contaminated with HMs may result in direct access of the HMs to humans which causes several physiological disorders (Shen et al., 2019). Several studies report that crops and plants growing in HM polluted areas get contaminated with HMs. Table 2 represents the contamination of HMs in selected crops/plants around the globe.

5. Physiological effect of heavy metal contamination in plants

In higher plants, the HMs are classified into necessary and additional elements; where the former are Iron, manganese, zinc, copper molybdenum, and nickel, while cadmium, lead, mercury fall in the latter category (Marschner, 1995) and are known to be either neutral or toxic. Lethal HMs mostly inhibit physiological and developmental processes such as growth, photosynthesis, membrane damage, oxidative stress, genomic and proteomic profile, crop yields, soil biomass, fertility, nutrient loss, ion/water uptake, nitrate assimilation, etc.; however, their magnitude is largely dependent on two main factors, i.e., concentration of the HMs and the physiological importance of the plant (Salinitro et al., 2019). Many models are reported on the composition of HMs toxicity in advanced plants as the impact of HM on enzyme function or specific tolerance mechanisms or metal, etc. Here, we overview the current updates on the physiological effect of HM contamination in plants.

5.1 Uptake of heavy metals

In plants, there are mainly two sources of HM contamination-natural and anthropogenic sources. The former consists of breaking of volcanic rocks such as Augite, Homblende and, Olivine, that generates good amounts of HMs, while sedimentary rocks contribute minimally (Nagajyoti et al., 2010); volcanoes, air-blown dust, forest fires, ocean sprays, and aerosols (Lin et al., 2016); and geothermal sources like volcanic eruptions (Ojuederie and Babalola, 2017). The latter source consists of agricultural (Nriagu, 1989; Yanqun et al., 2005), industrial (Tchounwou et al., 2012; Guan et al., 2018); and domestic effluents (Azizi et al., 2016). In aquatic plants, the entire surface is accessible to HMs and elements, while in land species, intake is mainly through the roots that are assisted by some protein transfer, chelating agents produced in the root microbiome, and pH-induced by changes in plants (Tangahu et al., 2011); by stomata, lesions, lenticels, etc. (Shahid et al., 2017); or by direct adsorption on the foliar surface.

5.2 Effects of heavy metals on plant's growth and development

In plants, the effect of HMs depends on the capacity to take up and amass nutrients from the soil. Heavy metals can also function as a stressor, resulting in physiological restrains like decreased vigour and growth inhibition (Singh et al., 2016). The most observant evidence of HMs toxicity in plants is reduced plant growth

Table 2: Heavy metal contamination in selected crops/plants.

S. No.	Locations	Source of contamination	Heavy metal(s) contaminated	Plant affected	References
Asia					
1.	Guizhou Province, China	Zn smelter	Cd, Cu, Pb, Zn	Rice, Corn, Wheat	Chen et al., 2017
2.	Guangxi Province, China	Pingle Mn mine	Cd, Cu, Pb, Mn, Zn	Vegetables, fruits and cereals	Liu et al., 2018
3.	Southwest China	Pb-Zn Mining area	As, Cd, Cr, Pb, No, Zn	Corn	Zhou et al., 2020
4.	Tongling, Eastern China	The non-ferrous metal mining area	Cr, Cd, Cu, Pb, Hg, Zn	Corn, Rice, vegetables, Fengdan, *Paeoniaostii*	Wang et al., 2019
5.	Tongling city, China	Fenghuangshan mining tailings	As, Cu, Cd, Fe,Cr, Pb, Zn, Mn	Paeoniaostii	Shen et al., 2017
6.	Guangdong Province, South China	Lead (Pb)-Zn (Zn) mining areas	Cd, Cr, Cur, Pb, Mn, Nii, Zn	High concentration of HMs in paddy soils	Xu et al., 2017
7.	Human Province, China	HMs of Hunan province	Sb, As, Cd, Pb, Hg	Rice	Fan et al., 2017
8.	Shimen, Fenghuang, and Xiangtan, Hunan Province, China	Mining area	As, Cd, Cr, Pb, Mn, Ni,	Brown rice	Zeng et al., 2015
9.	Tangshi village, Jiyuan, Henan Province, China	Lead smelting area	As, Cd, Pb	Wheat	Guo et al., 2018
10.	Esfahan, Iran	Drip irrigation with treated municipal wastewater	Cd, Cr	Wheat and Corn	Asgari and Cornelis, 2015
11.	Urmia City, western Azerbaijan Province, northwestern Iran	Irrigation with treated wastewater	Cd, Cu, Pb, Ni, Zn	Wheat	Rezapour et al., 2019a
12.	Urmia Plain, western Azerbaijan Province, northwestern Iran	Irrigation with treated wastewater	Cd	Wheat	Rezapour et al., 2019b

No.	Location	Source/cause	Heavy metals	Plants	Reference
13.	Sargodha City, Pakistan	Treatment with wastewater (experiment)	Cd, Fe, Cr, Mn, Pb, Ni	Wheat	Ahmad et al., 2019
14.	KP Province, Pakistan	Use of agriculture related insecticidal, herbiciadal and fungicidal chemicals, sewage sludge, irrigation through wastewater	Cr	Fruits and vegetables	Ur Rehman et al., 2018
15.	Faisalabad city, Pakistan	Irrigation by wastewater	Cr, Cu, Mn, Ni, Zn	Rice, Wheat, Corn, sugarcane, Millet	Mahfooz et al., 2020
16.	Barapukuria mine, Bangladesh	Coal mine area	As, Cu, Cr, Mn, Pb, Ni, Zn	Paddy	Halim et al., 2014
17.	Industrial Areas of Savar, Bangladesh	Industrial region	Pb, Cd, Cr, Co	Jute, Red amaranth, Okra, Zucchini, Stem amaranth	Al Amin et al., 2020
18.	Savar Upazila, Dhaka District, Bangladesh	Waste-water-Irrigated Site	Pb, Ni, As	Vegetables	Hossain et al., 2015
19.	Mining areas Thai Nguyen Ha Nam provinces, Vietnam	Mining area	As, Hg, Cd, Mn, Fe, Zn, Cu, Ni, Cr, Co, Pb	Paddy	Chu et al., 2021
20.	Huong River and Red Rivers, Vietnam	Brick factory, Fertilizer factory, Chemical factory	Cd	Paddy	Nguyen et al., 2020a
21.	Mekong River Delta, Vietnam	Contamination from sediments of rivers	As, Pb, Cd	Paddy	Nguyen et al., 2020b
America					
22.	Cacao plantation region in North, Central, and South Peru	Presence of HMs in the cultivated soil	Cd, Cr, Cu, Fe, Mn, Ni, Pb, Zn	Cocoa	Arévalo-Gardini et al., 2017

Table 2 contd. ....

...*Table 2 contd.*

S. No.	Locations	Source of contamination	Heavy metal(s) contaminated	Plant affected	References
23.	Orellana, Sucumbios and Morona-Santiago, in Amazon area	Occurrence of HMs in soil	As, Cr, Co, Mo, Cu, Mn, Pb, Ni, V, Zn	Cocoa	Barraza et al., 2021
24.	Orellana and Sucumbios in the northern Amazon region	Presence of oil refinery	Cd	Cocoa	Barraza et al., 2017
25.	Pilcomayo River watershed near Potosí', Bolivia	Acid mine drainage, solid mine waste leaching, and runoff	As, Cd, Pb, Zn	Potato	Garrido et al., 2017
26.	Lake Titicaca and Lake UruUru, Bolivia	Contamination due to mining and smelting activities	As	Totora plants (Schoenoplectus californicus)	Sarret et al., 2019
27.	La Serena, Los Tilos, Chillán, Chile	Presence of Cd in soil	Cd	Maize	Retamal-Salgado et al., 2017
28.	municipality of Sibaté, Cundinamarca, Columbia	Contamination from industrial complexes	As, Cr, Co, Pb, Cu, Zn	Daucus carota, Cynarascolymus, and Petroselinum crispum	Lizarazo et al., 2020
29.	mining zone of Zimapa'n, Hidalgo, Mexico	Contamination from mine tailings	Zn, Fe, Pb, As, Cd	Maize	Armienta et al., 2020
Europe					
30.	Kosovska Mitrovica, Kosovo	Contamination from industries and smelter	As, Co, Cd, Cu, Pb, Sb, U, Zn	Maize	Nannoni et al., 2016
31.	Mitrovica, Kosovo	Contamination from the mining area	Zn, Pb, Ni, Cd, Cu, As, Co, Cr	Salix purpurea	Zabergja-Ferati et al., 2021
32.	Warta Bolesławiecka district, Lower Silesia province, SW Poland	Contamination from mine tailings	Cu	Cerastiumarvense, Polygonum aviculare,	Kasowska et al., 2018
33.	Baia Mare area, Romania	Contamination from mine tailings	Zn, Cu,Pb, Cd	Fruits and vegetables	Roba et al., 2016

	Location	Source/Cause	Heavy metals	Plant	Reference
34.	La Rochelle, France	Irrigation water	Ni, Zn, Cu	Vegetables	Cherfi et al., 2016
35.	Dunaújváros, Hungary	Sediments of rivers around the industrial location	Cd, Zn, Cu	Radish, Pea	Kovács-Bokor et al., 2021
36.	Danube river-basin, Hungary, Romania, Serbia	Contamination due to mining, anthropogenic activity, processing of metal order	Cd, Zn	*Salix sp.*	Pavlović et al., 2016
37.	Pljačkovica, Bratoselce, Borovac, Reljan, Serbia	Bombing during the Yugoslavian war	Radioactive nucleotides Ra, Th, K, Cs, U, Be, Pb	Plant samples	Sarap et al., 2014
Africa					
38.	Ofla, Tigray, Ethiopia	Brewery sludge	Pb, Cr,Cd, Ni, Cu, Mn	Wheat	Tesfahun et al., 2021
39.	Kombolcha town in northeastern Ethiopia	brewery diatomite waste sludge	Cu, Zn, Pb, Ni	Wheat	Dessalew et al., 2018
40.	Mojo, Central Ethiopia	Release of industrial effluents in mojo river	As, Pb, Cd, Cr, Hg	Cabbage, Tomato	Gebeyehu and Bayissa, 2020
41.	Kogi State in north-central Nigeria	Irrigation and application of fertilizer	Ni, Cd, Cu	Pumpkin, Passion fruit, maize, sugarcane, cassava	Emurotu and Onainwa, 2017
42.	Boumerdes, Algeria	Pollution due to irrigation by treated wastewater	Cu, Zn, Pb, Cr	Fruits and vegetables	Cherfi et al., 2014
43.	Tarkwa, Southwest Ghana	Industrial processes	Pb, As, Cu	Cassava and Plantain	Bortey-Sam et al., 2015
44.	Kabwe town, central Zambia	Kabwe Mine in Zambia	Pb, Zn, Cd	Maize	Mwilola et al., 2020

(Sharma and Dubey, 2007; Gill, 2014) including chlorosis, reduction in turgor, slowing of germination, disrupted photosynthesis, and progressed senility or death of plant (Dalcorso et al., 2010; Singh et al., 2016). Cumulatively, these aspects are associated with biochemical, molecular, and ultrastructural alterations in plant tissues and cells brought about by HMs (Gamalero et al., 2009).

5.3 Effects of heavy metals on secondary metabolism

An enviable result of a detrimental environmental aspect has been proposed by therapeutic, aromatic, and non-food crops in HM contaminated soils as an alternative method to elicit the secondary metabolite biosynthesis by using various strategies to cope with the HM interference in their plant cells; and specific HM tolerant is managed by an intricate unified array of biochemical, genetic, morphological, and physiological mechanisms like HM ions-protein group binding, replacing specific cation in binding sites resulting in enzymes inactivation and generation of free radicals, that may induce membrane lipid peroxidation, break ribonucleic acid (RNA) and DNA, deactivation of vital enzymes, and oxidation of proteins and amino acids (Mahmood et al., 2017).

5.4 ROS production and scavenging

ROS within plants can be foraged by both nonenzymatic and enzymatic antioxidants. The former like phenolics, ascorbates, carotenoids and glutathione, bind HMs and neutralize free radicals generated in plants; whereas the commencement of the latter like superoxide dismutase (SOD), catalase (CAT), ascorbate peroxidase (APx), glutathione S-transferase (GST), glutathione reductase (GR) and dehydroascorbate reductase (DHAR), uses the inherent defense approach to regulate the oxidative stress of cells according to their metabolic need at a definite time. The ROS includes various molecules like superoxide anion, singlet oxygen species, alkoxyl and peroxyl radicals,and organic hydroperoxides (Mahmood et al., 2017).

5.5 Plant strategies to overcome heavy metal stress

HMs interact with biomolecules (nuclear proteins/DNA), forms ROS which in turn generates oxidative stress and affects the physiological and genetic structure of a plant (Manara, 2012); and thus have devised several mechanisms to counteract stress and can be broadly categorised into two groups namely avoidance and tolerance. The first line of defense against HM contamination are physical barriers like thick cell wall (CW), cuticle, mycorrhiza, trichomes, and plasma membrane (PM) (Emamverdian et al., 2015; Harada et al., 2010). Once, these barriers are overcome by HMs, several biomolecules such as PCs, metallothionein, organic acids, cellular by-products comprising of phenolic compounds, amino acids, and hormones come into play and initiate the process of detoxification (Dalvi and Bhalerao, 2013; Viehweger, 2014).

Various strategies have been evolved by plants to combat HMs stress at the genetic and biochemical levels. For example, the brassinosteroids (BRs) group's plant hormone 28-homobrassinolide (28-HBL) group is known to overcome the lethal effects of Cr responsible for a declined growth, chlorophyll, proline contents, enhanced MDA content, and metal intake (Sharma et al., 2011). Over secreted glyoxyalase enzymes detoxifies methyl-glyoxal and is reported to be tolerant toward salinity and HM stresses in *Arabidopsis* transgenic plants (Mustafiz et al., 2011); the CDR3 gene extracted from cadmium-resistant *Arabidopsis* species modulates HM tolerance seed development and the expression of GSH1 resulting in improved GSH levels (Wang et al., 2011). Lead toxicity in *Arabidopsis* is overcome by the expression of the ACBP1 gene (Xiao et al., 2008), while Cd toxicity by a change in the expression of a gene that controls sulphur metabolism, namely ATP sulfurylase (APS) and adenosine 5' phosphor sulfate reductase (APR), up-regulated expression of Ser-acetyl transferase (SAT) and O-acetyl-ser (thiol)-lyase (OASTL); and by the higher expression of Glutamyl cysteine synthetase (GCS) and glutathione synthetase (GS) that catalyze GSH production; Phytochelatin synthase (PCS), activation of antioxidant machinery and metal transporters (Zhang and Shu, 2006).

6. Economic loss due to heavy metal contamination

Contamination of HMs in soil affects the economy and human health. The HMs exposure and bioaccumulation are a threat to humans (Wuana and Okieimen, 2011) and great economic losses. China, for instance, estimates mining-induced HM contamination created 1.5 lakh hectares of discarded land at a rate of approximately fifty thousand hectares per annum (Li et al., 2014). The soil loses a part of its function, once it is contaminated by HMs as it maintains its functionality up to a certain concentration of the elements. Wuana and Okieimen, 2011, mentioned the importance of intervention and target values of soil HMs, where the former indicates the soil quality for human, flora, and fauna is crucially damaged and surplus value specify serious contamination; while the latter aims at soil quality and indicate the quality of soil needed for sustainability or expressed in terms of remedial policy, and required for the complete restitution of the soil's efficiency for the living beings. Soil is polluted by various elements, however, the soil value loss is calculated based on the highest metal extent of pollution (Abrahim and Parker, 2008; Duong and Lee, 2011).

Basta and Mcgowen (2004) reported that soil stabilization/solidification, excavation, transport, chemical extraction, and land-filling induced soil remediation strategies are highly efficient (with less risk) but expensive to implement; whereas, *in situ* phytoremediation (physicochemical remediation mechanism) is considered more cost-effective, environment-friendly and promising technology (Xiao et al., 2018); the cost is around 30 percent of physicochemical process, such as the washing of soil (EPA, 2000).

While dealing with somewhat polluted contaminated soils, reports have indicated that phytoextraction is cheaper but for higher contamination soil washing is preferred (Chen and Li, 2018). Soil HM contamination treatment requires extensive Cost-Effectiveness Analysis (CEA) between physicochemical methods and phytoremediation; which includes several steps like a clear understanding of the ecologicalgoal involved, the degree to which the goal has been met, identification of sources of contamination, stresses, and the effects of the current and future time horizons, a measure to connect the gap between the current condition and the desired outcome, assess the efficiency of these measures in accomplishing the goal, and calculate the cost, grading concerning enhanced unit costs and analysing the least-cost strategy (Chen and Li, 2018). However, the ideal and cost-effective alternative to treat soil contaminated by HM is a combination of both physicochemical and phytoremediation treatments.

7. Phytoremediation of heavy metals

During the last century, industrialization has grown at a rapid pace globally, resulting in increased demand for the exploitation of earth's resources in an unplanned manner. This has resulted in the menace of environmental pollution (Briffa et al., 2020). At present, the earth is polluted by several pollutants—toxic gases (Manisalidis et al., 2020), sludge (Wu et al., 2019), organometallic compounds (Babayigit et al., 2016; Fitella and Bonet, 2017), radioactive isotopes (Bjørklund et al., 2020) and HMs (Briffa et al., 2020). This has led to some serious environmental issues concerning the adverse effect on human health (Manisalidis et al., 2020; Bălă et al., 2021) and agricultural productivity (Toumisto et al., 2017; Sharma et al., 2020b). Thus, decontaminating the environment has become a matter of priority among the scientific community (Guillerm and Cesari, 2015; Moustafa, 2017). At present, the scientific community is focussing on the use of biological agents for cleaning environmental pollution through a process termed bioremediation. The bioremediation process uses organisms to clean up the contaminations. The major advantage of bioremediation is its cheap cost which gives it an edge over other conventional physicochemical methods (van Dillewijn et al., 2009). The process of bioremediation is also eco-friendly, sustainable, renewable, and can be easily adopted (Deniz, 2019). Phytoremediation is one subset of bioremediation that includes the eco-friendly approach to remediate soil and water using plants. Phytoremediation has two components, one being remediation with the assistance of root colonizing microbes while the other being remediation through the plant itself (Suresh and Ravishanker, 2004). In this section, the potential of plants to remediate HMs contaminated substrates is discussed. Plants have unique characteristics in the sense that they continuously interact with the environment. In the process, the roots of the plants transform the soil through several processes including uptake of water, minerals, and release of organic compounds through exudates (Morel et al., 1999). The presence of plant exudates in the soil attracts various microflora which results in the formation of a rhizosphere microflora favourable for plant growth (Doornbos et al., 2012). The root exudates also change the mobility and uptake of minerals and ions by the plants (Luo et al., 2014). Thus, plants are instrumental in altering the fate of ionic pollutants present in the soil and thus qualify as a suitable candidate for bioremediation. In this section, the importance of plants in the bioremediation process is discussed in detail. Plants have the potential to remediate a wide variety of pollutants. However, in this chapter, only the remediation of HMs is discussed.

Since plants are immobile and remain fixed to the substratum, they have developed very sophisticated processes to cope with the HMs stress or high HM concentration. The very first step which is related to HMs toxicity is metal avoidance, in which a plant restricts the entry of HMs (Emamverdian et al., 2015). The metal avoiders are not of much interest in phytoremediation as they don't contribute to the uptake process. However, metal hyperaccumulators are suitable candidates for phytoremediation as they uptake the contaminants (Sarma et al., 2011). Moreover, a good candidate for phytoremediation should possess an efficient translocation system for metallic contaminants, or in other words should have a high translocation factor (Girdhar et al., 2014). In general, the Metal hyperaccumulator species can accumulate extremely large levels of metals in their aerial parts without showing any signs of toxicity (Balafrej et al., 2020). A HM hyperaccumulator can accumulate HMs 100–1000 times more than that of a non-hyperaccumulating plant (Suman et al., 2018). To qualify as a HM accumulator, a plant should have the following qualities.

- The shoot to root HM ratio should be more than one which indicates efficient transportation of HMs to the aboveground parts. This is accompanied by efficient overexpression of transport systems required for transportation and sequestration of HMs (Viehweger, 2014).
- The shoot to soil HM ratio should be greater than one which indicates a higher capability of uptake of HMs from the soil by the roots (Wei et al., 2008).
- The concentration of accumulated metals in the shoots should be 10 mg per kg in case of Hg, 100 mg per kg for Cd and Se, 1000 mg per kg for Co, Cr, Cu, Ni, and Pb, and 10000 mg per kg for Zn and Mn (Yan et al., 2020).

Several strategies are adopted for the remediation of HMs by the plants. They are as follows:

- **Phytostabilization**

  In this process, a vegetation cover is established over a contaminated region to trap the contaminants within the vadose region through accumulation in the root system and immobilization within the rhizosphere (Bolan et al., 2011).

- **Phytoextraction**

  In this process, the metals from the contaminated sites are absorbed by the plants and translocated in a harvestable part. The main motive of phytoextraction is the reduction in the concentration of metals in contaminated sites (Nascimento and Xing, 2006).

- **Phytovolatilization**

  In this case, the contaminants are taken up by the root system and are converted into a gaseous state by plants and thereby released through the process of evapotranspiration (Greipsson, 2011).

- **Phytofiltration**

  Phytofiltration is the process through which plants are used to clean up contamination from an aqueous environment. When rooted plants are used to clean up the contaminants, the process is called rhizofiltration, while blastofiltration is the process of using seeds to clean up the contaminants. When shoots are involved, the process is called caulofiltration (Mesjasz-Przybyłowicz et al., 2004).

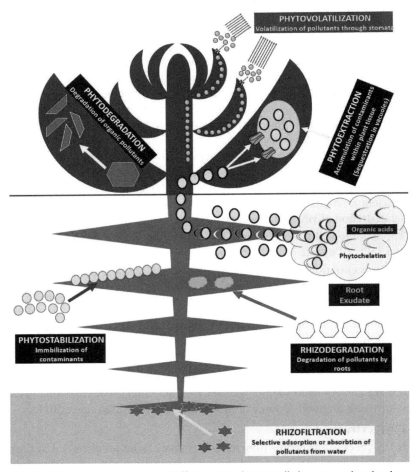

Figure 3: Schematic representation of different types phytoremediation process done by plants.

The plants involved in phytoremediation adapted to several physiological processes to sequester the contaminants. These include an efficient system for uptake, translocation of the contaminants, and final sequestration of the contaminants in the above-ground part. At present, 450 metal hyperaccumulators have been found in 45 angiosperm families (Praveen and Pandey, 2020). A lot of variation is observed in the accumulation pattern by the plants. A few species can accumulate a single element whereas certain species are reported to accumulate multiple elements. Table 3 represents the phytoremediation potential of selected plants.

8. Future prospects

The last twentieth and current century have witnessed a fast industrial revolution, speedy modernization, terrain utility changes particularly in populated developing countries like India and China. This has contributed to HM pollution over the globe, disquieting the environment, has adversely affected the growth of the plant and

Table 3: Phytoremediation potential of heavy metals by selected plants.

S. No.	Heavy metals remediated	Name of the Plant	Site of remediation	Beneficiary plant	Study outcome	Reference
Single metal						
1.	Pb	*Sinapis arvensis*	Pot experiment		Accumulation of lead in the roots in the roots.	Saghi et al., 2016
		Rapistrumrugosum				
2.	Pb	*Sonchus arvensis*	Hydroponic experiment		The highest accumulation of 849 mg per kg of lead was observed in the shoot.	Surat et al., 2008
			Pot experiment		The highest accumulation of 1397 mg per kg of lead was observed in the case of soil amended with organic fertilizer and ethylene diamine tetraacetic acid (EDTA).	
			Field trial		Accumulation of lead upto 3664 mg per kg in shoots.	
3.	Pb	*Sesuvium portulacastrum*	Experiment-soil		Accumulation of 3772ppm of lead in shots of plants exposed to exogenous lead.	Zaier et al., 2014
4.	Pb	*Mirabilis jalapa*	Experiment-soil		Coefficients and translocation factor increased for the plant with an increase in the concentration of chelating agents especially EDTA and NTA.	Yan et al, 2017
5.	Cr-benzo[a]pyrene	*Zea mays*	Experiment-soil		Enhancement of Cr accumulation upto 79% in shoots in the presence of benzo[a]pyrene	Chigbo and Batty, 2013
6.	Cr	*Pistiastratiotes*	Experiment-pond water		Improvement of Cr removal and translocation in plants upon supplementation with phosphorus and nitrogen.	Di Luca et al., 2014

Table 3 contd. ...

Table 3 contd. ....

S. No.	Heavy metals remediated	Name of the Plant	Site of remediation	Beneficiary plant	Study outcome	Reference
7.	Cr	*Ipomoea aquatica*	Hydroponic experiment		High accumulation of Cr in roots in presence of EDTA.	Chen et al., 2010
8.	Cr	*Polygonum coccineum*	Experiment-Soil/rock		Transfer of Cr from wastewater to plant body.	Kassaye et al., 2017
		Brachiara mutica	Synthetic solution		High translocation factor and removal efficiency of *Brachiaramutica*.	
		Cyprus papyrus	Tannery effluent		High translocation factor and removal efficiency of *Cyprus papyrus*.	
9.	Cd	*Brassica juncea*	Pot experiment		Large quantity of Cd accumulation in roots and shoots with translocation factor greater than 1.	Siddiqui et al., 2020
10.	Cd	*Parthenium hysterophorus*	Pot experiment		Increase in concentration, accumulation, and bioconcentration upon treatment with GA3 and EDTA.	Ali and Hadi, 2015
11.	Cd	*Carduusnutans, Phlomis* sp.	Soil-Field experiment		Accumulation of cd in shoots and roots.	Palutoglu et al., 2018
12.	Cd	*Typha latifolia*	Greenhouse pot culture		A high accumulation of cd was observed in the roots. Cd concentration in tissues ranged from 77.0 and 410.7 mg/kg.	Yang and Shen, 2020
13.	Cd	*Solanum nigrum*	Soil		*Solanum nigrum*: 36.9mg per kg of Cd was accumulated in leaves. *Lobelia chinensis*: 141 mg per kg of Cd in shoots.	Peng et al., 2009
	Cd	*Lobelia chinensis*	Hydroponic culture		*Solanum nigrum*: 1,110 mg per kg of Cd was accumulated in leaves. *Lobelia chinensis*: 414 mg per kg of Cd in shoots.	

14.	Hg	*Jatropha curcas*	Experiment-Soil from mining sites		Greater accumulation of Hg in the roots. Aerial parts accumulated a lesser amount of metals. High bioconcentration factor and low translocation factor.	Marrugo-Negrete et al., 2015
15.	Hg and methyl Hg	*Eichhornia crassipes*	Hydroponic experiment		The concentration of Hg and methyl Hg in the roots with minimal translocation in aerial parts.	Chattopadhyay et al., 2012
16.	Zn	*Salix pedicellata*	Hydroponic experiment		High concentration of Zn in SP-K20 clones in comparison to the control.	Amdoun et al., 2020
17.	Zn	*Salix matsudana × alba*	Field trial		High accumulation of Zn in plant tissues	Labrecque et al., 2020
18.	Mn	*Paulownia fortunei*	Experiment-soil		Increase in accumulation of Mn by 408.54% upon addition of spent mushroom compost.	Zhang et al., 2020
19.	Sr	*Sorghum bicolor*	Pot experiment		Accumulation of Sr in roots, stems and leaves. The concentration of Sr in the tissue decreased in the order leaves > roots > stems.	Wang et al., 2017
20.	As	*Pteris vittata*	Experiment-Soil	*Oryza sativa*	Accumulation of As and reduction in As content in rice grain.	Mandal et al., 2012
21.	As	*Pteris vittata*	Soil-Pot experiment	*Oryza sativa*	Removal of 3.5–11.4% of total As in the soil. Reduction in As content in grains and straw of rice plant growing on Pteris remediated soil.	Ye et al., 2011
22.	As	*Pteris vittata*	As contaminated site.	n/a	16.57% removal of As from the soil.	Lei et al., 2018
23.	As	*Pityrogramma calomelanos var. austroamericana* ; *Pteris vittata*	As contaminated cattle-dip site		Reduction of As content in the soil by (49– 63) % by using *P. calomelanos* var. *austroamericana* and (17–15) % by *Pteris vittata*.	Niazi et al., 2012
24.	As	*Lemna minor*	Hydroponic culture		70% removal of As.	Goswami et al., 2014

Table 3 contd. ...

Table 3 contd. ...

S. No.	Heavy metals remediated	Name of the Plant	Site of remediation	Beneficiary plant	Study outcome	Reference
25.	Ni	*Salvinia minima*	Hydroponic culture		Accumulation of 16.3 mg g($^{-1}$) dry weight of Ni.	Fuentes et al., 2014
Multiple Metals						
26	Co and Ni	*Berkheyacoddii*	Pot experiment		Co is retained in basal leaves whereas Ni is translocated to the top of the plant. The yield of Ni and Co were 77 kg and 16.5 kg per hectare respectively.	Rue et al., 2020
27.	Cr and Mn	*Ipomoea aquatica*	Hydroponic experiment		High efficiency in removal of chromium and Mn in both microcosms and mesocosms.	Haokip and Gupta, 2020
28.	Cd and Zn	*Ricinus communis*	Pot experiment		Maximum accumulation of Cd: 175.3 mg/kg Maximum accumulation of Zn: 386.8 mg/kg	He et al., 2020
29.	Cd and Pb	*Alternanthera bettzickiana*	Pot experiment		Uptake of metals in roots and shoots.	Tauqeer et al., 2016
30.	Cr and Cu	*Pistiastratioes*	Experiment-Water container		85 mg and 56 mg of chromium and 96 mg and 70 mg of Cu were accumulated on an average in roots and leaves respectively.	Tabinda et al., 2020
		Eichhornia crassipes			90 mg and 53 mg of chromium and 86mg and 50 of Cu were accumulated on an average in roots and leaves respectively.	
31.	Ni, Cd, Pb	*Azolla filiculoides*	Hydroponic experiment		Increase in removal efficiency of the metals from 40% to 70% in 10 days along with an increase in biomass from 2 g to 8 g.	Naghipour et al., 2018
32.	Cd, Cu, Zn	Intercropping of *Phyllostachys pubescens* with *Sedum plumbizincicola*	Field experiment		Available Cu, Zn, and Cd contents in the soil having intercropped plantation were 65.0, 28.7, and 48.4% lower than uncultivated control.	Bian et al., 2017

No.	Metals	Plant species	Experiment	Findings	Reference
33.	As, Cd, Pb, and Zn	*Pteris vittata*, co-planted with *Morus alba* or *Broussonetia papyrifera*	Pot experiment	The uptake of metals by *Pteris vittata* was increased by 80.0% and 64.2% when co-planted with *Morus alba* and *Broussonetia papyrifera*.	Zeng et al.,2019
34.	As, Cd, Pb, Zn	Intercropping of *Arundo donax* with *Broussonetia papyrifera* or *Morus alba*	Greenhouse experiment	Accumulation of Pb and Zn in shoots of combined plants increased by 171% and 124% through intercropping approach in comparison to monoculture treated *Arundo donax*. The accumulation of As and Pb in shoots also increased in intercropped plants.	Zeng et al., 2018
35.	Cd, Ni, Pb, Zn	*Salvinia minima*	Experiment	Absorption of Cd, Ni, Pb, Zn	Iha and Bianchini, 2015
36.	Cd, Cu, Pb, Ni	*Lemna minor*	Experiment	Removal efficiency for all metals ranged between 80–99%.	Bokhari et al., 2016
37.	Pb, Cu, Hg, Zn	*Eichhornia crassipes* *Myriophyllum aquaticum*		Removal of Cu PbHgZn	Romero-Hernández et al., 2017
38.	Zn, Cu, Pb, and Cd	*Ceratophyllum demersum* *Echinochloa pyramidalis* *Eichhornia crassipes* *Myriophyllum spicatum* *Phragmites australis* *Typha domingensis*	Water sediments	Higher accumulation of Cu and Zn in the shoots whereas roots accumulated higher concentration of Pb.	Fawzy et al., 2012
39.	Cd, Pb, Cu, and Zn	*Rhazya stricta*	Pot experiment	Highest bioconcentration of the HMs in plant roots.	Azab and Hegazy, 2020
40.	Pb, Cd, Zn, Cu, Iron, As	*Helianthus annuus*	Pot experiment	Accumulation of HMs in the plant	Chauhan and Mathur, 2020

increased human health hazards. With enhanced industrialization, the contamination of the environment has increased manifold. Thus, issues of soil contamination are presently a matter of global concern, particularly their intimate association with plants and animal.

Although, certain metals in trace amounts are important for growth, however, their higher values resulting deleterious effects. The release of industrial effluent and various types of sewage (unprocessed/processed) on soil are the main sources of HM pollution, though there are several other ways. Therefore, the discarding of industrial effluents (both raw and treated) on agricultural and forest land needs to be avoided. It thus becomes essential to avert the discard of processed waste as they are dangerous to plants and animals if assimilated into the system. Quick efficient mapping of land contamination is required to avoid the transport of HM pollutants into the food web and to devise effective remediation methods—specific microbial strains may be used for HMs degradation; optimize different conditions like biotechnological interventions to accelerate the HMs degradation/transformation processes. Therefore, HM remediation of soil could avoid the movement of HMs in the soil-crop system; this is an intricate process and uses versatile techniques. These translocation mechanisms of HMs from soil to crops are well known and attempts for remediation are directed toward decreasing metal concentrations in the land to reduce the successive transfer to crops. Remediation strategies include biological, chemical, ecological, and physical approaches. Biological remediation (phytoremediation and PGPR) may be applied as environment-friendly and economically feasible strategies for less polluted soils; also, eco-friendly inventions like the use of nano-tools could assist in the remediation of metallic contaminants, and farmer's awareness may improve socio-economic livelihoods. The Biochar application, which is derived from waste and effectively sequesters HMs by altering the physicochemical conditions of soil and reducing the phytoavailability of hazardous elements, is a powerful eco-remediation strategy to alleviate HM contamination pollution in soil and convergent multifaceted benefits. Biotechnological and molecular biology insights may increase our understanding of HM uptake in plants, and molecular biology techniques may have assisted in the elucidation of HM toxicity processes. Therefore, to enhance the effectiveness of any remediation mechanisms, physicochemical and biological features of polluted soil and concentrations of HM in soil are required to be investigated.

The majority of plant species are resistant to HM toxicity, with some being hyperaccumulative with several tolerance reactions compared to non-accumulative ones; nonetheless, these mechanisms are largely reliant on exclusion, chelation, and sequestration processes (Goyal et al., 2020). Also, metal-induced signal transduction pathways are triggered by stress, and these triggers need exhaustive study for understanding metal homeostasis. Therefore, the analysis of multiple stress factors under real environmental conditions remains a matter of challenge and hence, the focus should be based on low and environmentally pertinent HM concentrations. As a result, there is a wide range of tools to build holistic sensitivity detection approaches, such as tools that provide new insights into metalloproteases and their interactions, known as metallomics interactions.

9. Conclusion

Increased industrial-urban-agricultural activities have resulted in mismanagement of pooling and utilization of natural resources and results are seen as environmental pollution on a massive scale. This has led to several deleterious effects among plants, animals, and humans. The toxicity of HMs in the plants has resulted in a decrease in growth and overall productivity and this has also become a threat to the human population greatly impacting the sustainable food supply. Additionally, there have been numerous reports of accumulation of HMs within the plants used by people for regular consumption that is visible as biomagnification of HMs in trophic levels of ecosystem thereby increasing the chances of HM-related toxicity in the living organisms. Thus, the need of the hour is to minimize the environmental contamination of HMs. To achieve this, scientists have adopted a green and biological-based approach to decontaminate the HMs from the surrounding environment through the process of bioremediation. Plants have played a significant role in the bioremediation process largely aided by their unique physiological machinery to adsorb or sequester HMs. Bioremediation is of great advantage as it is eco-friendly, cost-effective, and sustainable. This has motivated the scientists to concentrate on the plants which are metal hyper accumulators that are fortified with unique physiological and genetic machinery to cope up with HM stress. Rapid advancements have been achieved in the field of transgenic technology to develop plants expressing genes related to HM accumulation. The decontamination of HMs needs to be set on top priority and further elaboration, emphasis, and exploration on the sustainable process to remediate the same needs to be done to achieve a healthy future.

References

Abedin MA, and Shaw R. 2013. Safe water adaptability for salinity, arsenic and drought risks in Southwest of Bangladesh. *Risk, Hazards Crisis Public Policy* 4: 62–82.

Abrahim GMS, and Parker RJ. 2008. Assessment of heavy metal enrichment factors and the degree of contamination in marine sediments from Tamaki estuary, Auckland, New Zealand. *Environ Monit Assess* 136: 227–238.

Ahmad K, Wajid K, Khan ZI, Ugulu I, Memoona H, Sana M, Nawaz K, Malik IS, Bashir H, and Sher M. 2019. Evaluation of potential toxic metals accumulation in wheat irrigated with wastewater. *Bull Environ Contam Toxicol* 102(6): 822–828.

Al Amin M, Rahman ME, Hossain S, Rahman M, Rahman MM, Jakariya M, and Sikder MT. 2020. Trace metals in vegetables and associated health risks in industrial areas of savar, Bangladesh. *J Health Pollut* 10(27): 200905.

Alengebawy A, Abdelkhalek ST, Qureshi SR, and Wang MQ. 2021. Heavy metals and pesticides toxicity in agricultural soil and plants: ecological risks and human health implications. *Toxics* 9(3): 42.

Ali H, Khan E, and Ilahi I. 2019. Environmental chemistry and ecotoxicology of hazardous heavy metals: environmental persistence, toxicity, and bioaccumulation. *J Chem* 2019: Article ID 6730305.

Ali N, and Hadi F. 2015. Phytoremediation of cadmium improved with the high production of endogenous phenolics and free proline contents in *Parthenium hysterophorus* plant treated exogenously with plant growth regulator and chelating agent. *Environ Sci Pollut Res Int* 22(17): 13305–18.

Amdoun R, Bendifallah N, Sahli F, Moustafa K, Hefferon K, Makhzoum A, and Khelifi L. 2020. Improving zinc phytoremediation characteristics in *Salix pedicellata* with a new acclimation approach. *Int J Phytoremediation* 22(7): 745–754.

Andersen EJ, Ali S, Byamukama E, Yen Y, and Nepal MP. 2018. Disease resistance mechanisms in plants. *Genes* 9(7): 339.

Aponte H, Medina J, Butler B, Meier S, Cornejo P, and Kuzyakov Y. 2020. Soil quality indices for metal(loid) contamination: An enzymatic perspective. *Land Degrad Dev* 31: 2700–2719.

Arévalo-Gardini E, Arévalo-Hernández CO, Baligar VC, and He ZL. 2017. Heavy metal accumulation in leaves and beans of cacao (*Theobroma cacao* L.) in major cacao growing regions in Peru. *Sci Total Environ* 605-606: 792–800.

Armienta MA, Beltrán M, Martínez S, and Labastida I. 2020. Heavy metal assimilation in maize (*Zea mays* L.) plants growing near mine tailings. *Environ Geochem Health* 42(8): 2361–2375.

Asgari K, and Cornelis WM. 2015. Heavy metal accumulation in soils and grains, and health risks associated with use of treated municipal wastewater in subsurface drip irrigation. *Environ Monit Assess* 187(7): 410.

Awa SH, and Hadibarata T. 2020. Removal of heavy metals in contaminated soil by phytoremediation mechanism: a review. *Water Air Soil Pollut* 231: 47.

Aydinalp C, and Marinova S. 2003. Distribution and forms of heavy metals in some agricultural soils. *Polish Journal of Env Studies* 12(5): 629–633.

Azab E, and Hegazy AK. 2020. Monitoring the efficiency of *Rhazya stricta* L. plants in phytoremediation of heavy metal-contaminated soil. *Plants (Basel)* 9(9): 1057.

Azizi S, Kamika I, and Tekere M. 2016. Evaluation of heavy metal removal from wastewater in a modified packed bed biofilm reactor. *PLoS One* 11(5): e0155462.

Babayigit A, Ethirajan A, Muller M, and Conings B. 2016. Toxicity of organometal halide perovskite solar cells. *Nat Mater* 15(3): 247–51.

Bălă GP, Râjnoveanu RM, Tudorache E, Motişan R, and Oancea C. 2021. Air pollution exposure-the (in) visible risk factor for respiratory diseases. *Environ Sci Pollut Res Int* 28(16): 19615–19628.

Balafrej H, Bogusz D, Triqui ZA, Guedira A, Bendaou N, Smouni A, and Fahr M. 2020. Zinc Hyperaccumulation in Plants: A Review. *Plants (Basel)* 9(5): 562. doi: 10.3390/plants9050562. PMID: 32365483; PMCID: PMC7284839.

Barraza F, Schreck E, Lévêque T, Uzu G, López F, Ruales J, Prunier J, Marquet A, and Maurice L. 2017. Cadmium bioaccumulation and gastric bioaccessibility in cacao: a field study in areas impacted by oil activities in Ecuador. *Environ Pollut* 229: 950–963.

Barraza F, Schreck E, Uzu G, Lévêque T, Zouiten C, Boidot M, and Maurice L. 2021. Beyond cadmium accumulation: distribution of other trace elements in soils and cacao beans in Ecuador. *Environ Res* 192: 110241.

Basta NT, and McGowen SL. 2004. Evaluation of chemical immobilization treatments for reducing heavy metal transport in a smelter-contaminated soil. *Environ Pollut* 127(1): 73–82.

Bian F, Zhong Z, Zhang X, and Yang C. 2017. Phytoremediation potential of moso bamboo (Phyllostachys pubescens) intercropped with *Sedum plumbizincicola* in metal-contaminated soil. *Environ Sci Pollut Res Int* 24(35): 27244–27253.

Bjørklund G, Semenova Y, Pivina L, Dadar M, Rahman MM, Aaseth J, and Chirumbolo S. 2020. Uranium in drinking water: a public health threat. *Arch Toxicol* 94(5): 1551–1560.

Bokhari SH, Ahmad I, Mahmood-Ul-Hassan M, and Mohammad A. 2016. Phytoremediation potential of *Lemna minor* L. for heavy metals. *Int J Phytoremediation* 18(1): 25–32.

Bolan NS, Park JH, Robinson B, Naidu R, and Huh KY. 2011. Phytostabilization: a green approach to contaminant containment. *Advances in Agronomy* 112: 145–204.

Bortey-Sam N, Nakayama SM, Akoto O, Ikenaka Y, Fobil JN, Baidoo E, Mizukawa H, and Ishizuka M. 2015. Accumulation of heavy metals and metalloid in foodstuffs from agricultural soils around Tarkwa area in Ghana, and associated human health risks. *Int J Environ Res Public Health* 12(8): 8811–27.

Briffa J, Sinagra E, and Blundell R. 2020. Heavy metal pollution in the environment and their toxicological effects on humans. *Heliyon* 6(9): e04691.

Brunet J, Varrault G, Zuily-Fodil Y, and Repellin A. 2009. Accumulation of lead in the roots of grass pea (*Lathyrus sativus* L.) plants triggers systemic variation in gene expression in the shoots. *Chemosphere* 77: 1113–1120.

CERCLA Priority List of Hazardous Substances. Agency for Toxic Substances and Disease Registry, USA. 2017. Available online: https://www.atsdr.cdc.gov/spl/(accessed on 20 September 2019).

Chabukdhara M, Munjal A, Nema AK, Gupta SK, and Kaushal RK. 2016. Heavy metal contamination in vegetables grown around peri-urban and urban-industrial clusters in Ghaziabad, India. *Hum Ecol Risk Assess* 22(3): 736–752.

Chang TC, You SJ, Yu BS, Chen CM, and Chiu YC. 2009. Treating high-mercury-containing lamps using full-scale thermal desorption technology. *J Hazard Mater* 162(2-3): 967–972.

Charfeddine M, Charfeddine S, and Bouaziz D. 2017. The effect of cadmium on transgenic potato (*Solanum tuberosum*) plants overexpressing the StDREB transcription factors. *Plant Cell Tiss Organ Cult* 128: 521–541.

Chattopadhyay S, Fimmen RL, Yates BJ, Lal V, and Randall P. 2012. Phytoremediation of mercury- and methyl mercury-contaminated sediments by water hyacinth (*Eichhornia crassipes*). *Int J Phytoremediation* 14(2): 142–61.

Chauhan P, and Mathur J. 2020. Phytoremediation efficiency of *Helianthus annuus* L. for reclamation of heavy metals-contaminated industrial soil. *Environ Sci Pollut Res* 27: 29954–66.

Chen F, Dong ZQ, Wang CC, Wei XH, Hu Y, and Zhang LJ. 2017. Heavy metal contamination of soils and crops near a zinc smelter. *Huan Jing Ke Xue* 38(10): 4360–4369. Chinese.

Chen JC, Wang KS, Chen H, Lu CY, Huang LC, Li HC, Peng TH, and Chang SH. 2010. Phytoremediation of Cr(III) by Ipomonea aquatica (water spinach) from water in the presence of EDTA and chloride: effects of Cr speciation. *Bioresour Technol* 101(9): 3033–9.

Chen W, and Li H. 2018. Cost-effectiveness analysis for soil heavy metal contamination treatments. *Water Air Soil Pollut* 229: 126.

Cheng JH. 2015. Background of land development and opportunity of land use transition. *Asian Agri Res* 7: 45–48.

Cherfi A, Abdoun S, and Gaci O. 2014. Food survey: levels and potential health risks of chromium, lead, zinc and Copper content in fruits and vegetables consumed in Algeria. *Food Chem Toxicol* 70: 48–53.

Chigbo C, and Batty L. 2014. Phytoremediation for co-contaminated soils of chromium and benzo[a] pyrene using *Zea mays* L. *Environ Sci Pollut Res Int* 21(4): 3051–9.

Chu DB, Duong HT, NguyetLuu MT, Vu-Thi HA, Ly BT, and Loi VD. 2021. Arsenic and heavy metals in Vietnamese Rice: assessment of human exposure to these elements through rice consumption. *J Anal Methods Chem* 2021: 6661955.

Coakley S, Cahill G, Enright AM, O'Rourke B, and Petti C. 2019. Cadmium hyperaccumulation and translocation in Impatiens glandulifera: from foe to friend? *Sustainability* 11(18): 5018.

Csuros M, and Csuros C. 2002. Environmental Sampling and Analysis for Metals, Lewis Publishers, Boca Raton, FL, USA.

Dalcorso G, Farinati S, and Furini A. 2010. Regulatory networks of cadmium stress in plants. *Plant. Signaling and Behavior* 5(6): 1–5.

Dalvi A, and Bhalerao SA. 2013. Response of plants towards heavy metal toxicity: an overview of avoidance, tolerance and uptake mechanism. *Ann Plant Sci* 2: 362–368.

Deng THB, van der Ent A, and Tang YT. 2018. Nickel hyperaccumulation mechanisms: a review on the current state of knowledge. *Plant Soil* 423: 1–11.

Deniz F. 2019. Bioremediation potential of waste biomaterials originating from coastal *Zostera marina* L. meadows for polluted aqueous media with industrial effluents. *Prog Biophys Mol Biol* 145: 78–84.

Dessalew G, Beyene A, Nebiyu A, and Astatkie T. 2018. Effect of brewery spent diatomite sludge on trace metal availability in soil and uptake by wheat crop, and trace metal risk on human health through the consumption of wheat grain. *Heliyon* 4(9): e00783.

Di Luca GA, Hadad HR, Mufarrege MM, Maine MA, and Sánchez GC. 2014. Improvement of Cr phytoremediation by Pistiastratiotes in presence of nutrients. *Int J Phytoremediation* 16(2): 167–78.

Dinu C, Gheorghe S, Tenea AG, Stoica C, Vasile GG, Popescu RL, Serban EA, and Pascu LF. 2021. Toxic Metals (As, Cd, Ni, Pb) Impact in the most common medicinal plant (Mentha piperita). *Int J Environ Res Public Health* 18(8): 3904.

Doornbos RF, van Loon LC, and Bakker PAHM. 2012. Impact of root exudates and plant defense signaling on bacterial communities in the rhizosphere. A review. *Agron Sustain Dev* 32: 227–243.

Duffus JH. 2002. "Heavy metals" a meaningless term? (IUPAC Technical Report). *Pure Appl Chem* 74(5): 793–807.

Duong TT, and Lee BK. 2011. Determining contamination level of heavy metals in road dust from busy traffic areas with different characteristics. *Journal of Environmental Management* 92: 554–562.

Emamverdian A, Ding Y, Mokhberdoran F, and Xie Y. 2015. Heavy metal stress and some mechanisms of plant defense response. *Sci World J* 2015: 756120.

Emurotu JE, and Onianwa PC. 2017. Bioaccumulation of heavy metals in soil and selected food crops cultivated in Kogi State, north central Nigeria. *Environ Syst Res* 6: 21.

EPA U. 2000. Introduction to phytoremediation. EPA/600/R-99/107.

Ertani A, Mietto A, Borin M, and Nardi S. 2017. Chromium in agricultural soils and crops: A review. *Water Air Soil Pollut* 228: 190.

Fan Y, Zhu T, Li M, He J, and Huang R. 2017. Heavy metal contamination in soil and brown rice and human health risk assessment near three mining areas in central China. *J Healthc Eng*: 4124302.

Fawzy MA, BadrNel-S, El-Khatib A, and Abo-El-Kassem A. 2012. Heavy metal biomonitoring and phytoremediation potentialities of aquatic macrophytes in River Nile. *Environ Monit Assess* 184(3): 1753–71.

Field RW, and Withers BL. 2012. Occupational and environmental causes of lung cancer. *Clin Chest Med* 33(4): 681–703.

Filella M, and Bonet J. 2017. Environmental impact of Alkyl Lead(IV) Derivatives: perspective after their phase-out. *Met Ions Life Sci* 17:/books/9783110434330/9783110434330-014/9783110434330-014.xml.

Fuentes II, Espadas-Gil F, Talavera-May C, Fuentes G, and Santamaría JM. 2014. Capacity of the aquatic fern (*Salvinia minima* Baker) to accumulate high concentrations of nickel in its tissues, and its effect on plant physiological processes. *Aquat Toxicol* 155: 142–50.

Gamalero E, Lingua G, Berta G, and Glick BR. 2009. Beneficial role of plant growth promoting bacteria and arbuscular mycorrhizal fungi on plant responses to heavy metal stress. *Canadian. J. Microbiol* 55(5): 501–514.

Garg N, and Singla P. 2011. Arsenic toxicity in crop plants: physiological effects and tolerance mechanisms. *Environ Chem Lett* 9: 303–321.

Garg VK, Yadav P, Mor S, Singh B, and Pulhani V. 2014. Heavy metals bioconcentration from soil to vegetables and assessment of health risk caused by their ingestion. *Biol Trace Elem Res* 157(3): 256–265.

Garrido AE, Strosnider WHJ, Wilson RT, Condori J, and Nairn RW. 2017. Metal-contaminated potato crops and potential human health risk in Bolivian mining highlands. *Environ Geochem Health* 39(3): 681–700.

Gebeyehu HR, and Bayissa LD. 2020. Levels of heavy metals in soil and vegetables and associated health risks in Mojo area, Ethiopia. *PLoS One* 15(1): e0227883.

Genchi G, Carocci A, Lauria G, Sinicropi MS, and Catalano A. 2020. Nickel: Human health and environmental toxicology. *Int J Environ Res Public Health* 17(3): 679.

Gichner T, Znidar I, and Záková J. 2008. Evaluation of DNA damage and mutagenicity induced by lead in tobacco plants. *Mutat Res Genet Toxicol Environ Mutagen* 652: 186–190.

Gill M. 2014. Heavy metal stress in plants: a review. *International Journal of Advanced Research* 2(6): 1043–1055.

Girdhar M, Sharma NR, Rehman H, Kumar A, and Mohan A. 2014. Comparative assessment for hyperaccumulatory and phytoremediation capability of three wild weeds. *3 Biotech* 4(6): 579–589.

Goss MJ, Tubeileh A, and Goorahoo D. 2013. A review of the use of organic amendments and the risk to human health. *Adv Agron* 120: 275–379.

Goswami C, Majumder A, Misra AK, and Bandyopadhyay K. 2014. Arsenic uptake by Lemna minor in hydroponic system. *Int J Phytoremediation* 16(7-12): 1221–7.

Goyal D, Yadav A, Prasad M, Singh TB, Shrivastav P, Ali A, Dantu PK, and Mishra S. 2020. Effect of Heavy metals on plant growth: an overview. *Contaminants in Agriculture*, pp: 79–101.

Greipsson S. 2011. Phytoremediation. Nature Education Knowledge 3(10): 7.

Grobelak A, and Napora A. 2015. The chemo phyto stabilisation process of heavy metal polluted soil. *PLoS One* 10(6): e0129538.

Guan Q, Wang F, Xu C, Pan N, Lin J, Zhao R, and Luo H. 2018. Source apportionment of heavy metals in agricultural soil based on PMF: a case study in Hexi Corridor, northwest China. *Chemosphere* 193: 189–197.

Guillerm N, and Cesari G. 2015. Fighting ambient air pollution and its impact on health: from human rights to the right to a clean environment. *Int J Tuberc Lung Dis* 19(8): 887–97.

Guo G, Lei M, Wang Y, Song B, and Yang J. 2018. Accumulation of As, Cd, and Pb in sixteen wheat cultivars grown in contaminated soils and associated health risk assessment. *Int J Environ Res Public Health* 15(11): 2601.

Gupta AK, Sinha S, Basant A, and Singh KP. 2007. Multivariate analysis of selected metals in agricultural soil receiving UASB treated tannery effluent at Jajmau, Kanpur (India). *Bull Environ Contam Toxicol* 79: 577582.

Gupta N, Ram H, and Kumar B. 2016. Mechanism of Zinc absorption in plants: uptake, transport, translocation and accumulation. *Rev Environ Sci Biotechnol* 15: 89–109.

Halim MA, Majumder RK, and Zaman MN. 2014. Paddy soil heavy metal contamination and uptake in rice plants from the adjacent area of Barapukuria coal mine, northwest Bangladesh. *Arab J Geosci* 8: 3391–3401.

Hanson PJ, Lindberg SE, Tabberer TA, Owens JA, and Kim KH. 1995. Foliar exchange of mercury vapor: evidence for a compensation point Springer: Netherlands, In Mercury as a Global Pollutant 373–382.

Haokip N, and Gupta A. 2020. Phytoremediation of chromium and Manganese by *Ipomoea aquatica Forssk.* from aqueous medium containing chromium-Manganese mixtures in microcosms and mesocosms. *Water Environ J* doi: 10.1111/WEJ.12676.

Harada E, Kim JA, Meyer AJ, Hell R, Clemens S, and Choi YE. 2010. Expression profiling of tobacco leaf trichomes identifies genes for biotic and abiotic stresses. *Plant Cell Physiol* 51: 1627–1637.

He C, Zhao Y, Wang F, Oh K, Zhao Z, Wu C, Zhang X, Chen X, and Liu X. 2020. Phytoremediation of soil heavy metals (Cd and Zn) by castor seedlings: tolerance, accumulation and subcellular distribution. *Chemosphere* 252: 126471.

Hossain M, Karmakar D, Begum SN, Ali SY, and Patra PK. 2021. Recent trends in the analysis of trace elements in the field of environmental research: a review. *Microchem J* 165: 106086.

Hossain MS, Ahmed F, Abdullah ATM, Akbor MA, and Ahsan MA. 2015. Public health risk assessment of heavy metal uptake by vegetables grown at a waste-water-irrigated site in Dhaka, Bangladesh. *J Health Pollut* 5(9): 78–85.

Hu HY, Lin H, Zheng W, Tomanicek S J, Johs A, Feng X, Elias DA, Liang L, and Gu B. 2013. Oxidation and methylation of dissolved elemental mercury by anaerobic bacteria. *Nature Geoscience* 6: 751–754.

Iha DS, and Bianchini I Jr. 2015. Phytoremediation of Cd, Ni, Pb and Zn by Salvinia minima. *Int J Phytoremediation* 17(10): 929–35.

International Agency for Research on Cancer. 1987. Overall evaluations of carcinogenicity: And updating of IARC monographs, vol. 1 to 42. IARC monographs on the evaluation of the carcinogenic risk of chemicals to humans: Suppl 7. *IARC* 7: 1–440.

Iwona M, Woźniak A, Mai VC, Rucińska-Sobkowiak R, and Jeandet P. 2018. The role of heavy metals in plant response to biotic stress. *Molecules* 23(9): 2320.

Jaishankar M, Tseten T, Anbalagan N, Mathew BB, and Beeregowda KN. 2014. Toxicity, mechanism and health effects of some heavy metals. *Interdiscip Toxicol* 7(2): 60–72.

Jan AT, Azam M, Siddiqui K, Ali A, Choi I, and Haq QM. 2015. Heavy metals and human health: mechanistic insight into toxicity and counter defense system of antioxidants. *Int J Mol Sci* 16(12): 29592–29630.

Jiang W, and Liu D. 2010. Pb-induced cellular defense system in the root meristematic cells of *Allium sativum* L. *BMC Plant Biol* 10: 1–40.

Joutey NT, Sayel H, Bahafid W, and El Ghachtouli N. 2015. Mechanisms of hexavalent chromium resistance and removal by microorganisms. In Reviews of Environmental Contamination and Toxicology; Springer: Berlin/Heidelberg, Germany 233: 45–69.

Junhe LU, Xinping YA, Xuchao ME, Guoqing WA, Yusuo LI, Yujun WA, and Fangjie ZH. 2017. Predicting cadmium safety thresholds in soils based on cadmium uptake by Chinese cabbage. *Pedosphere* 27(3): 475–81.

Kasowska D, Gediga K, and Spiak Z. 2018. Heavy metal and nutrient uptake in plants colonizing post-flotation Copper tailings. *Environ Sci Pollut Res Int* 25(1): 824–835.

Kassaye G, Gabbiye N, and Alemu A. 2017. Phytoremediation of chromium from tannery wastewater using local plant species. *Water Practice & Technology* (4): 894–901.

Kovács-Bokor É, Domokos E, and Biró B. 2021. Toxic metal phytoextraction potential and health-risk parameters of some cultivated plants when grown in metal-contaminated river sediment of Danube, near an industrial town. *Environ Geochem Health* 2021.

Kubier A, Wilkin RT, and Pichler T. 2019. Cadmium in soils and groundwater: A review. *Appl Geochem* 108: 1–16.

Kumar A, Kumar A, Chaturvedi AK, Shabnam AA, Subrahmanyam G, and Mondal R. 2020. Lead toxicity: Health hazards, influence on food chain, and sustainable remediation approaches. *Int J Environ Res Public Health* 17(7): 2179.

Kumar V, Sharma A, Kaur P, Sidhu GPS, Bali AS, Bhardwaj R, Thukral AK, and Cerda A. 2019. Pollution assessment of heavy metals in soils of India and ecological risk assessment: A state-of-the-art. *Chemosphere* 216: 449–462.

Labrecque M, Hu Y, Vincent G, and Shang K. 2020. The use of willow microcuttings for phytoremediation in a Copper, zinc and lead contaminated field trial in Shanghai, China. *Int J Phytoremediation* 22(13): 1331–1337.

Lei M, Wan X, Guo G, Yang J, and Chen T. 2018. Phytoextraction of arsenic-contaminated soil with Pteris vittata in Henan Province, China: comprehensive evaluation of remediation efficiency correcting for atmospheric depositions. *Environ Sci Pollut Res Int* 25(1): 124–131.

Li R, Wu H, and Ding J. 2017. Mercury pollution in vegetables, grains and soils from areas surrounding coal-fired power plants. *Sci Rep* 7: 46545.

Li Z, Ma Z, Jan TK, Yuan Z, and Huang L. 2014. A review of soil heavy metal pollution from mines in China: pollution and health risk assessment. *Science of The Total Environment* 468-469: 843–853.

Lin YC, Hsu SC, Chou CC, Zhang R, Wu Y, Kao SJ, Luo L, Huang CH, Lin SH, and Huang YT. 2016. Winter time haze deterioration in Beijing by industrial pollution deduced from trace metal fingerprints and enhanced health risk by heavy metals. *Environ Pollut* 208(Pt A): 284–293.

Liu Y, Liu K, Li Y, Yang W, Wu F, Zhu P, Zhang J, Chen L, Gao S, and Zhang L. 2016. Cadmium contamination of soil and crops is affected by intercropping and rotation systems in the lower reaches of the Minjiang River in south-western China. *Environ Geochem Health* 38(3): 811–820.

Liu K, Fan L, Li Y, Zhou Z, Chen C, Chen B, and Yu F. 2018. Concentrations and health risks of heavy metals in soils and crops around the Pingle Manganese (Mn) mine area in Guangxi Province, China. *Environ Sci Pollut Res Int* 25(30): 30180–30190.

Lizarazo MF, Herrera CD, Celis CA, Pombo LM, Teherán AA, Piñeros LG, Forero SP, Velandia JR, Díaz FE, Andrade WA, and Rodríguez OE. 2020. Contamination of staple crops by heavy metals in Sibaté, Colombia. *Heliyon* 6(7): e04212.

Lokeshwari H, and Chandrappa GT. 2006. Impact of heavy metal contamination of Bellandur Lake on soil and cultivated vegetation. *Current Science* 91(5): 622627.

Luo Q, Sun L, Hu X, and Zhou R. 2014. The variation of root exudates from the hyperaccumulator Sedum alfredii under cadmium stress: metabonomics analysis. *PLoS One* 9(12): e115581.

Mahfooz Y, Yasar A, Guijian L, Islam QU, Akhtar ABT, Rasheed R, Irshad S, and Naeem U. 2020. Critical risk analysis of metals toxicity in wastewater irrigated soil and crops: a study of a semi-arid developing region. *Sci Rep* 10(1): 12845.

Mahmood M, Mansour G, and Khalil K. 2017. Physiological and antioxidative responses of medicinal plants exposed to heavy metals stress. *Plant Gene* 11(B): 247–254.

Małecka A, Piechalak A, Morkunas I, and Tomaszewska B. 2008. Accumulation of lead in root cells of *Pisum sativum*. *Acta Physiol Plant* 30: 629–637.

Manara A. 2012. Plant responses to heavy metal toxicity. pp. 27–53. *In*: Plants and heavy metals. Springer, Dordrecht.

Mandal A, Purakayastha TJ, Patra AK, and Sanyal SK. 2012. Phytoremediation of arsenic contaminated soil by *Pteris vittata* L. II. Effect on arsenic uptake and rice yield. *Int J Phytoremediation* 14(6): 621–8.

Manisalidis I, Stavropoulou E, Stavropoulos A, and Bezirtzoglou E. 2020. Environmental and health impacts of air pollution: a review. *Front Public Health* 20;8: 14.

Mao QQ, Guan MY, Lu KX, Du ST, Fan SK, Ye YQ, Lin XY, and Jin CW. 2014. Inhibition of nitrate transporter 1.1-controlled nitrate uptake reduces cadmium uptake in Arabidopsis. *Plant Physiol* 166(2): 934–944.

Marrugo-Negrete J, Durango-Hernández J, Pinedo-Hernández J, Olivero-Verbel J, and Díez S. 2015. Phytoremediation of mercury-contaminated soils by *Jatropha curcas*. *Chemosphere* 127: 58–63.

Marschner H. 1995. Mineral Nutrition of Higher Plants. Second edition. pp: 889. London: Academic Press.

Martínez CE, and Motto HL. 2000. Solubility of lead, zinc and Copper added to mineral soils. *Environ Pollutn* 107(1): 153–158.

Mesjasz-Przybyłowicz JO, Nakonieczny MI, Migula PA, Augustyniak MA, Tarnawska MO, Reimold WU, Koeberl CH, Przybyłowicz WO, and Głowacka EL. 2004. Uptake of cadmium, lead nickel and zinc from soil and water solutions by the nickel hyperaccumulator Berkheyacoddii. *Acta Biol CracoviensiaSerBot* 46: 75–85.

Metanat K, Ghasemi-Fasaei R, Ronaghi A, and Yasrebi J. 2019. Lead phytostabilization and cationic micronutrient uptake by maize as influenced by Pb levels and application of low molecular weight organic acids. *Commun Soil Sci Plant Anal* 50: 1–10.

Minkina T, Rajput V, and Fedorenko G. 2020. Anatomical and ultrastructural responses of *Hordeum sativum* to the soil spiked by Copper. *Environ Geochem Health* 42: 45–58.

Mishra A, and Tirpathi BD. 2008. Heavy metal contamination of soil, and bioaccumulation in vegetables irrigated with treated wastewater in the tropical city of Varanasi, India. *Toxicological & Environmental Chemistry* 1–11. Retrieved on 24.06.2008 at www.informaworld.com.

Morel JL, Chaineau CH, Schiavon M, and Lichtfouse E. 1999. The role of plants in the remediation of contaminated soils. *In*: Baveye P, Block JC, and Goncharuk VV (eds.). *Bioavailability of Organic Xenobiotics in the Environment*. NATO ASI Series (Series 2: Environment), vol. 64. Springer, Dordrecht.

Mou H, Liu W, and Zhao L. 2021. Stabilization of hexavalent chromium with pretreatment and high temperature sintering in highly contaminated soil. *Front Environ Sci Eng* 15: 61.

Moustafa K. 2017. A clean environmental week: Let the nature breathe. *Sci Total Environ* 598: 639–646.

Mustafa K, Sukru D, Celalettin O, and Mehmet EA. 2006. Heavy metal accumulation in irrigated soil with waste water. *Ziraat Fakultesi Dergisi* 20(38): 6467.

Mustafiz A, Singh AK, Pareek A, Sopory SK, and Singla-Pareek SL. 2011. Genome-wide analysis of rice and *Arabidopsis* identifies two glyoxalase genes that are highly expressed in abiotic stresses. *Funct Integr Genomics* 11: 293–305.

Mwilola PN, Mukumbuta I, Shitumbanuma V, Chishala BH, Uchida Y, Nakata H, Nakayama S, and Ishizuka M. 2020. Lead, zinc and cadmium accumulation, and associated health risks, in maize grown near the Kabwe Mine in Zambia in response to organic and inorganic soil amendments. *Int J Environ Res Public Health* 17(23): 9038.

Nagajyoti PC, Lee KD, and Sreekanth TVM. 2010. Heavy metals, occurrence and toxicity for plants: a review. *Environ Chem Lett* 8: 199–216.

Naghipour D, Ashrafi SD, Gholamzadeh M, Taghavi K, and Naimi-Joubani M. 2018. Phytoremediation of heavy metals (Ni, Cd, Pb) by *Azolla filiculoides* from aqueous solution: A dataset. *Data Brief* 21: 1409–1414.

Nannoni F, Rossi S, and Protano G. 2016. Potentially toxic element contamination in soil and accumulation in maize plants in a smelter area in Kosovo. *Environ Sci Pollut Res Int* 23(12): 11937–46.

Nascimento CW, and Xing B. 2006. Phytoextraction: a review on enhanced metal availability and plant accumulation. *Scientia Agricola* 63(3): 299–311.

Nguyen TP, Ruppert H, Sauer B, and Pasold T. 2020a. Harmful and nutrient elements in paddy soils and their transfer into rice grains (*Oryza sativa*) along two river systems in northern and central Vietnam. *Environ Geochem Health* 42(1): 191–207.

Nguyen TP, Ruppert H, Pasold T, and Sauer B. 2020b. Paddy soil geochemistry, uptake of trace elements by rice grains (*Oryza sativa*) and resulting health risks in the Mekong River Delta, Vietnam. *Environ Geochem Health* 42(8): 2377–2397.

Niazi NK, Singh B, Van Zwieten L, and Kachenko AG. 2012. Phytoremediation of an arsenic-contaminated site using *Pteris vittata* L. and *Pityrogramma calomelanos* var. *austro americana*: a long-term study. *Environ Sci Pollut Res Int* 19(8): 3506–15.

Nkwunonwo UC, Odika PO, and Onyia PI. 2020. A review of the health implications of heavy metals in food chain in Nigeria. *Sci World J* 2020: Article ID 6594109, 11 pages.

Nriagu JO. 1989. A global assessment of natural sources of atmospheric trace metals. *Nature* 338: 47–49.

Ojuederie OB, and Babalola OO. 2017. Microbial and plant-assisted bioremediation of heavy metal polluted environments: a review. *Int J Environ Res Public Health* 14(12): 1504.

Olaniran AO, Balgobind A, and Pillay B. 2013. Bioavailability of heavy metals in soil: impact on microbial biodegradation of organic compounds and possible improvement strategies. *Int J Mol Sci* 14(5): 10197–10228.

Onakpa MM, Njan AA, and Kalu OC. 2018. A review of heavy metal contamination of food crops in Nigeria. *Ann Glob Health* 84(3): 488–494.

Oorts K. 2013. Copper. *In*: Alloway B (eds.). *Heavy Metals in Soils. Environmental Pollution*, 22: Springer, Dordrecht.

Palutoglu M, Akgul B, Suyarko V, Yakovenko M, Kryuchenko N, and Sasmaz A. 2018. Phytoremediation of cadmium by native plants grown on mining soil. *Bull Environ Contam Toxicol* 100(2): 293–297.

Pavlović P, Mitrović M, Đorđević D, Sakan S, Slobodnik J, Liška I, Csanyi B, Jarić S, Kostić O, Pavlović D, Marinković N, Tubić B, and Paunović M. 2016. Assessment of the contamination of riparian soil and vegetation by trace metals—A Danube River case study. *Sci Total Environ* 540: 396–409.

Peng KJ, Luo CL, Chen YH, Wang GP, Li XD, and Shen ZG. 2009. Cadmium and other metal uptake by *Lobelia chinensis* and *Solanum nigrum* from contaminated soils. *Bull Environ Contam Toxicol* 83(2): 260–4.

Praveen A, and Pandey VC. 2020. Pteridophytes in phytoremediation. *Environ Geochem Health* 42: 2399–2411.

Rafati RM, Kazemi S, and Moghadamnia AA. 2017. Cadmium toxicity and treatment: An update. *Caspian J Intern Med* 8(3): 135–145.

Rai PK, and Tripathi BD. 2008. Heavy metals in industrial wastewater, soil and vegetables in Lohta village, India. *Toxicol. Environ Chem* 90(2): 247–257.

Rai PK, Lee SS, Zhang M, Tsang YF, and Kim K. 2019. Heavy metals in food crops: Health risks, fate, mechanisms, and management. *Environ Int* 125: 365–385.

Rasul SB, Munir AKM, Hossain ZA, Khan AH, Alauddin M, and Hussam A. 2002. Electrochemical measurement and speciation of inorganic arsenic in groundwater of Bangladesh. *Talanta* 58: 33–43.

Rattan RK, Datta SP, Chhonkar PK, Suribabu K, and Singh AK. 2005. Long term impact of irrigation with sewage effluents on heavy metal content in soils, crops and groundwater—a case study. *Agriculture, Ecosystem and Environment* 109: 310–322.

Rea AW, Lindberg SE, and Keeler GJ. 2001. Dry deposition and foliar leaching of mercury and selected trace elements in deciduous forest throughfall. *Atmos Environ* 35: 3453–3462.

Resaee A, Derayat J, Mortazavi SB, Yamini Y, and Jafarzadeh MT. 2005. Removal of Mercury from chlor-alkali industry wastewater using *Acetobacter xylinum* cellulose. *Am J Environ Sci* 1(2): 102–105.

Retamal-Salgado J, Hirzel J, Walter I, and Matus I. 2017. Bioabsorption and bioaccumulation of cadmium in the straw and Grain of Maize (*Zea mays* L.) in Growing soils contaminated with cadmium in different environment. *Int J Environ Res Public Health* 14(11): 1399.

Rezapour S, Atashpaz B, Moghaddam SS, and Damalas CA. 2019a. Heavy metal bioavailability and accumulation in winter wheat (*Triticum aestivum* L.) irrigated with treated wastewater in calcareous soils. *Sci Total Environ* 656: 261–269.

Rezapour S, Atashpaz B, Moghaddam SS, Kalavrouziotis IK, and Damalas CA. 2019b. Cadmium accumulation, translocation factor, and health risk potential in a wastewater-irrigated soil-wheat (*Triticum aestivum* L.) system. *Chemosphere* 231: 579–587.

Roba C, Roşu C, Piştea I, Ozunu A, and Baciu C. 2016. Heavy metal content in vegetables and fruits cultivated in Baia Mare mining area (Romania) and health risk assessment. *Environ Sci Pollut Res Int* 23(7): 6062–73.

Romero-Hernández JA, Amaya-Chávez A, Balderas-Hernández P, Roa-Morales G, González-Rivas N, and Balderas-Plata MÁ. 2017. Tolerance and hyperaccumulation of a mixture of heavy metals (Cu, Pb, Hg, and Zn) by four aquatic macrophytes. *Int J Phytoremediation* 19(3): 239–245.

Rue M, Paul ALD, Echevarria G, van der Ent A, Simonnot MO, and Morel JL. 2020. Uptake, translocation and accumulation of nickel and Cobalt in Berkheyacoddii, a 'metal crop' from South Africa. *Metallomics* 12(8): 1278–1289.

Rydh CJ, and Svärd B. 2003. Impact on global metal flows arising from the use of portable rechargeable batteries. *Sci Total Environ* 302(1-3): 167–84.

Saghi A, Rashed, Mohassel MH, Parsa M, and Hammami H. 2016. Phytoremediation of lead-contaminated soil by *Sinapis arvensis* and *Rapistrumrugosum*. *Int J Phytoremediation* 18(4): 387–92.

Saha S, Hazra GC, Saha B, and Mandal B. 2015. Assessment of heavy metals contamination in different crops grown in long-term sewage-irrigated areas of Kolkata, West Bengal, India. *Environ Monit Assess* 187(1): 1–2.

Salinitro M, Tassoni A, Casolari S, de Laurentiis F, Zappi A, and Melucci D. 2019. Heavy metals bio indication potential of the common weeds *Senecio vulgaris* L., *Polygonum aviculare* L. and *Poa annua* L. *Molecules* 24(15): 2813.

Sarap NB, Janković MM, Todorović DJ, Nikolić JD, and Kovačević MS. 2014. Environmental radioactivity in southern Serbia at locations where depleted uranium was used. *ArhHigRadaToksikol* 65(2): 189–97.

Sarma H. 2011. Metal hyperaccumulation in plants: a review focusing on phytoremediation technology. *J Environ Sci Technol* 4(2): 118–38.

Sarret G, Guédron S, Acha D, Bureau S, Arnaud-Godet F, Tisserand D, Goni-Urriza M, Gassie C, Duwig C, Proux O, and Aucour AM. 2019. Extreme arsenic bioaccumulation factor variability in lake titicaca, Bolivia. *Sci Rep* 9(1): 10626.

Selvi A, Theertagiri J, Ananthaselvam A, Kumar KS, Madhavan J, and Rahman P. 2019. Integrated remediation processes towards heavy metal removal/recovery from various environments-a review. *Frontiers Environ* Sci 7: 66.

Shahid M, Dumat C, Khalid S, Schreck E, Xiong T, and Niazi NK. 2017. Foliar heavy metal uptake, toxicity and detoxification in plants: a comparison of foliar and root metal uptake. *J Hazard Mater* 325: 36–58.

Shankar S, Shanker U, and Shikha. 2014. Arsenic contamination of groundwater: a review of sources, prevalence, health risks, and strategies for mitigation. *Sci World J* 2014: 304524.

Sharma A, Kapoor D, Wang J, Shahzad B, Kumar V, Bali AS, Jasrotia S, Zheng B, Yuan H, and Yan D. 2020a. Chromium bioaccumulation and its impacts on plants: An overview. *Plants* 9(1): 100.

Sharma A, Shukla A, Attri K, Kumar M, Kumar P, Suttee A, Singh G, Barnwal RP, and Singla N. 2020b. Global trends in pesticides: A looming threat and viable alternatives. *Ecotoxicol Environ Saf* 201: 110812.

Sharma I, Pati PK, and Bhardwaj R. 2011. Effect of 28-homobrassinolide on antioxidant defence system in *Raphanus sativus* L. under chromium toxicity. *Ecotoxicology* 20: 862–74.

Sharma P, and Dubey RS. 2007. Involvement of oxidative stress and role of antioxidative defense system in growing rice seedlings exposed to toxic concentrations of aluminum. *Plant Cell Reports* 26(11): 2027–2038.

Shen X, Chi Y, and Xiong K. 2019. The effect of heavy metal contamination on humans and animals in the vicinity of a zinc smelting facility. *PLoS One* 14(10): e0207423.

Shen ZJ, Xu C, Chen YS, and Zhang Z. 2017. Heavy metals translocation and accumulation from the rhizosphere soils to the edible parts of the medicinal plant Fengdan (Paeoniaostii) grown on a metal mining area, China. *Ecotoxicol Environ Saf* 143: 19–27.

Siddiqui H, Ahmed KBM, Sami F, and Hayat S. 2020. Phytoremediation of cadmium contaminated soil using *Brassica juncea*: Influence on PSII Activity, Leaf Gaseous Exchange, Carbohydrate Metabolism, Redox and Elemental Status. *Bull Environ Contam Toxicol* 105(3): 411–421.

Singh R, Gautam N, Mishra A, and Gupta R. 2011. Heavy metals and living systems: an overview. *Ind J Pharmacol* 43(3): 246–253.

Singh S, Parihar P, Singh R, Singh VP, and Prasad SM. 2016. Heavy metal tolerance in plants: role of transcriptomics, proteomics, metabolomics, and ionomics. *Front Plant Sci* 26: 1143.

Stambulska UY, Bayliak MM, and Lushchak VI. 2018. Chromium (VI) toxicity in legume plants: Modulation effects of rhizobial symbiosis. *BioMed Res Int* 2018: 1–13.

Suman J, Uhlik O, Viktorova J, and Macek T. 2018. Phytoextraction of heavy metals: a promising tool for clean-up of polluted environment? *Front Plant Sci* 9: 1476.

Surat W, Kruatrachue M, Pokethitiyook P, Tanhan P, and Samranwanich T. 2008. Potential of *Sonchus arvensis* for the phytoremediation of lead-contaminated soil. *Int J Phytoremediation* 10: 325–42.

Suresh B, and Ravishankar GA. 2004. Phytoremediation—a novel and promising approach for environmental clean-up. *Crit Rev Biotechnol* 24(23): 97–124.

Tabinda AB, Irfan R, Yasar A, Iqbal A, and Mahmood A. 2020. Phytoremediation potential of *Pistiastratiotes* and *Eichhornia crassipes* to remove chromium and Copper. *Environ Technol* 41(12): 1514–1519.

Tangahu BV, Abdullah S, Rozaimah S, Basri H, Idris M, Anuar N, and Mukhlisin M. 2011. A review on heavy metals (As, Pb, and Hg) uptake by plants through phytoremediation. *Int J Chem Eng* 939161: 1–31.

Tauqeer HM, Ali S, Rizwan M, Ali Q, Saeed R, Iftikhar U, Ahmad R, Farid M, and Abbasi GH. 2016. Phytoremediation of heavy metals by *Alternanthera bettzickiana*: Growth and physiological response. *Ecotoxicol Environ Saf* 126: 138–146.

Tchounwou PB, Yedjou CG, Patlolla AK, and Sutton DJ. 2012. Heavy metal toxicity and the environment. *In*: Luch A (eds.). Molecular, *Clinical and Environmental Toxicology. Experientia Supplementum*, vol 101. Springer, Basel.

Tesfahun W, Zerfu A, Shumuye M, Abera G, Kidane A, and Astatkie T. 2021. Effects of brewery sludge on soil chemical properties, trace metal availability in soil and uptake by wheat crop, and bioaccumulation factor. *Heliyon* 7(1): e05989.

Tuomisto HL, Scheelbeek PF, Chalabi Z, Green R, Smith RD, Haines A, and Dangour AD. 2017. Effects of environmental change on agriculture, nutrition and health: A framework with a focus on fruits and vegetables. *Wellcome Open Res* 2: 21.

Ur Rehman Z, Khan S, Tahir Shah M, Brusseau ML, Akbar Khan S, and Mainhagu J. 2018. Transfer of heavy metals from soils to vegetables and associated human health risks at selected sites in Pakistan. *Pedosphere* 28(4): 666–679.

VanDillewijn P, Nojiri H, Van der Meer JR, and Wood TK. 2009. Bioremediation, a broad perspective. *Microb Biotechnol* 2(2): 125–7.

Varrica D, Aiuppa A, and Dongarrà Gaetano. 2000. Volcanic and anthropogenic contribution to heavy metal content in lichens from Mt. Etna and Vulcano island (Sicily). *Environmental pollution (Barking, Essex: 1987)* 108: 153–62.

Vaverková MD, Maxianová A, Winkler J, Adamcová D, and Podlasek A. 2019. Environmental consequences and the role of illegal waste dumps and their impact on land degradation. Land Use Policy 89.

Viehweger K. 2014. How plants cope with heavy metals. *Bot Stud* 55: 1–12.

Wang J, Su J, Li Z, Liu B, Cheng G, Jiang Y, Li Y, Zhou S, and Yuan W. 2019. Source apportionment of heavy metal and their health risks in soil-dustfall-plant system nearby a typical non-ferrous metal mining area of Tongling, Eastern China. *Environ Pollut* 254(Pt B): 113089.

Wang X, Chen C, and Wang J. 2017. Phytoremediation of Strontium contaminated soil by *Sorghum bicolor* (L.) Moench and soil microbial community-level physiological profiles (CLPPs). *Environ Sci Pollut Res Int* 24(8): 7668–7678.

Wang Y, Zong K, Jiang L, Sun J, Ren Y, Sun Z, Wen C, Chen X, and Cao S. 2011. Characterization of an *Arabidopsis* cadmium-resistant mutant cdr3-1D reveals a link between heavy metal resistance as well as seed development and flowering. *Planta* 233: 697–706.

Wani AL, Ara A, and Usmani JA. 2015. Lead toxicity: a review. *Interdiscip Toxicol.* 28(2): 55–64.

Wei SH, Yang CJ, and Zhou QX. 2008. Hyperaccumulative characteristics of 7 widely distributing weed species in composite family especially Bidenspilosa to heavy metals. *Huan Jing Ke Xue* 29(10): 2912–8. Chinese.

Wu Q, Liu Z, Liang J, Kuo DTF, Chen S, Hu X, Deng M, Zhang H, and Lu Y. 2019. Assessing pollution and risk of polycyclic aromatic hydrocarbons in sewage sludge from wastewater treatment plants in China's top coal-producing region. *Environ Monit Assess* 191(2): 102.

Wuana RA, and Okieimen FE. 2011. Heavy metals in contaminated soils: a review of sources, chemistry, risks and best available strategies for remediation. *Int Sch Res Notices* 2011: Article ID 402647, 20 pages.

Xiao R, Shen F, Du J, Li R, Lahori AH, and Zhang Z. 2018. Screening of native plants from wasteland surrounding a Zn smelter in Feng County China, for phytoremediation. *Ecotoxicol Environ Saf* 162: 178–183.

Xiao S, Gao W, Chen QF, Ramalingam S, and Chye ML. 2008. Overexpression of membrane-associated acyl-CoA-binding protein ACBP1 enhances lead tolerance in *Arabidopsis*. *Plant J* 54: 141–51.

Xu DM, Yan B, Chen T, Lei C, Lin HZ, and Xiao XM. 2017. Contaminant characteristics and environmental risk assessment of heavy metals in the paddy soils from lead (Pb)-zinc (Zn) mining areas in Guangdong Province, South China. *Environ Sci Pollut Res Int* 24(31): 24387–24399.

Yan A, Wang Y, Tan SN, MohdYusof ML, Ghosh S, and Chen Z. 2020. Phytoremediation: a promising approach for revegetation of heavy metal-polluted land. *Front Plant Sci* 11: 359.

Yan L, Li C, Zhang J, Moodley O, Liu S, Lan C, Gao Q, Zhang W. 2017. Enhanced Phytoextraction of Lead from Artificially Contaminated Soil by *Mirabilis jalapa* with Chelating Agents. *Bull Environ Contam Toxicol* 99(2): 208–212.

Yan ZZ, Ke L, and Tam NFY. 2010. Lead stress in seedlings of Avicennia marina, a common mangrove species in South China, with and without cotyledons. *Aquat Bot* 92: 112–118.

Yang Y, and Shen Q. 2020. Phytoremediation of cadmium-contaminated wetland soil with *Typha latifolia* L. and the underlying mechanisms involved in the heavy-metal uptake and removal. *Environ Sci Pollut Res Int* 27(5): 4905–4916.

Yanqun Z, Yuan L, Jianjun C, Haiyan C, Li Q, and Schratz C. 2005. Hyper accumulation of Pb, Zn and Cd in herbaceous grown on lead-zinc mining area in Yunnan, China. *Environ Int* 31: 755–762.

Ye WL, Khan MA, McGrath SP, and Zhao FJ. 2011. Phytoremediation of arsenic contaminated paddy soils with *Pteris vittata* markedly reduces arsenic uptake by rice. *Environ Pollut* 159(12): 3739–43.

Zabergja-Ferati F, Mustafa MK, and Abazaj F. 2021. Heavy metal contamination and accumulation in soil and plant from mining area of Mitrovica, Kosovo. *Bull Environ Contam Toxicol* 2021.

Zaier H, Ghnaya T, Ghabriche R, Chmingui W, Lakhdar A, Lutts S, and Abdelly C. 2014. EDTA-enhanced phytoremediation of lead-contaminated soil by the halophyte *Sesuvium portulacastrum*. *Environ Sci Pollut Res Int* 21(12): 7607–15.

Zeng F, Wei W, Li M, Huang R, Yang F, and Duan Y. 2015. Heavy metal contamination in rice-producing soils of hunan province, china and potential health risks. *Int J Environ Res Public Health* 12(12): 15584–93.

Zeng P, Guo ZH, Xiao XY, Peng C, and Huang B. 2018. Intercropping Arundo donax with Woody Plants to Remediate Heavy Metal-Contaminated Soil. *Huan Jing Ke Xue* 39(11): 5207–5216. Chinese.

Zeng P, Guo Z, Xiao X, Peng C, Feng W, Xin L, and Xu Z. 2019. Phytoextraction potential of *Pteris vittata* L. co-planted with woody species for As, Cd, Pb and Zn in contaminated soil. *Sci Total Environ* 650(Pt 1): 594–603.

Zhang J, and Shu WS. 2006. Mechanisms of heavy metal cadmium tolerance in plants. *Zhi Wu Sheng Li Yu Fen Zi Sheng Wu Xue Xue Bao* 32: 1–8.

Zhang M, Chen Y, Du L, Wu Y, Liu Z, and Han L. 2020. The potential of *Paulownia fortunei* seedlings for the phytoremediation of Manganese slag amended with spent mushroom compost. *Ecotoxicol Environ Saf* 196: 110538.

Zhang X, Zhong T, Liu L, and Ouyang X. 2015. Impact of soil heavy metal pollution on food safety in China. *PLoS One* 10(8): e0135182.

Zhou Y, Wan JZ, Li Q, Huang JB, Zhang ST, Long T, and Deng SP. 2020. Heavy metal contamination and health risk assessment of corn grains from a Pb-Zn mining area. *Huan Jing Ke Xue* 41(10): 4733–4739. Chinese.

3

Heavy Metal Contamination in Plants
Sources and Effects

Savita Bhardwaj,[1,#] *Sadaf Jan,*[2,#] *Dhriti Sharma,*[1,#]
Dhriti Kapoor,[1,*] *Rattandeep Singh*[2] *and Renu Bhardwaj*[3]

ABSTRACT

The predicament of heavy metal contamination is progressively becoming ubiquitous. Heavy metals including chromium, cadmium, mercury, copper, and lead are perilous environmental contaminants, notably in those realms with major anthropogenic activities. Heavy metal accretion in the soil is a major issue for agricultural yield due to its detrimental impact on food safety and market potential and phytotoxicity induced in crops. The ascendancy of plants along with their metabolic actions influences the geological and biological re-distribution of heavy metals via air, soil, and water pollution. The usual aftermath of heavy metal toxicity is immoderate accretion of reactive oxygen species as well as methylglyoxal, together lead to lipid peroxidation, enzyme inactivation, nucleic acid damage, and protein oxidation in plants. The current chapter constitutes the array of heavy metals, their prevalence, and toxicity induced in plants. Heavy metal toxicity has eminent reverberation to plants and subsequently affecting the ecosystem, where the flora is an elemental component. Plants thriving in metal contaminated soil evince altered metabolism, reduced growth, and less biomass

---

[1] Department of Botany, School of Bioengineering and Biosciences, Lovely Professional University, Phagwara, 144411, Punjab, India.
[2] Department of Biotechnology, School of Bioengineering and Biosciences, Lovely Professional University, Phagwara, 144411, Punjab, India.
[3] Department of Botanical and Environmental Sciences, Guru Nanak Dev University, Amritsar, 143005, Punjab, India.
* Corresponding author: dhriti405@gmail.com
# Contributed equally

yield. The contemporary analysis for toxicity and resistance in metal stressed plants actuated by the metal-contaminated environment. This study emphasizes on impact of heavy metals on plant proliferation, production, and mode of toxicity within plants.

1. Introduction

Heavy metals are metallic elements possessing high density. These are toxic even at smaller concentrations (Diaconu et al., 2020). The atomic density of heavy metal is greater than 4 g/cm^3 (Singh et al., 2016). Heavy metals comprise arsenic (As), iron (Fe), cadmium (Cd), copper (Cu), chromium (Cr), mercury (Hg), lead (Pb), platinum (Pt), and silver (Ag). Heavy metal pollution is a major environmental concern and is conceivably perilous due to bioaccumulation via the food chain (Aycicek et al., 2008), which emerges from expeditious industrialization, progresses in the utilization of farming synthetics, and urbanization. This has prompted the distribution of substantial metals in the environment, bringing about debilitated health of the populace, particularly by consumption of food crops adulterated by these toxic components (Zukowska et al., 2008). Assimilation of heavy metals by plants via absorption and accretion throughout the food chain is life-threatening to humans as well as animals (Jordao et al., 2006; Sprynskyy et al., 2007).

Heavy metals have the highest accessibility in soils, water bodies and to a considerably lesser extent in the atmosphere as vapors/particulate. Heavy metal noxiousness in plants fluctuates with plant varieties, particular metal, amount, compound structure, and soil constituents/pH. Few heavy metals are indispensable for plant growth and development among which copper and zinc act as cofactors and activators of various enzymatic reactions (Arif et al., 2016). These fundamental elements participate in redox reactions, electron transports, and function in nucleic acid metabolism. Several heavy metals, for example, cadmium, mercury, and arsenic are explicitly toxic to metal-sensitive enzymes, ensuing impediments in the growth and development of organisms. Another categorization of metals is derived from their coordination chemistry, and are known as class B metals. These are considered as nonessential trace components, which are exceptionally poisonous components, for example, Hg, Pb, Ni (Selvi et al., 2019). Among these, few are bio-accumulative and they neither degrade nor are metabolized. Aforesaid metals aggregate in the food chain during assimilation and consumption. Plants are sessile, and roots form an elementary contact site for metal particles. In an aquatic environment, the entire plant is susceptible to these particles. Heavy metals are directly absorbed by leaves because of ion deposition on foliar areas.

2. Source of Contamination

Various wellsprings of heavy metals exist in the ecosystem for example, (a) natural, (b) agrarian, (c) industrialization, (d) other sources (Fig. 1). Heavy metal contamination can arise from anthropogenic and natural activities. Processes like mining, refining, and agribusiness have defiled broad regions of the world, for instance, China, Japan, and Indonesia generally by substantial metals, viz, cadmium,

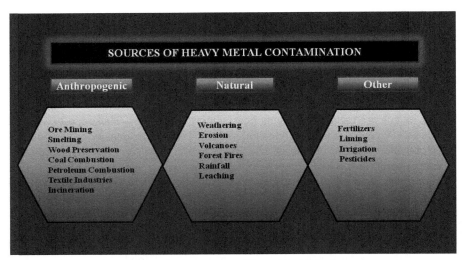

Figure 1: Various sources of Heavy Metal Pollution.

copper and zinc, copper, cadmium, and lead in North Greece, in Albania also, copper, lead, nickel, zinc, and cadmium in Australia (Nazli et al., 2020). Heavy metals are present inside the Earth's crust; henceforth their presence in soil is merely a result of weathering.

2.1 Natural source

The most significant natural genesis of heavy metals is geologic parent material/rock outcroppings. The amount and composition of substantial metals are determined by rock variety and natural conditions, triggering the weathering cycle. The geologic plant materials usually possess a major amount of Cu, manganese (Mn), cobalt (Co), nickel (Ni), zinc (Zn), selenium (Se), mercury (Hg), and lead (Pb). Howbeit, the concentration of heavy metals varies among the rocks. Sedimentary rocks are primarily the sources of soil formation, though they do not get weathered easily, they ares the source of heavy metals, albeit in minute quantity. Nonetheless, numerous magmatic rocks, such as olivine, augite, and hornblende offer a high quantity of Mn, Co, Ni, Cu, and Zn to the soil. Among the class of sedimentary rocks, shale has the most elevated amounts of Cr, Mn, Co, Ni, Cu, Zn, Cd, Se, Hg, and Pb, lastly by limestone and sandstone (Nagajyoti et al., 2010).

Volcanoes have been accounted to emanate elevated levels of aluminum, zinc, manganese, lead, nickel, copper, and mercury and some poisonous and hazardous gases (Edwards et al., 2020). Global information on the emanation of heavy metals from raw materials is inadequate. Wind dust, emerging from the desert area, for example, Sahara, consists of elevated levels of iron and a low quantity of manganese, zinc, chromium, nickel, and lead (Blonder et al., 2017). Marine vapors and woodland fires likewise elicit the distribution of few heavy metals in numerous conditions. Some significant volcanic eruptions have a prolonged impact, for example, an emanation from Mount Etna, Sicily comprises 10×10^6 kg/yr of cadmium, and chromium, copper, manganese, and zinc. This volcano remarkably increased the mercury level within plants and soils of nearby regions (Gworek et al., 2020). Airborne emanations of substantial metals arise from woods, grassland fires, and prairie fires. Unstable heavy metals, for example, mercury and selenium are the components of carbonaceous substances generated during the fire.

2.2 Agricultural sources

Inorganic and natural fertilizers are a major source of heavy metals in the agrarian realm while other sources, for instance liming, irrigation, pesticides, and sewage sludge also contribute to heavy metals in farming lands. Cd is accumulated in plant leaves at exceptionally higher levels, which is devoured by humans/animals. Cd enhancement likewise happens by using sewage sludge, compost, and limes (Yanqun et al., 2005). Manure enhances the soil with manganese, zinc, copper, and cobalt while sewage sludge adds zinc, chromium, lead, nickel, cadmium, and copper to the soil (Wuana and Okieimen, 2011). Heavy metal accretion in the soil also occurs by amending soil with compost and fertilizers. Pesticides, for instance, lead arsenate were utilized in Canada for agricultural purposes for nearly six years which caused amassing of lead, arsenic, and zinc in the soil thereby leading to food defilement. Frequent irrigation also results in heavy metal contamination viz, using water sources like deep wells, irrigation canals, rivers, and lakes causes an accmilation of lead and cadmium in the soil.

2.3 Industrial sources

Anthropogenic activities like mining, smelting, metal recycling, and finishing contribute to heavy metal accumulation in the environment. Mining activity produces diverse heavy metals that are determined by the type of mining. For instance, coalmines give rise to arsenic, cadmium, and iron etcetera which augment soil either directly or indirectly. The use of mercury in gold mining and the transport of mercury is also becoming a source of contamination (Esdaile and Chalker, 2018). High temperature processes viz., smelting emanate metals in the form of particulate and vapors. Vapor forms of metals, for instance, arsenic, cadmium, copper, lead, selenium, and zinc amalgamate with water to produce aerosols. Aerosols are disseminated by wind or rainfall resulting in water bodies and soil pollution. Pollution of soil and water bodies by heavy metals also ensues by runoff from crude ores mine industries, corrosion, and leaching into groundwater and soils. Soil pollution by heavy metals also arises from refinery plants. Petroleum combustion, coal combustion power industries, and nuclear power plants promote heavy metal pollution such as selenium, cadmium, copper, zinc, and nickel (Kaur et al., 2019). Various other industries including textiles, wood preservation, micro-electronics, paper, and plastic processing also contribute to heavy metal pollution (Gill, 2014).

3. Effect of heavy metals on plants

Whenever plants get a chance to grow in contaminated areas, they readily get exposed to high concentrations of heavy metals which are otherwise non-essential for their survival except for a few acting as micronutrients in minimal concentrations. These heavy metals are amassed in different parts of the plant once they enter the plant species from contaminated soil, water, or air. Toxic aftereffects of these heavy metals on various aspects of the life cycle of plants have been reported (Shamshed et al., 2018), some of which are discussed as follows:

3.1 Impact of heavy metals on seed germination

Seed germination is a vital phenomenon in the life of flowering plants which is banefully affected by heavy metals. Heavy metal toxicity in seeds reduces their productivity by decreasing the rate of germination along with causing oxidative harm to their membranal integrity, irregularities in sugar, and protein metabolism leading to nutrient loss (Wang et al., 2003; Ahmad and Ashraf, 2011). In germinating seeds, the accumulation of heavy metals such as Cd^{2+}, Ni^{2+}, and Pb^{2+} is more in radicles as compared to plumules. They have been found to reduce the rate of seed germination and radicle emergence in several plant species via inhibiting activities of hydrolyzing enzymes like α-amylases, β-amylases, acid invertases, acid phosphatases, proteases, and ribonucleases so that supply of reserve food materials in the form of carbohydrates and proteins is hampered to the developing embryo (Rahoui et al., 2010; Ashraf et al., 2011; Singh et al., 2011). Pb even modifies the genomic DNA profile of the seeds (Mohamed et al., 2011). Cd hyperaccumulation in seeds is reported to result in the piling of lipid peroxidation products, loss of nutrients due to membrane leakage, reduced embryo growth, and overall percentage of seed germination (Ahsan et al., 2007; Sfaxi-Bousbih et al., 2010; Smiri et al., 2011). However, this decrease in the total percentage of seed germination is associated with relatively higher concentrations of these heavy metals whereas their low concentrations are responsible for only causing a delay in the process of seed germination. A study by Munzuroglu and Geckil (2002) revealed that Cu lies next to Hg in impairing germination of seeds, elongation of radicle, hypocotyl and, epicotyl growth in wheat (*Triticum aestivum*) and cucumber (*Cucumis sativus*). In *Arabidopsis* plants, the reduction in the rate of seed germination due to heavy metal toxicity has been found in the order of Hg > Cd > Pb > Cu (Li et al., 2005).

3.2 Impact of heavy metals on plant growth

The physiological and biochemical functioning of plants is adversely affected by heavy metal toxicity resulting in impaired growth and development. For instance, Copper toxicity leads to a reduction in root growth along with severe malformities in *Chloris gayana* (Rhodes grass) and *Phaseolus vulgaris* (kidney bean) (Sheldon and Menzis, 2005; Katare et al., 2015; Pichhode and Nikhil, 2015). This is accounted for by the fact that Cu in excess becomes cytotoxic, induces chlorosis, and generates reactive oxygen species causing metabolic imbalance plus oxidative damage leading to retarded plant growth. Cd and Zn induced phytotoxicity has also been reported to bring about modulations in catalytic efficacy of enzymes, stoppage of several metabolic functions, oxidative harm, reduced growth and senescence in *Phaseolus vulgaris* (kidney bean), *Brassica juncea* (brown mustard), and *Pisumsativum* (pea) plants (Devries, 2002; Romero-Puertas et al., 2004). Cd excess lowers down nutrient uptake, thus affecting the proper growth of shoots in *Allium sativum* (garlic) and both root and shoot growth in *Triticum* sp. (wheat), *Zea mays* (maize) (Jiang et al., 2001; Wang et al., 2007; Ahmed et al., 2012a; Yourtchi and Bayat, 2013).

Hg, on the other hand, gets piled up in roots and shoots followed by an exhibition of toxicity symptoms in the form of stunted growth, less tillering, and inflorescence formation in *Oryza sativa* (rice) effectuating into a reduction in yield (Du et al., 2005; Kibra, 2008). Further, plant growth, nutrient uptake, biomass in *Zea mays* (maize) and plant height, leaf number, and leaf area in *Thespesia populnea* (portia tree) are drastically lowered down by Pb excess (Kabir et al., 2009; Hussain et al., 2013); whereas inhibition of leaf expansion, root and shoot growth, reduction in biomass, lowering of quantities of protein, sugar, and other nutrients in *Brassica napus* (oilseed rape), *Hordeum vulgare* (barley), and *Raphanus sativus* (radish) are owed due to the presence of Co beyond optimal concentrations (Jayakumar et al., 2007; Li et al., 2009).

3.3 Impact of heavy metals on photosynthesis

The inhibitory effect of heavy metals on plant metabolism is well documented but still, some of the heavy metals, being part of several enzymes and proteins, are found to be essential for keeping different metabolic reactions in their optimum state. For example, in the case of the most important metabolic process, i.e., photosynthesis, heavy metals are found to affect its light and dark phases directly and indirectly. Some of the heavy metals as Fe, Mn, and Cu are considered indispensable in photosynthesis because Fe is an integral part of redox enzymes involved in electron transport and also required for chlorophyll synthesis; Mn is a cofactor of several photosynthetic enzymes and has a crucial role in evolving molecular oxygen during photosynthesis whereas Cu is an important constituent of Plastocyanins of PSI involved in electron transport chains in light reactions. Their low concentrations or absence lead to deficiency symptoms however, they turn detrimental to the overall wellbeing of the plant when present in excess. It can be substantiated by citing toxicity symptoms produced in plants due to Cu in excess when it modulates the photosynthetic apparatus; alters the photosynthetic membranes in terms of their protein and pigment composition; distort the chlorophyll structure and impairs its functioning by replacing Mg from it (Küpper et al., 2003).

Besides these three, other heavy metals have also been reported to interfere with photosynthesis like Ni which cast an impact on the synthesis of photosynthetic pigments leading to a reduction in yield in *Vigna mungo* (mung bean) (Ahmad et al., 2007). Pb toxicity stops chlorophyll synthesis, modifies the structure of chloroplast, hampers the working of electron transport systems, inhibits enzymes of the C3 cycle, reduces uptake of essential nutrients, and decreases CO_2 content by closing stomata (Pourrut et al., 2011). In a study on the seedlings of *Pisum sativum* (pea) plants which when treated with heavy metals, also exhibited a substantial decline in chlorophyll and carotenoid content with Cd being most detrimental to their synthesis followed by Ni and Pb (Muradoglu et al., 2015; Paunov et al., 2018; Yousefi et al., 2018).

4. Heavy metals induced oxidative stress and antioxidant defense system

Heavy metal stress results in the generation of a plethora of ROS, for instance, superoxide radicals ($O_2{}^{\cdot-}$), hydroxyl radicals ($OH^{\cdot-}$), and hydrogen peroxide (H_2O_2); which are responsible for damaging cell wall lipids, nucleic acids and proteins (Cuypers et al., 2011). To decline this increased ROS level, plants have an antioxidant defense system comprised of enzymes like superoxide dismutase (SOD), catalase (CAT), glutathione reductase (GR), peroxidase (POD), ascorbate peroxidase (APX), and guaiacol peroxidases (GPX) and non-enzymatic antioxidants for example ascorbic acid (AsA), glutathione (GSH), and tocopherol (Sytar et al., 2013). For example, Cd toxicity reduced the functioning of cytosolic Cu, Zn-SOD genes, whereas the expression of Mn-SOD enhanced (Cuypers et al., 2011). Cd stress in *Solanum nigrum* exhibited enhanced functioning of SOD and APX enzymes, while CAT activity reduced after forty days of application (Fidalgo et al., 2011); where as functioning of SOD, CAT, APX, and DHAR antioxidative enzymes escalated under Cd toxicity in rice genotypes (Iqbal et al., 2010). Impact of Pb on growth and functioning of antioxidative enzymes in amassing ecotype (AE) and non-accumulating ecotype (NAE) of *Sedum alfredii* have exhibited that AE was more resistant to a plethora of Pb amounts, in which SOD and CAT amounts were increased in AE, whereas only SOD amount improved, while CAT activity reduced in NAE, after Pb application in comparison to control plants. Moreover, Pb toxicity elevated the level of MDA in both cultivars which ultimately, caused lipid peroxidation and membrane damage (Liu et al., 2008; Huang et al., 2012).

Cd, Pb, Cu, and Zn metals accumulated through roots caused oxidative stress in pea cultivars via excessive accretion of ROS, i.e., $O2^{\cdot-}$, H_2O_2 and also enhanced the action of SOD, CAT, and GR enzymes (Malecka et al., 2012). Heavy metal stress stimulates the different nonenzymatic antioxidants in plants, that are beneficial in heavy metal detoxification and diminish the level of ROS. Nonenzymatic antioxidants viz, phenolics, AsA, proline, tocopherol, flavonoid, and GSH, total ascorbate, have a promising appearance in metal detoxification. AsA and DHA are the crucial antioxidants, which are important for declining the plethora of ROS; ultimately protecting plants from metal-triggered oxidative damage (Mishra et al., 2014). Ni toxicity in *Oryza sativa* caused a reduction in plant growth, altered the membrane integrity through the production of excessive $O2^{\cdot-}$, H_2O_2, and $OH^{\cdot}$ radicals increased lipid peroxidation, and activities of SOD, CAT, and GPX antioxidative enzymes also got enhanced (Rajpoot et al., 2016). Treatment with Cd showed escalated amount of H_2O_2, which ultimately caused lipid peroxidation, electrolyte leakage, and improved the activities of SOD, APX and GSH in *Brassica juncea* genotypes (Ahmad et al., 2016).

5. Heavy metals mediated osmolytes and nutrient uptake

Osmo-protectants such as proline and glycine betaine (GB) play a major part in maintaining membrane integrity and provide tolerance to plants against stress conditions. The presence of significant levels of proline and GB does not alter enzyme function however, it hydrates enzymes; hence is helpful in the renewal of enzyme functions (Kishor et al., 2005). Cd treatment leads to the increased amount of proline and GB in comparison to control plants in faba bean genotypes (Ahmad et al., 2017); whereas in *Brassica juncea*, Cd triggered the level of proline (Ahmad et al., 2011, 2012b), pea plants (Pandey and Singh, 2012) and in wheat and maize (Zhao, 2011). A greater level of proline against stress environments regulates nutrient accumulation *via* H_2O passage, which is recognized to be a significant stress monitoring moiety (Ahmad et al., 2016). Proline has antioxidant characteristics and is helpful in defending tissues from ROS inducing oxidative stress (Ahmad et al., 2010; Jogaiah et al., 2013); whereas GB is beneficial to maintain the tissue osmoregulation progressions, obstruction of ROS production, defending photosynthetic apparatus stimulation of stress-related genes in response to stress conditions (Sakamoto and Murata, 2002; Chen and Murata, 2008; Ahmad et al., 2010, 2016). Treatment with Cd elevated the level of proline and GB, compared to the control plants (Ahmad et al., 2018).

Heavy metals like As hinder plant growth, alters micro and macronutrient accumulation by competing with plant nutrients for transporters—As is acknowledged to interfere with plant Phosphorus accumulation (Garg and Singla, 2011). Exposure to Hg, Pb, and As caused increment in Mg, K, and Ca level in stem portion while, Cu, Fe, and Zn levels increased in roots of *Pfaffia glomerate* (Gupta et al., 2013). Exposure to Cd and Cu metals resulted in similar alterations in the content of mineral nutrients in which both Cd and Cu reduced the amount of K, Fe, Zn, and Mn in a concentration-dependent manner, however, no noteworthy influence was noticed in Ca and Mg amount in *Brassica napus* (Mwamba et al., 2016). Cd treatment altered the level of nutrients like Zn, Cu, Mg, Mn, Fe, Ca and K of king grass in which the amount of root Zn, Mg and Ca enhanced expressively by elevating the soil Cd amount, whereas the level of other nutrients was considered stable. In addition to this, the amount of Zn, Cu, Mg, and Ca in the shoot was elevated expressively, and the amount of Fe and K remained stable, and the Mn amount was reduced (Zhang et al., 2014).

6. The defense mechanism of plants against heavy metals toxicity

Plants contain simple and well-organized protective mechanisms to evade or resist heavy metal stress in which morphological barricades like dense cuticles, physiologically active portions, i.e., trichomes, cell walls, and mycorrhizal symbiosis, are the plants' 1st line of protection in response to heavy metals tress (Harada et al., 2010). However, when heavy metals ions crossed morphological barricades and accumulated in various plants tissues and cells, plants trigger numerous protective strategies to alleviate the harmful impacts of heavy metals. Production of various biological moieties such as protein metallochaperones, chelators like organic acids, GSH, PCs, and MTs, plant exudates (flavonoid and phenolic compounds), phytohormones and many more, are the basic and most important assets to mitigate heavy metal stress (Dalvi and Bhalerao, 2013; Viehweger, 2014). However, when previously stated mechanisms do not alleviate heavy metal stress, then oxidative stress conditions arise due to the excessive generation of ROS (Mourato et al., 2012). Therefore, to reduce ROS level, plants antioxidant defense mechanism is stimulated, i.e., enzymatic and non-enzymatic antioxidants by performing the ROS scavenging activity (Sharma et al., 2012).

GSH plays an important role in providing defense against heavy metal stress through ROS quenching or the production of PCs. PCs, heavy metal-binding moieties are formed by exposure to heavy metals from GSH via phytochelatin synthase (PCS) enzyme that are regarded useful for metal remediation and regulation of cellular homeostasis (Yadav, 2010) (Fig. 2). As-phytochelatin synthase 1 (*AsPCS1*) and yeast cadmium factor 1 (*YCF1*) exhibited improved resistance for Cd and As and increased the potential of metal uptake in *Arabidopsis thaliana* (Guo et al., 2012). *Elsholtziahai chowensis* metallothionein type 1 (*EhMT1*) declined Cu stress by reducing H_2O_2 level and enhanced the peroxidase (POD) action, ultimately lowering the oxidative damage in tobacco genotypes (Xia et al., 2012). Whereas, *TaMT3*, isolated from *Tamarixand rossowii*, integrated into tobacco provides improved tolerance to Cd toxicity via declining the level of ROS by escalating the activity SOD enzyme, however, reduced POD functioning (Zhou et al., 2014).

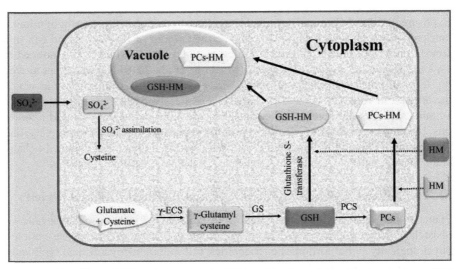

Figure 2: Glutathione (GSH) and phytochelatin (PCs) mediated metal detoxification in plants. GSH conjugates with metal ions by glutathione *S*-transferase enzyme to sequester metal ions into vacuole, whereas PCs synthesized from GSH, also make complexes with the metal ions to sequester them in the plant vacuole. Abbreviations used: SO_4^{2-}—sulphate; γ-ECS—γ-Glutamyl cysteine synthetase; GS—glutathione synthetase; PCS—phytochelatin synthase; HM—heavy metal (modified after Dubey et al., 2018).

7. Conclusion

Soil and H_2O pollution by heavy metals causes acute toxicity threats to plants and humans also. Heavy metal toxicity causes several adverse effects in plants such as altering plant growth, photosynthetic apparatus, antioxidative enzymes activities, osmolytes level, mineral nutrients, as well as gene expression. Plants resistance to heavy metals shows their potential to alleviate the adverse impact of metals via inhibiting accumulation of metals, compartmentation within the cell, compound formation, metal chelation, as well as by metal sequestration in the vacuole and they also induce the plant antioxidant defense system to diminish the oxidative damage caused by ROS. Glutathione is also beneficial in declining the level of ROS by the formation of PCs which sequester metal ions in the plant vacuoles. The multifaceted role of PCs and MTs for heavy metal detoxification is becoming a major focus of researchers. There is a high demand for research to discover how heavy metals influence plants in small input sustainable agronomic techniques.

References

Ahmad I, Akhtar MJ, Zahir ZA, and Jamil A. 2012a. Effect of cadmium on seed germination and seedling growth of four wheat (*Triticum aestivum* L.) cultivars. *Pak J Bot* 44: 1569–1574.

Ahmad MS, and Ashraf M. 2011. Essential roles and hazardous effects of nickel in plants. *Rev Environ Conta Sm Toxicol* 214: 125–67.

Ahmad MS, Hussain M, Saddiq R, and Alvi AK. 2007. Mungbean: A nickel indicator, accumulator or excluder? *Bull Environ Contam Toxicol* 78: 319–24.

Ahmad P, Abd Allah EF, Hashem A, Sarwat M, and Gucel S. 2016. Exogenous application of selenium mitigates cadmium toxicity in *Brassica juncea* L. (Czernand Cross) by up-regulating antioxidative system and secondary metabolites. *J Plant Growth Regul* 35: 936–950.

Ahmad P, Ahanger MA, Alyemeni MN, Wijaya L, and Alam P. 2018. Exogenous application of nitric oxide modulates osmolyte metabolism, antioxidants, enzymes of ascorbate-glutathione cycle and promotes growth under cadmium stress in tomato. *Protoplasma* 255: 79–93.

Ahmad P, Alyemeni MN, Wijaya L, Alam P, Ahanger MA, and Alamri SA. 2017. Jasmonic acid alleviates negative impacts of cadmium stress by modifying osmolytes and antioxidants in faba bean (*Viciafaba* L.). *Arch Agron Soil Sci* 63: 1889–1899.

Ahmad P, Jaleel CA, Salem MA, Nabi G, and Sharma S. 2010. Roles of enzymatic and nonenzymatic antioxidants in plants during abiotic stress. *Crit Rev Biotechnol* 30: 161–175.

Ahmad P, Nabi G, and Ashraf M. 2011. Cadmium-induced oxidative damage in mustard [*Brassica juncea* (L.) Czern. and Coss.] plants can be alleviated by salicylic acid. *South Afr J Bot* 77: 36–44.

Ahmad P, Ozturk M, and Gucel S. 2012b. Oxidative damage and antioxidants induced by heavy metal stress in two cultivars of mustard (*Brassica juncea* L.) plants. *Fresenius Environ Bull* 21: 2953–2961.

Ahsan N, Lee SH, Lee DG, Lee H, Lee SW, Bahk JD, and Lee BH. 2007. Physiological and protein profiles alternation of germinating rice seedlings exposed to acute cadmium toxicity. *C RBiol* 330: 735–746.

Arif N, Yadav V, Singh S, Singh S, Ahmad P, Mishra RK, Sharma S, Tripathi DK, Dubey NK, and Chauhan DK. 2016. Influence of high and low levels of plant-beneficial heavy metal ions on plant growth and development. *Front Environ Sci* 4: 69.

Ashraf MY, Sadiq R, Hussain M, Ashraf M, and Ahmad MSA. 2011. Toxic effect of nickel (Ni) on growth and metabolism in germinating seeds of sunflower (*Helianthus annuus* L.). *Biol Trace Elem Res* 143: 1695–1703.

Aycicek M, Kaplan O, and Yaman M. 2008. Effect of cadmium on germination, seedling growth and metal contents of sunflower (*Helianthus annus* L.). *Asian J Chem* 20: 2663.

Blonder B, Boyko V, Turchyn AV, Antler G, Sinichkin U, Knossow N, Klein R, and Kamyshny Jr. A. 2017. Impact of aeolian dry deposition of reactive iron minerals on sulfur cycling in sediments of the Gulf of Aqaba. *Front Microbiol* 8: 1131.

Chen TH, and Murata N. 2008. Glycinebetaine: an effective protectant against abiotic stress in plants. *Trends Plant Sci* 13: 499–505.

Cuypers A, Karen S, Jos R, Kelly O, Els K, Tony R, Nele H, Nathalie V, Yves G, Jan C, and Jaco V. 2011. The cellular redox state as a modulator in cadmium and copper responses in *Arabidopsis thaliana* seedlings. *J Plant Physiol* 168: 309–316.

Dalvi AA, and Bhalerao SA. 2013. Response of plants towards heavy metal toxicity: an overview of avoidance, tolerance and uptake mechanism. *Ann Plant Sci* 2: 362–368.

Devries W, Lofts S, Tipping E, Meili M, Groenenberg JE, and Schutze G. 2002. Impact of soil properties on critical concentrations of cadmium, lead, copper, zinc and mercury in soil and soil solution in view of ecotoxicological effects. *Rev Environ Contam Toxicol* 191: 47–89.

Diaconu M, Pavel LV, Hlihor RM, Rosca M, Fertu DI, Lenz M, Corvini PX, and Gavrilescu M. 2020. Characterization of heavy metal toxicity in some plants and microorganisms—A preliminary approach for environmental bioremediation. *New Biotechnol* 56: 130–139.

Du X, Zhu Y-G, Liu, W-J, and Zhao X-S. 2005. Uptake of mercury (Hg) by seedlings of rice (*Oryza sativa* L.) grown in solution culture and interactions with arsenate uptake. *Environ Exp Bot* 54: 1–7.

Dubey S, Shri M, Gupta A, Rani V, and Chakrabarty D. 2018. Toxicity and detoxification of heavy metals during plant growth and metabolism. *Environ Chem Lett* 16: 1169–1192.

Edwards BA, Kushner DS, Outridge PM, and Wang F. 2020. Fifty years of volcanic mercury emission research: knowledge gaps and future directions. *Sci Total Environ* 143800.

Esdaile LJ, and Chalker JM. 2018. The mercury problem in artisanal and small-scale gold mining. *Chemistry* 24: 6905.

Fidalgo F, Freitas R, Ferreira R, Pessoa AM, and Teixeira J. 2011. *Solanum nigrum* L. antioxidant defence system isozymes are regulated transcriptionally and post translationally in Cd-induced stress. *Environ Exp Bot* 72: 312–319.

Garg N, and Singla P. 2011. Arsenic toxicity in crop plants: physiological effects and tolerance mechanisms. *Environ Chem Lett* 9: 303–321.

Gill M. 2014. Heavy metal stress in plants: a review. *Int J Adv Res* 2: 1043–1055.

Guo J, Xu W, and Ma M. 2012. The assembly of metals chelation by thiols and vacuolar compartmentalization conferred increased tolerance to and accumulation of cadmium and arsenic in transgenic *Arabidopsis thaliana*. *J Hazard Mater* 199: 309–313.

Gupta DK, Huang HG, Nicoloso FT, Schetinger MR, Farias JG, Li TQ, Razafindrabe BHN, Aryal N, and Inouhe M. 2013. Effect of Hg, As and Pb on biomass production, photosynthetic rate, nutrients uptake and phytochelatin induction in *Pfaffia glomerata*. *Ecotoxicology* 22: 1403–1412.

Gworek B, Dmuchowski W, and Baczewska-Dąbrowska AH. 2020. Mercury in the terrestrial environment: a review. *Environ Sci Eur* 32: 1–19.

Harada E, Kim JA, Meyer AJ, Hell R, Clemens S, and Choi YE. 2010. Expression profiling of tobacco leaf trichomes identifies genes for biotic and abiotic stresses. *Plant Cell Physiol* 51: 1627–1637.

Huang H, Gupta DK, Tian S, Yang XE, and Li T. 2012. Lead tolerance and physiological adaptation mechanism in roots of accumulating and non-accumulating ecotypes of *Sedum alfredii*. *Environ Sci Pollut Res* 19: 1640–1651.

Hussain A, Abbas N, and Arshad F. 2013. Effects of diverse doses of lead (Pb) on different growth attributes of *Zea mays* L. *Agric Sci* 4: 262–265.

Iqbal N, Masood A, Nazar R, Syeed S, and Khan NA. 2010. Photosynthesis, growth and antioxidant metabolism in mustard (*Brassica juncea* L.) cultivars differing in cadmium tolerance. *Agric Sci China* 9: 519–527.

Jayakumar K, Jaleel CA, and Vijayarengan P. 2007. Changes in growth, biochemical constituents, and antioxidant potentials in radish (*Raphanus sativus* L.) under cobalt stress. *Turk J Biol* 31: 127–136.

Jiang W, Liu D, and Hou W. 2001. Hyperaccumulation of cadmium by roots, bulbs and shoots of garlic. *Bioresour Technol* 76: 9–13.

Jogaiah S, Govind SR, and Tran LSP. 2013. Systems biology-based approaches toward understanding drought tolerance in food crops. *Crit Rev Biotechnol* 33: 23–39.

Jordao CP, Nascentes CC, Cecon PR, Fontes RLF, and Pereira JL. 2006. Heavy metal availability in soil amended with composted urban solid wastes. *Environ Monit Assess* 112: 309–326.

Kabir M, Iqbal MZ, and Shafiq M. 2009. Effects of lead on seedling growth of *Thespesia populnea* L. *Adv Environ Biol* 3: 184–190.

Katare J, Pichhode M, and Nikhil K. 2015. Growth of *Terminalia bellirica* [(gaertn.) roxb.] on the malanjkhand copper mine overburden dump spoil material. *Int J Res* GRANTHAALAYAH 3: 14–24.

Kaur R, Sharma S, and Kaur H. 2019. Heavy metals toxicity and the environment. *J Pharmacogn Phytochem* SP1: 247–249.

Kibra MG. 2008. Effects of mercury on some growth parameters of rice (*Oryza sativa* L.). *Soil Environ* 27: 23–28.

Kishor PK, Sangam S, Amrutha RN, Laxmi PS, Naidu KR, Rao KS, Rao S, Reddy KJ, Theriappan P, and Sreenivasulu N. 2005. Regulation of proline biosynthesis, degradation, uptake and transport in higher plants: its implications in plant growth and abiotic stress tolerance. *Curr Sci* 88: 424–438.

Küpper H, Šetlík I, Šetliková E, Ferimazova N, Spiller M, and Küpper FC. 2003. Copper-induced inhibition of photosynthesis: limiting steps of *in vivo* copper chlorophyll formation in *Scenedesmus quadricauda. Funct Plant Biol* 30: 1187–1196.

Li HF, Gray C, Mico C, Zhao FJ, and McGrath SP. 2009. Phytotoxicity and bioavailability of cobalt to plants in a range of soils. *Chemosphere* 75: 979–986.

Li W, Mao R, and Liu X. 2005. Effects of stress duration and non-toxic ions on heavy metals toxicity to *Arabidopsis* seed germination and seedling growth. Chinese. *J Appl Ecol* 16: 1943–1947.

Liu D, Li TQ, Yang XE, Islam E, Jin XF, and Mahmood Q. 2008. Effect of Pb on leaf antioxidant enzyme activities and ultrastructure of the two ecotypes of *Sedum alfredii* Hance. *Russ J Plant Physiol* 55: 68–76.

Malecka A, Piechalak A, Mensinger A, Hanć A, Baralkiewicz D, and Tomaszewska B. 2012. Antioxidative defense system in *Pisum sativum* roots exposed to heavy metals (Pb, Cu, Cd, Zn). *Pol J Environ Stud* 21(6).

Mishra B, Sangwan RS, Mishra S, Jadaun, JS, Sabir F, and Sangwan NS. 2014. Effect of cadmium stress on inductive enzymatic and nonenzymatic responses of ROS and sugar metabolism in multiple shoot cultures of Ashwagandha (*Withania somnifera* Dunal). *Protoplasma* 251: 1031–1045.

Mohamed HI. 2011. Molecular and biochemical studies on the effect of gamma rays on lead toxicity in cowpea (*Vigna sinensis*) plants. *Biol Trace Elem Res* 144: 1205–1218.

Mourato M, Reis R, and Martins LL. 2012. Characterization of plant antioxidative system in response to abiotic stresses: a focus on heavy metal toxicity. *In*: Montanaro G (eds.). *Advances in Selected Plant Physiology Aspects* 12: 1–17.

Munzuroglu O, and Geckil H. 2002. Effects of metals on seed germination, root elongation, and coleoptile and hypocotyls growth in *Triticumaestivum* and *Cucumissativus. Arch Environ Cont Tox* 43: 203–213.

Muradoglu F, Gundogdu M, Sezai E, Tarik E, Balta F, Jaafar HZE, and Ziaul-Haq M. 2015. Cadmium toxicity affects chlorophyll a and b content, antioxidant enzyme activities and mineral nutrient accumulation in strawberry. *Biol Res* 48: 1–7.

Mwamba TM, Ali S, Ali B, Lwalaba JL, Liu H, Farooq MA, Shou J, and Zhou W. 2016. Interactive effects of cadmium and copper on metal accumulation, oxidative stress, and mineral composition in *Brassica napus. Int J Environ Science Technol* 13: 2163–2174.

Nagajyoti PC, Lee KD, and Sreekanth TVM. 2010. Heavy metals, occurrence and toxicity for plants: a review. *Environ Chemlett* 8(3): 199–216.

Nazli F, Mustafa A, Ahmad M, Hussain A, Jamil M, Wang X, Shakeel Q, Imtiaz M, and El-Esawi MA. 2020. A review on practical application and potentials of phytohormone-producing plant growth-promoting rhizobacteria for inducing heavy metal tolerance in crops. *Sustainability* 12: 9056.

Pandey N, and Singh GK. 2012. Studies on antioxidative enzymes induced by cadmium in pea plants (*Pisum sativum*). *J Environ Biol* 33: 201.

Paunov M, Koleva L, Vassilev A, Vangronsveld J, and Goltsev V. 2018. Effects of different metals on photosynthesis: Cadmium and Zinc affect chlorophyll fluorescence in Durum Wheat. *Int J Mol Sci* 19: 787.

Pichhode M, and Nikhil K. 2015. Effect of copper dust on photosynthesis pigments concentrations in plants species. *Int J Eng Res Manag* 2: 63–66.

Pourrut B, Shahid M, Dumat C, Winterton P, and Pinelli E. 2011. Lead uptake, toxicity, and detoxification in plants. *Rev Environ Contam Toxicol* 213: 113–136.

Rahoui S, Chaoui A, and El Ferjani EJ. 2010. Membrane damage and solute leakage from germinating pea seed under cadmium stress. *Hazard Mater* 178: 1128–1131.

Rajpoot R, Rani A, Srivastava RK, Pandey P, and Dubey RS. 2016. *Terminalia arjuna* bark extract alleviates nickel toxicity by suppressing its uptake and modulating antioxidative defence in rice seedlings. *Protoplasma* 253: 1449–1462.

Romero-Puertas MC, Rodriquez-Serrano M, Corpas FJ, Gomez M, Del Rio LA, and Sandalio LM. 2004. Cadmium-induced subcellular accumulation of O2 and H2O2 in pea leaves. *Plant Cell Env* 27: 1122–1134.

Sakamoto A, and Murata N. 2002. The role of glycine betaine in the protection of plants from stress: clues from transgenic plants. *Plant Cell Environ* 25: 163–171.

Selvi A, Rajasekar A, Theerthagiri J, Ananthaselvam A, Sathishkumar K, Madhavan J, and Rahman PK. 2019. Integrated remediation processes toward heavy metal removal/recovery from various environments-a review. *Front Environ Sci* 7: 66.

Sfaxi-Bousbih A, Chaoui A, and El Ferjani E. 2010. Cadmium impairs mineral and carbohydrate mobilization during the germination of bean seeds. *Ecotoxicol Environ Saf* 73: 1123–1129.

Shamshed S, Shahis M, Ratia M, Khalid S, Dumat C, Sabir M, Farooq ABU, and Shah NS. 2018. Effect of organic amendment on cadmium stress to pea: A multivariate comparison of germination vs young seedlings and younger vs older leaves. *Ecotoxicol Environ Saf* 151: 91–97.

Sharma P, Jha AB, Dubey RS, and Pessarakli M. 2012. Reactive oxygen species, oxidative damage, and antioxidative defense mechanism in plants under stressful conditions. *J Bot*, 1–26.

Sheldon AR, and Menzies NW. 2005. The effect of copper toxicity on the growth and root morphology of Rhodes grass (*Chloris gayana* Knuth.) in resin buffered solution culture. *Plant Soil* 278: 341–349.

Singh HP, Kaur G, Batish DR, and Kohli RK. 2011. Lead (Pb)-inhibited radicle emergence in *Brassica campestris* involves alterations in starch-metabolizing enzymes. *Biol Trace Elem Res* 144: 1295–1301.

Singh S, Parihar P, Singh R, Singh VP, and Prasad SM. 2016. Heavy metal tolerance in plants: role of transcriptomics, proteomics, metabolomics, and ionomics. *Front Plant Sci* 6: 1143.

Smiri M, Chaoui A, Rouhier N, Gelhaye E, Jacquot JP, and El Ferjani E. 2011. Cadmium affects the glutathione/glutaredoxin system in germinating pea seeds. *Biol Trace Elem Res* 142: 93–105.

Sprynskyy M, Kosobucki P, Kowalkowski T, and Buszewski B. 2007. Influence of clinoptilolite rock on chemical speciation of selected heavy metals in sewage sludge. *J Hazardous Mat* 149: 310–316.

Sytar O, Kumar A, Latowski D, Kuczynska P, Strzałka K, and Prasad MNV. 2013. Heavy metal-induced oxidative damage, defense reactions, and detoxification mechanisms in plants. *Acta Physiol Plant* 35: 985–999.

Viehweger K. 2014. How plants cope with heavy metals. *Botanical Stud* 55: 35.

Wang M, Zou J, Duan X, Jiang W, and Liu D. 2007. Cadmium accumulation and its effects on metal uptake in maize (*Zea mays* L.). *Bioresour Technol* 98: 82–88.

Wang W, Vinocur B, and Altman A. 2003. Plant responses to drought, salinity and extreme temperatures: Towards genetic engineering for stress tolerance. *Planta* 218: 1–14.

Wuana RA, and Okieimen FE. 2011. Heavy metals in contaminated soils: a review of sources, chemistry, risks and best available strategies for remediation. *Int Sch Res Notices* 2011.

Xia Y, Qi Y, Yuan Y, Wang G, Cui J, Chen Y, Zhang H, and Shen Z. 2012. Overexpression of *Elsholtziahai chowensis* metallothionein 1 (EhMT1) in tobacco plants enhances copper tolerance and accumulation in root cytoplasm and decreases hydrogen peroxide production. *J Hazard Mater* 233: 65–71.

Yadav SK. 2010. Heavy metals toxicity in plants: an overview on the role of glutathione and phytochelatins in heavy metal stress tolerance of plants. *South Afr J Bot* 76: 167–179.

Yanqun Z, Yuan L, Jianjun C, Haiyan C, Li Q, and Schvartz C. 2005. Hyperaccumulation of Pb, Zn and Cd in herbaceous grown on lead–zinc mining area in Yunnan, China. *Environ Int* 31: 755–762.

Yourtchi MS, and Bayat HR. 2013. Effect of cadmium toxicity on growth, cadmium accumulation and macronutrient content of durum wheat (Dena CV.). *Int J Agric Crop Sci* 6: 1099–1103.

Yousefi Z, Lolahi M, Majd A, and Jonoubi P. 2018. Effect of cadmium on morphometric traits, antioxidant enzyme activity and phytochelatins synthase gene expression (SoPCS) of *Saccharum officinarum* var. pp. 48–103 *in vitro*. *Ecotoxicol Environ Saf* 157: 472–481.

Zhang, X, Zhang, X, Gao, B, Li, Z, Xia, H, Li, H, and Li, J. 2014. Effect of cadmium on growth, photosynthesis, mineral nutrition and metal accumulation of an energy crop, king grass (*Pennisetum americanum× P. purpureum*). *Biomass Bioenergy*, 67: 179–187.

Zhao Y. 2011. Cadmium accumulation and antioxidative defenses in leaves of *Triticumaestivum* L. and *Zea mays* L. *Afr J Biotechnol* 10: 2936–2943.

Zhou B, Yao W, Wang S, Wang X, and Jiang T. 2014. The metallothionein gene, TaMT3, from Tamarix androssowii confers Cd2+ tolerance in tobacco. *Int J Mol Sci* 15(6): 10398–10409.

Żukowska J, and Biziuk M. 2008. Methodological evaluation of method for dietary heavy metal intake. *J Food Sci* 73: R21–R29.

4

Heavy Metal Contamination in Plants
Present and Future

Mamta Pujari,[1,*] *Savita Bhardwaj,*[1] *Dhriti Sharma,*[1] *Anju Joshi,*[2]
Shivani Kotwal[1] and *Dhriti Kapoor*[1]

ABSTRACT

Owing to increasing contamination in the environment across the globe, mainly in developing nations, pollutants always have been a widely discussed topic and it will continue to be so far in several coming years. Heavy metal pollution is emphasized in particular, taking into consideration the variability and the robustness of contamination processes plus its harmful impacts on soil, crops, and human health. Many aspects of the environment, biotic as well as abiotic, are mostly affected by heavy metal toxicity. Urbanization, agricultural practices, absorption of hazardous waste in the atmosphere along inability to break the cycle of pollution are all major factors that make these metal pollutants threatening by nature. Metals exist in a stable system in the environment as a natural element of the surface of the earth, with a variety of habitats thriving. They are involved in many physiochemical, as well as metabolic responses, and functions in biological processes. Metals are, in a way, of cellular existence, as protein-stabilizing elements, enzymes, and cofactors of essential constituents of biomolecules, and their abundance or scarcity affects metabolic processes, resulting in a variety of ailments. This implementation has encouraged us to reconsider how we use the natural resources as well as how we work on them, and how we might minimize or eradicate the harm we have created, and concentrate on sustainable development, for the sake of coming generations.

[1] Department of Botany, School of Bioengineering and Biosciences, Lovely Professional University, Phagwara (Punjab) India.
[2] Department of Botany, Govt. Degree College, Sitarganj U. S. Nagar (Uttarakhand).
* Corresponding author: mamta.21901@lpu.co.in

1. Introduction

The elements of metallic nature with quite high density and molecular weight are generally referred to as heavy metals. They might be found in differing amounts in the entirety of our ecosphere but can cast their deleterious effects even in the lowest of concentrations owing to their high persistence and bioaccumulative potential. The likes of arsenic (As), cadmium (Cd), chromium (Cr), lead (Pb), mercury (Hg), Nickel (Ni), and thallium (Tl) are some prominent examples of heavy metals which naturally exist either in elemental forms or complexed with different chemical substances. However, accurate characterization of heavy metals by taking into account their physical and chemical attributes has to be performed quite cautiously as some of the heavy metals tend to share their properties with lighter ones and likewise. Heavy metals do have impressive applications. For example, copper (Cu) and iron (Fe) are the integral parts of respiratory pigments; cobalt (Co) is the central constituent of metabolically efficient vitamin cyanocobalamin or B12; gold (Au), silver (Ag), platinum (Pt) are precious metals; manganese (Mn), vanadium (V), and zinc (Zn) act as inorganic cofactors in complex enzymes (Rao and Reddi, 2000; Feldmann, 2009; Butcher, 2010; Stowers, 2014). But heavy metals such as arsenic (As), cadmium (Cd), lead (Pb), and mercury (Hg) definitively symbolize the downside of chemistry by being extremely toxic, even in trace quantities (Duruibe et al., 2007).

Both the natural as well as human inputs have resulted in the addition of heavy metals into our immediate surroundings throwing the biogeochemical cycles off-balance, polluting the soil, affecting the optimal growth and yield of plants, food scarcity, and human existence thereof. Various sources which can be held responsible for this inclusion of heavy metals in our lives are listed as activities involving natural disintegration of earth's crustal plate, volcanic eruptions, quarrying, industrial effluents, wastewater sludge, automobile exhausts, mobile phones, eye cosmetics, fertilizers, paints, pesticides, and some agricultural malpractices (Ming-Ho, 2005). Vaporous heavy metals and the ones which get readily carried over minute particulate matter are extensively transported to large distances.

Alteration in physiochemical and biological characteristics of soil due to the presence of heavy metals, which affect the performance of crops, plants, in particular, pressing the need of looking out for effective measures of remediation. Procedures based on the principles of adsorption and filtration such as ion exchange (with adsorbents like kaolinite, bio-char, or agri-waste), nano-carriers (carbon, metal-oxide, or simply metal-based nanoparticles), photocatalysis, advanced membrane filtration, electrodialysis, etc., have been quite popular lately for the expedite recovery of heavy metals from contaminated soils (Wang, 2005; Rasheed et al., 2018; Ullah et al., 2018). Other modern approaches are centered around using either plants (phytoremediation) or microorganisms (microbial remediation) or both in combination for decontaminating the heavy metal polluted soils; collectively falling under the umbrella term of bioremediation. Since bioremediation is an eco-friendly, cost-efficient, and more effective measure, it reclaims the metal-polluted soil for the successful establishment or reestablishment of affected plant species in a more natural manner. This chapter discusses the mechanism of entry and storage of heavy metals from the soil into the plants, negative aftereffects of heavy metal-induced toxicity on different aspects of plant life, activation of innate defence system to mitigate this particular stress, and adoption of biological corrective procedures in particular.

2. Mechanism of uptake and accumulation of heavy metals in plants

The mechanism by which plant roots uptake heavy metals from their immediate environment is influenced by exudates discharged from roots along with a multitude of soil attributes such as its water and organic matter content, pH, and occurrence of heavy metals. The heavy metals found in free ionic nature or combined with organic/ inorganic substances are available for uptake in their soluble form whereas oxides, hydroxides, or silicates of heavy metals are absorbed by plant roots via chelant induced process (Abollino et al., 2006). Several research works have proven the chelating nature of root exudates such as for increased accumulation of heavy metals like Cd and Zn in *Sedum alfredii* and Pb, Cd, and Cu in *Echinochloa crus-galli* (Li et al., 2005). Root exudates help in reducing the organically complex heavy metals found in the rhizosphere and make them readily available to the plant roots in free ionic forms or simple compounds (Quartacci et al., 2009). Other such natural chelators include phytosiderophores, organic acids of low molecular weight, and soil microbes such as bacteria and mycorrhizal fungi which influence the availability of metals by solubilizing them (Neubauer et al., 2000; Yang et al., 2005). The proton transport pumps located in cell membranes of root cells help in maintaining the pH of the soil by acidifying the soils with the release of H^+ ions as observed in the case of Cu uptake by *Elsholtzia splendens* (Peng et al., 2005).

Heavy metals penetrate the roots via both apoplastic (Cell wall and intercellular spaces) and symplastic (plasmodesmata connections) pathways with the help of different transport proteins. Further, if after uptake, the heavy metals reside mainly in roots or other below-ground organs barring restricted translocation, then such plant species are known as non-hyperaccumulators but if these are translocated from roots to above-ground parts as well, i.e., shoots and leaves, then these plant species fall in the category of hyperaccumulators (Memon and Schroder, 2009). Heavy metals, after translocation in an upward direction, are finally stockpiled in leaves in case of hyperaccumulator plant species (Mahmood, 2010). During the process of translocation, these heavy metals first enter the water transport channels of shoots via xylem loading and are then transported along with water and dissolved salts. Plants tend to detoxify themselves by combining heavy metals with chelating agents or sequestering them inside the cell wall or mainly the vacuolar region of leaf cells (Bhargava et al., 2012). The uptake, transport, and accumulation mechanism of heavy metals have been summarised in Fig. 1.

The uptake of heavy metals and then their accumulation depends upon the hardiness of such plants to be able to sustain in contaminated soils along with the bioavailability of these metals. Furthermore, these metals accumulate due to their transport coupled with sugars and minerals, a befitting balance of heavy metals between roots and shoots is needed to be maintained which is known as translocation

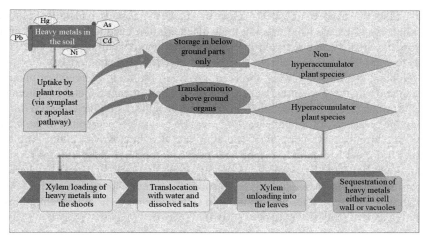

Figure 1: Uptake, transport, and accumulation mechanism of heavy metals inside plants.

factor (TF) and for hyperaccumulators, the value of TF has to be more than 1 (Tangahu et al., 2011). One more element that helps in knowing the heavy metal accumulation capacity of plants is the proportion of heavy metals absorbed by roots in comparison to their presence in soil which is termed as bioconcentration factor (BC), the value of which again exceeds 1 for hyperaccumulators (Ahmadpour et al., 2012).

3. Heavy metal toxicity and its impact on plants

Stress-induced due to amassing of heavy metals in plants has contributed to affecting agricultural productivity by interfering with the major processes invariably linked with requisite growth and development as that of photosynthesis, water, and nutrient uptake. Besides this, one of the principal aftermaths of heavy metal exposure is the production of ROS, i.e., reactive oxygen species in the plants consequently leading to oxidative stress (Mithofer et al., 2004). This disturbs the otherwise naturally operational antioxidant system and vital metabolic activities of essential nutrients (Srivastava et al., 2004; Dong et al., 2006). Furthermore, plant cell membranes undergo lipid peroxidation disrupting their integrity resulting in ion leakage through them. The extensive occurrence coupled with a multitude of unpropitious effects has put the heavy metals in the category of worst soil pollutants which in turn interrupts principal metabolic reactions in the plants upon entry. All these consequences of heavy metal toxicity are discussed in detail under the following heads:

3.1 Plant growth

Heavy metals are strikingly baneful in manifesting detrimental aftermaths on different plant growth parameters such as length of roots plus shoots, cell expansion, biomass, vital metabolic activities, and absolute inhibition of growth even in minimal amounts (Soares et al., 2001). This can be attributed to a reduction in leaf area, obstruction of photosynthetic mechanism, respiration, protein synthesis along with disturbing the balance between essential components which might prove fatal to cultivars (Astolfi et al., 2004; Chaves et al., 2011; Khan et al., 2015). Lipids constituents of cell membranes get peroxidized causing their disintegration eventually resulting in ion leakage through themas reported in exceedingly high concentrations of Cu, which leads to complete stoppage of root growth as well (Bouazizi et al., 2010).

Gopal and Khurana (2011) observed a decrease in the total height of plants in the range of 18–77% when they chance to grow upon in heavy metal contaminated soils. Similar results were obtained in the stem length of Lettuce (*Lactuca sativa*) due to the presence of Cd in the growing medium and this is accredited to the metal-induced chromosomal anomalies (Seregin and Kozhevnikova, 2006; Monteiro et al., 2007). Cd is also held responsible to reduce optimal elongation of root and shoot in wheat (*Triticum aestivum*) in quantities barely reaching 5 mg L^{-1} whereas a major drop in all-inclusive plant growth and inflorescence formation in rice (*Oryza sativa*) was reported if these plants grow in soils containing Hg in concentrations of 1 mg/Kg (Kibra, 2008; Ahmad et al., 2012).

3.2 *Photosynthetic activity*

The process of photosynthesis is of prime importance in the lives of plants whereby they not only transduce radiant energy into chemical energy but also synthesize their food. Several photosynthetic pigments are crucially involved in carrying out photosynthesis through the formation as well as physiological aspects of these key drivers get adversely affected in the proximity of heavy metals. In a sort of cascading effect, the presence of heavy metals destroys chlorophyll (chlorosis), hampering pathways synthesizing photosynthetic pigments leading to underdeveloped plastids, less photosynthetic action, and upsetting of numerous metabolic functions (Rai et al., 2016). Cytological studies on plants treated with Cd revealed damage to the photosynthetic apparatus, i.e., chloroplasts at the structural level and reduction in D1 protein of photosystem II of photochemical phase and RuBP-carboxylase enzyme of the thermochemical phase of photosynthesis, however, converse results concerning these very proteins and chlorophyll content were obtained for Pb treated maize plants (Figlioli et al. 2019). Immature leaves of Soybean (*Glycine max*) exhibit reduced concentrations of Chlorophyll, the performance of PSII, and net rate of photosynthesis, consequently inhibiting the very process vis-à-vis mature leaves when given the enhanced dosage of Cd in the growing media (Xue et al., 2014). An experiment conducted on *Brassica juncea* plants when subjected to varying concentrations of heavy metals like Cd, Cr, Hg, Ni, and Pb showed a significant reduction in amounts of Chl *a* plus Chl *b*, photosynthetic efficiency, and biomass upon increasing the metal content in comparison to plants grown under control (Sheetal et al., 2016).

3.3 *Antioxidative defence system*

Oxidative harm in the form of excessive accumulation of reactive oxygen species (ROS) like OH^{-} (hydroxyl radical), O$_2$$^{-}$ (superoxide), RO$^{\bullet}$ (alkoxyl), ROO$^{\bullet}$ (peroxyl), H$_2$O$_2$ (hydrogen peroxide), ^{1}O$_2$ (singlet oxygen), etc., is also one of the harmful impacts of heavy metal generated stress in the plants. These ROS hamper different growth and developmental advancements at the cellular and tissue level finally disturbing the complete metabolism. Plants innate immune system get activated to combat this overaccumulation of ROS which involves both enzymatic [ascorbate peroxidase (APX), catalase (CAT), glutathione reductase (GR), superoxide dismutase (SOD), etc.] and non-enzymatic antioxidants [ascorbic acid (AsA or Vitamin C), glutathione (GSH) and tocopherol or Vitamin E] (Sytar et al., 2013). *Brassica juncea* plants were shown to have enhanced production of ROS such as hydrogen peroxide (H_2O_2) under cadmium (Cd) toxicity which caused lipid peroxidation led to breaking down of membranes resulting in leakage of electrolytes through them followed by activation of antioxidative enzymes as the likes of APX, GSH, and SOD (Ahmad et al., 2016). On the very similar lines results were obtained in the case of *Oryza sativa*, where the application of Ni altered morphological growth, brought about overaccumulation of ROS, oxidative stress, ultimately leading to an escalation in the activities of antioxidative enzymes (Rajpoot et al., 2016).

4. Analysis of heavy metals in the environment

Metalliferous mining, smelting, phosphatic fertilizers, fossil fuel combustion, sewage refuse or municipal composts, incinerator discharge, and industrial activities continuously add heavy metals into our environment. Since heavy metals qualify as chief pollutants in the present times, constant checking and analysis are quite essential for their evaluation and regulation in the environment (Elzwayie et al., 2017). To keep a proper track of all the heavy metals and metalloids of toxic nature, environmental media ranging from water, soil, deposits, and local biota, need to be carefully and regularly monitored. Procedures like wet chemical methods (gravimetric, titrimetric, colorimetric, etc.). Inductively coupled plasma mass spectrometry (ICP-MS), Inductively coupled plasma optical emission (ICP-OES), and Atomic absorption spectroscopy (AAS) are variously used for the analytical studies of heavy metals in the environment (Hossain and Brennan, 2011; Salimi et al., 2018). Furthermore, frequent use of diverseion-selective electrodes has been reported for the quantification of heavy metals (Verma and Singh, 2005; Ghica et al., 2013). In the present times, all new, highly effective, sensitive, accurate, cost-effective and rapid, optical (Zhylyak et al., 1995; Ghica and Bret, 2008), chemical (Aragay and Merkoçi, 2012), and biological sensory systems are presently in different stages of development. These progressive pursuits in the field of analytical chemistry are now intricately linked to nanotechnology. Such analytical studies will certainly provide accurate facts and figures regarding the origin, distribution, permissible limits, and potential role of heavy metals in the ecosphere and their trophic level transfer inside food chains.

5. Bioremediation of heavy metals

Bioremediation or employment of microorganisms/plants to rectify contamination of heavy metals from the soil has been widely acknowledged to be an efficient and inexpensive approach as compared to conventional remedial measures. About one acre of Pb polluted land was decontaminated by 50% to 65% after being subjected to bioremediation process vis-à-vis conventional techniques of landfilling or excavation (Blaylock et al., 1997). This process either removes the toxic substances altogether or simply degrade them into less harmful ones, however, factors like the nature of heavy metal pollutants, the extent of pollution, type of soil, temperature, humidity, and topographical location do affect this very process.

5.1 *Phytoremediation*

Phytoremediation is a biological corrective technique in which the plants with high tolerance and accumulation capacity are made to grow in heavy metal contaminated soils whereby they pile up these heavy metal pollutants in their different parts. Up to now, more than 400 plants, mostly belonging to the Brassicaceae family and genera like *Alyssums, Brassica*, and *Thlaspi*, have been reported to possess this remediation potential (Xin et al., 2003). This remedial measure works by resorting to different mechanisms of extraction, stabilization or immobilization and volatilization carried out with the help of plants explained as follows:

(i) Phytoextraction: After accumulating heavy metals in sufficient quantities from the contaminated soil or after maturity, the plant species can be simply harvested, incinerated, or cured of their toxicity and this is termed as phytoextraction. The accumulation capability of these plants can allow them to pile up about 10 to 500 fold more heavy metals than other ordinary plants. However, different plant parts exhibit varied accumulation potential as reported in the case of sunflower (*Helianthus annuus*) where seeds and leaves can stock 1% and 59% Pb respectively from the polluted soil (Angelova et al., 2016). Further, the specificity of these plants for each heavy metal also differs as rightly observed in plant *Micranthemum umbrosum* of ornamental importance which accumulates more As as compared to Cd (Islam et al., 2013).

(ii) Phytostabilization: It is another variant of phytoremediation technique whereby plants are utilized to negatively alter the bioavailability of heavy metals and contain their further spread via simply immobilizing the latter in the soil layer itself preventing their erosion and leaching in the environment (Lone et al., 2008). Heavily contaminated soils where phytoextraction will take longer than usual time, such a technique is highly recommended and it operates through precipitation, complex formation, sorption, and metal valency reduction (Jadia and Fulekar, 2009). Since plant roots play a major role in this process, so the root system has to be extensive and it should permit only limited translocation of metals from roots to stem (Islam et al., 2013). The remedial potential of these plants can be further improved by adopting certain soil amending procedures like the addition of lime, phosphates, agricultural compost, or biochars which in turn modulate the soil pH causing escalation of the metal inactivation process.

(iii) Phytovolatilization: This particular process converts soil pollutants into volatile form first followed by their release into the atmosphere, mostly resorted to in case of heavy metals like Hg and Se polluted soils. Due to the rare occurrence of plants with high Hg accumulation potential, genetically modified plants such as *Arabidopsis thaliana, Nicotiana tabacum,* and *Liriodendron tulipifera* having genes for mercury reduction, i.e., merA and merB are employed for detoxification of Hg (Meagher et al., 2000). On the other hand, plants such as *Brassica napus* and *B. juncea* have been reported to effectively volatilize Se from the polluted soil (Banuelos et al., 1997).

5.1.1 Mechanism of heavy metal remediation in plants

The metabolic malfunctioning and developmental aberrations are the common consequences of heavy metal toxicity in plants. For the instance, heavy metals disposition in the amino acids by combining with—SH groups plus active and integral components of some major macronutrients are prevented from operating normally leading to breaking down of plasma membranes, stoppage of action of vital enzymes catalysing metabolic processes of prime importance like photosynthesis, respiration and transportation, etc. (Dixit et al., 2015).

However, plants try to battle out heavy metal toxicity with the help of their innate immune network in which the first line of defence involves the prevention of heavy metals from entering the plant bodies. And for this, several morphological adaptations on the exterior level like resilient cell wall, thick impassable cuticle, specializations in the form of tough trichomes, or mycorrhizal alliances are present (Emamverdian et al., 2015). Still, if metal ions remain successful in crossing all these physical barriers, then the second line of defence operates at the cellular level so that baneful impacts of heavy metal toxicity could subside (Silva and Matos, 2016). The otherwise naturally present defence system of enzymatic and non-enzymatic antioxidants gets activated on the behest of heavy metals which mitigate metal-generated oxidative harm by lowering down the measures of deleterious Reactive Oxygen Species (Skórzynska-Polit et al., 2010; Sytar et al., 2013).

Other safeguarding approaches adopted by plants include the production of phytochelatins (PCs) (catalyzed by the enzyme phytochelatin synthase) which get readily combined with heavy metals for their sequestration in vacuoles and further action revolves around their transformation into metallothioneins (MTs) or enhancing the production of stress tolerance providing amino acid proline (Ehsanpour et al., 2012). Xia et al. in 2012 obtained the relevant results in *Elsholtzia haichowensis* metallothionein type 1 (EhMT1) which declines the toxicity brought about by Cu by simply reducing the measures of H_2O_2 (a type of ROS) through enhancing the functioning of peroxidase (POD) enzyme to combat oxidative stress in *Nicotiana tabacum* (Tobacco) genotypes. On similar lines, another metallothionein TaMT3, obtained from *Tamarix androssowii*, suffused into *Nicotiana tabacum* plants confer more resistance potential to Cd toxicity by lowering down the levels of ROS with the help of SOD enzyme, though limiting the POD enzyme from action (Zhou et al., 2014). Another warrior from the antioxidative defence squad, GSH, also provides a shielding effect against heavy metal toxicity by quashing ROS or by simply synthesizing phytochelatins (PCs), all these activities not only mitigate the metal-induced stress but also maintain the cellular homeostatic levels (Yadav, 2010). Cd and As tolerance potential of plants are elevated by As-phytochelatin synthase 1 (*AsPCS1*) and yeast cadmium factor 1 (YCF1) in *Arabidopsisthaliana* along with enhancing the uptake and storage of heavy metals (Guo et al., 2012).

5.2 Microbial remediation

The usage of microorganisms especially bacteria to rectify the presence of heavy metals in the soils by precipitating, inactivating, oxidizing, or reducing these pollutants is called microbial remediation (Su, 2014). The soil enriched with waste deposits of metal industry or soil tainted with Cd can be cured by using strains of *Bacillus* bacteria like *B. cereus* and *B. thuringiensis* which escalate uptake of Cd and Zn (Ajaz et al., 2010). Heavy metals with higher oxidation state are downgraded to their low oxidation state of less toxic nature as reported in Cr^{+6} to Cr^{+3} by microbes like *Bacillus subtilis, Enterobacter cloacae*, and *Pseudomonas putida* (Garbisu et al., 1998). This can be reasoned with the enhanced production of siderophores (iron-chelating compounds) by these bacteria in association with heavy metals themselves which increase both bioavailability and uptake of heavy metals, as substantiated by *Azotobacter vinelandii* which stimulates the generation of siderophores in the presence of Zn metal. The heavy metals are found to modify the physiological activities of siderophore-inducing bacteria which in turn boosts mobilization and removal of heavy metals from the soil (Chibuike and Obiora, 2014). Creating the suitable conditions for this microbial action to take place is called bio-stimulation which involves the addition of organic matter or nutrients so that the extraction of heavy metals from the soils can be enhanced by either altering the pH of the soil, improving solubility, or availability of heavy metals for the plants (Karaca, 2004; McCauley et al., 2009).

6. Conclusion

Heavy metals enter the food chain after polluting the atmosphere, hydrosphere, or cultivated lands, and are ingested by humans and other species. The hazardous load on the entire ecosystem is rising as a result of industrial development and environmentally harmful farming practices. Heavy metal is more prevalent in urban areas. Air quality has deteriorated dramatically as a result of vehicle exhausts and industrial emissions into the environment, particularly in urban areas. Humans are compelled to inhale polluted air, breathing a variety of toxins, some of which are extremely toxic and cancer-causing. There are several mitigation procedures. Phytoremediation and other forms of bioremediation tend to be more eco-sustainable. Many metals and plants have had their accumulation pathways exposed. The process of having genetically modified plants with increased toxic metal bioaccumulation is beginning to yield results. By simulating gene variants in metal uptake and transport, several transgenic plants have been developed. Consequently, removing all pollutants from the soil with a single remediation method would be inadequate. For the effective removal of harmful toxic substances and several dangerous substances from soil, it is often suggested to be using a combination of remedial techniques.

References

Abollino O, Giacomino A, Malandrino M, Mentasti E, Aceto M, and Barberis R. 2006. Assessment of metal availability in a contaminated soil by sequential extraction. *Water, Air, and Soil Pollution* 173(1): 315–338.

Ahmad P, Abd Allah EF, Hashem A, Sarwat M, and Gucel S. 2016. Exogenous application of selenium mitigates cadmium toxicity in *Brassica juncea* L. (Czern and Cross) by up-regulating antioxidative system and secondary metabolites. *J Plant Growth Regul* 35(4): 936–950.

Ahmad, I, Akhtar MJ, Zahir ZA, and Jamil A. 2012. Effect of cadmium on seed germination and seedling growth of four wheat (*Triticum aestivum* L.) cultivars. *Pak J Bot* 44: 1569–1574.

Ahmadpour P, Ahmadpour F, Mahmud TMM, Abdu A, Soleimani M, and Tayefeh FH. 2012. Phytoremediation of heavy metals: A green technology. *African Journal of Biotechnology* 11(76): 14036–14043.

Ajaz HM, Arasuc RT, Narayananb VKR, and Zahir HMI. 2010. Bioremediation of heavy metal contaminated soil by the exigobacterium and accumulation of Cd, Ni, Zn and Cu from soil environment. *Inter J Biol Technol* 1: 94–101.

Angelova VR, Perifanova-Nemska M, Uzunova G, Ivanov K, and Lee H. 2016. Potential of sunflower (*Helianthus annuus* L.) for phytoremediation of soils contaminated with heavy metals. *World J Sci Eng Technol* 10: 1–11.

Aragay G and Merkuci A. 2012. Nanomaterials application in electrochemical detection of heavy metals. *Electrochimica Acta* 84: 49–61.

Astolfi S, Zuchi S and Passera C. 2004. Effects of cadmium on the metabolic activity of *Avena sativa* plants grown in soil or hydroponic culture. *Biologia Plantarum* 48(3): 413–418.

Banuelos GS, Ajwa HJ, Mackey B, Wu L, Cook C, Akohoue S, and Zambruzuski S. 1997. Evaluation of different plant species used for phytoremediation of high soil selenium. *Journal of Environmental Quality* 26(3): 639–646.

Bhargava A, Carmona FF, Bhargava M, and Srivastava S. 2012. Approaches for enhanced phytoextraction of heavy metals. *Journal of Environmental Management* 105: 103–120.

Blaylock MJ, Salt DE, Dushenkov S, Zakharova O, Gussman C, Kapulnik Y. Ensley BD, and Raskin, I. 1997. Enhanced accumulation of Pb in Indian mustard by soil-applied chelating agents. *Environ Sci Tech* 31: 860–865.

Bouazizi H, Jouili H, Geitmann A, and Ferjani EEI. 2010. Copper toxicity in expanding leaves of *Phaseolus vulgaris* L.: antioxidant enzyme response and nutrient element uptake. *Ecotoxicol Environ Saf* 73: 1304–1308.

Butcher DJ. 2010. Advances in inductively coupled plasma optical emissions pectrometry for Environmental analysis. *Instrumentation Science and Technology* 38(6): 458–469.

Chaves LHG, Estrela MA, and Sena de Souza R. 2011. Effect on plant growth and heavy metal accumulation by sunflower. *J Phytol* 3: 04–09.

Chibuike GU, and Obiora SC. 2014. Heavy metal polluted soils: effect on plants and bioremediation methods. *Appl Environ Soil Sci* 2014: 1–12.

Dixit R, Malaviya D, Pandiyan K, Singh UB, Sahu A, Shukla R, Singh BP, Rai JP, Sharma PK, Lade H, and Paul D. 2015. Bioremediation of heavy metals from soil and aquatic environment: an overview of principles and criteria of fundamental processes. *Sustainability* 7(2): 2189–2212.

Dong J, Wu F, and Zhang G. 2006. Influence of cadmium on antioxidant capacity and four microelement concentrations in tomato seedlings (*Lycopersicon esculentum*). *Chemosphere* 64(10): 1659–1666.

Duruibe JO, Ogwuegbu MOC, and Egwurugwu JN. 2007. Heavy metal pollution and human bio-toxic effects. *International Journal of Physical Sciences* 2(5): 112–118.

Ehsanpour AA, Zarei S, and Abbaspour J. 2012. The role of over expression of p5cs gene on proline, catalase, ascorbate peroxidase activity and lipid peroxidation of transgenic tobacco (*Nicotiana tabacum* L.) plant under in vitro drought stress. *J Cell Mol Res* 4: 43–49.

Elzwayie A, Afan HA, Allawi MF, and El-Shafie A. 2017. Heavy metal monitoring, analysis and prediction in lakes and rivers: state of the art. *Environmental Science and Pollution Research* 24(13): 12104–12117.

Emamverdian A, Ding Y, Mokhberdoran F, and Xie Y. 2015. Heavy metal stress and some mechanisms of plant defense response. *Sci World J*, 756120.

Feldmann J, Salaün P, and Lombi E. 2009. Critical review perspective: Elemental speciation analysis methods in environmental chemistry—Moving towards methodological integration. *Environmental Chemistry* 6(4): 275–289.

Figlioli F, Sorrentino MC, Memoli V, Arena C, Maisto G, Giordano S, Capozzi F, and Spagnuolo V. 2019. Overall plant responses to Cd and Pb metal stress in maize: Growth pattern, ultrastructure, and photosynthetic activity. *Environ Sci Pollut Res* 26(2): 1781–1790.

Garbisu C, Alkorta I, Llama MJ and Serra JL. 1998. Aerobic chromate reduction by *Bacillus subtilis*. *Biodegradation* 9(2): 133–141.

Ghica ME, and Bret CM. 2008. Glucose oxidase inhibition in poly (neutral red) mediated enzyme Biosensors for heavy metal determination. *Microchimica Acta* 163(3-4): 185–193.

Ghica ME, Carvalho RC, Amine A and Bret CM. 2013. Glucose oxidase enzyme inhibition sensors. For heavy metals at carbon film electrodes modified with cobalt or copper hexacyanofer Rate. *Sensors and Actuators B: Chemical* 178: 270–278.

Gopal R, and Khurana N. 2011. Effect of heavy metal pollutants on sunflower. *African J Plant Sci* 5(9): 531–536.

Guo J, Xu W, and Ma M. 2012. The assembly of metals chelation by thiols and vacuolar compartmentalization conferred increased tolerance to and accumulation of cadmium and arsenic in transgenic *Arabidopsis thaliana*. *J Hazard Mater* 199: 309–313.

Hossain SZ, and Brennan JD. 2011. B-Galactosidase-based colorimetric paper sensor for determination of heavy metals. *Analytical Chemistry* 83(22): 8772–8778.

Islam MS, Ueno Y, Sikder MT, and Kurasaki M. 2013. Phytofiltration of arsenic and cadmium from the water environment using *Micranthemum umbrosum* (jf GMEL) sf blake as a hyperaccumulator. *Int J Phytoremed* 15: 1010–1021.

Jadia CD, and Fulekar MH. 2009. Phytoremediation of heavy metals: recent techniques. *African Journal of Biotechnology* 8(6): 921–928.

Karaca A. 2004. Effect of organic wastes on the extractability of cadmium, copper, nickel, and zinc in soil. *Geoderma* 122: 297–303.

Khan A, Khan S, Khan MA, Qamar Z, and Waqas M. 2015. The uptake and bioaccumulation of heavy metals by food plants, their effects on plants nutrients, and associated health risk: a review. *Environ Sci Pollut Res* 22(18): 13772–13799.

Kibra MG. 2008. Effects of mercury on some growth parameters of rice (*Oryza sativa* L.). *Soil Environ* 27: 23–28.

Li TQ, Yang XE, Jin X, He ZL, Stoffella PJ, and Hu QH. 2005. Root responses and metal accumulation in two contrasting ecotypes of Sedum alfredii Hance under lead and zinc toxic stress. *Journal of Environmental Science and Health, Part A* 40(5): 1081–1096.

Lone MI, He ZL, Stoffella PJ, and Yang XE. 2008. Phytoremediation of heavy metal polluted soils and water: Progresses and perspectives. *J Zhejiang Univ Sci B* 9: 210–220.

Mahmood T. 2010. Phytoextraction of heavy metals-the process and scope for remediation of contaminated soils. *Soil Environ* 29(2): 91–109.

McCauley A, Jones C, and Jacobsen J. 2009. Soil pH and organic matter. In Nutrient Management Module, 8, Montana State University Extension, Bozeman, Mont, USA.

Meagher RB, Rugh CL, Kandasamy MK, Gragson G, and Wang NJ. 2000. Engineered phytoremediation of mercury pollution in soil and water using bacterial genes. pp. 201–219. *In*: Terry N, and Banuelos G (eds.). *Phytoremediation of Contaminated Soil and Water*. Lewis Publishers, Boca Raton, Fla, USA.

Memon AR, and Schröder P. 2009. Implications of metal accumulation mechanisms to phytoremediation. *Environmental Science and Pollution Research* 16(2): 162–175.

Ming-Ho, Y. 2005. Environmental Toxicology: Biological and Health Effects of Pollutants, Chap. 12, 2nd Edition, CRC Press LLC, Boca Raton.

Mithofer A, Schulze B, and Boland W. 2004. Biotic and heavy metal stress response in plants: evidence for common signals. *FEBS Letters* 566: 1–5.

Monteiro M, Santos C, Mann RM, Soares AMVM, and Lopes T. 2007. Evaluation of cadmium genotoxicity in *Lactuca sativa* L. using nuclear microsatellites. *Environ Exp Bot* 60: 4 21–427.

Neubauer U, Furrer G, Kayser A, and Schulin R. 2000. Siderophores, NTA, and citrate: potential soil amendments to enhance heavy metal mobility in phytoremediation. *Int J Phytoremediation* 2(4): 353–368.

Peng HY, Yang XE, Jiang LY, and He ZL. 2005. Copper phytoavailability and uptake by Elsh {o} ltzia splendens from contaminated soil as affected by soil amendments. *Journal of Environmental Science and Health* 40(4): 839–856.

Quartacci MF, Irtelli B, Gonnelli C, Gabbrielli R, and Navari-Izzo F. 2009. Naturally-assisted metal phytoextraction by *Brassica carinata*: Role of root exudates. *Environmental Pollution* 157(10): 2697–2703.

Rai R, Agrawal M, and Agrawal SB. 2016. Impact of heavy metals on physiological processes of plants: with special reference to photosynthetic system. pp. 127–140. *In*: Singh A, Prasad S, and Singh R (eds.). *Plant Responses to Xenobiotics*. Springer, Singapore.

Rajpoot R, Rani A, Srivastava RK, Pandey P, and Dubey RS. 2016. Terminalia arjuna bark extract alleviates nickel toxicity by suppressing its uptake and modulating antioxidative defence in rice seedlings. *Protoplasma* 253(6): 1449–1462.

Rao CRM, and Reddi GS. 2000. Platinum group metals (PGM); occurrence, use and recent trends in their determination. *TrAC Trends in Analytical Chemistry* 19(9): 565–586.

Rasheed T, Bilal M, Nabeel F, Iqbal HM, Li C, and Zhou Y. 2018. Fluorescent sensor-based models for the detection of environmentally-related toxic heavy metals. *Science of the Total Environment* 615: 476–485.

Salimi F, Kiani M, Karami C, and Taher MA. 2018. Colorimetric sensor of detection of Cr (III) and Fe (II) ions in aqueous solutions using gold nanoparticles modified with methylene blue. *Optik-International Journal for Light and Electron Optics* 158: 813–825.

Seregin IV, and Kozhevnikova AD. 2006. Physiological role of nickel and its toxic effects on higher plants. *Russ J Plant Physiol* 53: 257–277.

Sheetal KR, Singh SD, Anand A, and Prasad S. 2016. Heavy metal accumulation and effects on growth, biomass and physiological processes in mustard. *Ind J Plant Physiol* 21: 219–223.

Silva P, and Matos M. 2016. Assessment of the impact of aluminum on germination, early growth and free proline content in *Lactuca sativa* L. *Ecotoxicol Environ Saf* 131: 151–156.

Skórzynska-Polit E, Dra̜z̓kiewicz M, and Krupa Z. 2010. Lipid peroxidation and antioxidative response in arabidopsis thaliana exposed to cadmium and copper. *Acta Physiol Plant* 32: 169.

Soares CR, Grazziotti FS, Siquaira PH, Carvalno JO, and De JH. 2001 Zinc toxicity on growth and nutrition of *Eucalyptus muculata* and *Eucalyptus urophylla*. *Pesq Agrop Brasileira* 36: 339–348.

Srivastava S, Tripathi RD, and Dwivedi UN. 2004. Synthesis of phytochelatins and modulation of antioxidants in response to cadmium stress in Cuscuta reflexa–an angiospermic parasite. *Journal of Plant Physiology* 161(6): 665–674.

Stowers CC, Cox BM, and Rodriguez BA. 2014. Development of an industrializable fermentation process for propionic acid production. *Journal of Industrial Microbiology & Biotechnology* 41(5): 837–852.

Su C. 2014. A review on heavy metal contamination in the soil worldwide: Situation, impact and remediation techniques. *Environmental Skeptics and Critics* 3(2): 24.

Sytar O, Kumar A, Latowski D, Kuczynska P, Strzałka K, and Prasad MNV. 2013. Heavy metal-induced oxidative damage, defense reactions, and detoxification mechanisms in plants. *Acta Physiol Plant* 35(4): 985–999.

Tangahu BV, Sheikh Abdullah SR, Basri H, Idris M, Anuar N, and Mukhlisin M. 2011. A review on heavy metals (As, Pb, and Hg) uptake by plants through phytoremediation. *International Journal of Chemical Engineering* 2011: 1–31.

Ullah N, Mansha M, Khan I, and Qurashi A. 2018. Nanomaterial-based optical chemical sensor for the detection of heavy metals in water: Recent advances and challenges. TrAC Trends. *In Analytical Chemistry* 100: 155–166

Verma N, and Singh M. 2005. Biosensors for heavy metals. *Biometals* 18(2): 121–129.

Wang J. 2005. Stripping analysis of bismuth electrodes: a review. *Electroanalysis* 17(15-16): 1341–1346.

Xia Y, Qi Y, Yuan Y, Wang G, Cui J, Chen Y, Zhang H, and Shen Z. 2012. Overexpression of Elsholtzia haichowensis metallothionein 1 (EhMT1) in tobacco plants enhances copper tolerance and accumulation in root cytoplasm and decreases hydrogen peroxide production. *J Hazard Mater* 233: 65–71.

Xin QG, Pan WB, and Zhang TP. 2003. On phytoremediation of heavy metal contaminated soils. *Ecologic Sci* 22(3): 275–279.

Xue Z, Gao H, and Zhao S. 2014. Effects of cadmium on the photosynthetic activity in mature and young leaves of soybean plants. *Environ Sci Pollut Res* 21(6): 4656–4664.

Yadav SK. 2010. Heavy metals toxicity in plants: an overview on the role of glutathione and phytochelatins in heavy metal stress tolerance of plants. *S Afr J Bot* 76(2): 167–179.

Yang X, Feng Y, He Z, and Stoffella PJ. 2005. Molecular mechanisms of heavy metal hyperaccumulation and phytoremediation. *Journal of Trace Elements in Medicine and Biology* 18(4): 339–353.

Zhou B, Yao W, Wang S, Wang X, and Jiang T. 2014. The metallothionein gene, TaMT3, from Tamarix and rossowii confers Cd2+ tolerance in tobacco. *Int J Mol Sci* 15(6): 10398–10409.

Zhylyak GA, Dzyadevich SV, Korpan YI, Soldatkin AP, and El'Skaya AV. 1995.Application of urease conductometric biosensor for heavy-metal ion determination. *Sensors and Actuators B: Chemical* 24(1-3): 145–148.

5

Heavy Metal Contamination in Crop Plants

Naziya Tarannum[1] and *Nivedita Chaudhary*[1,*]

ABSTRACT

Heavy metals can be essential, expensive, and radioactive, and their influence on humans and the environment depends upon their properties and exposure. Essential heavy metals play a crucial role in physiological and biochemical functions in living organisms. Heavy metals are found naturally; however, they may be released due to various natural actions in the environment. In recent years, anthropogenic activities caused the undesirable release of heavy metals into the environment affecting both humans and vegetation. Major impacts in the environment can be considered in terms of phytotoxicity due to the excess formation of reactive oxygen species (ROS) causing oxidative stress. They disrupt various metabolic processes of the plants. Generally, heavy metals are considered to be stable and bioaccumulate, causing more adverse effects. Plants can uptake heavy metals majorly from the soil, contaminated irrigation water, and foliar deposition. The excess formation of ROS affects the plants' physiological, biochemical and metabolic activities, and ultimately the yield and biomass of major crops. Various mitigating measures such as the development of transgenic and phytoremediation techniques may reduce heavy metal stress.

1. Introduction

'Heavy metals' is the set of metals encompassing a density of more than 4 gcm^{-1}, for example, Mn, Cr, Cu, Co, Zn, Mo, Ni, Cd, Sn, Pb, Hg, Sb, etc. (Hawkes, 1997). Certain heavy metals such as Pb, Cr, Hg, Cu, Ni, Zn, Cd, As, Sn, and Co are

---
[1] Department of Environmental Science, School of Earth Sciences, Central University of Rajasthan, Ajmer, Rajasthan, India.
* Corresponding author: nivedita@curaj.ac.in

categorized as toxic; some are expensive metals, such as Ag, Au,Pd, Pt, Ru, etc., and some are radioactive for example, Th, U, Ra, Am, etc. Due to the essential functions such as contribution in redox reactions and role as a component for enzymes, Zn, Cu, Mn, Mo, and Fe are reflected as essential heavy metals and display physiological and biochemical functions in living beings. The endurance and stable characteristics of these heavy metals cannot quickly be deteriorated or eliminated. Therefore, they can also be bio-accumulative and may gradually enter the plants and animals via air, water, and the food chain extension (Nagajyoti et al., 2010). Activities such as extensive agriculture, industrialization, and burning of fossils fuels release heavy metals in the environment and cause detrimental effects on plants and animals in one way or another (Shahid et al., 2017).

Heavy metals come to a specific site by diffusion primarily during natural catastrophes (earthquakes, landslides, or floods) or by accelerated anthropogenic activities (Wuana and Okieimen, 2011). However, anthropogenic activities enhance the levels of heavy metals in the environment, such as chemical manures fertilizers, herbicides, pesticides, and weedicides in agriculture and military operations. Other sources are municipal waste, energy production, fuel burning, mining activities, irrigation with wastewater, power production and industrial effluents, organic waste manure, land application of sewage sludge, by-products from industries, etc. (Woldetsadik et al., 2017). Transportation of heavy metals from contaminated soil affects plants. In this regard, soil texture and pH also participate in the accumulation and manifestation of heavy metal toxicity (Hassan et al., 2017). Phytotoxic effects are observed in various plants triggering the generation of ROS, which causes oxidative stress. Nagajyoti et al. (2010) reviewed the impact of heavy metals on plants, such as visible physiological effects caused by Cu, Zn, Hg, electron transport chain, photophosphorylation, and enzymatic activities, and CO_2 fixation by Cr-stress on plants. Ni can cause nutrient imbalance and reduced cell membrane functions in some plant species, and chlorosis caused by Mn toxicity. Fe causes the generation of excessive free radical that permanently destroys the cellular structure and membranes and affects DNA and proteins (Nagajyoti et al. 2010). Foliar injuries in the form of wilting and necrosis in leaves, subsequently leading to retardation in shoot and root growth is caused by As phytotoxicity (Woolson et al., 1971). Heavy metal absorption in plants occurs via roots and leaves, resulting in the dysfunction of plant metabolism that later causes severe risk potential health of humans (Shahid et al., 2017). Schreck et al. (2012) implied that plants use several approaches to confront absorption of heavy metals in roots and foliar cells involving different mechanisms within the plants.

Some conventional transgenic and breeding technologies are in use to develop heavy metal stress-resistant cultivars and varieties. Also, certain microbes are in use for alleviating the metal resilience of plants. Similarly, microbes associated with plants reduce metal accumulation in plant tissues and minimize the bioavailability of metals in soil. Additionally, in the foliage of several aromatic and medicinal plants, heavy metals get absorbed and accumulated and, hence, become a possible substitute for remediation of contaminated sites. Largely, crop plants' growth and development play a significant role in the heavy metal affecting plants and is a leading cause of detrimental effect on agricultural productivity. Therefore, the objective of the present chapter is to impart an insight into the sources of heavy metals, the present scenario of heavy metal contamination affecting plants, mechanism of action, and mitigative perspectives.

2. Sources of heavy metals

2.1 Natural sources

Naturally, rock weathering is considered the prominent source of heavy metal present in the soil; in general, the levels and composition of heavy metals are subject to the kind of rock and the environmental situations (Abdu et al., 2011). Heavy metals inputs in the soil, for example, Co, Mn, Cu, Ni, and Zn are from igneous rocks and Cd, Sn, Hg, Pb, and Cr from shale, which is a sedimentary rock from the geologic origin (Sandeep et al., 2019); volcanos are also the source of Al, Mn, Pb, Zn, Ni, Cu, and Nagajyoti et al. (2010) suggested that Hg dust particles blown by winds from the desert contribute to high concentrations of Fe and some quantity of Zn, Cr, Mn, Ni, and Pb. Apart from weathering, vegetation also retrieves heavy metals from the environment by decomposition and discharging via volatilization (Cuypers et al., 2013). Oceanic events release sea sprays and aerosols into coastal areas containing heavy metals (Monge et al., 2015). In groundwater, heavy metals are naturally present by their release from aquifer rocks. Globally, groundwater contamination is observed majorly in India, Thailand, Vietnam, Greece, Hungary, inner parts of Mongolia, the USA, Ghana, Chile, Argentina, and Mexico (Battsengel et al., 2020).

2.2 Agricultural sources

Agriculture is the prime anthropogenic source of heavy metals, which are added to the soil by inorganic fertilizers that are found in the water used for irrigation and sewage sludge. Specific sources of heavy metals in the soil due to many agricultural activities are provided in Table 1. Mahfooz et al. (2020) reported higher heavy metals such as Fe, Mn, Cd, Zn levels in plants due to irrigation using municipal wastewater contaminated with heavy metals. Jaramillo and Restrepo (2017) have stated that sewage and irrigation disrupt crop growth and development. While Heavy metals

Table 1: Heavy metals due to agricultural activities.

Source	Heavy metal input	References
Fertilizers: Phosphate, Nitrate fertilizers, Potash, Lime	Cd, Cu, Cr, Zn, Mn, Ni, and Pb	Gimeno-García et al., 1996
Pesticides: Herbicides, Insecticides, Fungicides	Zn, Cd, and Cu	Nicholson et al., 2003
Biosolids and manure: Sewage sludge, Composts, Livestock manures, Fly ash	Cd, Zn, Pb, Cu, Ni, As, and Cr.	Chauhan et al., 2012; Srivastava et al., 2016
Wastewater: Irrigation with municipal wastewater and industrial wastewater	Cu, Ni, As, Cr, Zn, and Cd	Nicholson et al., 2003; Woldetsadik et al., 2017.
Atmospheric deposition, Metal smelting, mining, transport and waste incineration	Pb, Cu, Zn, Ni, Cd, Hg, and Cr	Liu et al., 2015

Table 2: Indian standards for heavy metals in soil suitable for agriculture.

Heavy metals (mg kg^{-1})	Indian Standards
Cr	-
Mn	-
Cu	135–270
Zn	300–600
Ni	75–150
Cd	3–6
Pb	250–500

content in agricultural soil is determined by the soil features and configuration, excessive use of inorganic fertilizers, pesticides, sewage sludge, and wastewater also contribute to the contamination, affecting soil activities and functions. Table 2 shows the Indian standards for heavy metals, particularly in soil dedicated to agricultural activities (Awasthi, 1998).

2.3 Industrial sources

Several activities, for instance, mining and alteration in the refinement activity, are the extensive cause of heavy metal contamination and are represented in Table 3. Vaporized forms of heavy metals, such as Zn, Cu, As, Pb, Cd, and Sn combined with water and created aerosol (Nagajyoti et al., 2010). Further, these heavy metals undergo dry or wet deposition contributing to soil and water contamination. Also, mines waste runoffs, transference of crude ores generated dust, deterioration, and discharge, polluting the water and soil (Rout et al., 2013).

Table 3: Sources of heavy metals from various industries.

Source	Heavy metal input	References
Coal mines	Cd, As, and Fe	Srivastava et al., 2016
Thermal power stations, Petroleum industries, and Nuclear power plants	Cu, Se, Zn, Cd, Ni, Cs	Zhu et al., 2016
Inefficient engines of automobiles	Zn, Pb, Cd, Ni, Hg, Cr	Srivastava et al., 2016
Lead containing gasoline combustion	Pb	Tangahu et al., 2011
Municipal solid waste incinerated products	Pb, Zn, Al, Sn, Fe, Cu	Srivastava et al., 2016

2.4 Domestic sources

Singh and Kumar (2017) suggested that runoffs are a vital contributor to the elevated levels of heavy metal in aquatic bodies. Effluents commonly comprise treated or untreated wastewater, biological treatment plants, and sewage waste matter outlet into water bodies, contributing to the severe issue of heavy metal pollution by urban effluent runoff.

3. Present scenario of heavy metal contamination

Soil contaminated with heavy metals and its negative effect has been assessed globally by various researchers suggesting that its accumulation subsequently causes an impact on plant products. In the present scenario, high levels of heavy metals and their bioaccumulation in the agricultural harvests, predominantly in vegetables, were observed by Mishra and Tripathi (2008) when intensive irrigation was done with wastewater in Varanasi. Similarly, irrigation with wastewater affects the quality of soil over time, and observation suggests significant increases in Ni, Mn, Zn, Cu, and Pb compared to the irrigation done with rainwater (Masto et al., 2009). Lokeshwari and Chandrappa (2006) reported that the heavy metal contamination in the Bellandurlake situated in Bengaluru affects soil and cultivated vegetation as its water is used for irrigation purposes. Due to automobiles and industries, urban areas add to heavy metal pollution in soils found in urban and suburban areas. These heavy metals are dissolved in the soil or present in organic and inorganic soil and deposits in the lattices of soil minerals and are readily available for plants (Aydinalp and Marinova, 2003). In India water bodies, especially rivers, are the foremost source of irrigation, however, it is reported that most of the rivers are contaminated with heavy metals well over the tolerable limit. Bhattacharya et al. (2015) reported Cu, Cd, Cr and Zn above the admissible limits in the river Yamuna at various study sites in Delhi.

Similarly, in the eastern part of the country Subarnarekha river is contaminated with Cu (Giri and Singh, 2014). Gupta et al. (2014) reported high Ni concentration in the various samples of water of the Gomti River. Cu and Cd were also observed to surpass permissible limits in the water of Jaikwadi Dam present in Maharashtra. Different reports suggested the higher levels of heavy metals in the water bodies such as Cu and Ni in various rivers located in Serbia (Dević et al., 2016), concentrations of Ni, Cr, and Cu reaching over permissible limits were found in the water samples of Dzindi River, South Africa (Edokpayi et al., 2014). High levels of Cd concentrations were observed in the River Nile and Ismailia canal, Egypt (Goher et al., 2014). Certain metals such as Zn, Ni, Cu, Cd, Pb, and Cr were also observed in high levels in Pardo River, Brazil (Alves et al., 2014). These reports suggest the high levels of heavy metals in aquatic bodies and the prolonged use of the water for irrigation lead to several impacts on plants.

Certain studies also suggested the negative effect on specific crops and vegetables grown with the substantial use of irrigated water contaminated with heavy metals. Punetha et al. (2015) reported high levels of Zn and Cr in *Brassica* sp., *Chenopodium,* and *Spinacia*. Similarly, Jia et al. (2013) suggested the existence of high levels of Cd, Zn, Cu, Pb, and Ni in various vegetables such as tomato, mint, parsley, coriander, leek, cabbage, garlic, and mustard. Rahman et al. (2014) also reported heavy metals in vegetables in Bangladesh. *Coriandrum sativum, Anethum graveolens*, and *Raphanus sativa* were also affected by heavy metals (Maleki et al., 2014). Apart from vegetables, the impact was also observed in rice and lentils as reported by Rahman et al. (2014), Goni et al. (2014), Chung et al. (2011). The impact of heavy metal contaminated water bodies on vegetables and crop plants subsequently led to the formation of policies required at a local level for ameliorating edaphic conditions and amendments in agricultural practices.

4. Mechanism of action of heavy metals

4.1 Mechanism of action of heavy metal affecting plants

Agricultural soil that is polluted with elevated levels of heavy metals affects plants' physiological and biochemical structure at the molecular level resulting in detrimental effects on growth, development and metabolism, fertility and biodiversity. Further, consumption of these plants affects organisms present at higher trophic levels (Zia-ur-Rehman et al., 2015). Heavy metals cause increased ROS production in plants, and these highly reactive species affect the electron transport activities in the mitochondrial membrane and chloroplast. ROS also disturb the redox condition of cells, causing leakage of ions in the membrane and also instigating lipid peroxidation (Anjum et al., 2017), and the biologically active breakdown of macromolecule (Venkatachalam et al., 2017). Physiological effects observed in plants due to the ROS formation and the anti-oxidative defence mechanism lead to oxidative stress in plants (Venkatachalam et al., 2017). Both redox-active (Cu, Fe, Cr, Co) and non-redox active (Ni, Cd, Zn) heavy metals cause direct production of ROS and its binding to the protein leading to redox; the metabolic imbalance and ionic disturbances result in damages to the plants and under optimum ROS production causes signalling for the repair mechanism. Therefore, the primary suggested mechanism of action is immediate interaction with ROS generation and driving essential cations from explicit binding sites and disturbing its roles (Sharma and Dietz, 2009).

Metal-dependent Haber-Weiss reactions involve certain uninhibited redox-active metals, for instance, Cr, Fe, and Cu, which instantaneously magnify ROS production in the plant (Das and Roychoudhury, 2014). In general, the plant's anti-oxidative defence enzymes for instance, superoxide dismutase, catalase, glutathione S-transferase, ascorbate peroxidase, and dehydroascorbate reductase are employed concurrently against ROS and oxidative stress generated as reported by Blokhina et al. (2003). Additionally, Shikimate dehydrogenase, Glutathione reductase, guaiacol peroxidase, polyphenol oxidase, cinnamyl alcohol dehydrogenase, and phenylalanine ammonia-lyase also participate against heavy metal stimulated oxidative stress (Jaskulak et al., 2018). Antioxidants (non-enzymatic) such as alpha-tocopherol, carotenoids, glutathione, proline, ascorbic acid, and flavonoids, tannins, lignins, anthocyanins, phenolic acid and associated compounds, coumarin, flavanol, cinnamic acid, and cinnamyl aldehyde, also work against heavy metal-initiated ROS in the plants,as reported by Singh et al. (2016).

4.1.1 Heavy Metals root uptake soil

An insoluble metal compound in soil readily available for uptake by the root comprises a minor portion of the metal content in the soil; therefore, soil characteristics are influenced by a heavy metal content in soil (Chaney, 1988). Certain microorganisms and root exudates also play a crucial role in metal availability and mobility, particularly in the rhizosphere, leading to metal uptake. Usually, metal ions are transported across the root cellular membrane, which enables its entry into plant tissues. Once inside, initially received by the apoplast, which is a free intercellular area leading towards the xylem, the metal is transported to the epidermis of the root, cortex and in root cells. Further, in the Casparian strip, metal reaches the xylem and this passage of metal ions accompanied by the Casparian strip takes place through a power-demanding active transport system (Cunningham and Berti, 1993). Sandeep et al. (2019) suggested that heavy metal is translocated from root to shoot by apoplast and symplast. Heavy metals are carried to leaves through the xylem loading within, and metal accumulation occurs in vacuoles from the shoot. The movement of metals in apoplast is achievable by non-cationic metal chelates, causing high exchange capability for cations through cell walls (Raskin et al., 1997). Therefore, metals remain immobilized in apoplast and symplast regions due to the formation of sulfate, carbonate, or precipitation of phosphate ions (Garbisu and Alkorta, 2001). Furthermore, the root–shoot metal translocation is mainly transported from the root symplast into the xylem apoplast (Marschner et al., 1996). The plasma membrane's negative potential encourages the deepest metal ions to travel due to the electrochemical gradient (Raskin et al., 1997). Usually, concentrations of heavy metals controlled by tonoplast and transporters specific for metals are present in the plasma membrane; also, specific membrane transport proteins regulated metal transport in the xylem (Thakur et al., 2016).

4.1.2 Leaf uptake

The plant's canopy is considered an efficient screen for the atmospheric deposition of heavy metals (Liu et al., 2012). Leaf uptake ensues from the stomata, cuticular splits, and aqueous pores (Fernández et al., 2013). Schreck et al. (2013) considered foliar uptake of metals was generally by adsorption and stomatal uptake. Studies suggested that small-sized particles having metals such as Cu penetrate easily and accumulate in plants such as in *Vicia faba*, as Eichert et al. (2008) reported. Similarly, Maiti et al. (2016) suggested that heavy metals accumulation in fruit plants such as *Artocarpus heterophyllus, Psidium guajava, Anacardium occidentale, Syzygium cumini,* and *Mangifera indica* first cultivated in mine soil then plants grown in garden soil.

4.2 Morphology of crop plants

The rate of cell division reduction due to heavy metal stress affects water and nutrient uptake by roots (Hu et al., 2013). Morphological characteristics of plants such as plant height reduction, leaf area, root growth, black blemishes in the cortex, and pericycle are affected due to Hg, As, Cu, Cd in rice plants (Kim et al., 2014). Similarly, in wheat plants there is, shoot/root length reduction and seed germination which is affected by Zn, Pb, Ni, Cd, Cr (Nagajyoti et al., 2010; Deshmukh et al., 2017), Zn and Cd affects maize plants root/shoot biomass; reduction in primary root length and dry weights of root/shoot (Hosseini et al., 2019). Barley, sorghum, and oats also depicted similar effects due to heavy metals (Deshmukh and Belanger, 2015; Zia-ur-Rehman et al., 2015). Leguminous plants such as mung bean, soybean, chickpea, pigeon pea, pea, and faba beans showed heavy metals accumulation (Hg, Cd, Co, Ni,

Pb, and Zn), affecting root and shoot growth, and seed germination (Foucault et al., 2013; Shi et al., 2018). Oil yielding plants (groundnut, canola, and *Brassica juncea*) also showed similar effects (Shi et al., 2010; Tripathi et al., 2012; Ding et al., 2013; Shim et al., 2014). Cd, Co, and Cr accumulation occur in garlic, radish, and onion (Li et al., 2012; Coskun et al., 2019). Ahmad et al. (2018) suggested stunted growth of cotton due to Cd toxicity. As and Pb causes plant and yield reduction (Lukačová et al., 2013).

4.3 Physiological and Biochemical Response

4.3.1 Effects of heavy metals on physiological characteristics of plants

Heavy metals coordinate with cells, tissue, and organs directly or indirectly by modifying cell signalling that may also cause changes in metabolic processes and may also be observed as visible injuries (Shahid et al., 2014). For instance, Cu accumulation causes loss of chlorophyll. Heavy metal also reduces the plant biomass and blocks the electron transport chain by overwhelming the plastocyanin protein pigment during photosynthetic processes (Wang et al., 2013). Similarly, Ni participates in various physiological disarrays in necrosis and chlorosis, and photosynthetic activity inhibition, consequently enhancing oxidative stress in plants (Yadav, 2010). Certain metals Zn, Fe, Ni, Cu, Co, and Mo, considered essential plant nutrients and optimum levels, are used as the enzymes that are cofactors for physiological activities (Kováčik et al., 2010); however, higher levels are detrimental for the same physiological activities. Sharma et al. (2019) reported the metal toxicity affecting electron transport between photosystem PS (II) and (I). Singh et al. (2019) and Per et al. (2016a) suggested Cd toxicity in tomato and mustard plants, causing a decrease in chlorophyll and a negative impact on PSII. Per et al. (2016b) reported stomatal closure, causing less gaseous exchange, uptake of water and its transportation by Cd toxicity further decreases photosynthetic activity, and enzymes of Calvin–Benson cycle (Mittler and Blumwald, 2010). Water content, stomatal conductance, transpiration rate, internal carbon dioxide concentrations, water-use efficiency, and rate of photosynthesis reported being reduced by increased Ni in plants (Khan and Khan 2014). Various effects on the physiological characteristics of crop plants are given below in Table 4.

5. Amelioration of heavy metals toxicity

Various methods are applied to plants to inhibit the penetration of harmful heavy metals into plant tissues by direct action on plants and by indirectly treating metals from the soil and irrigated water. One of the approaches is grafting vegetables, which helps in enhancing the tolerance levels in plants against heavy metals. Edelstein and Ben-Hur (2018) suggested that grafting of *Cucurbita* 'TZ-148' rootstock with melon cv. Arava increases tolerance for metals such as Cr, Cd, Mn, Pb, Ti, and Ni in the shoots and fruit. Similarly, in tomatoes, grafting limits the Cd transport (Kumar et al., 2015). The development of transgenic plants is another approach used to relieve the consequences of heavy metals. For elevating the impact of heavy metal by the action of transgenic plants which generated phytochelatins, such as in *Arabisopsis*

Table 4: Phytotoxicity of heavy metals on physiological characteristics of different crops.

Crop Species	Heavy Metal	Plant responses	References
Wheat	Pb, Ni, Zn	Decrease in photosynthetic pigments and increase in the chlorophyll *a*/*b* ratio	Nagajyoti et al. (2010); Deshmukh et al. (2017)
Barley	Cd, Cu	Diminution in chlorophyll a, b and carotenoids content	Fauteux et al. (2006)
Sorghm	Cd	Reduction in chlorophyll pigments, growth, and root characteristics	Ali et al. (2011)
Oat	Pd	Affected CO_2 fixation affected by inhibition of enzyme activity	Zia-ur-Rehman et al. (2015)
Soyben	Cd	Reduction in chlorophyll synthesis and Mg uptake declined	Wu et al. (2012)
Bean	Zn	Reduction of photosynthetic pigments; Assimilation and translocation of Fe and Mg into the chloroplast	Greger et al. (2018)
Pigeonpea	Ni, Pd	Reduction in stomatal conductance and chlorophyll content; enzymatic activities affects the CO_2 fixation; photosynthetic activity reduction; decrease in chlorophyll content	Shi et al. (2005)
Fababean	Mn	Decrease in photosynthetic pigments	Arya and Roy (2011)
Pea	Mn, Zn	Reduction in photosynthetic pigments; reduction in O_2 evolution and photosystem II activity; change in the structure of chloroplast	Li et al. (2012); Liang et al. (2005)
Groundnut	Cr	Decrease in photosynthetic pigments	Shim et al. (2014)
Cotton	Cd, Pb	Reduction of pigments, stomatal conductance, rate of transpiration and photosynthesis, water use efficiency	Ahmad et al. (2018); Shi et al. (2005)
Tomato	Hg	Chlorosis	Lukačová et al. (2013)
Canola	As	Chlorosis	Tripathi et al. (2012)
Brassiajuncea	Cd, Pb	Reduction in photosynthetic pigments	Shi et al. (2010)
Radish	Cd, Co	Reduction in photosynthetic pigments	Chen et al. (2003)

thaliana enzyme phytochelatin synthase (AtPCS1), and *Triticum aestivum* L. enzymes phytochelatin synthase (TaPCS1) worked against heavy metal toxicity (Bohra et al., 2015). Numerous transgenic plants are created by the change in the primary genes/proteins; for instance, *Arabidopsis thaliana* develop Hg tolerance when encoding mercuric ion reductase and against Cd produced in *Nicotiana tabacum* by the yeast metallothionein-encoding gene (Rugh et al., 1998). AtPCS1 gene participates in the As tolerance in *Nicotiana tabacum* (Zanella et al., 2016). A similar gene can also provide Cd accretion and tolerance in *Arabidopsis* (Soda et al., 2016). Cd hypersensitivity can be reduced in rice plants by the TaPCS1 gene (Mayerová et al., 2017).

Certain bacteria also reduce the harmful effects of the heavy metal ion by restraining, uptake, and transformation (Hassan et al., 2017). Microbes can transport metals across the cytoplasmic membrane, participate in the bioaccumulation and biosorption to the cell wall; entrapping metals in the extra-cellular capsules causes heavy metals precipitation process of oxidation and reduction causes metal detoxification (Zubair et al., 2016). Bacillus, Pseudomonas, Streptomyces, and Methylobacterium can ameliorate the effect of heavy metals in crops (Sessitsch et al., 2013). Similarly, Co, Cu, Cd, Cr, Ni, Mn, and Pb effects can be ameliorated by *Pseudomonas moraviensis* and *Bacillus cereus* in wheat plants (Hassan et al., 2017). *Klebsiella pneumonia is* associated with rice plants to reduce the effect of Cd (Pramanik et al., 2017). *Enterobacter, Leifsonia, Klebsiella,* and *Bacillus* help to mitigate the impact of Cd in Zea mays (Ahmad et al., 2016). Similarly, *Rhodococcus* sp., *Variovorax paradoxus*, and *Flavobacterium* sp. work against Cd in mustard (Adediran et al., 2015). In *Solanum nigrum*, Cu, Cd, and Zn toxicity can be reduced by *Pseudomonas* sp. (Chen et al., 2015).

Apart from bacteria, fungi also take part in the mitigation of heavy metals from plants and soil. Particularly the genera Penicillium, Trichoderma, Aspergillus, and Mucor, are filamentous to reduce heavy metal stress (Oladipo et al., 2016). Cell walls of fungus also have prominent metal-binding capacities due to the negative charge of the functional groups such as carboxylic, phosphate, amine, or sulfhydryl (Ong et al., 2017). Tripathi et al. (2012, 2017) suggested that *Trichoderma* species causes the reduction in As induced stress in chickpea. Essential soil microorganism, arbuscular mycorrhizal fungi (AMF), promotes nutrient absorption from soil and mitigates heavy metal stress in the host plant (Saxena et al., 2017). AMF helps to inbound the heavy metals in the cell wall and deposit them in the AMF vacuoles by siderophores, causing metal sequestration in the soil and also into root apoplast; usually, metals conjoined to metallothioneins or phytochelatins inside the plant and fungal cells, other metal transporters present in the tonoplast help in catalyzing the metals moving from the cytoplasm (Jan and Parray, 2016).

Specific plants (medicinal) also help in the remediation progression of soil polluted with heavy metals. Phytoremediation processes involve connecting microorganisms in the rhizosphere to eradicate, depreciate, or assemble pollutants from the environment (Ouyang, 2002). For tackling heavy metal phytoremediation contamination, using woody plant species, high biomass plants including crop plants, medicinal, and aromatics plants are recommended (Zheljazkov et al., 2008). The consumption of crops contaminated by phytoremediation is not acceptable because of a chance of the heavy metal entering the food chain both via intake by humans and animals (Gupta et al., 2014). Due to the advanced capability of certain medicinal plants to accumulate metals in their edible parts, the cultivation and usage of these plants in contaminated soil require strict regulation for use as recommended by WHO (2005).

6. Conclusion

Heavy metals such as Pd, Cr, Cd, Hg, Ag, and As are considered to be phytotoxic; however, it is important to note that some heavy metals (Zn, Cu, Mn, Fe, Ni, Mo, and Co) act as micronutrients due to their participation in the various redox reaction involved in the cellular activities. Deficiency may disrupt plant metabolic activity. However, both essential and non-essential heavy metals can cause growth diminution and different biochemical and physiological alterations in higher concentrations. Various anthropogenic activities are liable for the increasing intensity of heavy metal in the soil, emission in the atmosphere, and deposition on soil, water bodies, and water uptake, as well as accumulation, and translocation to plant parts. The present chapter provides an insight into the various sources of heavy metals involving both natural and anthropogenic activities, mechanism of action of metals on plants, uptake and its effects at the different levels of plants resulting in changes in the morphological physiological, and biochemical characteristics. Specific mitigation strategies also suggested the advancement in the development of the transgenic plant, and the phytoremediation use of bacterial and fungi. Heavy metals in the environment above the permissible limits, directly and indirectly, affect plants. Therefore, in the present scenario, the investigation of pathways and mitigation strategies can prevent the detrimental effects of heavy metals on humans and vegetation.

References

Abdu N, Agbenin JO, and Buerkert A. 2011. Geochemical assessment, distribution, and dynamics of trace elements in urban agricultural soils under long-term wastewater irrigation in Kano, northern Nigeria. *J Plant Nutr Soil Sci* 174(3): 447–58.

Adediran GA, Ngwenya BT, Mosselmans JF, Heal KV, and Harvie BA. 2015. Mechanisms behind bacteria induced plant growth promotion and Zn accumulation in *Brassica juncea*. *J Hazard Mater* 283: 490–9.

Ahmad B, Jaleel H, Sadiq Y, Khan MMA, and Shabbir A. 2018. Response of exogenous salicylic acid on cadmium induced photosynthetic damage, antioxidant metabolism and essential oil production in peppermint *Plant Growth Regul* 86: 273–286.

Ahmad I, Akhtar MJ, Asghar HN, Ghafoor U, and Shahid M. 2016. Differential effects of plant growth-promoting rhizobacteria on maize growth and cadmium uptake. *J Plant Growth Regul* 35(2): 303–15.

Ali S, Bai P, Zeng F, Cai S, Shamsi IH, Qiu B, Wu F, and Zhang G. 2011. The ecotoxicological and interactive effects of chromium and aluminum on growth, oxidative damage and antioxidant enzymes on two barley genotypes differing in Al tolerance. *Environ Exp Bot* 70(2-3): 185–91.

Alves RI, Sampaio CF, Nadal M, Schuhmacher M, Domingo JL, and Segura-Muñoz SI. 2014. Metal concentrations in surface water and sediments from Pardo River, Brazil: human health risks. *Environ Res* 133: 149–55.

Anjum SA, Ashraf U, Imran KH, Tanveer M, Shahid M, Shakoor A, and Longchang WA. 2017. Phyto-toxicity of chromium in maize: oxidative damage, osmolyte accumulation, anti-oxidative defense and chromium uptake. *Pedosphere* 27(2): 262–73.

Arya SK, and Roy BK. 2011. Manganese induced changes in growth, chlorophyll content and antioxidants activity in seedlings of broad bean (*Viciafaba* L.). *J. Environ. Biol.* 32(6): 707.

Awasthi, S. 1998. Prevention of food adulteration Act (Act No. 37 of 1954), along with central and state rules, as amended in 1997 and 1998. *International Law Publishing Company Pvt.* Limited.

Aydinalp C, and Marinova S. 2003. Distribution and forms of heavy metals in some agricultural soils. *Pol J Environ Stud* 12(5).

Battsengel E, Murayama T, Fukushi K, Nishikizawa S, Chonokhuu S, Ochir A, Tsetsgee S, and Davaasuren D. 2020. Ecological and human health risk assessment of heavy metal pollution in the soil of the ger district in Ulaanbaatar, Mongolia. *Int J Environ Res Public Health* 17(13): 4668.

Bhattacharya A, Dey P, Gola D, Mishra A, Malik A, and Patel N. 2015. Assessment of Yamuna and associated drains used for irrigation in rural and peri-urban settings of Delhi NCR. *Environ Monit Assess* 187(1): 1–3.

Blokhina O, Virolainen E, and Fagerstedt KV. 2003. Antioxidants, oxidative damage and oxygen deprivation stress: a review. *Ann Bot* 91(2): 179–94.

Bohra A, Sanadhya D, and Chauhan R. 2015. Heavy metal toxicity and tolerance in plants with special reference to cadmium: A Review. *J Plant Sci Res* 31(1).

Chaney RL. 1988. Metal speciation and interaction among elements affect trace element transfer in agricultural and environmental food-chains. *Metal Precipitation: Theory, Analysis, and Application.* pp. 219–59.

Chauhan PS, Singh A, and Singh RP. 2012. Ibrahim MH. Environmental impacts of organic fertilizer usage in agriculture. Organic fertilizers: types, production and environmental impact. Nova Science Publisher, Hauppauge, pp. 63–84.

Chen M, Xu P, Zeng G, Yang C, Huang D, and Zhang J. 2015. Bioremediation of soils contaminated with polycyclic aromatic hydrocarbons, petroleum, pesticides, chlorophenols and heavy metals by composting: applications, microbes and future research needs. *Biotechnol Adv* 33(6): 745–55.

Chen YX, He YF, Luo YM, Yu YL, Lin Q, and Wong MH. 2003. Physiological mechanism of plant roots exposed to cadmium. *Chemosphere* 50(6): 789–93.

Chung BY, Song CH, Park BJ, and Cho JY. 2011. Heavy metals in brown rice (*Oryza sativa* L.) and soil after long-term irrigation of wastewater discharged from domestic sewage treatment plants. *Pedosphere* 21(5): 621–7.

Coskun D, Deshmukh R, Sonah H, Menzies JG, Reynolds O, Ma JF, Kronzucker HJ, and Bélanger RR. 2019. The controversies of silicon's role in plant biology. *New Phytolo.* 221(1): 67–85.

Cunningham SD, and Berti WR. 1993. Remediation of contaminated soils with green plants: an overview. *In Vitro Cellular & Developmental Biology-Plant* 29(4): 207–12.

Cuypers A, Remans T, Weyens N, Colpaert J, Vassilev A, and Vangronsveld J. 2013. Soil-plant relationships of heavy metals and metalloids. pp. 161–193. *In Heavy Metals in Soils.* Springer, Dordrecht.

Das K, and Roychoudhury A. 2014. Reactive oxygen species (ROS) and response of antioxidants as ROS-scavengers during environmental stress in plants. *Front Environ Sci* 2: 53.

Deshmukh R, and Bélanger RR. 2015. Molecular evolution of aquaporins and silicon influx in plants. *Funct Ecol* 30(8): 1277–85.

Deshmukh RK, Ma JF, and Bélanger RR. 2017. Role of silicon in plants. *Front Plant Sci* 25(8): 1858.

Dević G, Sakan S, and Đorđević D. 2016. Assessment of the environmental significance of nutrients and heavy metal pollution in the river network of Serbia. *Environ Sci & Pollut* 23(1): 282–97.

Ding X, Zhang S, Li S, Liao X, and Wang R. 2013. Silicon mediated the detoxification of Cr on Pakchoi (*Brassica Chinensis* L.) in Cr-contaminated Soil. *Procedia Environ Sci* 18: 58–67.

Edelstein M and Ben-Hur M. 2018. Heavy metals and metalloids: sources, risks and strategies to reduce their accumulation in horticultural crops. *Sci Hortic* 234: 431–44.

Edokpayi JN, Odiyo JO, and Olasoji SO. 2014. Assessment of heavy metal contamination of Dzindiriver, in Limpopo Province, South Africa. *Int J Nat Sci Res* 2(10): 185–94.

Eichert T, Kurtz A, Steiner U, and Goldbach HE. 2008. Size exclusion limits and lateral heterogeneity of the stomatal foliar uptake pathway for aqueous solutes and water-suspended nanoparticles. *Physiol Plant* 134(1): 151–60.

Fauteux, F, Chain F, Belzile F, Menzies JG, and Bélanger RR. 2006. The protective role of silicon in the Arabidopsis–powdery mildew pathosystem. *Proc Natl Acad Sci U S A* 103(46): 17554–17559.

Fernández V, and Brown PH. 2013. From plant surface to plant metabolism: the uncertain fate of foliar-applied nutrients. *Front Plant Sci* 4: 289.

Foucault Y, Lévèque T, Xiong T, Schreck E, Austruy A, Shahid M, and Dumat C. 2013. Green manure plants for remediation of soils polluted by metals and metalloids: Ecotoxicity and human bioavailability assessment. *Chemosphere* 93(7): 1430–5.

Garbisu C, and Alkorta I. 2001. Phytoextraction: a cost-effective plant-based technology for the removal of metals from the environment. *Bioresour Technol* 77(3): 229–36.

Gimeno-García E, Andreu V, and Boluda R. 1996. Heavy metals incidence in the application of inorganic fertilizers and pesticides to rice farming soils. *Environ Pollut* 92(1): 19–25.

Giri S, and Singh AK. 2014. Assessment of surface water quality using heavy metal pollution index in Subarnarekha River, India. *Water Qual Expo Health* 5(4): 173–82.

Goher ME, Hassan AM, Abdel-Moniem IA, Fahmy AH, and El-sayed SM. 2014. Evaluation of surface water quality and heavy metal indices of Ismailia Canal, Nile River. *Egypt Egypt J Aquat Res* 40(3): 225–33.

Goni MA, Ahmad JU, Halim MA, Mottalib MA, and Chowdhury DA. 2014. Uptake and translocation of metals in different parts of crop plants irrigated with contaminated water from DEPZ area of Bangladesh. *Bull Environ Contam Toxicol* 92(6): 726–32.

Greger M, Landberg T, and Vaculík M. 2018. Silicon influences soil availability and accumulation of mineral nutrients in various plant species. *Plants* 7(2): 41.

Gupta SK, Chabukdhara M, Kumar P, Singh J, and Bux F. 2014. Evaluation of ecological risk of metal contamination in river Gomti, India: a biomonitoring approach. *Ecotoxicol Environ Saf* 110: 49–55.

Hassan TU, Bano A, and Naz I. 2017. Alleviation of heavy metals toxicity by the application of plant growth promoting rhizobacteria and effects on wheat grown in saline sodic field. *Int J phytoremediation* 19(6): 522–9.

Hawkes SJ. 1997. What is a "heavy metal"? *J Chem Educ* 74(11): 1374.

Hosseini SA, Naseri Rad S, Ali N, and Yvin JC. 2019. The ameliorative effect of silicon on maize plants grown in Mg-deficient conditions. *Int J Mol Sci* 20(4): 969.

Hu J, Wu F, Wu S, Cao Z, Lin X, and Wong MH. 2013. Bioaccessibility, dietary exposure and human risk assessment of heavy metals from market vegetables in Hong Kong revealed with an *in vitro* gastrointestinal model. *Chemosphere* 91(4): 455–61.

Jan S, and Parray JA. 2016. Use of mycorrhiza as metal tolerance strategy in plants. pp. 57–68. *In Approaches to Heavy Metal Tolerance in Plants*. Springer, Singapore.

Jaramillo MF, and Restrepo I. 2017. Wastewater reuse in agriculture: a review about its limitations and benefits. *Sustainability* 9(10): 1734.

Jaskulak M, Rorat A, Grobelak A, and Kacprzak M. 2018. Antioxidative enzymes and expression of rbcL gene as tools to monitor heavy metal-related stress in plants. *J Environ Manage* 218: 71–78.

Jia-Wen WU, Yu SH, Yong-Xing ZH, Yi-Chao WA, and Hai-Jun GO. 2013. Mechanisms of enhanced heavy metal tolerance in plants by silicon: a review. *Pedosphere* 23(6): 815–25.

Khan MI, and Khan NA. 2014. Ethylene reverses photosynthetic inhibition by nickel and zinc in mustard through changes in PS II activity, photosynthetic nitrogen use efficiency, and antioxidant metabolism. *Protoplasma* 251(5): 1007–19.

Kim YH, Khan AL, Kim DH, Lee SY, Kim KM, Waqas M, Jung HY, Shin JH, Kim JG, and Lee IJ. 2014. Silicon mitigates heavy metal stress by regulating P-type heavy metal ATPases, Oryza sativa low silicon genes, and endogenous phytohormones. *BMC Plant Biology* 14(1): 1–3.

Kováčik J, Klejdus B, Hedbavny J, and Bačkor M. 2010. Effect of copper and salicylic acid on phenolic metabolites and free amino acids in *Scenedesmus quadricauda* (Chlorophyceae). *Plant Sci* 178(3): 307–11.

Kumar P, Lucini L, Rouphael Y, Cardarelli M, Kalunke RM, and Colla G. 2015. Insight into the role of grafting and arbuscular mycorrhiza on cadmium stress tolerance in tomato. *Front Plant Sci* 6: 477.

Li L, Zheng C, Fu Y, Wu D, Yang X, and Shen H. 2012. Silicate-mediated alleviation of Pb toxicity in banana grown in Pb-contaminated soil. *Biol Trace Elem Res* 145(1): 101–8.

Liang Y, Wong JW, and Wei L. 2005. Silicon-mediated enhancement of cadmium tolerance in maize (*Zea mays* L.) grown in cadmium contaminated soil. *Chemosphere* 58(4): 475–83.

Liu L, Guan D, and Peart MR. 2012. The morphological structure of leaves and the dust-retaining capability of afforested plants in urban Guangzhou, South China. *Environ Sci Pollut Res* 19(8): 3440–9.

Liu M, Yang Y, Yun X, Zhang M, and Wang J. 2015. Concentrations, distribution, sources, and ecological risk assessment of heavy metals in agricultural topsoil of the Three Gorges Dam region, China. *Environmental Monitoring and Assessment* 187(3): 1–1.

Lokeshwari H, and Chandrappa GT. 2006. Impact of heavy metal contamination of Bellandur Lake on soil and cultivated vegetation. *Curr Sci* 10: 622–7.

Lukačová Z, Švubová R, Kohanová J, and Lux A. 2013. Silicon mitigates the Cd toxicity in maize in relation to cadmium translocation, cell distribution, antioxidant enzymes stimulation and enhanced endodermal apoplasmic barrier development. *Plant Growth Regul* 70(1): 89–103.

Mahfooz Y, Yasar A, Guijian L, Islam QU, Akhtar AB, Rasheed R, Irshad S, and Naeem U. 2020. Critical risk analysis of metals toxicity in wastewater irrigated soil and crops: a study of a semi-arid developing region. *Sci Rep* 10(1): 1–10.

Maiti SK, Kumar A, and Ahirwal J. 2016. Bioaccumulation of metals in timber and edible fruit trees growing on reclaimed coal mine overburden dumps. *Int J Min Reclam Environ* 30(3): 231–44.

Maleki A, Amini H, Nazmara S, Zandi S, and Mahvi AH. 2014. Spatial distribution of heavy metals in soil, water, and vegetables of farms in Sanandaj, Kurdistan, Iran. *J Environ Health Sci Eng* 12(1): 1–0.

Marschner H, Kirkby EA, and Cakmak I. 1996. Effect of mineral nutritional status on shoot—root partitioning of photoassimilates and cycling of mineral nutrients. *J Exp Bot* 1: 1255–63.

Masto RE, Chhonkar PK, Singh D, and Patra AK. 2009. Changes in soil quality indicators under long-term sewage irrigation in a sub-tropical environment. *Environ Geology* 56(6): 1237–43.

Mayerová M, Petrová Š, Madaras M, Lipavský J, Šimon T, and Vaněk T. 2017. Non-enhanced phytoextraction of cadmium, zinc, and lead by high-yielding crops. *Environ Sci Pollut Res* 17: 14706–16.

Mishra A, and Tripathi BD. 2008. Heavy metal contamination of soil, and bioaccumulation in vegetables irrigated with treated waste water in the tropical city of Varanasi, India. *Toxicol Environ Chem* 90(5): 861–71.

Mittler R, and Blumwald E. 2010. Genetic engineering for modern agriculture: challenges and perspectives. *Annu Rev Plant Biol* 61: 443–62.

Monge G, Jimenez-Espejo FJ, García-Alix A, Martínez-Ruiz F, Mattielli N, Finlayson C, Ohkouchi N, Sánchez MC, de Castro JM, Blasco R, Rosell J, Carrión J, Rodríguez-Vidal J, and Finlayson G. 2015. Earliest evidence of pollution by heavy metals in archaeological sites. *Sci Rep* 5: 14252.

Nagajyoti PC, Lee KD, and Sreekanth TV. 2010. Heavy metals, occurrence and toxicity for plants: a review. *Environ Chem Lett* 8(3): 199–216.

Nicholson FA, Smith SR, Alloway BJ, Carlton-Smith C, and Chambers BJ. 2003. An inventory of heavy metals inputs to agricultural soils in England and Wales. *Sci Total Environ* 205–19.

Oladipo OG, Awotoye OO, Olayinka A, Ezeokoli OT, Maboeta MS, and Bezuidenhout CC. 2016. Heavy metal tolerance potential of *Aspergillus* strains isolated from mining sites. *Bioremediat J* 20(4): 287–97.

Ong GH, Ho XH, Shamkeeva S, Manasha Savithri Fernando AS, and Wong LS. 2017. Biosorption study of potential fungi for copper remediation from Peninsular Malaysia. *Remediation Journal* 27(4): 59–63.

Ouyang Y. 2002. Phytoremediation: modeling plant uptake and contaminant transport in the soil–plant–atmosphere continuum. *J Hydrol* 266(1-2): 66–82.

Per TS, Khan NA, Masood A, and Fatma M. 2016a. Methyl jasmonate alleviates cadmium-induced photosynthetic damages through increased S-assimilation and glutathione production in mustard. *Front Plant Sci* 7: 1933.

Per TS, Khan S, Asgher M, Bano B, and Khan NA. 2016b. Photosynthetic and growth responses of two mustard cultivars differing in phytocystatin activity under cadmium stress. *Photosynthetica* 54(4): 491–501.

Pramanik K, Mitra S, Sarkar A, Soren T, and Maiti TK. 2017. Characterization of cadmium-resistant *Klebsiella pneumoniae* MCC 3091 promoted rice seedling growth by alleviating phytotoxicity of cadmium. *Environ Sci Pollut Res* 24(31): 24419–37.

Punetha D, Tewari G, Pande C, Kharkwal GC, and Tewari K. 2015. Investigation on heavy metal content in common grown vegetables from polluted sites of Moradabad district, India. *J Indian Chem Soc* 92(1): 97–103.

Rahman MA, Rahman MM, Reichman SM, Lim RP, and Naidu R. 2014. Heavy metals in Australian grown and imported rice and vegetables on sale in Australia: health hazard. *Ecotox Environ Safe* 100: 53–60.

Raskin I, Smith RD, and Salt DE. 1997. Phytoremediation of metals: using plants to remove pollutants from the environment. *Curr Opin Biotechnol* 8(2): 221–6.

Rout TK, Masto RE, Ram LC, George J, and Padhy PK. 2013. Assessment of human health risks from heavy metals in outdoor dust samples in a coal mining area. *Environ Geochem Health* 35(3): 347–56.

Rugh CL, Senecoff JF, Meagher RB, and Merkle SA. 1998. Development of transgenic yellow poplar for mercury phytoremediation. *Nat Biotechnol* 16(10): 925–8.

Sandeep G, Vijayalatha KR, and Anitha T. 2019. Heavy metals and its impact in vegetable crops. *Int J Chem Stud* 7(1): 1612–21.

Saxena B, Shukla K, and Giri B. 2017. Arbuscular mycorrhizal fungi and tolerance of salt stress in plants. *Arbuscular Mycorrhizas and Stress Tolerance of Plants*, pp. 67–97.

Schreck E, Foucault Y, Sarret G, Sobanska S, Cécillon L, Castrec-Rouelle M, Uzu G, and Dumat C. 2012. Metal and metalloid foliar uptake by various plant species exposed to atmospheric industrial fallout: mechanisms involved for lead. *Sci Total Environ* 427: 253–62.

Schreck E, Laplanche C, Le Guédard M, Bessoule JJ, Austruy A, Xiong T, Foucault Y, and Dumat C. 2013. Influence of fine process particles enriched with metals and metalloids on *Lactuca sativa* L. leaf fatty acid composition following air and/or soil-plant field exposure. *Environ Pollut* 179: 242–9.

Sessitsch A, Kuffner M, Kidd P, Vangronsveld J, Wenzel WW, Fallmann K, and Puschenreiter M. 2013. The role of plant-associated bacteria in the mobilization and phytoextraction of trace elements in contaminated soils. *Soil Biol Biochem* 60: 182–94.

Shahid M, Pourrut B, Dumat C, Nadeem M, Aslam M, and Pinelli E. 2014. Heavy-metal-induced reactive oxygen species: phytotoxicity and physicochemical changes in plants. *Rev Environ Contam Toxicol* 232: 1–44.

Shahid M, Dumat C, Khalid S, Schreck E, Xiong T, and Niazi NK. 2017. Foliar heavy metal uptake, toxicity and detoxification in plants: A comparison of foliar and root metal uptake. *J Hazard Mater* 325: 36–58.

Sharma A, Kumar V, Shahzad B, Ramakrishnan M, Sidhu GP, Bali AS, Handa N, Kapoor D, Yadav P, Khanna K, and Bakshi P. 2019. Photosynthetic response of plants under different abiotic stresses: a review. *J Plant Growth Regul* 19: 1–23.

Sharma SS, and Dietz KJ. 2009. The relationship between metal toxicity and cellular redox imbalance. *Trends Plant Sci* 14(1): 43–50.

Shi G, Cai Q, Liu C, and Wu L. 2010. Silicon alleviates cadmium toxicity in peanut plants in relation to cadmium distribution and stimulation of antioxidative enzymes. *Plant Growth Regul* 61(1): 45–52.

Shi Q, Bao Z, Zhu Z, He Y, Qian Q, and Yu J. 2005. Silicon-mediated alleviation of Mn toxicity in Cucumissativus in relation to activities of superoxide dismutase and ascorbate peroxidase. *Phytochemistry* 66(13): 1551–9.

Shi Z, Yang S, Han D, Zhou Z, Li X, Liu Y, and Zhang B. 2018. Silicon alleviates cadmium toxicity in wheat seedlings (*Triticumaestivum* L.) by reducing cadmium ion uptake and enhancing antioxidative capacity. *Environ Sci Pollut Res* 25(8): 7638–46.

Shim J, Shea PJ, and Oh BT. 2014. Stabilization of heavy metals in mining site soil with silica extracted from corn cob. *Water Air Soil Pollut* 225(10): 1–2.

Singh S, Parihar P, Singh R, Singh VP, and Prasad SM. 2016. Heavy metal tolerance in plants: role of transcriptomics, proteomics, metabolomics, and ionomics. *Front Plant Sci* 6: 1143.

Singh S, Singh VP, Prasad SM, Sharma S, Ramawat N, Dubey NK, Tripathi DK, and Chauhan DK. 2019. Interactive effect of silicon (Si) and salicylic acid (SA) in maize seedlings and their mechanisms of cadmium (Cd) toxicity alleviation. *J Plant Growth Regul* 38(4): 1587–97.

Singh UK, and Kumar B. 2017. Pathways of heavy metals contamination and associated human health risk in Ajay River basin, India. *Chemosphere* 174: 183–99.

Soda N, Sharan A, Gupta BK, Singla-Pareek SL, and Pareek A. 2016. Evidence for nuclear interaction of a cytoskeleton protein (OsIFL) with metallothionein and its role in salinity stress tolerance. *Sci Rep* 6(1): 1–4.

Srivastava V, De Araujo AS, Vaish B, Bartelt-Hunt S, Singh P, and Singh RP. 2016. Biological response of using municipal solid waste compost in agriculture as fertilizer supplement. *Rev Environ Sci Biotechnol* 15(4): 677–96.

Tangahu BV, Sheikh Abdullah SR, Basri H, Idris M, Anuar N, and Mukhlisin M. 2011. A review on heavy metals (As, Pb, and Hg) uptake by plants through phytoremediation. *Int J Chem Eng* 939161.

Thakur S, Singh L, Ab Wahid Z, Siddiqui MF, Atnaw SM, and Din MF. 2016. Plant-driven removal of heavy metals from soil: uptake, translocation, tolerance mechanism, challenges, and future perspectives. *Environ Monit Assess* 188(4): 206.

Tripathi DK, Singh VP, Kumar D, and Chauhan DK. 2012. Impact of exogenous silicon addition on chromium uptake, growth, mineral elements, oxidative stress, antioxidant capacity, and leaf and root structures in rice seedlings exposed to hexavalent chromium. *Acta Physiol Plant* 34(1): 279–89.

Tripathi P, Singh PC, Mishra A, Srivastava S, Chauhan R, Awasthi S, Mishra S, Dwivedi S, Tripathi P, Kalra A, and Tripathi RD. 2017. Arsenic tolerant *Trichoderma* sp. reduces arsenic induced stress in chickpea (*Cicerarietinum*). *Environ Pollut* 223: 137–45.

Venkatachalam P, Jayalakshmi N, Geetha N, Sahi SV, Sharma NC, Rene ER, Sarkar SK, and Favas PJ. 2017. Accumulation efficiency, genotoxicity and antioxidant defense mechanisms in medicinal plant *Acalyphaindica* L. under lead stress. *Chemosphere* 171: 544–53.

Wang CQ, Tao WA, Ping MU, LI ZC, and Ling YA. 2013. Quantitative trait loci for mercury tolerance in rice seedlings. *Rice Sci* 20(3): 238–42.

WHO. 2005. National policy on traditional medicine and regulations of herbal medicines. Geneva.

Woldetsadik D, Drechsel P, Keraita B, Itanna F, and Gebrekidan H. 2017. Heavy metal accumulation and health risk assessment in wastewater-irrigated urban vegetable farming sites of Addis Ababa, Ethiopia. *Int J Food Contam* 4(1): 1–3.

Woolson EA, Axley JH, and Kearney PC. 1971. The chemistry and phytotoxicity of arsenic in soils: I. Contaminated field soils. *Soil Sci Soc Am J* 35(6): 938–43.

Wu Y, Zhang D, Chu JY, Boyle P, Wang Y, Brindle ID, De Luca V, and Després C. 2012. The Arabidopsis NPR1 protein is a receptor for the plant defense hormone salicylic acid. *Cell Rep* 1(6): 639–47.

Wuana, RA, and Okieimen, FE. 2011. Heavy metals in contaminated soils: a review of sources, chemistry, risks and best available strategies for remediation. International Scholarly Research Notices, 2011.

Yadav SK. 2010. Heavy metals toxicity in plants: an overview on the role of glutathione and phytochelatins in heavy metal stress tolerance of plants. *S Afr J Bot* 76(2): 167–79.

Zanella L, Fattorini L, Brunetti P, Roccotiello E, Cornara L, D'Angeli S, Della Rovere F, Cardarelli M, Barbieri M, Di Toppi LS and Degola F. 2016. Overexpression of AtPCS1 in tobacco increases arsenic and arsenic plus cadmium accumulation and detoxification. *Planta* 243(3): 605–22.

Zheljazkov VD, Jeliazkova EA, Kovacheva N, and Dzhurmanski A. 2008. Metal uptake by medicinal plant species grown in soils contaminated by a smelter. *Environ Exp Bot* 64(3): 207–16.

Zhu C, Tian H, Cheng K, Liu K, Wang K, Hua S, Gao J, and Zhou J. 2016. Potentials of whole process control of heavy metals emissions from coal-fired power plants in China. *J Clean Prod* 114: 343–51.

Zia-ur-Rehman M, Sabir M, and Nadeem M. 2015. Remediating cadmium-contaminated soils by growing grain crops using inorganic amendments. pp. 367–396. *In: Soil Remediation and Plants: Prospects and Challenges.* Elsevier Inc., Academic Press.

Zubair M, Shakir M, Ali Q, Rani N, Fatima N, Farooq S, Shafiq S, Kanwal N, Ali F, and Nasir IA. 2016. Rhizobacteria and phytoremediation of heavy metals. *Environ Technol Rev* 5(1): 112–9.

6

Heavy Metal Perception in Plants

Dwaipayan Sinha,[1,*] *Arun Kumar Maurya,*[2] *Shilpa Chatterjee,*[3]
Priyanka De,[4] *Moumita Chetterjee*[5] and *Junaid Ahmad Malik*[6]

ABSTRACT

Heavy metals (HMs) are metals that have a density greater than that of water by multiple times and are present in the form of ores and minerals in the earth's crust. However, due to natural processes and anthropogenic activities, HMs are causing pollution on the earth. This pollution negatively impacts almost all living beings including the plants by affecting their yield, growth, reproduction, and physiological processes. However, certain plants can resist HM induced stress through different physio-biochemical mechanisms. Biomolecules, particularly the proteins (PCs, MTs, transporters), secondary metabolites like proline, NO, osmolytes, etc., are involved in conferring resistance from HMs in the plants. The chelators and transporters and other chelating biomolecules play distinct roles in providing resistance. This chapter gives an overview of how the plants recognise the HMs present in their environment, particularly soil. Attempts have also been made to illustrate the mechanisms adopted by the plants during HM stress.

[1] Department of Botany, Government General Degree College, Mohanpur, Paschim Medinipur, West Bengal, India-721436.
[2] Department of Botany, Multanimal Modi College, Modinagar, Ghaziabad, Uttar Pradesh, India-201204.
[3] Master of Research (MRes) Student, Faculty of Medical Sciences, Newcastle University, UK.
[4] Post-Graduate Department of Biotechnology, St Xavier's College (Autonomous), Kolkata, West Bengal, India-700016.
[5] Institute of Wood Science and Technology, 18th Cross, Malleswaram, Bangalore, Karnataka, India-560003.
[6] Department of Zoology, Government Degree College, Bijbehara, Kashmir (J&K)-192124, India.
* Corresponding author: dwaipayansinha@hotmail.com
ORCIDiD: https://orcid.org/0000-0001-7870-8998.

1. Introduction

Heavy metals (HMs) are prevalent in the lithosphere, albeit in comparatively low concentrations. HMs are the group of naturally occurring elements that encompass various metallic elements and metalloids possessing comparatively higher density (5 times) in comparison to water. In the earth's crust, HMs are largely prevalent in bound states either in the form of salts or as free elements. With time, geogenic sources such as weathering and erosion clubbed with various anthropogenic activities cause HMs leaching and their entry into the soil and aquatic bodies including groundwater (Dytłow and Górka-Kostrubiec, 2021).

In the field of botany, HMs encompasses three subgroups, namely,

- Subgroup 1 includes the middle block of the transition elements (transition metals) excluding Lanthanum (La) and Actinium (Ac)
- Subgroup 2 includes the bottom rare earth elements (lanthanide series, actinide series)
- Subgroup 3 includes the right-hand side lead group elements in the periodic table.

Figure 1 displays the distribution of HMs in the typical periodic table of elements.

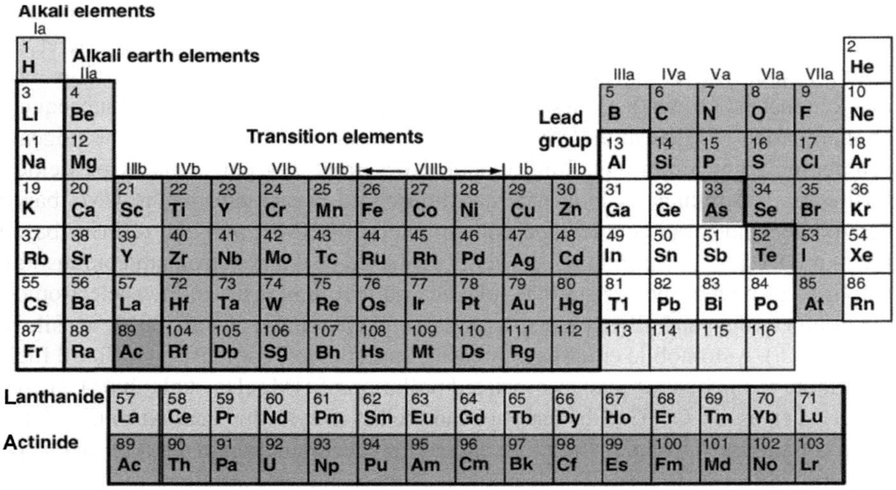

Figure 1: Periodic table showing the position of heavy metals.

1.1 Properties of heavy metals

The properties of HMs are vital in understanding the environmental impact, both on plants and animals. They have a high atomic weight and are denser than water, having a density of at least 5 gcm^{-3} differentiating them from other so-called 'light metals'. According to Tchounwou et al. (2012), HMs are called trace elements due to less availability in nature (ranging from parts per billion, ppb to 10 parts per million, ppm). A few HMs namely copper (Cu), cobalt (Co), zinc (Zn), etc., act as essential trace elements since they play crucial roles in various metabolic pathways with daily requirements ranging in amounts from 50 µg to 18 mg (Mertz, 1981). Iron is central to cytochromes of the electron transport chain (ETC) while selenium acts as a powerful antioxidant. Zinc is a central component of Zn-finger transcription factors in plants with Zn finger proteins instrumental in stress responses (Han et al., 2020). Many HMs have exceptional technological implications and act as catalysts in transformation reactions (Terfassa et al., 2014). Heavy metals such as mercury (Hg), arsenic (As), lead (Pb), chromium (Cr), and cadmium (Cd) are toxic (Kaiser, 1998; Rao and Reddi, 2000; Duruibe et al., 2007) and the magnitude of toxicity depends upon the oxidation state and the ambient chemical condition.

1.2 Sources and contamination of heavy metals

HMs mediated contamination of the environment is a serious global concern of ecological and physiological well-being. As evident from humus geochemistry, HMs contamination of the hydrosphere, lithosphere, biosphere, and atmosphere as well as their interface can be both natural and anthropogenic. The natural sources consist of rock weathering, volcanic outbreaks, sea-salt sprays, forest fires, and many such natural occurrences, leaching out these heavy metals from their endemic sites to the external environment. There is an increasing trend of HMs exposure due to their augmented applications in the domains of agriculture, industry, and technology, in addition to geogenic, pharmaceutical, or domestic sources (He et al., 2005; Tchounwou et al., 2012). Moreover, present-day agricultural activities and quick industrial growth have significantly added to the accumulation and subsequent contamination of HMs in the environment (Gu et al., 2019). Various anthropogenic sources including mining, smelting, metallurgical processes, power plants, pesticides, fertilizers, and numerous industrial and household wastes add to the HMs based ecotoxicity. Industrial sources and contamination of HMs comprise refinery based metal processing, electroplating, power plant-based coal and petroleum combustion, activities of nuclear power stations, plastic and textile industries, microelectronics, paper processing, and allied plants (Nriagu, 1989; Tchounwou et al., 2012; Musilova et al., 2016). Automobile emission, smelting, insecticides, combustion of fossil fuels contribute to the bulk of human-mediated pollution of HMs (He et al., 2005). Heavy metals such as As, Cd, Cr, Pb, and Hg head the list of systemic ecotoxicants, even at a low dose of exposure, and therefore undergo bioaccumulation in the food chain and food web. Since HMs are not biodegradable, their rising concentrations over time, can modify soil properties and reduce nutrient availability. Toxic HMs from wastewater irrigated agricultural fields may get stored in the plant tissues or hyperaccumulators, especially the edible parts of leafy vegetables (Hossain et al., 2015; Sharma and Nagpal, 2020). The hazard index (HI) is one of the key parameters utilized to assess human health hazards for any environmental pollutant. Ecotoxicological studies indicate that Cr is the most hazardous HM followed by Pb, Copper (Cu), Ni, Cd, and Zn in children as well as adults (Qing et al., 2015). Concerning ecotoxicity and toxicogenomics, certain HMs are carcinogenic, mutagenic, teratogenic, allergenic, endocrine-disruptors, neurotoxic, hepatotoxic and/or nephrotoxic, cardiotoxic, and having various other toxicological impacts (Tchounwou et al., 2012; Koedrith et al., 2013).

Plants are frequently exposed to HM due to contamination of soil by the same. Although known for playing an imperative role in physiological processes, unbalanced doses of HMs like Cd, Pb, Zn, Ni, Cr, Co, etc., induces both cytotoxic and genotoxic effects in plants leading to genome instability, plant growth inhibition, and subsequent reduction in crop yield (Choudhary et al., 2020). Soil contamination with these HMs generates important stress conditions for plants. The HMs compete with the essential nutrients for absorption in the plant roots to induce genotoxic and cytotoxic effects.

Heavy metals are first encountered by the root system (Gu et al., 2019) (Fig. 2) and the cell plasma membrane (PM) of the root is instrumental in regulating the access of HMs in plants. Heavy metal-mediated toxic responses in plants are perceived and responded to in coordinated and complex interlinked short and long-term mechanisms. The short-term mechanisms comprise the rapid changes in the transcriptome of the responsive genes through the regulation of gene expression whereas, the long-term ones have an association with genetic and epigenetic modifications (Dutta et al., 2018). The HMs storing plants called metallophytes, especially the obligate types are known to develop diverse strategies including the development of specific ecotypes to survive in HM soil conditions (Bothe and Słomka, 2017). In this chapter, an attempt would be taken to overview the perception of HMs by the plants. Efforts are taken to illustrate the effects of HMs and the roles of various proteins which act as transporters in the transportation of the metal elements inside the plant body.

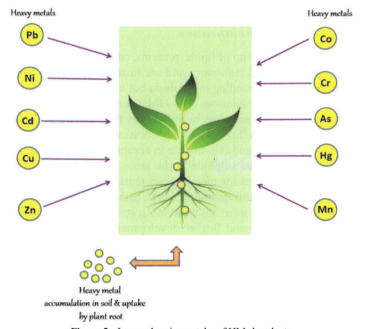

Figure 2: Image showing uptake of HMs by plants.

2. Effect of heavy metals on the physiology of plants

The middle of the twentieth century witnessed a population rise and urbanization across the globe which led to an increase in agricultural and industrial activities thereby escalating pollution levels. Among diverse environmental pollutants, HMs are placed on the top list that mainly affects soil and consequently plants and humans' health due to their toxic effects (Mishra et al., 2020; Aslam et al., 2021). They induce several physiological effects in plants which are discussed in the following section (Fig. 3).

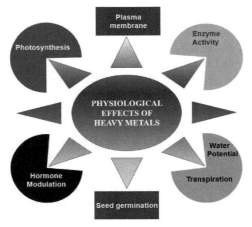

Figure 3: Physiological effects of heavy metals in plants.

2.1 Effect of HMs on membranes

Plasma membranes are made up of lipids, proteins, carbohydrates and are fluid. The membranes are dynamic and regulatory and are instrumental in the functioning of cells by ensuring not only bounding protoplasm but also of recognition, appropriate transport, and signalling mechanisms. The HMs cause a disturbance in the nature and stability of membranes and thereby influences their function. Cadmium is an HM absorbed by plants from the soil and translocates freely in plants. Cd is present in a divalent inorganic state and is available in aerobic soil rather than in anaerobic soil due to its precipitation as sulphide in the latter. The effect of Cd and Ni was seen on the rice plant as the resting membrane potential (EM) change and membrane permeability and was found differential in nature (Sanz et al., 2009). The membrane proteins and lipids oxidation were observed in Zn stress that caused solute removal, failure of seed germination, and further development in the absence of nutrients (Rahoui et al., 2010). HMs like Co and Cu occur together in soil and show toxic effects in plants when they are bioavailable in excess. Transcriptomic studies reveal the expression of 358 genes upon Co and Cu treatments. Among them, the genes, perhaps, whose role in membrane synthesis is important (fatty acid elongation) were expressed and augmented tolerance in the barley plant (Lwalaba et al., 2021). The acquisition of phosphorus by the roots can be hampered by that high concentration of Cu through a direct effect on uptake mechanisms, particularly on transporters present in PM (Feil et al., 2020).

2.2 *Effect of heavy metals on photosynthesis*

HMs actions are visible on the photosynthetic machinery of plants and take place by affecting directly binding sites at photosystems (PS I and PSII, pigments, or associated machinery) that works for capturing light and converting it into chemical energy or the enzymes participating in CO_2 fixation. The toxic effect of HMs on phototropic organisms is possibly related to increased lipid peroxidation and carbonylation of proteins (El-Amier et al., 2019).

Excess Cu distorts the internal structure of chloroplast (Baszynski et al., 1988), degrades chlorophyll, and blocks the photosynthetic electron transport (Pätsikkä et al., 2002). The Cd effects are widespread and visible on photosynthetic machinery by affecting pigments, carotenoids, and enzymes associated with photosynthesis as suppression of ribulose 1,5 dicarboxylase/oxygenase (RuBisCo), affecting the biosynthesis of δ-aminolevulinic acid (ALA), leading to the inhibition of chlorophyll synthesis that ultimately resulting in the onset of chlorotic and necrotic processes (Qadir et al., 2013). Increased concentration of Cd decreased photosynthetic pigments in *Satureja hortensis* (Azizi et al., 2020). The effects HMs were observed on soybean and mung bean grown in a hydroponic system. Both the plants exhibited reduced chlorophyll content resulting in chlorosis and consequently plant growth suppression (Mao et al., 2018).

The plant *Salvinia* is an HM accumulator with differential ranges. The decline in photosystem II and RuBisCo activity were observed in Ni, Co, Cd, Pb, Zn, and Cu treatment but contrary to this, photosystem I activity got enhanced in comparison to control. The enhancement was correlated with a build-up of the trans-thylakoidal proton gradient that supported the maintenance of the photophosphorylation potential suggesting the effects of these HMs on carbon assimilation efficiency and related process (Dhir et al., 2011). The transcriptomic studies suggested that the barley plant grown in Co and Cu contaminated soils showed expression of photosynthesis-related genes providing them resistance (Lwalaba et al., 2021).

2.3 *Effects of HMs on seed germination*

Germination of seed is the beginning of new plant life that is catalyzed by integrated actions of hormones, favorable environmental conditions, and nutrient availability. HMs influence the availability of these factors causing negative impacts on seed germination. The inhibition is concentration-dependent and a decline in nutrient availability was observed in the presence of Ni in *Cajanus cajan* (Rao and Sresty, 2000) and *Lotus corniculatus* (Bae et al., 2016). Pb caused an effect on inhibition of root volume, poor germination, and poor root growth by suppressing cell division and enlargement (Ali et al., 2014b).

2.4 Effect of HMs on transpiration and water relation

HMs cause impediments in water movement from the root to shoot (Rucińska-Sobkowiak et al., 2016). Such impediments are due to diverse effects of HMs on plants ranging from inhibition of stomatal function, affecting transpiration, and water movement from root systems to shoots as seen in Ni stressed plants (Molas, 1997). It also induces the accumulation of hormones like ABA helping in stomatal closure (Seregin and Ivanov, 2001) and enhanced oxidative damage and lipid peroxidation in response to Ni toxicity (Rao and Sresty, 2000). Exposure of excess Zn or Cu in Scot's pine seedlings for a short duration showed a great influence on water status and nutrient balance (Ivanov et al., 2021). Cd significantly inhibited mulberry plant growth and primarily accumulated in mulberry roots (Dai et al., 2020).

2.5 Heavy metals on plant hormones

Plant hormones help in the growth processes of plants and act as the first messenger. HMs disturb their normal functioning. *Lolium perenne* has a great capacity to grow in Cd-contaminated soil with superior tillering capacity. Cd causes inhibition in tiller production which is linked with the axillary bud rather than bud initiation. It is observed that Cd upregulated genes are linked with the dormancy of axillary bud and down-regulated genes, related to bud activity strigolactone biosynthesis and signalling and transportation of auxin and signalling. In addition to it, there was an enhanced degradation of cytokinin and decreased expression of cytokinin biosynthetic gene with no change in concentration of indole-3-acetic acid (IAA) (Niu et al., 2021).

Jasmonic acid (JA) can influence plant growth through crosstalk with other signalling molecules namely brassinosteroids (BRs) and ethylene (ET). Exposure to Cd results in upregulation of ET biosynthesis and induces oxidative stress, which in turn affects ET signalling, indicating crosstalk between two pathways (Schellingen et al., 2015). Exposure to Cd results in the activation of signalling pathways related to jasmonate and ethylene in *Arabidopsis* for mediating stress-induced NO_3^- allocation in roots for enhancing tolerance towards Cd (Zhang et al., 2014).

Copper (Cu) interferes with the growth process in partial association with plant growth regulators. Exposure to Cu induces oxidative stress thereby altering hormonal homeostasis in roots primarily in the apical region where drastic reductions in hormonal levels were observed (Matayoshi et al., 2020).

Salicylic acid (SA) and gibberellin (GA) are instrumental in the tolerance of abiotic stress in plants (Emamverdian et al., 2020). The role of SA in providing metal tolerance was observed for Hg in alfalfa and *Medicago sativa* (Zhou et al., 2009) and Pb for pea and Indian mustard (Boroumand et al., 2011; Ghani et al., 2015) and Cd in *Thlaspi* and maize (Gondor et al., 2016; Llugani et al., 2013). In *Chlorella vulgaris*, GA3 increased tolerance to low levels of Pb and Cd (Falkowska et al., 2011).

3. Perception of HMs by the plant

Plants have a very intricate uptake mechanism of HMs from the soil. The uptake mechanisms can be divided into two parts. The first is making the metal or nutrient in a bioavailable or readily available form and the second is the actual uptake mechanism. The roots uptake nutrients from the rhizosphere. In this case, the root exudates are crucial in the uptake of HMs. Plant metabolites that are released by the roots into the surrounding soil to enhance the nutrient absorption capacity are known as root exudates (Luo et al., 2014). The root exudates are of two types namely, high molecular weight (HMW) and low molecular weight (LMW) exudates. The HMW exudates include mucilage which is comprised of polysaccharides and polyuronic acids and enzymes while LMW exudates comprise amino and organic acids (OAs), phenols, and siderophores (Bais et al., 2008). Among the root exudates, LMWOAs have the potential to modify HMs bioavailability (Montiel-Rozas et al., 2016). The LMWOAs promote the soil phase dissolution and aids in the uptake of dissolved compounds from the soil by chelating them and also by proton-promoted reactions which can be donated by the carboxyl groups present in their molecule (Zhao et al., 2017). In addition, root exudates are involved in attracting beneficial microbes (PGPR) which in turn help in the growth promotion of the plant (Vives-Peris et al., 2020). For example, nonspecific acid phosphatases secreted by the microbes help to solubilize phosphate salts and make them available to the plant among which the most elaborately studied enzymes are the phosphomonoesterases (Alori et al., 2017). These actions are largely centered on the uptake of beneficial elements or more appropriately the mineral nutrients. The scenario varies for ions that are not beneficial to the plants. Some root secretions released into the soil confers protection from HMs. They possess chelating properties and prevent the entry of HMs in the root. For example, histidine and citrates present in root exudates inhibit entry of Ni in the roots (Salt et al., 2000). However, the entry of HMs within the plant cell depends on their concentration in soil and the plant's physiological stature (accumulators or tolerants). All these play the determining role in the passage of HMs within a plant cell.

4. Entry of HMs in the root system

In a case where the concentration of HMs is high (in case of pollution mediated toxicity), the cell wall (CW) is first line of fortification which an HM needs to cross. The root CW possesses cation binding sites and modulates the availability of ions for uptake in the apoplastic region. It is widely believed that pectic substances of the CW are responsible for the binding of cations (Szatanik-Kloc et al., 2017). The binding of Cd to pectic substances in the root CW of *Oryza sativa* showed greater pectin methylesterase activities (Yu et al., 2020). The hemicellulose present in CW is a significant site of aluminium (Al) adsorption in *Arabidopsis* (Yang et al., 2011). Another report states the involvement of hemicellulose in binding to Cd in the roots of *Sedum alfredii* (Guo et al., 2019). The pectic sites, histidyl groups of CW along with callose and mucilage often provide an ideal site for immobilization of HMs prevents their entry into the cytosol (Jan and Parray, 2016). In *Silene vulgaris*, HMs are bound to the proteins and silicates in their CW and accounts for their metal tolerance (Jain et al., 2018). The CW acts not only as a site for HM adsorption but also acts as a structure for responding to HM stress. However, adsorption of HMs on the CW and further entry into the cytosol are also governed by a very intricate signalling process. In this case, receptors such as wall-associated kinases (WAKs) are capable of sensing the status of the cell concerning any external disturbance and transmiting the signal to the interior of the cell through their kinase domain located in the cytoplasm. The WAKs create a signalling link between the CW and membrane and respond to variations in turgor pressure which is the prime parameter in a metal stressed condition (Jain et al., 2018).

4.1 Movement of HMs across the cytoplasmic membrane

Under the condition of HMs stress, several membrane-bound transporters and proteins aid in the uptake and consequent translocation of HMs. In this section, we would briefly discuss the various proteins which are associated with the transportation process.

4.1.1 Proton pumps

A proton pump is a membrane-bound enzyme complex having the capacity to mobilize protons thereby generating a transmembrane proton gradient. This proton gradient constitutes the fundamental energy reservoir and is actively involved in respiration and the photosynthetic process in plants (Gomez, 2011). The PM H^+-ATPase is an electrogenic proton pump that exports cellular H^+ ions thereby generating a transmembrane proton gradient (ΔpH; acidic on the outside) and an electrical gradient where the potential is negative inside the membrane (Falhof et al., 2016). Ion transport across the PM is associated with the hydrolysis of ATP facilitated by P-type ATPases integral membrane proteins. The P-type ATPase contains three conserved domains namely (i) the transmembrane helix (TM) bundle composed of 6 to 10 transmembrane helix and acts as translocation pathway of the substrate, (ii) the ATP-binding domain which is composed of phosphorylation domain (P-domain) and nucleotide-binding domain (N-domain) and the phosphorylation domain (P-domain) consisting of an invariant and transiently phosphorylated Asp residue, (iii) the actuator domain (AD), which possibly transmits changes in the ATP-binding domain to the transmembrane region and propels dephosphorylation (Smith et al., 2014). The P1B-ATPases are commonly termed heavy metal ATPases (HMA) and transport HMs across the membranes. There are 9 and 8 members of the P1B-ATPase family in *Oryza sativa* and *Arabidopsis* respectively (Zhang et al., 2018). Out of the various types of HMAs, of these, AtHMA5, AtHMA7, AtHMA6, and AtHMA8 are responsible for the homeostasis of Cu ions (Burkhead et al., 2009). PAA2 (AtHMA8) and PAA1 (AtHMA6) deliver Cu to the chloroplast whereas PAA1 (AtHMA6) is responsible for the transportation of Cu across the plastid-envelope, while PAA2 (AtHMA8) transports Cuacross the membrane of thylakoid (Shikanai et al., 2003; Abdel-Ghany et al., 2005). In *Oryza sativa*, OsHMA4–OsHMA9 functions as Cu^+/Ag^+ transporters (Deng et al., 2013; Huang et al., 2016).

4.1.2 The ZIP family of proteins

The zinc-regulated andiron-regulated transporter-like protein (ZIP) family of proteins are instrumental in the transportation and homeostasis of Zn and other transition metal cations in the plant system (González-Guerrero et al., 2016; Ajeesh Krishna et al., 2020). These proteins have transmembrane metal coordination sites which enable them to capture the transition elements (Antala et al., 2015). The iron-regulated transporters (IRT) also belong to the ZIP family and transport Fe (Krohling et al., 2016). In *Arabidopsis*, the AtIRT1 (Varotto et al., 2002), AtIRT2 (Vert et al., 2009), and AtIRT3 (Lin et al., 2009) uptake and transport Fe within the plant. In addition, IRTs also transport several divalent cations (Bowers and Srai, 2018). The genes for IRTs are expressed in a different metal stressed condition and increase with an increase in deficiency of Fe (Ajeesh Krishna et al., 2020).

4.1.3 The NRAMP family of proteins

Natural Resistance-Associated Macrophage Proteins (NRAMPS) are a family of proton/metal transporter proteins that help in the uptake of metal ions such as Cd^{2+}, Fe^{2+}, Mn^{2+}, and Zn^{2+} (Ullah et al., 2018). In *Oryza sativa*, seven members of NRAMP have been identified namely OsNRAMP1, OsNRAMP2, OsNRAMP3, OsNRAMP4, OsNRAMP5, OsNRAMP6 and OsNRAMP7 which transport Cd^{2+}, Fe^{2+}, Zn^{2+}, and Mn^{2+} (Mani and Sankaranarayanan, 2018). In *Arabidopsis*, the NRAMP gene encodes a family of highly hydrophobic membrane proteins and transfers Cd^{2+} and Fe^{2+} (Thomine et al., 2000). In *Arabidopsis*, there are six NRAMPS proteins. AtNRAMP1 controls Fe homeostasis (Curie et al., 2000) and also acts as a transporter for manganese (Mn) uptake due to transcriptional upregulation of the same Mn deficiency and its PM localization (Socha and Guerinot, 2014). AtNRAMP4 and AtNRAMP3 present in the vacuolar membrane (VM) mobilize Fe from the vacuoles during the germination of seeds (Lanquar et al., 2005). Through the vesicular-shaped endomembrane compartment, AtNRAMP6 transports metals within the cell (Chen et al., 2017).

4.1.4 Copper transporter family of proteins

They uptake Cu from the soil and are composed of metal-binding sites and transmembrane domains (Jain et al., 2018). In *Arabidopsis*, the high-affinity transporters (COPT) family consists of 6 members which are located in the PM (COPT6, COPT1, and COPT2) and internal membrane (COPT5 and COPT3) (Sanz et al., 2019). The COPT1 is a PM-localized transporter, involved in Cu influx *Arabidopsis* roots whereas COPT5 is present in the VM and pre-vacuolar vesicles and is responsible for mobilization of stored Cu and export out of the organelles during times of Cu deficiency (Tiwari et al., 2017). COPT2 is a high-affinity Cu and Fe transport protein and is responsible for Cu acquisition and distribution during the Cu-deficient condition (Perea-García et al., 2013).

4.2 Movement of metals from roots to shoots

Vascular bundles transport HMs from the roots to aboveground parts (Álvarez-Fernández et al., 2014). During the transportation process, they are initially carried by specific transporters in the form of metal conjugates, i.e., they bind or complex with metals and aids in the transportation process (González-Guerrero et al., 2016). Table 1 illustrates the various molecules involved in the transportation of metals.

Table 1: Various molecules involved in transportation.

Transporters	Nature/Properties	Specific Transporters/ proteins	Metal ions transported/Loaded	Functions	Reference
Heavy metal ATPase (HMA) family of transporters	P-type ATPase related to metal transport	HMA4	Zn	Loading in Xylem	Claus et al., 2013
		AtHMA4	Cd	Loading in Xylem	Zeng et al., 2017
		AtHMA2	Cd	Loading in Xylem	Zeng et al., 2017
		AtHMA5	Cu	Translocation from roots to shoots, Cu detoxification in roots.	Deng et al., 2013
		OsHMA5	Cu	Loading of Cu in the xylem of roots	Deng et al., 2013
Multidrug And Toxic Compound Extrusion (MATE) Family of Efflux Proteins	Multidrug efflux transporters (Kuroda and Tsuchiya, 2009)	Ferric reductase defective 3 (FRD3)	Fe	Fe translocation from roots to shoots through complex formation with citrate	Krohling et al., 2016
Oligopeptide Transporters Family (OPT)	Small gene family whose products are involved in the transportation of substrates synthesized from amino acids that can complex with metals (Lubkowitz, 2011)	Yellow stripe 1 protein (ZmYS1)	Fe	Uptake of Fe-Phytosiderophores in the root cells.	Zhang et al., 2019
		AtYSL2	Fe/Zn	Uptake of Fe and Zn in roots.	Schaaf et al., 2005

Phytochelatins	Metal-binding peptides consisting of repetitive γ-glutamylcysteine units with a carboxyl-terminal glycine (Grill et al., 1987)	AtABCC1 and AtABCC2	Cd, Hg	Complexation with Cd and Hg and then sequestered in vacuoles through the ABC transporters.	Park et al., 2012
			Cd	Cd-Phytochelatin complex for long-distance transport in plants(source to sink)	Mendoza-Cózatl et al., 2008
			As/Cd	Accumulation and detoxification of As and Cd through overexpression of AtPCS1	Zanella et al., 2016
Metallothioneins	Intracellular cysteine-rich metal-binding proteins (Ruttkay-Nedecky et al., 2013)	AtMT1a, AtMT2a, AtMT2b and AtMT3	Cu	Tolerance of Cu^{+2} in the leaves through metallothionein-metal complexation	Guo et al., 2003; Benatti et al., 2014
		AtMT4	Zn	Modulation of Zn homeostasis in seeds.	Ren et al., 2012
Ferritin	Fe storage protein important for Fe homeostasis (Knovich et al., 2010)		Fe	Storage of Fe in cell organelles	Morrissey and Guerinot, 2009

4.3 Distribution and uptake of heavy metals within the intracellular organelles

The transportation of metal ions through the vascular system ultimately results in a gradual build-up of the HMs in the cells of the aboveground parts of the plant. They are then transported within the cell organelles to lessen their burden in the cytosol. Several proteins and transporters are involved in the process of transportation. In this section, the proteins involved in the transportation of HMs to the cell organelles would be discussed.

The CPx type of ATPases facilitates the movement of toxic metals through the PM by utilization of ATPP-1B type form of energy from roots to the organelles present within the cells of aboveground parts (Tong et al., 2002; Seigneurin-Berny et al., 2006; Argüello et al., 2007; Kim et al., 2009). In higher plants, 3 distinct transporter families namely the ATP-Binding Cassette (ABC) superfamily, the Major Facilitator Superfamily (MFS), and the Multidrug and Toxic Compound Extrusion (MATE) family together form Multiple Drug Resistance (MDR) transporters. The MDR transporters are responsible for the primary detoxification process through the extrusion of toxic elements from the cytosol and storing them in the vacuole which is seen in other organisms also (Remy and Duque, 2014).

In *Arabidopsis*, the ABC transporters can further be classified into three groups namely P-glycoproteins (PGP) or multidrug resistance (MDR) which falls into the ABCB subfamily of the multidrug resistance-associated protein (MRP) which belongs to pleiotropic drug resistance (PDR) and ABCC subfamily, which falls in the ABCG subfamily (Sánchez-Fernández et al., 2001). In *Arabidopsis*, 16 members of MRP transporters are present in the VM (Rea, 2007). In plants, the AtABCC1 and AtABCC2 are reported to act as phytochelatin transporter and confer tolerance to HMs (Song et al., 2010; Park et al., 2012). In addition to it, AtABCC3 also escalates phytochelatin-mediated sufferance towards Cd (Brunetti et al., 2015). In addition to it, MATE also helps in the translocation of HMs. FDR, a member of the MATE family is expressed in the roots of *Arabidopsis halleri* and *Thlaspi caerulescens*, and the gene encoding the protein is involved in the translocation of HMs (Singh et al., 2016). In *Cajanus cajan*, CcMATE4 was upregulated in the roots under Mn, Zn and Al stress hinting at their possible roles in translocation or detoxification of the toxic metals (Dong et al., 2019). A report states that AtNRAMP4 and AtNRAMP3 mobilize Fe in the vacuoles of a cell (Lanquar et al., 2005) while AtNRAMP6 is present in the Golgi and trans-Golgi networks and participates in intracellular Fe homeostasis (Li et al., 2019b). Another study states that NRAMP2 is localized in the trans-Golgi network and is involved in the mobilization of HMs in the roots of the *Arabidopsis* (Gao et al., 2018). The role of NRAMP6 in Cd toxicity is also established (Cailliaette et al., 2009).

The Cation Diffusion Facilitator (CDF) family or the metal transporter proteins (MTP) in higher plants helps in HM transport to the organelle (Ibout et al., 2020). The CDF protein consists of two modular architectures which include a C-terminal domain (CTD) and transmembrane domain (TMD) (Kolaj-Robin et al., 2015). The

CDF is also known as MTP in plants helps in ion homeostasis especially in the sequestration of Zn in the cell vacuoles of metal and nonmetal hyperaccumulating plants (Ricachenevsky et al., 2013). The AtMTP1 and AtMTP3 transport Zn in the vacuole of *Arabidopsis* (Kobae et al., 2004; Arrivault et al., 2006). It is also noted that AtMTP8 is responsible for the transportation of Fe and Mn in the vacuole (Alejandro et al., 2020). AtMTP11 is also involved in the transportation of Mn tolerance towards the metal (Delhaize et al., 2017). It is localized in prevacuolar Golgi-like compartments and confers the tolerance towards Mn through vesicular trafficking and exocytosis of excess Mn (Chu et al., 2017). In *Oryza sativa*, OsMTPs are involved in Mn tolerance and toxicity. OsMTP9 is present in PM and helps in the efflux of Mn into the root stele for consequent root shoot translocation (Ueno et al., 2015). OsMTP8.1 is localized in the tonoplast and functions to sequester Mn ions in the vacuole (Farthing et al., 2017). The Zip transporters transport HMs to and fro within the intracellular organelle. It is reported that ZIP1, which is a homolog to IRT1 is present in the vascular membrane stele cells of the roots and remobilizes Mn from the vacuole to cytosol, and facilitates radial movement of Mn within the root (Alejandro et al., 2017). AtZIP2 is also responsible for the transport of Mn in the vascular bundles of root for further movement into the shoots (Li et al., 2019a). Figure 4 depicts the various transporters involved in HM management in a plant cell.

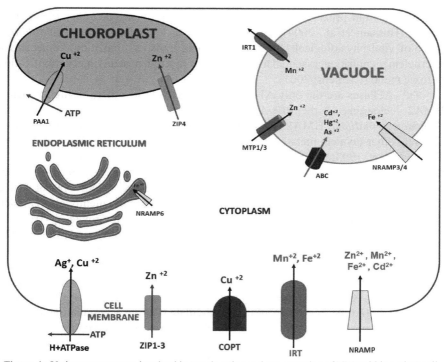

Figure 4: Various transporters involved in translocation and sequestration of HMs within a plant cell.

5. Metal ATPases and their physiological roles

P1B-ATPases are the universal cation transporters and regulate the fluxes of nutrients (Zn^{2+}, Cu^+) and also detoxify toxic cations (Co^{2+}, Pb^{2+}, Cd^{2+}, high Zn^{2+}). In plants, they are present in various cellular and intracellular membranes. P-type pumps are a kind of ATPases, which facilitate ions' active transport, are classified into multipass membrane protein groups. In *Arabidopsis thaliana* eight types of HM transporting genes of the P-type ATPases subfamily are present. The P-type ATPases helps in transporting several cations across PM which are further classified based on sequence and functional affinity. Axelsen and Palmgren (1998), among these type 1B protein subfamily of the P-type ATPases required for HM transportation (Rensing et al., 1999; Argüello, 2003). They have been known as HMA1 to HMA8, where HMA7 and HMA6 have been designated as RAN1 and PAA1, respectively (Baxter et al., 2003). Copper accumulation affects the biosynthetic pathway of chlorophyll by declining photosystem as the activity of protochlorophyllide reductase decreases. Excess Cu accumulation in plants ceased the Mg-chelatase activity which leads to the blocking of chlorophyll biosynthesis. Protochlorophyllide reductase and Mg-chelatase and present in bacteria, fungi, and plants are sensitive to oxygen.

Among these eight, HMA5, HMA8, HMA7 (RAN1), and HMA6 (PAA1), resemble Cu/Ag subclass. The remaining four types, i.e., HMA1, HMA2, HMA3, and HMA4 of 1B ATPases in *Arabidopsis*, are associated with the divalent cation transporters from prokaryotes and are not present in other eukaryotes (apart from plants) (Hussain et al., 2004). These 1B ATPases in *Arabidopsis* execute a wide range of vital physiological functions like metal transport transition, homeostasis, microelement nutrition, essential metals delivery to target proteins, and detoxification of toxic metal (Clemens, 2001).

Zn^{2+}-ATPases are the only ATPases that have been identified in higher plants. HMA2, belonging to the PIB-ATPase group has been successfully cloned in *Arabidopsis thaliana*. HMA2 is a Zn^{2+}-dependent ATPase that is also triggered by Cd^{2+} and other divalent metallic cations (Tsai et al., 1992; Okkeri and Haltia, 1999; Tsai and Linet, 1993; Sharma et al., 2000). A higher concentration of Zn^{+2} elicits the expression of HMA4 transcripts predominantly in roots (Mills et al., 2003). Promoters of HMA2 and HMA4 regulate the expression of reporter genes present inside the vascular bundle, HMA2 present at the *Arabidopsis* cells membrane (Hussain et al., 2004). Tolerance to Co^{2+}, Cd^{2+}, $Pb^{2+,}$ and Zn^{2+} can be improved by AtHMA3 overexpression (Clemens, 2001; Fraustro da Silva and Williams, 2001; Hall, 2002). All the P(1B)-type ATPases do not possess various cytoplasmic metal-binding domains (MBDs) adaptation, despite these differences C and N terminus of MBDs regulate the turnover rate of enzymes that are present in the membrane transport sites. In eukaryotes, multiple N-MBD regions of Cu^+-ATPases perform localization of the protein (Argüello et al., 2007). The modulation of the accumulation of HM in vivo for plant survival is intricate, in the metabolic process metal cation transporter has a vital role. To encounter several environmental stresses plants have adopted various defence mechanisms for protection. Heavy metal associated, i.e., HMA domain gene is vital for metal ions transportation in the cell. Six main cis-acting elements of HMA regulate abiotic stress-mediated adaptation in plants. HMA family gene expression varies in plant tissue types and also type of abiotic stress (He et al., 2020). P(1B)-type ATPases that transport copper is classified into two subfamilies, CopAs (P1B-1-ATPases) and CopBs (P1B-3-ATPases) which have a vital function in Cu homeostasis (Andersson et al., 2014).

Phosphate (play an important role in biological functions like cell energy metabolism, an important constituent of DNA, RNA and phospholipids), and Zn (Zn deficiency leads to biomass reduction, necrosis in the interveinal area, leaf deformities and chlorosis, yield depletion through a high concentration of Zn considered as toxic for plant metabolism) are two important inorganic nutrients required for the growth of plants. Although these two nutrients are present in soil in low quantity, sometimes they are absent. For that reason, during cultivation people choose fertilizer fortified with phosphate and Zn for better yield of the crop to overcome food scarcity. But this practice is not economically and ecologically suited for the long period as phosphate is a non-renewable element (Bouain et al., 2014). Phytic acid (PA) is present in grains and consists of 6 phosphate moieties joined to a myo-inositol molecule. It is also reported that a high amount of phosphate in phytic acid cause eutrophication (Bali and Satyanarayana, 2001). Root architecture will alter due to the low concentration of phosphate and Zn which leads to the escalating of important nutrients (Bayle et al., 2011; Jain et al., 2013). Element export from cells requires phosphate and Zn loading inside the xylem. As a result, nutrients will translocate from root to shoot (Hamburger et al., 2002; Hanikenne et al., 2008; Rouached et al., 2010; Sinclair and Kramer, 2012). Phosphate and Zn transportation in plants is regulated by several genes like phosphate transporter, i.e., PHT1 (Bayle et al., 2011), ZIP transporter, i.e., ZRT or IRT-like protein, phosphate exporter, i.e., PHO1 (Rouached et al., 2010), and HMA2 and HMA4 of P1B–ATPases (Sinclair and Kramer, 2012). Several morphological, transcriptional as well as metabolic modifications are influenced by phosphate deficit (Misson et al., 2005; Lan et al., 2012). There are 9 and 12 PHT1 genes present in *Arabidopsis* and rice (Paszkowsky et al., 2002). Soil pH regulates the Zn levels with higher concentrations in lower pH (Tagwira et al., 1992; Wang et al., 2006).

6. Heavy metal triggered oxidative stress in plants

Heavy metal stress inactivates and denatures several plant biomolecules, and inhibits subsequent substitution reactions in which essential metal ions are involved (Choudhary et al., 2020). Genotoxicity involves HMs entering the nucleus and cross-links deoxyribonucleic acid (DNA), modifies DNA, breaks DNA strands, removes purine molecules which eventually blocks transcription and translation processes (Dutta et al., 2018) (Fig. 5).

Heavy metals cause alteration in membrane properties of plants and change the overall physiology. It also triggers an oxidative burst resulting in the generation of superoxide (O_2^-) and hydroxyl (OH$^•$) radicals and also methylglyoxal (Gratão et al., 2019). Accumulation of these ROS leads to lipid peroxidation causing cellular membrane leakage, damaging of biomolecules, cleaving DNA, and culminating in the inhibition of downstream central dogma process of transcription and translation. These events are collectively termed oxidative stress (Dutta et al., 2018).

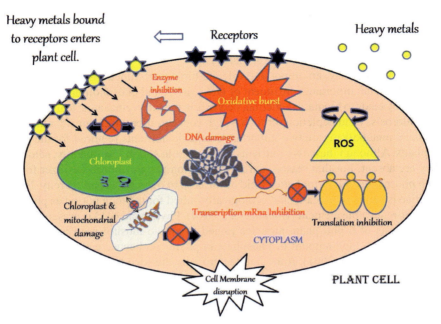

Figure 5: Pictorial representation of HM-induced oxidative stress in a plant cell.

6.1 Heavy metal stress-induced adaptive strategies

Several mechanisms are adopted by plants regarding increased HM levels in the environment (i.e., soil). They implement the same either through tolerance or through avoidance. Avoidance is the first defensive strategy of the plant to minimize HM uptake. It is accomplished by root cell and their exudates (Hasanuzzam et al., 2020). The root exudates restrict the metals in the apoplastic region and prevent entry into the plant cell by an increase of efflux or bio-absorption to the CWs (Ghori et al., 2019).

Some plants have mechanisms called tolerance mechanisms for HMs. It is achieved by components like amino acids, glutathione, superoxide dismutase (SOD), phytochelatins (chelation of HMs), transcription factors, etc. (Ghori et al., 2019). Their function is to sequester metal ions in the respective cell components protecting the sensitive plant components from their interaction with HMs. This is also termed as a "detoxification mechanism" acting as the second line of defense which takes place inside the cell or vacuole of the plants (Ghori et al., 2019).

6.1.1 Symbiotic association

The symbiotic association with vesicular-arbuscular mycorrhizal (VAM) fungi which immobilize HMs thereby reducing their uptake forms another defensive mechanism to tolerate stress induced by HM (Ghori et al., 2019; Sousa et al., 2019).

6.1.2 *Relevance of APX cycle in overcoming oxidative stress*

Ascorbate peroxidase (APX) converts harmful hydrogen peroxide (H_2O_2) into water generated by HMs (H_2O_2 + AsA2 H_2O + MDHA).

This detoxification occurs in chloroplast, peroxisome, cytosol, mitochondria, chloroplast, peroxisome, cytosol, mitochondria, and apoplast. Scavenging of H_2O_2 in the chloroplast is done by APX as Catalase (CAT) is absent in the chloroplast. The peroxisomal/glyoxysomal APX scavenges H_2O_2 and participate in AsA–GSH cycle through the production of mono dehydroascorbate (MDHA) which is then reduced to dehydroascorbate (DHA) and finally recycled to ascorbic acid (AsA) by glutathione (GSH) dependant dehydroascorbate reductase (DHAR) (Sousa et al., 2019; Giannakoula et al., 2021).

6.1.3 *Role of non-enzymatic antioxidants in overcoming oxidative stress*

Ascorbate and GSH are instrumental in the AsA-GSH cycle to scavenge free radicals. Other vital components (Ashraf et al., 2019) Antioxidant enzymes such as SOD, CAT and peroxidases are the frontline antioxidant defense system present in plants. SOD acts through the dismutation of superoxide (O_2^-) into H_2O_2 which is further taken over by CAT to generate water. Catalase is a heme-containing enzyme, is actively involved in the conversion of H_2O_2 to H_2O (26 million molecules/min). Peroxidase mainly oxidizes phenyl hydroxide (PhOH) for producing phenoxyl radical (PhO) (Giannakoula et al., 2021). These antioxidant catalytic reaction sites involve chloroplast, peroxisome, cytosol, mitochondria, apoplast (SOD), peroxisome (CAT and peroxidase), and other antioxidants such as polyphenol oxidase (PPO) glutathione reductase (GR), peroxidase transferase, and thioredoxin that exists as part of ROS scavenging antioxidant machinery (Ghori et al., 2019).

7. Strategies adopted by plants towards heavy metal tolerance

HMs are either not required by plants or required in a very minute amount for normal growth and development. When such HMs are encountered by plants, they show various strategies to overcome their deleterious effects. These strategies operate at two levels, one, by preventing entry into plants; two, if HMs enter inside plant then by detoxification mechanism, phytochelation (by complex formation or use of metallothionein). Plants use several genes for HMs tolerance. HMs tolerant plants show restricted uptake and transport of HMs in the following ways as described in this section.

7.1 *The cell wall*

The CW of the root system forms the first line of structure that comes in contact with HMs present in the soil. The CW decides the adsorption and entry of HMs in plants with pectins and histidines (Baetz and Martinoia, 2014). *Silene vulgaris* ssp. *humilis* accumulates diverse HMs through adsorption and binding in the silicates and proteins located in CW (Revathi and Subhashree 2013). Cadmiumtolerance in two wheat genotypes namely a low Cd accumulating and high Cd accumulating in grains in hydroponic culture treated with or without Cd for seven days was studied. The result showed that the tolerant genotype showed Cd bound to the pectin and hemicellulosemoieties of roots CW. Metabolomic profiling confirmed that upregulated CW biosynthesis is one of the mechanisms providing tolerance to HM stress (Lu et al., 2021).

7.2 Root exudates

Root exudates form the initial defensive barrier against HMs. Histidine and citrates are present in RE and are involved in the uptake of Ni (Nishida et al., 2008). Extracellular carbohydrates such as mucilage and callose are instrumental in the capture of HMs (Revathi and Venugopal, 2013). The accumulation of Cr in tomato cultivars through changes in pH of the rhizosphere, secretion of OAs along with the ionic composition and morpho-physiological responses showed enhanced Cr accumulation in the root and shoot in both cultivars. Greater retention of Cr and other mineral nutrients were also observed, which may be due to the basification of the growth medium (Javed et al., 2021).

7.3 By plasma membrane

Plasma membrane forms the second barrier and very sensitive biological structure of a plant that responds with HMs. The manifestation of HM stress is through ion leakage, oxidation of protein thiols and membrane lipids, inhibition H^+-ATPase, alteration of membrane fluidity, and integrity (Quartacci et al., 2001). HMs induced damage caused increased leakage of solutes from cells (Meharg, 1993). Contrarily, Cu, Zn has a protective effect on membrane leakage (Cakmak, 2000), through the involvement of heat shock proteins (HSP) and metallothioneins (Salt et al., 1998). Plants also take the help of diverse transporters (the HM P1B-ATPase, the NRAMP and the ZIP families), for metal uptake and homeostasis and present on PM and tonoplast. The complementary role for H_2S and GSH is to confer strength to the membrane during Cr stress in maize (Kharbech et al., 2020).

7.4 By detoxification mechanism

The detoxification mechanisms in plants operate by synthesizing specific low molecular weight chelators that reduce the chance to bind the key proteins. Along with this, PM exclusion strategy and vacuolar sequestration with the help of membrane transporters are additional strategies to minimize the effects of HMs. Soybean plant under Cd stress shows upregulation of glutamine synthetase (GS) and metabolites related to the biosynthesis of glutathione that results in increased GSH formation that helped more metal binding capacity and enhanced protection against oxidative stress. Similarly, cysteine synthase (CS) and GSH, involved in Al tolerance are up-regulated in Al-stressed soybean. Cd and Al-induced heat shock proteins (like HSP70) and LMW-HSP and three DnaJ-like proteins are reported in the leaves of soybean. DnaJ-like proteins are molecular chaperones and modulate Hsp70 -ATPase activity during folding, assembly and disassembly of proteins (Frugis et al., 1999).

7.5 By phytochelatins and metallothioneins

Plants detoxify HMs through the production of LMW thiol-rich peptides synthesized by phytochelatin synthase (PCS) from sulfur-rich glutathione (GSH) (Hassinen et al. 2011) (Fig. 6). PCS chelate HMs in the vacuole thereby reducing cell damage and playing a great role in phytoremediation (Fig. 7). *IpPCS1* is identified in Cd treated *Ipomoea pes-caprae* and confers Cd tolerance. The Cd activation site is located in the C-terminal of *IpPCS1* (Su et al., 2020). Two PCS genes (glutathione gamma-glutamyl-cysteinyl transferase 1 and glutathione gamma-glutamyl-cysteinyl transferase) and 4 proteins, four FC genes, and 4 mRNA were detected in chickpeas under Cd stress (Mohajel Kazemi et al., 2020). MYB4 is a transcription factor that belongs to the R2R3-subfamily of MYB domain protein and is instrumental in Cd-stress tolerance in *Arabidopsis*. The MYB4 protein binds directly to PCS1 phytochelatin synthase 1 and metallothionein 1C helps *in vivo* to control their transcriptional expression (Agarwal et al., 2020). Cd tolerance in presence of exogenous GSH is accomplished by scavenging stress-induced ROS and the biosynthesis of PCs in maize (Wang et al., 2021). Cadmium stress tolerance in two varieties of *citrus* showed Cd exclusion as a strategy to minimize HM accumulation in photosynthetic organs (López-Climent et al., 2014).

Cysteine-rich Metallothioneins (MTs), LMW cytoplasmic metal-binding proteins playing crucial roles in allocation of metal and homeostasis distributed across all kingdoms (Hassinen et al., 2011; Rono et al., 2021) and diverse affinity to

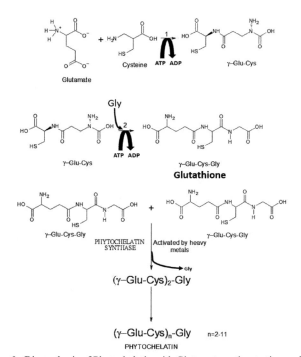

Figure 6: Biosynthesis of Phytochelatin with Glutamate as the starting molecule. 1-γ-glutamyl-cysteine-synthetase, 2-glutathione synthetase.

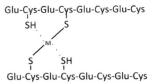

Figure 7: Mechanism of chelation of HMs with phytochelatins. Heavy metals form complexes with thiol groups of the cysteine residues in a Phytochelatin molecule.

HMs (Pomponi et al. 2006). The OsMT1b is more involved in the chelation of Cr in roots instead of shoots, while OsMT2c is more functional in neutralizing conditions and accumulated H_2O_2 in shoots than roots (Yu et al., 2019). The MYB4 protein binds directly to MT1C (metallothionein 1C) promoters *in vivo* and positively controls their expression at the transcriptional level, suggesting that MT1C are the key targets of MYB4 (Agarwal et al., 2020).

7.6 Genes involved in metal tolerance

Genes are the key to all cellular activities and are responsible for evolution. A moss (*Physcomitrella patens*) contains four metallothionein-like genes. MTs are involved in detoxification of metals, mobilization of nutrients, scavenging of free radicals, tolerance towards stress, and developmental processes. Protonema treated with Cd exhibited up-regulation of PpMT1.1a and PpMT1.1b and H_2O_2 was probably associated with HM accumulation and tolerance (Pakdee et al., 2019).

HMs stress inducts plants to resort to varied and complicated mechanisms of regulation of genes. Gene expression at the post-transcriptional level is modulated by MicroRNAs. The plant possesses various HM-responsive microRNAs target genes. They constitute a portion of the complex regulatory network that controls HM uptake, transport, folding and assembly of proteins, chelation of metals, free radical scavenging, signaling of growth regulators, and biogenesis of microRNA (Ding et al., 2020). An artificial microRNA (amiRNA) line targeting closely homologous ERF and CBF transcription factors showed that the CBF1, 2, and 3 are associated with the regulation of sensitivity towards arsenite. Furthermore, resistance towards Cd requires the presence of ERF34 and ERF35 transcription factors (Xie et al., 2021).

Excess of Cu and/or Cd in *Zostera marina* showed upregulation of GSH, PCs, and MT levels along with enzymes like GR, APX, and CAT. Both metals also showed upregulation of the DNA methyltransferases DRM2 and CMT3 (Greco et al., 2019). Cr exposures (Cr IV and Cr III) developed MTs enhancement in plant tissues and ten specific *OsMT* genes in rice tissues. The *OsMT1b* gene showed a better chelating response during Cr stress in roots in comparison to shoots. The *OsMT2c* gene functions in eliminating accumulated H_2O_2 in shoots than in roots (Yu et al., 2019).

Mercury and selenium (Se) are one of the major pollutants. Their separate and combined action caused inhibited growth of *S. salsa*, increased antioxidant enzyme activities, and disturbed osmotic regulation through the genes of betaine aldehyde dehydrogenase and choline monooxygenase (Liu et al., 2021). A RING E3 ligase gene (*SlRING1*) is associated with the positive regulation of Cd tolerance in tomatoes. The product of SlRING1 is present in PM and nucleus and its overexpression increased the net photosynthetic rate, chlorophyll content and photochemical efficiency of photosystem II (Fv/Fm), and reduction in levels of ROS by increased transcript level of CAT, MDHAR, GSH, PCs, and relative less electrolyte leakage under Cd stress (Ahammed et al., 2020). Cd tolerant variety consumed more carbohydrates to overcome Cd stress rather than to support growth. It is reported that Cd-sensitive plants utilize more carbohydrates to counter Cd stress at the expense of supporting growth than the Cd-tolerant variety (Li et al., 2020). The kinase CIPK11 responses were observed to Cd stress by the ABA signalling pathway (Gu et al., 2021).

Arsenic stress tolerance in *Ricinus communis* showed enhanced activity of CAT, antioxidant enzymes, the content of proline, and expression of nicotianamine synthase genes (*RcNAS1*, *RcNAS2*, and *RcNAS3*). NAS gene encodes an enzyme that catalyzes trimerization of S-adenosylmethionine to form nicotianamine, which is instrumental in metal chelation and tolerance of HMs (Singh et al., 2021).

Chromium stress showed blockage of the citric acid cycle, glutamine synthetase/glutamate synthase (GS/GOGAT) cycle, and partial amino acid metabolic pathways that inhibited normal growth and development of the bokchoy plant (Zhou et al., 2021). Nitric oxide partly operates through post translational modification of proteins, notably through S-nitrosylation in response to HM stress, involves in the function of various phytohormones and S-nitrosylation during plant responses to HM stress (Wei et al., 2020).

Several transcription factors (TFs) upregulation of the ERF family are important in response to Cd, Cu, and Zn by the plant. It was observed that Cd, Zn, and Cu induced the expression of TdSHN1 in durum wheat seedlings that provided tolerance in transgenic tobacco lines through the formation of longer roots, greater biomass, increased chlorophyll, and reduced ROS in comparison to wild type plants subjected to HM stress (Djemal Khoudi, 2021). Figure 8 is the pictorial overview perception of HMs by the plants and the sequence of events that occurs under metal stressed conditions.

8. Future prospect and conclusive remarks

Heavy metals occur naturally in the soil. However, due to indiscriminate anthropogenic activities, there is increasing pollution of HMs in the environment. These HMs are emitted largely through industrial processes as sludge or as particulate matter through smokes. The accumulation of HMs in the soil and water beyond a certain limit results in toxic effects on plants. The plants have devised several mechanisms through which they can counteract the HM toxicity. However, if the concentration of the HMs is more, the defensive mechanism often tends to break apart resulting in the manifestation of toxic effects. These effects include overall retardation of growth, decrease in yield, disruption of respiratory processes, and most importantly, induction of oxidative stress which in turn results in a cascade of events either directly or through signal transduction pathways. However, before the manifestation of the toxic responses, the metals get translocated from the roots to the parts where they express their toxic effects. On the other hand, some plants are efficient in metal absorption and store them in their intracellular organelles or vacuoles or even adsorb them in their robust architecture of the cell wall. All these processes involve the

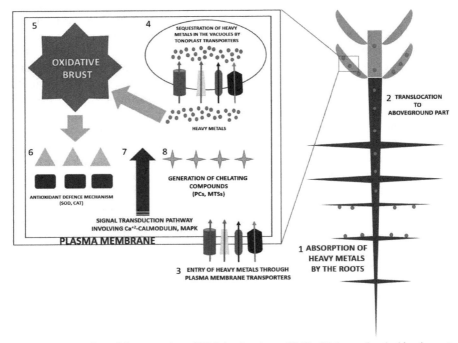

Figure 8: An overview of the perception of HMs by the plants. (1) The HMs are absorbed by the root system. (2) They are then translocated into the above-ground part through the vascular bundles with many proteins helping in the process. (3) Once they reach the above-ground parts, they are housed in the cell with the help of a wide array of PM transporters. (4) Inside the cell the metals are sequestered into the vacuoles through the transporters present in the tonoplast. Concomitantly they also result in (5) oxidative burst (6) upregulation of antioxidant defense system (7) Initiation of signal transduction pathway involving second messengers and (8) Generating of chelating compounds such as metallothioneins and phytochelatins.

action of specialized proteins which act as transporters. In addition, translocation of the HMs also involves the action of chelating agents namely the phytochelatins and metallothioneins. Thus, these biomolecules are instrumental in the management of HMs within a plant body and have been elaborately discussed in the chapter. It is has been observed that genes of several proteins get upregulated during HM stress in metal tolerant plants. Thus the study of these genes is extremely suitable to understand the mechanism of metal tolerance. A lot of avenues are there to devise biotechnological approaches to insert the genes of metal tolerance in susceptible plants and validate their performance in field conditions. This will pave the way for the generation of biofortified HM tolerant plants. Additionally, further studies are required in the field of signal transduction pathways as some of them are common to other stresses also. Thus proper fine-tuning of the study of signal transduction pathways concerning HM stress is also relevant to fully understanding metal homeostasis. This will again open up the avenue of exploring multiple stress responses at the same time due to similarities in signal transduction pathways. Further studies are also required for those HMs that might be required in fewer quantities by the plants. The scope also lies at the other end of the table. Till now the discussion stressed the study of metal tolerance in the plant and how they can be made more tolerant or biofortified to combat HM stress. However its equally important to study the health risk of consumption of metal hyperaccumulators especially by herbivores and insects. The proper study of environmental impact and biomagnification also requires to be done. Though metal hyperaccumulators act as an efficient tool for remediation of HMs, efficient formulations are also required for proper disposal of the same to avoid consumption by animals or insects. In this regard, the involvement of policymakers is relevant for efficient implementation. The HM perception and its consequent effects on the plants can thus be utilized as an efficient model system for the study of plant physiology which can then be imparted as bioremediation in the practical field keeping into consideration the environmental regulations.

References

Abdel-Ghany SE, Müller-Moulé P, Niyogi KK, Pilon M, and Shikanai T. 2005. Two P-type ATPases are required for copper delivery in *Arabidopsis thaliana* chloroplasts. *Plant Cell* 17(4): 1233–1251.

Agarwal P, Mitra M, Banerjee S, and Roy S. 2020. MYB4 transcription factor, a member of R2R3-subfamily of MYB domain protein, regulates cadmium tolerance via enhanced protection against oxidative damage and increases expression of PCS1 and MT1C in *Arabidopsis*. *Plant Sci* 297: 110501.

Ahammed GJ, Li CX, Li X, Liu A, Chen S, and Zhou, J. 2020. Overexpression of tomato RING E3 ubiquitin ligase gene SlRING1 confers cadmium tolerance by attenuating cadmium accumulation and oxidative stress. *Physiol Plant* 10.1111/ppl.13294. Advance online publication.

Ajeesh Krishna TP, Maharajan T, Victor Roch G, Ignacimuthu S, and Antony Ceasar S. 2020. Structure, function, regulation and phylogenetic relationship of ZIP family transporters of plants. *Front Plant Sci* 11: 662.

Alejandro S, Cailliatte R, Alcon C, Dirick L, Domergue F, Correia D, Castaings L, Briat JF, Mari S, and Curie C. 2017. Intracellular distribution of manganese by the trans-golgi network transporter NRAMP2 is critical for photosynthesis and cellular redox homeostasis. *Plant Cell* 29(12): 3068–3084.

Alejandro S, Höller S, Meier B, and Peiter E. 2020. Manganese in plants: from acquisition to subcellular allocation. *Front Plant Sci*, 11.

Ali B, Mwamba TM, Gill RA, Yang C, Ali S, Daud MK, Wu Y, and Zhou W. 2014b. Improvement of element uptake and antioxidative defense in *Brassica napus* under lead stress by application of hydrogen sulfide. *Plant Growth Regul* 74: 261–273.

Alori ET, Glick BR, and Babalola OO. 2017. Microbial phosphorus solubilization and its potential for use in sustainable agriculture. *Front Microbiol* 8: 971.

Alvarez-Fernández A, Díaz-Benito P, Abadía A, López-Millán AF, and Abadía J. 2014. Metal species involved in long distance metal transport in plants. *Front Plant Sci* 5: 105.

Andersson M, Mattle D, Sitsel O, Klymchuk T, Nielsen AM, Møller LB, White SH, Nissen P, and Gourdon P. 2014. Copper-transporting P-type ATPases use a unique ion-release pathway. *Nat Struct Mol Biol* 21: 43–48.

Antala S, Ovchinnikov S, Kamisetty H, Baker D, and Dempski RE. 2015. Computation and functional studies provide a model for the structure of the zinc transporter hZIP4. *J Biol Chem* 290(29): 17796–17805.

Argüello JM. 2003. Identification of ion selectivity determinants in heavy metal transport P1B-type ATPases. *J Membr Biol* 195: 93–108.

Argüello JM, Eren E, and González-Guerrero M. 2007. The structure and function of heavy metal transport P1B-ATPases. *Biometals* 20(3-4): 233–248.

Arrivault S, Senger T, and Krämer U. 2006. The *Arabidopsis* metal tolerance protein AtMTP3 maintains metal homeostasis by mediating Zn exclusion from the shoot under Fe deficiency and Zn oversupply. *Plant J* 46(5): 861–879.

Ashraf MA, Riaz M, Arif MS, Rasheed R, Iqbal M, Hussain I, and Salman M. 2019. The role of non-enzymatic antioxidants in improving abiotic stress tolerance in plants CRC Press: Boca Raton, FL, USA (pp. 129–143).

Aslam M, Aslam A, Sheraz M, Ali B, Ulhassan Z, Najeeb U, Zhou W, and Gill RA. 2021. Lead toxicity in Cereals: mechanistic insight into toxicity, mode of action and management. *Front Plant Sci* 11: 587785.

Axelsen KB and Palmgren MG. 1998. Evolution of substrate specificities in the P-type ATPase superfamily. *J Mol Evol* 46: 8–101.

Azizi I, Esmaielpour B, and Fatemi H. 2020. Effect of foliar application of selenium on morphological and physiological indices of savory (*Satureja hortensis*) under cadmium stress. *Food Sci Nutr* 8(12): 6539–6549.

Bae J, Benoit LD, and Watson AK. 2016. Effect of heavy metals on seed germination and seedling growth of common ragweed and roadside ground cover legumes. *Environ Pollut* 213: 112–8.

Baetz, U. and Martinoia E. 2014. Root exudates: the hidden part of plant defense. *Trends Plant Sci* 19: 90–8.

Bais HP, Broeckling CD, and Vivanco JM. 2008. Root exudates modulate plant—microbe interactions in the rhizosphere. pp. 241–252. *In*: Karlovsky P (eds.). *Secondary Metabolites in Soil Ecology. Soil Biology*, vol. 14. Springer, Berlin, Heidelberg.

Bali A and Satyanarayana T. 2001. Microbial phytases in nutrition and combating phosphorus pollution. *Everyman's Sci* 4: 207–9.

Baszynski T, Tukendorf A, Ruszkowska M, Skorzynska E, and Maksymieci W. 1988. Characteristics of the photosynthetic apparatus of copper non-tolerant spinach exposed to excess copper. *J Plant Physiol* 132(6): 708–13.

Baxter I, Tchieu J, Sussman MR, Boutry M, Palmgren MG, Gribskov M, Harper JF, and Axelsen KB. 2003. Genomic comparison of P-Type ATPase ion pumps in *Arabidopsis* and rice. *Plant Physiol* 132: 618–628.

Bayle V, Arrighi JF, Creff A, Nespoulous C, Vialaret J, Rossignol M, Gonzalez E, Paz-Ares J, and Nussaume L. 2011. *Arabidopsis thaliana* high-affinity phosphate transporters exhibit multiple levels of post translational regulation. *The Plant Cell* 23: 1523–1535.

Benatti MR, Yookongkaew N, Meetam M, Guo WJ, Punyasuk N, AbuQamar S, and Goldsbrough P. 2014. Metallothionein deficiency impacts copper accumulation and redistribution in leaves and seeds of *Arabidopsis. New Phytol* 202(3): 940–951.

Boroumand JS, Lari YH, and Ranjbar M. 2011 Effect of salicylic acid on some plant growth parameters under lead stress in *Brassica napus* var. *Iran J Plant Physiol* 1: 177–185.

Bothe H and Słomka A. 2017. Divergent biology of facultative heavy metal plants. *J Plant Physiol* 219: 45–61.

Bouain N, Shahzad Z, Rouached A, Khan GA, Berthomieu P, Abdelly C, Poirier Y, and Rouached H. 2014. Phosphate and zinc transport and signalling in plants: toward a better understanding of their homeostasis interaction. *J Exp Bot* 65(20): 5725–5741.

Bowers Kand Srai S. 2018. The trafficking of metal ion transporters of the Zrt- and Irt-like protein family. *Traffic (Copenhagen, Denmark)* 19(11): 813–822.

Brunetti P, Zanella L, De Paolis A, Di Litta D, Cecchetti V, Falasca G, Barbieri M, Altamura MM, Costantino P, and Cardarelli M. 2015. Cadmium-inducible expression of the ABC-type transporter AtABCC3 increases phytochelatin-mediated cadmium tolerance in *Arabidopsis. J Exp Bot* 66(13): 3815–3829.

Burkhead JL, Reynolds KA, Abdel-Ghany SE, Cohu CM, and Pilon M. 2009. Copper homeostasis. *New Phytol* 182(4): 799–816.

Cailliatte R, Lapeyre B, Briat JF, Mari S, and Curie C. 2009. The NRAMP6 metal transporter contributes to cadmium toxicity. *Biochem J* 422(2): 217–228.

Cakmak I. 2000. Possible roles of zinc in protecting plant cells from damage by reactive oxygen species. *New Phytol* 146: 185–205.

Chen S, Han X, Fang J, Lu Z, Qiu W, Liu M, Sang J, Jiang J, and Zhuo R. 2017. Sedum alfredii SaNramp6 metal transporter contributes to cadmium accumulation in transgenic *Arabidopsis thaliana. Sci Rep* 7(1): 13318.

Choudhary A, Kumar A, and Kaur N. 2020. ROS and oxidative burst: Roots in plant development. *Plant Divers* 42(1): 33–43.

Chu HH, Car S, Socha AL, Hindt MN, Punshon T, and Guerinot ML. 2017. The *Arabidopsis* MTP8 transporter determines the localization of manganese and iron in seeds. *Sci Rep* 7(1): 1–10.

Claus J, Bohmann A, and Chavarría-Krauser A. 2013. Zinc uptake and radial transport in roots of *Arabidopsis thaliana*: a modelling approach to understand accumulation. *Ann Bot* 112(2): 369–380.

Clemens S. 2001. Molecular mechanisms of plant metal tolerance and homeostasis. *Planta* 212: 475–486.

Curie C, Alonso JM, Le Jean M, Ecker JR, and Briat JF. 2000. Involvement of NRAMP1 from *Arabidopsis thaliana* in iron transport. *Biochem J* 347: Pt 3(Pt 3): 749–755.

Dai F, Luo G, Li Z, Wei X, Wang Z, Lin S, and Tang C. 2020. Physiological and transcriptomic analyses of mulberry (*Morus atropurpurea*) response to cadmium stress. *Ecotox Environmental Safe* 205: 111298.

Delhaize E, Gruber BD, Pittman JK, White RG, Leung H, Miao Y, Jiang L, Ryan PR, and Richardson AE. 2007. A role for the AtMTP11 gene of *Arabidopsis* in manganese transport and tolerance. *Plant J* 51(2): 198–210.

Deng F, Yamaji N, Xia J, and Ma JF. 2013. A member of the heavy metal P-type ATPase OsHMA5 is involved in xylem loading of copper in rice. *Plant Physiol* 163(3): 1353–1362.

Dhir B, Sharmila P, Pardha Saradhi P, Sharma S, Kumar R, and Mehta D. 2011. Heavy metal induced physiological alterations in *Salvinia natans*. *Ecotox Environ Safe* 74(6): 1678–1684.

Ding Y, Ding L, Xia Y, Wang F, and Zhu C. 2020. Emerging roles of micrornas in plant heavy metal tolerance and homeostasis. *J Agric Food Chem* 68(7): 1958–1965.

Djemal R, and Khoudi H. 2021. The ethylene-responsive transcription factor of durum wheat, TdSHN1, confers cadmium, copper and zinc tolerance to yeast and transgenic tobacco plants. *Protoplasma*, 10.1007/s00709-021-01635-z. Advance online publication.

Dong B, Niu L, Meng D, Song Z, Wang L, Jian Y, Fan X, Dong M, Yang Q, and Fu Y. 2019. Genome-wide analysis of MATE transporters and response to metal stress in *Cajanus cajan*. *J Plant Interact* 14(1): 265–75.

Duruibe JO, Ogwuegbu MOC, and Egwurugwu JN. 2007. Heavy metal pollution and human biotoxic effects. *Int J Phys Sci* 2(5): 112–118.

Dutta S, Mitra M, Agarwal P, Mahapatra K, De S, Sett U, and Roy S. 2018. Oxidative and genotoxic damages in plants in response to heavy metal stress and maintenance of genome stability. *Plant Signal Behav* 13(8): e1460048.

Dytłow S and Górka-Kostrubiec B. 2021. Concentration of heavy metals in street dust: an implication of using different geochemical background data in estimating the level of heavy metal pollution. *Environ Geochem Health* 43(1): 521–535.

El-Amier Y, Elhindi K, El-Hendawy S, Al-Rashed S, and Abd-ElGawad A. 2019. Antioxidant system and biomolecules alteration in *Pisum sativum* under heavy metal stress and possible alleviation by 5-aminolevulinic acid. *Molecules* 24(22): 4194.

Emamverdian A, Ding Y, and Mokhberdoran F. 2020. The role of salicylic acid and gibberellin signalling in plant responses to abiotic stress with an emphasis on heavy metals. *Plant Signal Behav* 15(7): 1777372.

Falhof J, Pedersen JT, Fuglsang AT, and Palmgren M. 2016. Plasma membrane H+-ATPase regulation in the center of plant physiology. *Mol Plant* 9(3): 323–337.

Falkowska M, Pietryczuk A, Piotrowska A, Bajguz A, Grygoruk A, and Czerpak R. 2011. The effect of gibberellic acid (GA3) on growth, metal biosorption and metabolism of the green algae *Chlorella vulgaris* (chlorophyceae) beijerinck exposed to cadmium and lead stress. *Pol J Environ Stud* 20: 53–59.

Farthing EC, Menguer PK, Fett JP, and Williams LE. 2017. OsMTP11 is localised at the Golgi and contributes to Mn tolerance. *Sci Rep* 7(1): 1–13.

Feil SB, Pii Y, Valentinuzzi F, Tiziani R, Mimmo T, and Cesco S. 2020. Copper toxicity affects phosphorus uptake mechanisms at molecular and physiological levels in *Cucumis sativus* plants. *Plant Physiol Biochem* 157: 138–147.

Fraustro da Silva JJR, and Williams RJP. 2001. The Biological Chemistry of the Elements, Ed 2. Oxford University Press, New York.

Frugis G, Mele G, Giannino D, and Mariotti D. 1999. MsJ1, an alfalfa DnaJ-like gene, is tissue-specific and transcriptionally regulated during cell cycle. *Plant Mol Biol* 40(3): 397–408.

Gao H, Xie W, Yang C, Xu J, Li J, Wang H, Chen X, and Huang CF. 2018. NRAMP2, a trans-Golgi network-localized manganese transporter, is required for *Arabidopsis* root growth under manganese deficiency. *New Phytol* 217(1): 179–193.

Ghani A, Khan I, Ahmed I, Mustafa I, and Abd-Ur-Rehman MN. 2015. Amelioration of lead toxicity in *Pisum sativum* (L.) by foliar application of salicylic acid. *J Environ Anal Toxicol* 5(292): 2161–2525.

Ghori NH, Ghori T, Hayat MQ, Imadi SR, Gul A, Altay V, and Ozturk M. 2019. Heavy metal stress and responses in plants. *Int J Environ Sc Technol* 16(3): 1807–1828.

Giannakoula A, Therios I, and Chatzissavvidis C. 2021. Effect of lead and copper on photosynthetic apparatus in citrus (*Citrus aurantium* L.) plants. The role of antioxidants in oxidative damage as a response to heavy metal stress. *Plants* 10(1): 155.

Gomez F. 2011. Proton pump. pp. 1356–1356. *In*: Gargaud M et al. (eds.). *Encyclopedia of Astrobiology*. Springer, Berlin, Heidelberg.

Gondor OK, Pál M, Darko E, Janda T, and Szalai G. 2016. Salicylic acid and sodium salicylate alleviate cadmium toxicity to different extents in maize (*Zea mays* L.). *PLoS One* 11(8): e0160157.

González-Guerrero M, Escudero V, Saéz Á, and Tejada-Jiménez M. 2016. Transition metal transport in plants and associated endosymbionts: arbuscular mycorrhizal fungi and rhizobia. *Front Plant Sci* 7: 1088.

Gratão PL, Alves LR, and Lima LW. 2019. Heavy metal toxicity and plant productivity: role of metal scavengers. pp. 49–60. *In*: Srivastava S et al. (eds.). *Plant-metal Interactions*. Springer, Cham.

Greco M, Sáez CA, Contreras RA, Rodríguez-Rojas F, Bitonti MB, and Brown MT. 2019. Cadmium and/ or copper excess induce interdependent metal accumulation, DNA methylation, induction of metal chelators and antioxidant defences in the seagrass *Zostera marina*. *Chemosphere* 224: 111–119.

Grill E, Winnacker EL, and Zenk MH. 1987. Phytochelatins, a class of heavy-metal-binding peptides from plants, are functionally analogous to metallothioneins. *Proc Natl Acad Sci USA* 84(2): 439–443.

Gu S, Wang X, Bai J, Wei T, Sun M, Zhu L, Wang M, Zhao Y, and Wei W. 2021. The kinase CIPK11 functions as a positive regulator in cadmium stress response in *Arabidopsis*. *Gene* 772: 145372.

Gu YG, and Gao YP. 2019. An unconstrained ordination-and GIS-based approach for identifying anthropogenic sources of heavy metal pollution in marine sediments. *Mar Pollut Bull* 146: 100–105.

Guo WJ, Bundithya W, and Goldsbrough PB. 2003. Characterization of the *Arabidopsis* metallothionein gene family: tissue-specific expression and induction during senescence and in response to copper. *New Phytol* 159(2): 369–381.

Guo X, Liu Y, Zhang R, Luo J, Song Y, Li J, Wu K, Peng L, Liu Y, Du Y, and Liang Y. 2019. Hemicellulose modification promotes cadmium hyperaccumulation by decreasing its retention on roots in *Sedum alfredii*. *Plant Soil* 5: 1–5.

Hall JL. 2002. Cellular mechanisms for heavy metal detoxification and tolerance. *J Exp Bot* 53: 1–11.

Hamburger D, Rezzonico E, Petétot JM, Somerville C, and Poirier Y. 2002. Identification and characterization of the *Arabidopsis* PHO1 gene involved in phosphate loading to the xylem. *The Plant Cell* 14(4): 889–902.

Han G, Lu C, Guo J, Qiao Z, Sui N, Qiu N, and Wang B. 2020. C2H2 Zinc Finger Proteins: Master Regulators of Abiotic Stress Responses in Plants. *Front Plant Sci* 11: 115.

Hanikenne M, Talke IN, Haydon MJ, Lanz C, Nolte A, Motte P, Kroymann J, Weigel D, and Krämer U. 2008. Evolution of metal hyperaccumulation required cis-regulatory changes and triplication of HMA4. *Nature* 453(7193): 391–5.

Hasanuzzaman M, Bhuyan MHM, Zulfiqar F, Raza A, Mohsin SM, Mahmud JA, and Fotopoulos V. 2020. Reactive oxygen species and antioxidant defense in plants under abiotic stress: revisiting the crucial role of a universal defense regulator. *Antioxidants* 9(8): 681.

Hassinen VH, Tervahauta AI, Schat H, and Kärenlampi SO. 2011. Plant metallothioneins—metal chelators with ROS scavenging activity? *Plant Biol (Stuttg)* 13(2): 225–232.

He G, Qin L, Tian W, Meng L, He T, and Zhao D. 2020. Heavy metal transporters-associated proteins in *S. tuberosum*: genome-wide identification, comprehensive gene feature, evolution and expression analysis. *Genes (Basel)* 11(11): 1269.

He ZL, Yang XE, and Stoffella PJ. 2005. Trace elements in agroecosystems and impacts on the environment. *J Trace Elem Med Biol* 19(2-3): 125–140.

Hossain MS, Ahmed F, Abdullah A, Akbor MA, and Ahsan MA. 2015. Public health risk assessment of heavy metal uptake by vegetables grown at a waste-water-irrigated site in Dhaka, Bangladesh. *J Health Pollut* 5(9): 78–85.

Huang XY, Deng F, Yamaji N, Pinson SR, Fujii-Kashino M, Danku J, Douglas A, Guerinot ML, Salt DE, and Ma JF. 2016. A heavy metal P-type ATPase OsHMA4 prevents copper accumulation in rice grain. *Nat Commun* 7: 12138.

Hussain D, Haydon MJ, Wang Y, Wong E, Sherson SM, Young J, Camakaris J, Harper JF, and Cobbetta CS. 2004. P-Type ATPase heavy metal transporters with roles in essential zinc homeostasis in *Arabidopsis*. *The Plant Cell* 16: 1327–1339.

Ibuot A, Dean AP, and Pittman JK. 2020. Multi-genomic analysis of the cation diffusion facilitator transporters from algae. *Metallomics* 12(4): 617–630.

Ivanov YV, Ivanova AI, Kartashov AV, and Kuznetsov VV. 2021. Phytotoxicity of short-term exposure to excess zinc or copper in Scots pine seedlings in relation to growth, water status, nutrient balance and antioxidative activity. *Environ Sci Pollut Res Int* 28(12): 14828–14843.

Jain A, Sinilal B, Dhandapani G, Meagher RB, and Sahi SV. 2013. Effects of deficiency and excess of zinc on morphophysiological traits and spatiotemporal regulation of zinc-responsive genes reveal incidence of cross talk between micro-and macronutrients. *Environ Sci Technol* 47(10): 5327–35.

Jain S, Muneer S, Guerriero G, Liu S, Vishwakarma K, Chauhan DK, Dubey NK, Tripathi DK, and Sharma S. 2018. Tracing the role of plant proteins in the response to metal toxicity: a comprehensive review. *Plant Signal Behav* 13(9): e1507401.

Jan S, and Parray JA. 2016. Heavy metal uptake in plants. pp. 1–18. *In:* Jan S, and Parray JA (eds.). *Approaches to Heavy Metal Tolerance in Plants*. Springer, Singapore.

Javed MT, Tanwir K, Abbas S, Saleem MH, Iqbal R, and Chaudhary HJ. 2021. Chromium retention potential of two contrasting *Solanum lycopersicum* Mill. cultivars as deciphered by altered pH dynamics, growth and organic acid exudation under Cr stress. *Environ Sci Pollut Res Int* 10.1007/s11356-020-12269-8. Advance online publication.

Kaiser J. 1998. Toxicologists shed new light on old poisons. *Science* 279(5358): 1850–1851.

Kharbech O, Sakouhi L, Ben Massoud M, Jose Mur LA, Corpas FJ, Djebali W, and Chaoui A. 2020. Nitric oxide and hydrogen sulfide protect plasma membrane integrity and mitigate chromium-induced methylglyoxal toxicity in maize seedlings. *Plant Physiol Biochem* 157: 244–255.

Kim YY, Choi H, Segami S, Cho HT, Martinoia E, Maeshima M, and Lee Y. 2009. AtHMA1 contributes to the detoxification of excess Zn(II) in *Arabidopsis*. *Plant J* 58(5): 737–753.

Knovich MA, Storey JA, Coffman LG, Torti SV, and Torti FM. 2009. Ferritin for the clinician. *Blood Rev* 23(3): 95–104.

Kobae Y, Uemura T, Sato MH, Ohnishi M, Mimura T, Nakagawa T, and Maeshima M. 2004. Zinc transporter of *Arabidopsis thaliana* AtMTP1 is localized to VM and implicated in zinc homeostasis. *Plant Cell Physiol* 45(12): 1749–1758.

Koedrith P, Kim H, Weon JI, and Seo YR. 2013. Toxicogenomic approaches for understanding molecular mechanisms of heavy metal mutagenicity and carcinogenicity. *Int J Hyg Environ Health* 216(5): 587–598.

Kolaj-Robin O, Russell D, Hayes KA, Pembroke JT, and Soulimane T. 2015. Cation diffusion facilitator family: Structure and function. *FEBS Lett* 589(12): 1283–1295.

Krohling CA, Eutrópio FJ, Bertolazi AA, Dobbss LB, Campostrini E, Dias T, and Ramos AC. 2016. Ecophysiology of iron homeostasis in plants. *Soil Sci Plant Nutr* 62(1): 39–47.

Kuroda T and Tsuchiya T. 2009. Multidrug efflux transporters in the MATE family. *Biochimica et Biophysica Acta* 1794(5): 763–768.

Lan P, Li W, and Schmidt W. 2012. Complementary proteome and transcriptome profiling in phosphate-deficient *Arabidopsis* roots reveals multiple levels of gene regulation. *Mol Cell Proteomics* 11(11): 1156–66.

Lanquar V, Lelièvre F, Bolte S, Hamès C, Alcon C, Neumann D, Vansuyt G, Curie C, Schröder A, Krämer U, Barbier-Brygoo H, and Thomine S. 2005. Mobilization of vacuolar iron by AtNRAMP3 and AtNRAMP4 is essential for seed germination on low iron. *EMBO J* 24(23): 4041–4051.

Li C, Liu Y, Tian J, Zhu Y, and Fan J. 2020. Changes in sucrose metabolism in maize varieties with different cadmium sensitivities under cadmium stress. *PloS one* 15(12): e0243835.

Li J, Jia Y, Dong R, Huang R, Liu P, Li X, Wang Z, Liu G, and Chen Z. 2019a. Advances in the mechanisms of plant tolerance to manganese toxicity. *Int J Mol Sci* 20(20): 5096.

Li J, Wang Y, Zheng L, Li Y, Zhou X, Li J, Gu D, Xu E, Lu Y, Chen X, and Zhang W. 2019b. The intracellular transporter AtNRAMP6 is involved in Fe homeostasis in *Arabidopsis*. *Front Plant Sci* 10: 1124.

Lin YF, Liang HM, Yang SY, Boch A, Clemens S, Chen CC, Wu JF, Huang JL, and Yeh KC. 2009. *Arabidopsis* IRT3 is a zinc-regulated and plasma membrane localized zinc/iron transporter. *New Phytol* 182(2): 392–404.

Liu T, Chen Q, Zhang L, Liu X, and Liu C. 2021. The toxicity of selenium and mercury in *Suaeda salsa* after 7-days exposure. *Comparative Biochemistry and Physiology. Toxicology & Pharmacology* 244: 109022.

Llugany M, Martin SR, Barceló J, and Poschenrieder C. 2013. Endogenous jasmonic and salicylic acids levels in the Cd-hyperaccumulator *Noccaea* (Thlaspi) *praecox* exposed to fungal infection and/or mechanical stress. *Plant Cell Rep* 32(8): 1243–1249.

López-Climent MF, Arbona V, Pérez-Clemente RM, Zandalinas SI, and Gómez-Cadenas A. 2014. Effect of cadmium and calcium treatments on phytochelatin and glutathione levels in citrus plants. *Plant Biol (Stuttg)* 16(1): 79–87.

Lu M, Yu S, Lian J, Wang Q, He Z, Feng Y, and Yang X. 2021. Physiological and metabolomics responses of two wheat (*Triticum aestivum* L.) genotypes differing in grain cadmium accumulation. *Sci Total Environ* 769: 145345.

Lubkowitz M. 2011. The oligopeptide transporters: a small gene family with a diverse group of substrates and functions? *Mol Plant* 4(3): 407–415.

Luo Q, Sun L, Hu, and Zhou, R. 2014. The variation of root exudates from the hyperaccumulator *Sedum alfredii* under cadmium stress: metabonomics analysis. *PloS one* 9(12): e115581.

Lwalaba J, Zvobgo G, Gai Y, Issaka JH, Mwamba TM, Louis LT, Fu L, Nazir MM, Ansey Kirika B, Kazadi Tshibangu A, Adil MF, Sehar S, Mukobo RP, and Zhang G. 2021. Transcriptome analysis reveals the tolerant mechanisms to cobalt and copper in barley. *Ecotox Environ Safe* 209: 111761.

Mani A and Sankaranarayanan K. 2018. *In Silico* analysis of natural resistance-associated macrophage protein (NRAMP) family of transporters in rice. *Protein J* 37(3): 237–247.

Mao F, Nan G, Cao M, Gao Y, Guo L, Meng X, and Yang G. 2018. The metal distribution and the change of physiological and biochemical process in soybean and mung bean plants under heavy metal stress. *Int J Phytoremediation* 20(11): 1113–1120.

Matayoshi CL, Pena LB, Arbona V, Gómez-Cadenas A, and Gallego SM. 2020. Early responses of maize seedlings to Cu stress include sharp decreases in gibberellins and jasmonates in the root apex. *Protoplasma* 257(4): 1243–56.

Meharg AA. 1993. The role of plasmalemma in metal tolerance in angiosperm. *Physiol Plantarum* 88: 191–8.

Mendoza-Cózatl DG, Butko E, Springer F, Torpey JW, Komives EA, Kehr J, and Schroeder JI. 2008. Identification of high levels of phytochelatins, glutathione and cadmium in the phloem sap of *Brassica napus*. A role for thiol-peptides in the long-distance transport of cadmium and the effect of cadmium on iron translocation. *Plant J* 54(2): 249–259.

Mertz W. 1981. The essential trace elements. *Science (New York, N.Y.)* 213(4514): 1332–1338.

Mills RF, Krijger GC, Baccarini PJ, Hall JL, and Williams LE. 2003. Functional expression of AtHMA4, a P1B-type ATPase of the Zn/Co/Cd/Pb subclass. *Plant J* 35: 164–176.

Mishra AK, Singh J, and Mishra PP. 2020. *Toxic metals in crops: a burgeoning problem*. pp. 273–301. *In*: Mishra K, Tandon P and Srivastava S (eds.). Sustainable Solutions for Elemental Deficiency and Excess in Crop Plants (Singapore: Springer).

Misson J, Raghothama KG, Jain A, Jouhet J, Block MA, Bligny R, Ortet P, Creff A, Somerville S, Rolland N, and Doumas P. 2005. A genome-wide transcriptional analysis using *Arabidopsis thaliana* Affymetrix gene chips determined plant responses to phosphate deprivation. *Proc Natl Acad Sci* 102(33): 11934–9.

Mohajel Kazemi E, Kolahi M, Yazdi M, and Goldson-Barnaby A. 2020. Anatomic features, tolerance index, secondary metabolites and protein content of chickpea (*Cicer arietinum*) seedlings under cadmium induction and identification of PCS and FC genes. *Physiol Molecular Biol Plants* 26(8): 1551–1568.

Molas J. 1997. Changes in morphological and anatomical structure of cabbage (*Brassica oleracea* L.) outer leaves and in ultrastructure of their chloroplasts caused by an *in vitro* excess of nickel. *Photosynthetica* 34(4): 513–22.

Montiel-Rozas MM, Madejón E, and Madejón P. 2016. Effect of heavy metals and organic matter on root exudates (low molecular weight organic acids) of herbaceous species: An assessment in sand and soil conditions under different levels of contamination. *Environ Pollut (Barking, Essex : 1987)* 216: 273–281.

Morrissey J and Guerinot ML. 2009. Iron uptake and transport in plants: the good, the bad and the ionome. *Chem Rev* 109(10): 4553–4567.

Musilova J, Arvay J, Vollmannova A, Toth T, and Tomas J. 2016. Environmental contamination by heavy metals in region with previous mining activity. *Bull Environ Contam Toxicol* 97(4): 569–575.

Nishida ST Mizuno, and Obata H. 2008. Involvement of histidine-rich domain of ZIP family transporter TjZNT1 in metal ion specificity. *Plant Physiol Biochem* 46: 601–6.

Niu K, Zhang R, Zhu R, Wang Y, Zhang D, and Ma H. 2021. Cadmium stress suppresses the tillering of perennial ryegrass and is associated with the transcriptional regulation of genes controlling axillary bud outgrowth. *Ecotox Environ Safe* 212: 112002.

Nriagu JO. 1989. A global assessment of natural sources of atmospheric trace metals. *Nature* 338: 47–49.

Okkeri J and Haltia T. 1999. Expression and mutagenesis of ZntA, a zinc-transporting P-type ATPase from *Escherichia coli*. *Biochemistry* 38: 14109–14116.

Pakdee O, Songnuan W, Panvisavas N, Pokethitiyook P, Yokthongwattana K, and Meetam M. 2019. Functional characterization of metallothionein-like genes from *Physcomitrella patens*: expression profiling, yeast heterologous expression and disruption of PpMT1.2a gene. *Planta* 250(2): 427–443.

Park J, Song WY, Ko D, Eom Y, Hansen TH, Schiller M, Lee TG, Martinoia E, and Lee Y. 2012. The phytochelatin transporters AtABCC1 and AtABCC2 mediate tolerance to cadmium and mercury. *Plant J* 69(2): 278–288.

Paszkowski U, Kroken S, Roux C, and Briggs SP. 2002. Rice phosphate transporters include an evolutionarily divergent gene specifically activated in arbuscular mycorrhizal symbiosis. *Proc Natl Acad Sci, USA* 99: 13324–13329.

Pätsikkä E, Kairavuo M, Šeršen F, Aro EM, and Tyystjärvi E. 2002. Excess copper predisposes photosystem II to photo inhibition *in vivo* by out competing iron and causing decrease in leaf chlorophyll. *Plant Physiol* 129(3): 1359–67.

Perea-García A, Garcia-Molina Aandrés-Colás N, Vera-Sirera F, Pérez-Amador MA, Puig S, and Peñarrubia L. 2013. *Arabidopsis* copper transport protein COPT2 participates in the cross talk between iron deficiency responses and low-phosphate signalling. *Plant Physiol* 162(1): 180–194.

Pomponi M, Censi V, and Di Girolamo V. 2006. Overexpression of *Arabidopsis* phytochelatin synthase in tobacco plants enhances Cd^{2+} tolerance and accumulation but not translocation to the shoot. *Planta* 223: 180–190.

Qadir S, Jamshieed S, Rasool S, Ashraf M, Akram NA, and Ahmad P. 2013. Modulation of plant growth and metabolism in cadmium-enriched environments. *Rev Environ Contam Toxicol*, 51–88.

Qing X, Yutong Z, and Shenggao L. 2015. Assessment of heavy metal pollution and human health risk in urban soils of steel industrial city (Anshan), Liaoning, Northeast China. *Ecotoxicol Environ Saf* 120: 377–385.

Quartacci MF, Cosi E, and Navari-Izzo F. 2001. Lipids and NADPH-dependent superoxide production in plasma membrane vesicles from roots of wheat grown under copper deficiency or excess. *J Exp Bot* 52: 77–84.

Rahoui S, Chaoui A, and Ferjani Eel. 2010. Reserve mobilization disorder in germinating seeds of *Vicia faba* L. exposed to cadmium. *J Plant Nutr* 33: 809–17.

Rao CRM, and Reddi GS. 2000. Platinum group metals (PGM); occurrence, use and recent trends in their determination. *Trends Analyt Chem* 19(9): 565–586.

Rao KM and Sresty TV. 2000. Antioxidative parameters in the seedlings of pigeonpea (*Cajanus cajan* (L.) Millspaugh) in response to Zn and Ni stresses. *Plant Sci* 8; 157(1): 113–28.

Rea PA. 2007. Plant ATP-binding cassette transporters. *Annu Rev Plant Biol* 58: 347–75.

Remy E and Duque P. 2014. Beyond cellular detoxification: a plethora of physiological roles for MDR transporter homologs in plants. *Front Physiol* 5: 201.

Ren Y, Liu Y, Chen H, Li G, Zhang X, and Zhao J. 2012. Type 4 metallothionein genes are involved in regulating Zn ion accumulation in late embryo and in controlling early seedling growth in *Arabidopsis*. *Plant Cell Environ* 35(4): 770–789.

Rensing C, Ghosh M, and Rosen BP. 1999. Families of soft-metal-ion-transporting ATPases. *J Bacteriol* 181: 5891–5897.

Revathi S and Venugopal S. 2013. Physiological and biochemical mechanisms of heavy metal tolerance. *Int J Environ Sci* 3: 1339–54.

Ricachenevsky FK, Menguer PK, Sperotto RA, Williams LE, and Fett JP. 2013. Roles of plant metal tolerance proteins (MTP) in metal storage and potential use in biofortification strategies. *Front Plant Sci* 4: 144.

Rono JK, Le Wang L, Wu XC, Cao HW, Zhao YN, Khan IU, and Yang ZM. 2021. Identification of a new function of metallothionein-like gene OsMT1e for cadmium detoxification and potential phytoremediation. *Chemosphere* 265: 129136.

Rouached H, Arpat AB, and Poirier Y. 2010. Regulation of phosphate starvation responses in plants: signalling players and cross-talks. *Molecular Plant* 3(2): 288–299.

Rucińska-Sobkowiak R. 2016. Water relations in plants subjected to heavy metal stresses. *Acta Physiol Plant* 38(11): 1–3.

Ruttkay-Nedecky B, Nejdl L, Gumulec J, Zitka O, Masarik M, Eckschlager T, Stiborova M, Adam V, and Kizek R. 2013. The role of metallothionein in oxidative stress. *Int J Mol Sci* 14(3): 6044–6066.

Salt DE, Smith RD, and Raskin I. 1998. Phytoremediation. *Annu Rev Plant Physiol Plant Mol Biol* 49: 643–68.

Salt DE, Kato N, Kramer U, Smith RD, and Raskin I. 2000. The role of root exudates in nickel hyperaccumulation and tolerance in accumulator and nonaccumulator species of Thlaspi. pp. 189–200. *In*: Terry N, Banuelos G (eds.). *Phytoremediation of Contaminated Soil and Water*. Boca Raton, FL: Lewis Publishers Inc.

Sánchez-Fernández R, Davies TG, Coleman JO, and Rea PA. 2001. The *Arabidopsis thaliana* ABC protein superfamily, a complete inventory. *J Biol Chem* 276(32): 30231–30244.

Sanz A, Llamas A, and Ullrich CI. 2009. Distinctive phytotoxic effects of Cd and Ni on membrane functionality. *Plant Signal Behav* 4: 980–2.

Sanz A, Pike S, Khan MA, Carrió-Seguí À, Mendoza-Cózatl DG, Peñarrubia L, and Gassmann W. 2019. Copper uptake mechanism of *Arabidopsis thaliana* high-affinity COPT transporters. *Protoplasma* 256(1): 161–170.

Schaaf G, Schikora A, Häberle J, Vert G, Ludewig U, Briat JF, Curie C, and von Wirén N. 2005. A putative function for the *Aarabidopsis* Fe-Phytosiderophore transporter homolog AtYSL2 in Fe and Zn homeostasis. *Plant Cell Physiol* 46(5): 762–774.

Schellingen K, Van Der Straeten D, Remans T, Vangronsveld J, Keunen E, and Cuypers A.2015. Ethylene signalling is mediating the early cadmium-induced oxidative challenge in *Arabidopsis thaliana*. *Plant Sci* 239: 137–46.

Seigneurin-Berny D, Gravot A, Auroy P, Mazard C, Kraut A, Finazzi G, Grunwald D, Rappaport F, Vavasseur A, Joyard J, and Richaud P. 2006. HMA1, a new Cu-atpase of the chloro plast envelope, is essential for growth under adverse light conditions. *J Biol Chem* 281(5): 2882–2892.

Seregin IV and Ivanov VB. 2001. Physiological aspects of cadmium and lead toxic effects on higher plants. *Russ J Plant Physiol* 48(4): 523–44.

Sharma A and Nagpal AK. 2020. Contamination of vegetables with heavy metals across the globe: hampering food security goal. *Journal of Food Science and Technology* 57(2): 391–403.

Sharma R, Rensing C, Rosen BP, and Mitra B. 2000. The ATP hydrolytic activity of purified ZntA, a Pb(II)/Cd(II)/Zn(II)-translocating ATPase from *Escherichia coli*. *J Biol Chem* 275: 3873–3878.

Shikanai T, Müller-Moulé P, Munekage Y, Niyogi KK, and Pilon M. 2003. PAA1, a P-type ATPase of *Arabidopsis*, functions in copper transport in chloroplasts. *Plant Cell* 15(6): 1333–1346.

Sinclair SA and Kramer U. 2012. The zinc homeostasis network of land plants. *Biochimica et Biophysica Acta* 1823: 1553–1567.

Singh R, Misra AN, and Sharma P. 2021. Effect of arsenate toxicity on antioxidant enzymes and expression of nicotianamine synthase in contrasting genotypes of bioenergy crop *Ricinus communis*. *Environ Sci Pollut Res Int* 28(24): 31421–31430.

Singh S, Parihar P, Singh R, Singh VP, and Prasad SM. 2016. Heavy metal tolerance in plants: role of transcriptomics, proteomics, metabolomics and ionomics. *Front Plant Sci* 6: 1143.

Smith AT, Smith KP, and Rosenzweig AC. 2014. Diversity of the metal-transporting P1B-type ATPases. *J Biol Inorg Chem* 19(6): 947–960.

Socha AL and Guerinot ML. 2014. Mn-euvering manganese: the role of transporter gene family members in manganese uptake and mobilization in plants. *Front Plant Sci* 5: 106.

Song WY, Park J, Mendoza-Cózatl DG, Suter-Grotemeyer M, Shim D, Hörtensteiner S, Geisler M, Weder B, Rea PA, Rentsch D, Schroeder JI, Lee Y, and Martinoia E. 2010. Arsenic tolerance in *Arabidopsis* is mediated by two ABCC-type phytochelatin transporters. *Proc Natl Acad Sci USA* 107(49): 21187–21192.

Sousa RH, Carvalho FE, Lima-Melo Y, Alencar VT, Daloso DM, Margis-Pinheiro M, and Silveira JA. 2019. Impairment of peroxisomal APX and CAT activities increases protection of photosynthesis under oxidative stress. *J Exp Bot* 70(2): 627–639.

Su H, Zou T, Lin R, Zheng J, Jian S, and Zhang M. 2020. Characterization of a phytochelatin synthase gene from *Ipomoea pes-caprae* involved in cadmium tolerance and accumulation in yeast and plants. *Plant Physiol Biochem* 155: 743–755.

Szatanik-Kloc A, Szerement J, Cybulska J, and Jozefaciuk G. 2017. Input of different kinds of soluble pectin to cation binding properties of roots cell walls. *Plant Physiol Biochem* PPB 120: 194–201.

Tagwira F, Piha M, and Mugwira L. 1992. Effect of pH and phosphorus and organic matter contents on zinc availability and distribution in two Zimbabwean soils. *Communications in Soil Science and Plant Analysis* 23: 1485–1500.

Tchounwou PB, Yedjou CG, Patlolla AK, and Sutton DJ. 2012. Heavy metal toxicity and the environment. *Experientia Supplementum* 101: 133–164.

Terfassa B, Schachner JA, Traar P, Belaj F, and Zanetti NCM. 2014. Oxorhenium(V) complexes with naphtholate-oxazoline ligands in the catalytic epoxidation of olefins. *Polyhedron* 75: 141–145.

Thomine S, Wang R, War JM, Crawford NM, and Schroeder JI. 2000. Cadmium and iron transport by members of a plant metal transporter family in *Arabidopsis* with homology to Nramp genes. *Proc Natl Acad Sci* 97(9): 4991–4996.

Tiwari M, Venkatachalam P, Penarrubia L, and Sahi SV. 2017. COPT2, a plasma membrane located copper transporter, is involved in the uptake of Au in *Arabidopsis*. *Sci Rep* 7(1): 1–9.

Tong L, Nakashima S, Shibasaka M, Katsuhara M, and Kasamo K. 2002. A novel histidine-rich CPx-ATPase from the filamentous cyanobacterium *Oscillatoria brevis* related to multiple-heavy-metal cotolerance. *J Bacteriol* 184(18): 5027–5035.

Tsai KJ, Yoon KP, and Lynn AR. 1992. ATP-dependent cadmium transport by the cadA cadmium resistance determinant in everted membrane vesicles of *Bacillus subtilis*. *J Bacteriol* 174: 116–121.

Tsai KJ and Linet AL. 1993. Formation of a phosphorylated enzyme intermediate by the cadA Cd(2+)-ATPase. *Arch Biochem Biophys* 305: 267–270.

Ueno D, Sasaki A, Yamaji N, Miyaji T, Fujii Y, Takemoto Y, Moriyama S, Che J, Moriyama Y, Iwasaki K, and Ma JF. 2015. A polarly localized transporter for efficient manganese uptake in rice. *Nat Plants* 1(12): 1–8.

Ullah I, Wang Y, Eide DJ, and Dunwell JM. 2018. Evolution and functional analysis of Natural Resistance-Associated Macrophage Proteins (NRAMPs) from *Theobroma cacao* and their role in cadmium accumulation. *Sci Rep* 8(1): 1–15.

Varotto C, Maiwald D, Pesaresi P, Jahns P, Salamini F, and Leister D. 2002. The metal ion transporter IRT1 is necessary for iron homeostasis and efficient photosynthesis in *Arabidopsis thaliana*. *Plant J* 31(5): 589–599.

Vert G, Barberon M, Zelazny E, Séguéla M, Briat JF, and Curie C. 2009. *Arabidopsis* IRT2 cooperates with the high-affinity iron uptake system to maintain iron homeostasis in root epidermal cells. *Planta* 229(6): 1171–1179.

Vives-Peris V, de Ollas C, Gómez-Cadenas A, and Pérez-Clemente RM. 2020. Root exudates: from plant to rhizosphere and beyond. *Plant Cell Rep* 39(1): 3–17.

Wang AS, Angle JS, Chaney RL, Delorme TA, and Reeves RD. 2006. Soil pH effects on uptake of Cd and Zn by *Thlaspi caerulescens*. Plant Soil 281: 325–337.

Wei L, Zhang M, Wei S, Zhang J, Wang C, and Liao W. 2020. Roles of nitric oxide in heavy metal stress in plants: Cross-talk with phytohormones and protein S-nitrosylation. *Environ Pollut* 259: 113943.

Xie Q, Yu Q, Jobe TO, Pham A, Ge C, Guo Q, Liu J, Liu H, Zhang H, Zhao Y, Xue S, Hauser F, and Schroeder JI. 2021. An amiRNA screen uncovers redundant CBF and ERF34/35 transcription factors that differentially regulate arsenite and cadmium responses. *Plant Cell Environ* 10.1111/pce.14023. Advance online publication.

Yang JL, Zhu XF, Peng YX, Zheng C, Li GX, Liu Y, Shi YZ, and Zheng SJ. 2011. Cell wall hemicellulose contributes significantly to aluminum adsorption and root growth in *Arabidopsis*. *Plant Physiol* 155(4): 1885–1892.

Yu H, Wu Y, Huang H, Zhan J, Wang K, and Li T. 2020.The predominant role of pectin in binding Cd in the root cell wall of a high Cd accumulating rice line (*Oryza sativa* L.). *Ecotoxicol Environ Saf* 206: 111210.

Zanella L, Fattorini L, Brunetti P, Roccotiello E, Cornara L, D'Angeli S, Della Rovere F, Cardarelli M, Barbieri M, Sanità di Toppi L, Degola F, Lindberg S, Altamura MM, and Falasca G. 2016. Overexpression of AtPCS1 in tobacco increases arsenic and arsenic plus cadmium accumulation and detoxification. *Planta* 243(3): 605–622.

Zeng L, Zhu T, Gao Y, Wang Y, Ning C, Björn LO, Chen D, and Li S. 2017. Effects of Ca addition on the uptake, translocation and distribution of Cd in *Arabidopsis thaliana*. *Ecotoxicol Environ Saf* 139: 228–237.

Zhang GB, Yi HY, and Gong JM. 2014. The *Aarabidopsis* ethylene/jasmonic acid-nrt signalling module coordinates nitrate reallocation and the trade-off between growth and environmental adaptation. *Plant Cell* 26: 3984–3998.

Zhang X, Zhang D, Sun W, and Wang T. 2019. The adaptive mechanism of plants to iron deficiency via Iron Uptake, Transport and Homeostasis. *Int J Mol Sci* 20(10): 2424.

Zhang Y, Chen K, Zhao FJ, Sun C, Jin C, Shi Y, Sun Y, Li Y, Yang M, Jing X, and Luo J. 2018. OsATX1 interacts with heavy metal P1B-type ATPases and affects copper transport and distribution. *Plant Physiol* 178(1): 329–44.

Zhao HM, Xiang L, Wu XL, Jiang YN, Li H, Li YW, Cai QY, Mo CH, Liu JS, and Wong MH. 2017. Low-molecular-weight organic acids correlate with cultivar variation in ciprofloxacin accumulation in *Brassica parachinensis* L. *Sci Rep* 7(1): 10301.

Zhou C, Ma Q, Li S, Zhu M, Xia Z, and Yu W. 2021. Toxicological effects of single and joint sulfamethazine and cadmium stress in soil on pakchoi (*Brassica chinensis* L.). *Chemosphere* 263: 128296.

Zhou ZS, Guo K, Elbas AA, and Yang ZM. 2009. Salicylic acid alleviates mercury toxicity by preventing oxidative stress in roots of *Medicago sativa*. *Environ Exp Bot* 65(1): 27–34.

7

Plant Response to Heavy Metals (at the Cellular Level)

Arun Kumar Maurya,[1,*] *Dwaipayan Sinha,*[2] *Kamakshi*[3]
and *Suchetana Mukherjee*[4]

ABSTRACT

Plants, as other organisms require certain metals and metalloids at physiological concentrations for the proper functioning of cellular functions. Some metals and metalloids which belong to the category of heavy metals are not required in general by the plants. Some heavy metals are used by plants in small quantities for their physiological function but most of these are harmful and beyond certain limits become toxic to plants and show morpho-anatomical, physio-biochemical symptoms on various processes like seed germination, root-shoot growth, grain yield, induction of reactive oxygen and nitrogen species, etc. To counter the adverse effects induced by heavy metals, plants have evolved diverse mechanisms to minimize the damage caused, such as the exclusion, compartmentalization, uptake reduction, inhibition of long-distance transportation and induction of antioxidant system, detoxification mechanism (storage into the vacuole, chelation, trafficking), heat shock proteins, hormones, and cell signaling molecules. Sugar is one of the key biochemical products of a photosynthetic process that aids in counteracting mechanisms. Soil contaminated with heavy metals is uptaken by plants and affects biochemical processes such as respiration, sugar metabolism, uptake of essential metal ions, and replication. The heavy metals also affect symbiotic N fixation and N metabolism by affecting the bacterial consortium.

[1] Department of Botany, Multanimal Modi College, Modinagar, Ghaziabad, Uttar Pradesh-201204 India.
[2] Department of Botany, Government General Degree College, Mohanpur, Paschim Medinipur, West Bengal-721436, India.
[3] Department of S&H, SRMIST (NCR Campus) Modinagar, Ghaziabad, U.P.-201204-India.
[4] Department of Botany, Sripat Singh College, Jiaganj, Murshidabad, West Bengal-742123, India.
* Corresponding author: botany25@gmail.com
ORCID iD:https://orcid.org/0000-0002-6650-5576

Some heavy metals adversely affect enzyme activities related to protein metabolism, synthesis, modifications, and protease activity, enzymes involved in N assimilation like NR, NiR, GS, and GOGAT. Therefore, it is amply clear that heavy metals have a wide spectrum impact on plant growth and physiology that requires correction to maintain proper functioning of plant and ecosystem health to achieve sustainable growth. This chapter is an attempt to overview the impact of heavy metals on the various physiological processes of plants.

1. Introduction

All organism including plants require essential heavy metals (HMs) at physiological concentrations, which exhibits multiple cellular functions at the primary level or some at the secondary level while mitigating the adversity of non-essential HMs ions. Some mechanisms are involved to minimize the damage caused by HMs while other mechanisms involve the exclusion, compartmentalization, uptake reduction, and inhibition of long-distance transportation. Plant metal homeostasis is maintained through developing the metal-induced detoxification mechanism such as storage into the vacuole, chelation, trafficking, etc. In advanced stages, plants evolve secondary mechanisms namely heat shock proteins (HSPs), hormones, cell signaling molecules, reactive oxygen species, etc. (Manara, 2012). Therefore, the stress responses mitigated by proteins have resulted from gene expression. Therefore, the functional proteomes are critical in the maintenance of the detoxification mechanism of plants (Hasan et al., 2017).

While the presence of non-essential HMs ions and an excessive amount of essential HMs ions disrupt the structure of newly synthesized proteins through intervening in the folding process or misleading the folding of existing proteins, or inducing alteration in their native conformation by binding. HMs can cause damage in the overexpression of important enzymes or checking their activity.

2. Impact of heavy metals on cellular respiration

Metals are essential elements of innate ecology and ecosystem. They possess the capacity of high electrical conduction therefore readily lose electrons and eventually form cations. Although they are found almost everywhere some HMs as chromium (Cr), cobalt (Co), copper (Cu), iron (Fe), magnesium (Mg), manganese (Mn), molybdenum (Mo), nickel (Ni), selenium (Se), and zinc (Zn) are required by the plants and animals for the proper functioning of various physiological and biochemical pathways but lethal in excessive quantity. However, arsenic (As), cadmium (Cd), chromium (Cr), lead (Pb), mercury (Hg), and silver (Ag) exhibit deleterious effects even in lesser quantities (Godwill et al., 2019). Due to the increasing urbanization and industrialization, the high amount of metals accumulated in the cells cause abnormality in the functioning of primary metabolites or alters the structure of secondary metabolites.

Their accumulation induces the adverse change in biochemical behavior of cells through binding proteins, nucleic acids by changing their native conformation (Hossain and Komatsu, 2013) while upon transportation they damage the distant organelles. Long-term and continuous exposure to the HM results in the dismantling of cellular organelles and declining the processing rate reactions. The plants respond through metal sequestration or removal from the symplastic high metal concentrations (Denny and Wilkins, 1987; Ernst, 1969).

A high concentration of HMs influences partially the pathways of carbon (C) assimilation by affecting the enzymes involved in the citrate cycle, electron transport chain, and ATP synthase. The functioning of intermediate reactions of the pathway is restricted hence declined consumption of oxygen, CO_2 removal, and obtaining less number of ATPs (Pfeffer et al., 1986). The role played by vacuole in metal sequestration keeps the respiration rate increased or unchanged to protect the organelles (Fig. 1).

The increased respiration rate could be the result of high energy requirement for export or ion uptake (Meharg, 1993), the release of PC-metal complexes (Vogeli-Lange and Wagner, 1990), and the formation of polypeptides to bind metal

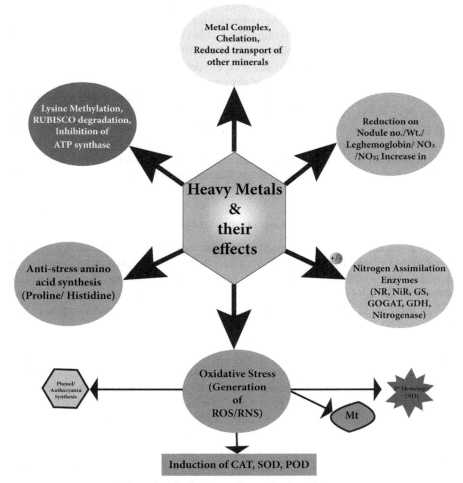

Figure 1: Metal sequestration mechanism of plants.

(Grill et al., 1985). The effect of some HMs such as Cu, Hg, and silver analyzed to increase the rate of carbon dioxide production initially but slow down gradually or caused to decline in *Aspergillus niger*, the reason might be a chemical combine with cell constituent induced metal activation which consecutively alters the velocity constants of respiratory reactions (Cook, 1926).

The experiments using Cu showed a unique pattern of the latent period in *Aspergillus niger* while the length of the latent period was found inverse to the Cu concentration (Cook, 1926). The exposure of various HMs such as Cd, Cr, Cu, Hg, Ni, and Zn affect the various degree of adverse effects depending on the period of exposure, metal concentration, and detoxification mechanism of the organism. Although, plants protect respiratory sites with barriers of cell wall's cation exchange potential and cell membrane.

In the cytoplasm, glycolytic enzymes are precipitated with phosphate anions (PO_4^-) or SH-rich molecules to mitigate the hindrance of metals. The transfer of metals is also checked partly by the inner mitochondrial membrane to encounter the effect of HM concentration Mycobacterium tuberculosis (Horio et al., 1955) and *Saccharomyces cerevisiae* (Murayama, 1961) activated more metal resistant isoenzymes but not in higher plants (Wu et al., 1975). The concentration of metal ions as mild metal stress increases but severely decreases the rate of respiration.

The high concentration of metals stresses the demand for more energy for the extrusion of metals. CAT is involved in H_2O_2 removal generated in photorespiration during HM stress (Mittler, 2006). The high accumulation of HMs induces the hyperproduction of ROS in the mitochondria due to the sharp deterioration in the electron transport chain (Davidson and Schiestl, 2001; Keunen et al., 2011). The ROS production is increased by HMs through Haber Weiss and Fenton reaction or inhibiting the enzymes involved in detoxification (Schut Zendubel and Polle, 2002; Halliwell, 2006; Keunen et al., 2013).

3. Impact of heavy metals on sugar metabolism

Sugars produced during photosynthesis in plants are readily available for primary metabolism. Sugar metabolism is an integrated enzyme-catalyzed pathway that provides the metabolites for related subsequent, immediate, or intermediate pathways such as synthesis of complex carbohydrates, some primary metabolites, energy production (Stein and Granot, 2019), signaling factors, stress response, and plant-microbe interactions (Young-Ling Ruan, 2014).

The accumulation of HMs imbalanced cell-ion homeostasis which triggers the plant detoxification strategies including chelation resulting in the production of Cys-rich peptides (PCs) as a marker of metal stress. Abiotic stress is responsible to increase the concentration of soluble sugar in contrast to HMs which decrease the concentration of soluble sugar (Rosa et al., 2009) depending on the genotype as well. Hence, the correlation between the concentration of soluble sugar and high-stress tolerance exists. The alteration in the soluble sugar contents is somewhat associated with the CO2 assimilation, processing enzymes, activation of relevant genes.

The soluble sugars as sucrose, glucose, and fructose are important substrates of cellular respiration or as osmolytes and formation of secondary metabolites consecutively. The high stress caused by HMs can damage chloroplast structure, inhibit electron transport in the chloroplast (Pego et al., 2000; Gratão et al., 2005) result in the functioning of antioxidant enzymes and oxidation of lipids (Apel and Hirt, 2004; Foyer and Noctor, 2005).

The HM stress can regulate the protein expression during carbohydrate metabolism involving the enzymes of starch biosynthesis such as ADP-Glcpyrophosphorylase AGPase and sucrose biosynthesis as sucrose synthase (SuSy), sucrose phosphate synthase (SPS), and Invertase (INV) (Wingler et al., 2000; Stitt and Hurry, 2002). The alteration of sucrose and hexose ratio due to stress modulates different pathways of signal transduction in the cell (Rosa et at., 2009). The study on *A. coniculatum* L. seedlings exposed to HMs showed yellow leaves and later defoliation (Guangqiu et at., 2007). It was also analyzed that upon increasing the concentration of HMs, sugar level enhanced but reverse effect on the starch level was noticed which might be an indicator of stressed carbohydrate metabolism (Kim et at., 2003).

4. Impact of HMs on uptake of mineral elements during metal stress

HM contamination in the soil affects the plant in physiological and genetic levels Plants are often compelled to absorb HMs from the soil where their concentration is high (Mishra et al., 2017). The HMs are taken up by the plants due to differences in concentration gradient guided by selective uptake of ions by roots or through the process of diffusion in the soil (Peralta-Videa et al., 2009).

An intricate process involving adsorption, complexation, and action of various transporters are involved in the entry and transportation of HMs from the environment to the plant cell (Angulo-Bejarano et al., 2021). At high concentrations, the HMs are uptaken by the plant and result in the manifestation of several responses, the earliest of which is possibly oxidative stress (Seneviratne et al., 2019; Kolahi et al., 2020; Rizvi et al., 2020). The oxidative stress in turn unleashes a cascade of events involving signal transduction pathways through secondary messengers (Dvorák et al., 2021). All these events result in altered gene expression to counteract the stress response.

In addition, the HMs also directly act upon essential biomolecules of plants rendering them ineffective or nonfunctional (Shahid et al., 2014; Lukowski and Dec, 2018). In most cases, the HMs compete with the essential nutrients during the uptake process. For example, As (V) competes with the P in root uptake and interferes with ATP synthesis and oxidative phosphorylation (Peralta-Videa et al., 2009). Other reports state that Pb and Cd compete with essential nutrients and result in the inhibition of uptake of potassium (K), calcium (Ca), Mg, Mn, Zn, Cu, and Fe ultimately resulting in the deficiency of the elements (Ouariti et al., 1997; Godwin et al., 2001). Nickel results in Fe deficiency either by slowing the absorption or immobilization of the latter in the roots (Mysliwa-Kurdziel et al., 2004; Nishida et al., 2012). Another report states that Ni may accumulate under Zn-deficient conditions in *Arabidopsis thalliana* (Nishida et al., 2015). Several mechanisms are involved in the reduced uptake of essential elements in presence of HMs. One mechanism is possibly due to comparable ionic radii which results in ready uptake of

HMs if present in greater concentrations in the surrounding atmosphere in place of the essential elements (Sreekanth et al., 2013).

The decline in nutrient uptake is also due to HMs induced impairment of enzymes bound to the cell membrane. For example, Cd toxicity results in the inactivation of metal ATPases which are largely involved in the transportation of elements across the membranes (Gallego et al., 2012; Genchi et al., 2020). Another report states that Cd inhibits plasma membrane H+-ATPase and inhibits the uptake of nitrate (Rizzardo et al., 2012). Table 1 illustrates alterations in mineral uptake by plants in presence of HMs.

5. Impact of HMs on uptake of nutrient balance and nutrient assimilation

In the previous section, the impact of HMs on nutrient uptake has been discussed. It is now a well-established fact that HMs alter the uptake pattern of the essential elements in a plant due to their comparable ionic radii and capacity to deactivate or inhibit plasma membrane-bound channels. These aspects have already been discussed in the previous section. In the current section, the impact of HMs on the balance and assimilation of nutrients would be discussed.

The essential elements perform a variety of physiological processes in plants (Maathuis and Diatloff, 2013). Some elements such as N and P are directly utilized in the synthesis of proteins and other biomolecules (Młodzińska and Zboińska, 2016; Perchlik and Tegeder, 2018). Sulfur is the major constituent of two amino acids namely cysteine and methionine which are involved in the primary and secondary metabolic processes of the plant (Droux et al., 2004). Cysteine acts as an S-donor for all S-containing compounds including methionine, glutathione, Fe-S cluster, Mo cofactor, vitamins (coenzyme A, lipoic acid, thiamine, biotin), and secondary compounds namely camalexin and glucosinolates (Hell and Wirtz, 2011). Other essential elements such as Mo act as a cofactor of a large number of plant enzymes (Llamas et al., 2017; Tejada-Jimenez et al., 2018). Other essential elements also have their specific role in the physiological processes of the plants. Table 2 illustrates the effect of HMs in the assimilation of elemental nutrients in plants.

However, the behavior of some elements is also very interesting in the fact that they also help in the amelioration of HM stress. A lot of investigations have been made to explore the ameliorating role of HMs. Silicon is one such element that is extensively used for neutralizing the toxicity of metals.

A study reports that silicon is responsible for reducing Al toxicity through the formation of Si-Al complexes in mucigel and outer cellular tissues. This decreases binding Al to the cell wall of the plants where it inhibits wall loosening required for the growth and elongation process (Kopittke et al., 2017). Another study reports that silicon reduced the Al content in shoots and roots of *Solanum tuberosum*.

In addition, silicon also ameliorated the toxic effect of Al with respect to the number of branches of roots and the number of leaves (Dorneles et al., 2016). In a recent study, it is reported that silicon dioxide nanoparticles can induce antioxidant defense mechanisms in maize plants subjected to Al stress. In addition to it, the metal detoxification process and accumulation of organic acids are also initiated

Table 1: Effect of Heavy metals of uptake of essential elements.

S. No.	Heavy Metal	Concentration applied	Plant tested	Effect on uptake of other minerals		Reference
				Increase	Decrease	
1.	Cd	$0–0.5\ \mu g\ mL^{-1}$	*Brassica campestris* L. ssp. *chinensis* var. *communis*	K, Mg, phosphorus(P), Mn, Boron(B), Fe concentration in roots. Ca, Sulfur(S), and Zn in shoots and roots.	K, Mg, P, Mn, B, Fe concentration in shoots.	Zhu et al., 2004
2.		$10–190\ mg\ Cd\ kg^{-1}$	*Brassica juncea*	K and P concentration in roots.	Zn concentration in roots.	Jiang et al., 2004
3.		0, 50, 100, 200 and $400\ \mu M$	*Atriplex halimus* subsp. *schweinfurthii*		Ca and K in shoots and roots.	Nedjimi and Daoud, 2009
4.		$10\ \mu mol\ L^{-1}$	*Solanum lycopersicum*		Zn concentration in the whole plant.	Cherif et al., 2011
5.		1 mM	*Solanum lycopersicum*	Fe, Cu, Mn, Zn uptake in the whole plant.	K, Mg, Ca, and S uptake in whole plants.	Gratão et al., 2015
6.	Cr	$0–300\ \mu M$	*Brassica juncea*		Sodium (Na), K, Ca, Mg, C, Hydrogen, and N content.	Handa et al., 2018
7.		0, 50, 150 and $200\ mg\ L^{-1}$	*Lactuca sativa*		K, Mg, Fe, and Zn contents in roots and leaves.	Dias et al., 2016
8.		0.05–0.4 mM	*Raphanus sativus*		Fe content in the leaves.	Tiwari et al., 2013
9.		$2.5–200\ mg\ L^{-1}$	*Oryza sativa*		Uptake of NP and K, Uptake of Mn, Cu, Zn, and Fe.	Sundaramoorthy et al., 2010
11.	Cu	$0–60\ mmol\ L^{-1}$	*Limoniastrum monopetalum*	S content In root at a concentration of $35\ mmol\ L^{-1}$.	Leaf S content at a concentration of 2 and $9\ mmol\ L^{-1}$. Mg concentration in leaf at an exposure of 15 and $35\ mmol\ L^{-1}$.	Cambrollé et al., 2013

Table 1 contd. ...

...Table 1 contd.

S. No.	Heavy Metal	Concentration applied	Plant tested	Effect on uptake of other minerals	Reference
12.		100, 400 and 600 mg L^{-1}	*Spinacea oleracea*	Foliar N, K, Ca, Mg, and Fe content at 400 mg L^{-1} concentration of Cu. Foliar N content at 600 mg L^{-1} concentration of Cu. Foliar P content at 400 mg L^{-1} concentration of Cu. Foliar P, K, Ca, Mg, and Fe content at 600 mg L^{-1} concentration of Cu. Root P, K, Ca, Mg, and Fe content at 600 mg L^{-1} concentration of Cu.	Gong et al., 2019
13.		0, 2, 9, 15 and 30 mmol L^{-1}	*Atriplex halimus*	Foliar Na, P, and N content Root P, N, K, and Mg content.	Mateos-Naranjo et al., 2013
14.	Pb	0.5, 1, 2, 3 and 4 mM	*Jatropha curcas*	Uptake of Zn. Uptake of Mn, Cu, and Fe.	Shu et al., 2014
15.		1.5, 3, and 15 mM	*Spinacia oleracea* *Triticum aestivum*	Mn uptake. Na, K, Ca, P, Mg, Fe, Cu, and Zn uptake.	Lamhamdi et al., 2013
16.		0, 30, 60 and 90 mg kg^{-1}	*Brassica napus*	N, P, K, Ca, Mg, Zn, Cu, and Mn uptake in shoots. N, Fe, Zn, and Cu uptake in roots.	Ashraf et al., 2011
17.	Ni	0, 20, and 40 mg L^{-1}	*Zea mays*	N and P content in shoots. K, Ca, Zn, and Cu content in shoots and roots.	Amjad et al., 2019
18.	Al	1 mML^{-1}	*Symplocos paniculata*	Ca content in the whole plant.	Schmitt et al., 2016
19.		0.5, 1.0, 2.0, 5.0 mM	*Melastoma malabathricum*	Foliar P, K, Mg, and Ca.	Mahmud and Burslem, 2020

Table 2: Effect of selected heavy metals on mineral assimilation of plants by affecting biochemical pathways or key biomolecules.

S. No.	Essential element affected	Heavy metals	Plant studied	Biochemical pathway affected/biomolecules affected	References
1.	N	Cd	*Brassica juncea*	Inactivation of Nitrate Reductase.	Irfan et al., 2014
2.	N	Cd + Pb	*Oryza sativa*	Protein carbonylation.	Srivastava et al., 2014
3.	N	Cr	*Oryza sativa*	Inactivation Nitrate Reductase.	Singh et al., 2019
4.	N	Pb	*Brassica pekinensis*	Inactivation Nitrate Reductase.	Xiong et al., 2006
5.	N/Mo	Tungstate		Tungstate competes with molybdate for incorporation in MoCo resulting in inhibition of nitrate reductase.	Adamakis et al., 2012
6.	N/NH$_3$	Cd	*Lycopersicon esculentum*	Glutamine synthetase.	Chaffei et al., 2004
7.	N/NH$_3$		*Pisum sativum*	Glutamine synthetase and glutamate synthase.	Gangwar et al., 2011a
8.	N/NH$_3$		*Pisum sativum*	Glutamine synthetase and glutamate synthase.	Gangwar et al., 2011b
9.	Fe	Co, Cr	*Alfalfa*	Ferric chelate reductase.	Barton et al., 2000
10.	C	Zn	*Lolium perenne*	Reduced activity of RUBISCO.	Monnet et al., 2001
11.	C	Cd	*Erythrina variegata*	Reduced activity of RUBISCO.	Muthuchelian et al., 2001
12.	S	Cr	*Brassica juncea*	Repression of the root low-affinity sulfate transporter (BjST1).	Schiavon et al., 2008
13.	S	Cd	*Sedum alfredii*	Adenosinetriphosphate (ATP) sulfurylase (ATPS) and serine acetyltransferase (SAT). Activities were higher in hyperaccumulating ecotypes than in non hyperaccumulating ecotypes.	Guo et al., 2009
14.	S	Cd	*Arabidopsis thaliana*	Induced expression of serine acetyltransferase involved in cysteine biosynthesis.	Howarth et al., 2003
15.	P/(PO$_4$$^{-3}$)			Replacement of phosphate groups in various biochemical reactions.	Dixon, 1996

Table 2 contd. ...

...Table 2 contd.

S. No.	Essential element affected	Heavy metals	Plant studied	Biochemical pathway affected/biomolecules affected	References
16.	$P/(PO_4^{-3})$			Inhibition of ATP synthesis due to the formation by substitution of arsenate in place of phosphate in the glycolytic pathway.	Dixon, 1996
17.	$P/(PO_4^{-3})$		*Arabidopsis*	Competes with (PO_4^-) involving PHT transporter.	Catarecha et al., 2007
18.	$P/(PO_4^{-3})$		*Oryza sativa*	OsPT1 is involved in the uptake of As in place of (PO_4^-).	Kamiya et al., 2013
19.	$P/(PO_4^{-3})$		*Oryza sativa*	OsPT8 is involved in the uptake of As in place of (PO_4^-).	Wu et al., 2011

Table 3: Inhibition of uptake of selected heavy metals by silicon.

S. No.	Heavy metal	Plant system	Reference
1.	Cd	*Pennisetum* sp.	Dong et al., 2019
2.	Cu	*Cucumis sativus*	Bosnić et al., 2019
3.	Cr	*Triticum aestivum*	Sarkar et al., 2020
4.	Hg	*Glycine max*	Li et al., 2020
5.	As	*Oryza sativade*	Wu et al., 2011
6.	Antimony(Sb)	*Oryza sativa*	Zhang et al., 2017

in the roots of Al stressed plants upon treatment with silicon dioxide nanoparticles (de Sousa et al., 2019).

One of the mechanisms through which silicon exerts its protective effect is through the reduction of uptake of HMs by the plants. Table 3 illustrates the inhibitory activity of silicon towards the uptake of HMs by the plants.

This decrease in uptake of metals in presence of silicon can be explained in several ways. Firstly, silicon stimulates the production of root exudates which have the capacity of metal chelation and resulting in reduced uptake by roots. In addition to it, a physical barrier is formed due to the deposition of silicon into the endoderm which results in a reduction of porosity of walls of the cells in root tissue. This inhibits access of metals into the vascular bundle for further conduction. Silicon also reduces apoplasmic transport of HMs through the reduction in their concentration in the apoplasm (Adrees et al., 2015).

Apart from silicon, Ca is reported to ameliorate Cd toxicity *Matricaria chamomilla* through reduction of cell surface negativity and competing with Cd ion influx (Farzadfar et al., 2013). Similar reports have also been obtained by a recent study on *Vicia faba*. The result from the study showed that Ca reduced the accumulation of Cd in the plant tissue along with improvement in cell membrane integrity and increase in antioxidant defense system (Nouairi et al., 2019).

Cd toxicity in plants can also be ameliorated by the addition of K. One study reported that treatment of K to Cd stressed Nicotiana tabacum plant resulted in a significant reduction in uptake and translocation of Cd along with improvement in physiological parameters (Wang et al., 2017). In another study, it is reported that Zinc oxide nanoparticles alleviate toxicity induced by Pb and Cd in *Leucaena leucocephala* through modulation of physiological processes including antioxidant defense mechanism (Venkatachalam et al., 2017). Similar reports with zinc oxide nanocatalysts were also obtained from a more recent study on *Gossypium hirsutum* subjected to stress by Cd and Pb (Priyanka et al., 2021).

6. Impact of heavy metals on nitrogen fixation

Nitrogen (N) is the most abundant element found in the atmosphere and is an essential element for plants as it is crucial for their growth and development. Nitrogen is an essential component of plant protein, DNA, chlorophyll, ATP, and other biomolecules. However, the plant cannot utilize atmospheric N, but the reduced forms of this element.

Thus, plants are dependent on other organisms which may be symbiotic or free-living, that can convert elementary N into other reduced forms which are suitable for plant uptake by a process called N fixation (Wagner, 2011; Buren and Rubio, 2017; Klobus et al., 2002). Symbiotic N_2 fixation is carried out by rhizobia which are present in root and shoot nodules of legume plants. Rhizobia comprises *Alphaproteobacteria* and *Betaproteobacteria* which are gram-negative. *Rhizobium* being the largest rhizobia genus. *Frankia* and Cyanobacteria are other groups of bacteria that can fix N symbiotically (Lindstrom and Mousari, 2019). HM in low concentrations often play important role in enzymatic activities but at higher concentration become toxic (Ahmad et al., 2012; Haddad et al., 2015). Inhibition of N-fixation by HMs is caused by depletion of ATP and Hg^{2+}. Mercuric ions inhibit nitrogenase activity by lysis of vegetative cells (Singh et al., 1987; Stratton et al., 1979). HMs like Hg, Cd, and Ni have been shown to affect the symbiotic relation between Rhizobium and legume negatively.

They also exert a detrimental impact on plant metabolism, biochemistry of nodules, nodule number and dry weight, and nitrogenase activity. They also induced host plant morphology and yield adversely. The most potent HM posing adverse effects was Hg followed by Cd and Ni (Pal, 1996). Cd adversely affects enzyme activity, DNA-mediated transformation in microbes, symbiotic relationship between host and microbe, and increases the fungal threat. HMs also transform viable bacterial cells into non-culturable ones thus making the soil unsuitable for microbial growth and crop production (Ahmad et al., 2012).

HMs like Cu, Ni, Zn, Cd, As probably inhibit the growth, morphology, and activities of various groups of micro-organisms. Dehydrogenase activity is most affected by HMs (Lukowski and Dec 2018). HMs have been reported to negatively affect nodulation and N nutrition in legumes (Haddad et al., 2015). Cr was found to adversely affect nitrogenase, nitrate reductase, glutamate synthetase, and glutamate dehydrogenase activity in clusterbeans.

Specific nodule nitrogenase activity (SNA) decreases with increasing Cr (VI) concentration (Sangwan et al., 2014). High Cd concentration also decreases SNA by increasing thiobarbituric acid and decreasing leghemoglobin levels which resulted in oxidative stress (Balestrasse et al., 2003). In *Pisum sativum* N fixation decreases in response to Cr (Bishnoi et al., 1993).

7. Impact of heavy metals on DNA replication and the repair of DNA

Replication and repair are complex processes that occur in a cell in the presence of a large number of proteins and enzymes. DNA damage is caused by both endogenous and exogenous factors and if not repaired may result in cell death. The most common cytotoxic damage caused to the DNA helix is the introduction of double-stranded breaks (DSB's). HMs like Cd, Ar, and Ni are potent carcinogens that cause DNA damage by induction of double-stranded breaks as well as inhibit crucial proteins for DNA repair (Morales et al., 2016). HM (Zn, Cu, Cd) induced DNA damage in plants has been reported in *Arabidopsis thaliana*, where the increased concentration of the HMs posed greater damage. Cu^{2+} and Cd^{2+} caused greater damage than

Zn^{2+} at the same concentration (Li et al., 2008). Cu at low concentration induces chromosomal alterations and increases micronuclei formation (Marcato-Romaio et al., 2009). Structural damage (oxidative stress) of DNA in the fruit bodies of plants in response to HMs like Cd, Zn, Cu, and Hg have been reported in Boletus edulis (Collin-Hansen et al., 2005). HM-induced DNA damage was also reported in *Utrica dioica* (Gjorgieva et al., 2013).

HM-induced nucleic acid damage is caused either by direct binding and cleavage or indirectly by oxidative stress and the generation of ROS (Nada et al., 2007; Pandey et al., 2009). Inhibition of DNA replication by HMs is induced by the production of ROS (Dutta et al., 2018). ROS-induced strand break in DNA is caused due to both altered base and damaged sugar residues that undergo fragmentation. Studies have revealed that the DNA repair mechanism in higher plants is similar to other eukaryotic DNA repair mechanisms (Moura et al., 2012).

Cr (VI) and Cd^{2+} can inhibit progression through the cell cycle by inhibiting mitotic activity in root meristem and expression of S-phase-specific CDK's respectively (Dutta et al., 2018). Proteins participating in DNA repair (excision and mismatch) are sensitive to Cd toxicity. Exposure to Cd decreases the expression of MMR genes and adversely affects the mismatch recognition process of MutS? and MutS? complex and nucleotide excision process of the MMR system.

Thus, Cd immensely hampers DNA repair and acts as a carcinogen (Moura et al., 2012). As-induced DNA damage is caused by oxidative stress (Lin et al., 2008). Pb-induced DNA damage is caused by the production of ROS which induces oxidative stress (Sharma and Dubey, 2005; Moura et al., 2012). The genotoxicity of Hg is due to the increased production of ROS induced by oxidative stress. Zn and Se are also reported to exhibit genotoxicity through ROS production (Moura et al., 2012).

HM-induced molecular changes led to DNA damage which includes strand breaks (both SSB and DSB), DNA protein cross-links, base and sugar lesion, abasic site production, and DNA modification in plants either by directly interacting with DNA or by inhibiting DNA repair enzymes and production of ROS compounds (Agarwal and Khan, 2020).

8. Impact of heavy metals on protein metabolism, protein synthesis, modifications, and protease activity; nitrate assimilating enzymes: nitrate reductase (NR), nitrite reductase (NiR); ammonia assimilating enzymes: glutamine synthetase (GS), glutamine oxoglutarate aminotransferase (GOGAT) and glutamate dehydrogenase (GDH)

HMs ions usually enter plant cells non-selectively and induce a negative impact on diverse cellular and physiological process. Cation-efflux family proteins (CEFP), members of Group III of the cation-efflux transporters reported from *B. juncea* (BjCET3 and BjCET4) show similarity in amino acid sequence. Their transcripts levels substantially increased by Zn, Cd, NaCl, or PEG suggesting that CEFP may play roles in several stress conditions (Lang et al., 2011). LmSAP is a member of the stress-associated protein (SAP) gene family. The response of LmSAP obtained from a halophyte (*Lobularia maritima*) expressed in transgenic tobacco plants to see the effect of stresses induced by HMs like Cd, Cu, Mn, and Zn where expression of SAP increased after 12 h of treatment with these metals indicating linkage of HMs stress having role against oxidative stress and protecting the plant cells. SAP induces CAT, POD, and SOD thereby scavenging ROS consequently lowering the free availability of HMs through metal-binding proteins present in the cytosol. Many other gene transcriptions are enhanced in metal tolerant tobacco such as metallothioneins, CCH a Cu transport protein, a Cys, His-rich domain-containing protein RAR1 (Rar1), and a ubiquitin-like protein 5 (PUB1) (Saad et al., 2018).

Cd toxicity impairs N uptake, assimilation and affects normal plant growth. Molybdenum is a key element and participates in N metabolism through regulating N assimilatory enzyme activities and expressions in higher plants. Two fragrant rice cultivars, Guixiangzhan and Meixiangzhan-2 showed differential responses with Cd toxicity for N utilizing and assimilatory enzyme activities, 2-acetyl-1-pyrroline (2AP) content. The stress caused by Cd toxicity was alleviated by Mo application and improved N assimilatory pathway through efficient NO^{3-} utilization with higher NR, NiR, GS, and GOGAT activities and transcript levels (Imran et al., 2021).

The *Salvia sclarea* L. (clary sage) is known to accumulate Cd and cause increased phenolic and anthocyanin contents as well as increased accumulation of Fe in leaves during stress (Dobrikova et al., 2021). In comparison to inorganic N, organic N may modify HMs uptake and toxicity. It was seen elevated during growth, chlorophyll content, and photosynthetic activity under Cd stress. The stress also affected the protein content, amino acids, total N, and accumulation of mineral nutrients. Parallelly allantoin and partially urea content affected the accumulation of acids like citrate and tartrate (Dresler et al., 2021). Nitrogen metabolism plays an important role in responses to HMs toxicity (Andrade et al., 2010). Various nitrogenous metabolites, namely amino acids, amino acid-derived molecules, and polyamines can bind to and scavenge HMs-induced ROS. Exposure to a high concentration of HMs levels induces an increase in special amino acids such as proline or histidine and scavenging of ROS observed (Sharma and Dietz, 2006; Fariduddin et al., 2009). Prolines accumulated in plants in response to HMs stress help augment several mechanisms like a regulatory role or protecting certain plant enzymes, stabilizing subcellular components and structure, and maintaining osmoregulation (Sharma and Dietz, 2006). Proline, similar to GSH, ascorbic acid, or tocopherol reduces free radical levels generated from HMs toxicity. Cd shows a negative effect on NR activity and NO alleviates such effects by promoting ROS scavenging or antioxidant enzyme activation. NO observed in Cd toxicity influencing its uptake and reduction of root growth (Arasimowicz-Jelonek et al., 2011). NR enzyme is indicated to be a potential contributor to EBR-induced NO generation playing a key role in Cd stress tolerance in pepper plants by regulating the AsA-GSH cycle and antioxidant enzymes (Kaya et al., 2020).

Tomato seedlings stressed with Cd caused increased nitrate content, a decline in plant growth, and enzymes involved in primary N assimilation pathways such as NR, NiR, and GS. Such effects got inversed after Cd removal from the nutrient solution. It was found that $NAD^{(+)}$-dependent glutamate dehydrogenase (GDH-NAD^+) activity gradually increased during the recovery time but the cognate NADH-dependent glutamate dehydrogenase (GDH-NADH) activity decreased suggesting that stress-induced ammonia (NH_3) generated from protein catabolism is incorporated by detoxification and re-assimilation with the help of the GDH-NADH isoenzyme (Chaffei et al., 2003). Other characteristics such as a decline in leaves, root biomass, the rate of photosynthetic activity due to both Rubisco and chlorophyll degradation, stomata closure, nitrate content, NR, NiR, GS and ferredoxin-glutamate synthase also decreased but contrarily increase in NADH-glutamate synthase, NADH-glutamate dehydrogenase activity the total amino acid content in the phloem, maintaining Gln/Glu ratios and Gln represented the major amino acid transported through xylem sap in response to Cd. Due to loss of protein and accumulation of amino acids and ammonium was also observed in Cd-treated plants (Chaffei et al., 2004). Tomato seedlings were grown on nitric medium and treated with various Cd concentrations. It was found that Cd remains predominantly located in the roots that act as trap-organ. Decreased NO_3^-, glutamine synthetase with plastidic isoform ARNm (GS2) accumulation but increased the cytosolic isoform ARNm (GS1) was observed in medium concentrations of Cd. It was complemented by stimulations of ammonium accumulation, ARNm quantity, NADH-dependent glutamate synthase, NADH-dependent glutamate dehydrogenase, and protease activity. Similarly, stimulations were also observed for NAD^+ and $NADP^+$-dependent malate dehydrogenase and $NADP^+$-dependent isocitrate dehydrogenase suggesting their role in the plant-defense processes against Cd-induced stresses (Chaffei et al., 2006).

The groundnut (*Arachis hypogaea*) seedlings, treated with Cd showed a decline in NR and NiR activities and proline, POD, and CAT level upregulated (Dinkar et al., 2009). Cd influences antioxidative enzyme activities (GR, SOD, APX, POD, and CAT) and an increase in proline as compared to control (Dinakar et al., 2008).

Deleterious effects of Cd in *Phaseolus mungo* L. were on seed germination, germination relative index, length and dry weight of root and shoot, shoot-root ratio and seedling vigor index, plant height, phytomass, number of leaves and branches, leaf area, and chlorophyll contents while 10^{-8} M revealed slightly promotion effects. NR and NiR activities were markedly inhibited at higher concentrations (Siddhu and Khan, 2012). Some proteins are differentially methylated at lysine residues in response to Cd and that a few genes coding KMTs are regulated by Cd.

Cd also showed lysin methylation in non-histone proteins and is suggestive for many other methylation events modulating the Cd response in Arabidopsis (Serre et al., 2020). Cd stress caused changes in several proteins found in xylem sap and play role in Cd tolerance. These are associated with cell wall modifications, lipid and protein metabolism, and stress/oxidoreductase category (a defensin-like protein, BnPDFL) (Luo et al., 2019). Another protein OsHMA3, acting as Cd transporter, when overexpressed in Indica cultivar (Zhongjiazao 17), decreased root-shoot Cd translocation and provided tolerance. Similarly, overexpression of protein in brown rice grown in Cd-contaminated paddy soils decreased Cd concentration and showed no influence on other mineral elements like Zn, Fe, Cu, and Mn (Lu et al., 2019).

NR activity was inhibited by Cu and Se pollution either individually or combined in different degrees (Hu et al., 2014). Selenium (Se) is found to be very effective in alleviating oxidative damages. The Se influence was observed on carbohydrate and N metabolism in potato plants (*Solanum tuberosum* L. cv. Sante) grown under Cd and/or As toxicity. HMs toxicity caused a reduction in NO^{3-} and NO_2^- content, and NR and NiR enzymatic activity, and enhanced NH_4^+ content and GDH activity in leaves, roots, and stolons in N metabolism (Shahid et al., 2019). Cu-treated plants showed a reduction in protease, amylase, nitrate, NR, and elevation of peroxidase in root and leaves and caused the generation of ROS where the failure of enzymes playing a crucial role in the biochemical reaction of plants to grow (Azmat and Moin, 2017). Ascorbate, glutathione, and PCs like molecules are induced by Cu in *U. compressa* suggesting Cu tolerance (Mellado et al., 2012) and involvement of thiol-containing peptides and proteins may participate in Cu accumulation and detoxification working in a coordinated and complementary fashion (Navarrete et al., 2019).

The seedlings exposed to Ni resulted in a rapid accumulation in the shoots that caused a significant reduction in fresh weight. Tissue nitrate (NO^{3-}) and glutamate content of seedling decreased in response to Ni stress, whereas ammonium and proline concentration increased substantially. Enzymes like NR declined without altering their activation state, decrease NiR activity, Fd-GOGAT showed reduced activity in the shoots. Contrary to this, GS activity in wheat shoots remained uninfluenced by Ni application but NADH-GOGAT activity was enhanced in comparison to the control. The activity of both NADH-GDH and NAD-GDH activities got enhanced after one week of Ni exposure in wheat shoots. Contrastingly, both AlaAT and AspAT glutamate-producing activities got stimulated by Ni treatment. It suggests that increased AlaAT, AspAT, NADH-GOGAT, and NADH-GDH activities may compensate for the reduced Fd-GOGAT activity that could serve as an alternative route of glutamate synthesis in wheat shoots under Ni stress (Gajewska and Sklodowska, 2009).

Cr treatment to plants showed a negative impact on affect nitrogenase, NR, NiR, glutamine synthetase (GS), and glutamate dehydrogenase (GDH) in many plant organs at different growth stages and found to be lethal at higher concentrations to cluster bean plants (Sangwan et al., 2014).

The decrease in soybean biomass in the combined treatment with $Pb^{(2+)}$ showed decreased root growth, nitrate-N assimilation, and peroxidase activity (Wang et al., 2013). *Silene vulgaris* growth, cell ultrastructure, and element accumulations were observed in untreated and treated with Pb, Cd and Zn ions under *in vitro* conditions showed a drop in protein level compared to the control, enhanced activity of proteases, glutamate changes to proline, reduced glutathione, resulting in intensified growth and cell senescence (Muszynska and Labudda, 2020).

Application of Hg as HM to the bean leaf segments increased the NADH-GDH and NADH-GOGAT enzyme activity in presence of ammonium nitrate (NH_4NO_3). GDH activity was reduced by glutamine and glutathione and enhanced by Al suggests its possible role as the enzyme under Hg-stress (Gupta and Gadre, 2005).

9. Conclusion

HM stress is another great abiotic challenge faced by crop plants as well as plants in general growing across the globe. HM stress not only contributes to a decline in crop productivity but also leaves a great impact on living beings including human health. It is observed to be more severe in the areas which are close to highly urbanized or any industrial area or areas downstream to them. The HMs are usually harmful to plants directly but they also affect indirectly by hampering the uptake of essential minerals. Despite that, some HMs are required by plants in trace amounts for their normal plant growth and development. The impact of HMs is visible on plants ranging from seed germination to seed formation. HMs tolerant plants have evolved diverse strategies to overcome the ill effects induced by HMs by prevention of uptake, reduced uptake, compartmentalization, exclusion, prevention of transport, induction of various biochemicals like RNS/ROS that in turn activates antioxidant systems, hormones, and diverse signaling responses. The impact of HMs was also seen on various physio-biochemical pathways or mechanisms such as sugar metabolism, replication machinery, and associated enzymes and N metabolizing enzymes involved in N assimilation received from external sources or made available through symbiotic N fixation. Considering the vast impact of HMs on the health and productivity of plant and living beings, it is very pertinent to tackle the HMs challenges and device strategies to minimize its impact. Traditional approaches such as minimizing the use of HMs, removal from the polluted ecosystem by eco-friendly techniques such as phytoremediation, use of crops which HMs tolerant created by either classical or modern genetic engineering mechanisms. HM tolerant crops can be created by utilizing the current understanding related to HMs and biochemical-molecular mechanisms and signal transduction pathways to achieve greater yield and crop productivity without compromising the health of living beings and the ecosystem.

References

Adamakis ID, Panteris E, and Eleftheriou EP. 2012. Tungsten toxicity in plants. *Plants* (Basel, Switzerland) 1(2): 82–99.

Adrees M, Ali S, Rizwan M, Zia-ur-Rehman M, Ibrahim M, Abbas F, Farid M, Qayyum MF, and Irshad MK. 2015. Mechanisms of silicon-mediated alleviation of heavy metal toxicity in plants: a review. *Ecotoxicol Environ Saf* 119: 186–197.

Agarwal S and Khan S. 2020. Heavy metal phytotoxicity: DNA damage. *In*: Faisal M., Saquib Q., Alatar A.A, andAl-Khedhairy AA (eds.). *Cellular and Molecular Phytotoxicity of Heavy Metals*. Springer, Cham.

Ahmad E, Zaidi A, Khan, MS, and Oves, M. 2012. Heavy metal toxicity to symbiotic nitrogen-fixing microorganism and host legumes. *In*: Zaidi A, Wani P, and Khan M (eds.). *Toxicity of heavy metals to legumes and bioremediation*. Springer, Vienna.

Amjad M, Raza H, Murtaza B, Abbas G, Imran M, Shahid M, Naeem MA, Zakir A, and Iqbal MM. 2019. Nickel toxicity induced changes in nutrient dynamics and antioxidant profiling in two maize (*Zea mays* L.) Hybrids. *Plants* (Basel, Switzerland) 9(1): 5.

Andrade SAL, Gratão PL, Azevedo RA, Silveira APD, Schiavinato MA, and Mazzafera P. 2010. Biochemical and physiological changes in jack bean under mycorrhizal symbiosis growing in soil with increasing Cu concentrations. *Environ Exp Bot* 68: 198–207.

Angulo-Bejarano PI, Puente-Rivera J, and Cruz-Ortega R. 2021. Metal and metalloid toxicity in plants: an overview on molecular aspects. *Plants* 10(4): 635.

Apel K and Hirt H. 2004. Reactive oxygen species: Metabolism, oxidative stress and signal transduction. *Annu Rev Plant Biol* 55: 373–99.

Arasimowicz-Jelonek M, Floryszak-Wieczorek J and Gwóźdź EA. 2011. The message of nitric oxide in cadmium challenged plants. *Plant Sci* 181(5): 612–620.

Ashraf MY, Azhar N, Ashraf M, Hussain M, and Arshad M. 2011. Influence of lead on growth and nutrient accumulation in canola (*Brassica napus* L.) cultivars. *J Environ Biol* 32(5): 659–666.

Azmat R and Moin S. 2017. The monitoring of Cu contaminated water through potato peel charcoal and impact on enzymatic functions of plants. *J Environ Manage* 203(Pt 1): 98–105.

Balestrasse KB, Benavides MP, Gallego SM, and Tomaro ML. 2003. Effect of cadmium stress on nitrogen metabolism in nodules and roots of soybean plants. *Funct Plant Biol* 30(1): 57–64.

Barton LL, Johnson GV, O'Nan AG, and Wagener BM. 2000. Inhibition of ferric chelate reductase in alfalfa roots by cobalt, nickel, chromium, and copper. *J Plant Nutr* 23(11-12): 1833–1845.

Bishnoi NR, Dua A, Gupta VK, and Sawhney SK. 1993. Effect of chromium on seed germination, seedling growth and yield of peas. *Agric Ecosyst Environ* 47(1): 47–57.

Bosnić D, Bosnić P, Nikolić D, Nikolić M, and Samardžić J. 2019. Silicon and iron differently alleviate copper toxicity in cucumber leaves. *Plants (Basel, Switzerland)* 8(12): 554.

Burén S and Rubio LM. 2017. State of the art in eukaryotic nitrogenase engineering. *FEMS Microbiol. Lett* 365(2): fnx274.

Cambrollé J, Mancilla-Leytón JM, Muñoz-Vallés S, Figueroa-Luque E, Luque T, and Figueroa ME. 2013. Effects of copper sulfate on growth and physiological responses of *Limoniastrum monopetalum*. *Environ Sci Pollut Res Int* 20(12): 8839–8847.

Catarecha P, Segura MD, Franco-Zorrilla JM, García-Ponce B, Lanza M, Solano R, Paz-Ares J, and Leyva A. 2007. A mutant of the Arabidopsis phosphate transporter PHT1; 1 displays enhanced arsenic accumulation. *The Plant Cell* 19(3): 1123–1133.

Chaffei C, Gouia H, Masclaux C, and Ghorbel MH. 2003. Réversibilité des effets du cadmium sur la croissance et l'assimilation de l'azote chez la tomate (*Lycopersicon esculentum*) [Reversibility of the effects of cadmium on the growth and nitrogen metabolism in the tomato (*Lycopersicon esculentum*)]. *C R Biol* 326(4): 401–412.

Chaffei C, Pageau K, Suzuki A, Gouia H, Ghorbel MH, and Masclaux-Daubresse C. 2004. Cadmium toxicity induced changes in nitrogen management in *Lycopersicon esculentum* leading to a metabolic safeguard through an amino acid storage strategy. *Plant Cell Physiol* 45(11): 1681–1693.

Chaffei C, Suzuki A, Masclaux-Daubresse C, Ghorbel MH, and Gouia H. 2006. Implication du glutamate, de l'isocitrate et de la malate déshydrogénases dans l'assimilation de l'azote chez la tomate stressée par le cadmium [*Implication of glutamate, isocitrate and malate deshydrogenases in nitrogen assimilation in the cadmium-stressed tomato*]. *C R Biol* 329(10): 790–803.

Cherif J, Mediouni C, Ben Ammar W, and Jemal F. 2011. Interactions of zinc and cadmium toxicity in their effects on growth and in antioxidative systems in tomato plants (*Solanum lycopersicum*). *J Environ Sci (China)* 23(5): 837–844.

Collin-Hansen C, Andersen RA, and Steinnes E. 2005. Damage to DNA and lipids in *Boletus edulis* exposed to heavy metals. *Mycological Research* 109(12): 1386–1396.

Cook SF. 1926a. The effects of certain heavy metals on respiration. *J Gen Physiol* 9(4): 575–601.

Cook SF. 1926b. A latent period in the action of copper on respiration. *J Gen Physiol* 9(5): 631–50.

Davidson JF and Schiestl RH. 2001. Mitochondrial respiratory electron carriers are involved in oxidative stress during heat stress in *Saccharomyces cerevisiae*. *Mol Cell Biol* 21(24): 8483–9.

de Sousa A, Saleh AM, Habeeb TH, Hassan YM, Zrieq R, Wadaan M, Hozzein WN, Selim S, Matos M, and AbdElgawad H. 2019. Silicon dioxide nanoparticles ameliorate the phytotoxic hazards of aluminum in maize grown on acidic soil. *Sci Total Environ* 693: 133636.

Denny HJ and Wilkins DA. 1987. Zinc tolerance in *Betula* spp. II. Microanalytical studies of zinc uptake into root tissues. *New Phytol* 106: 525–534.

Dias MC, Moutinho-Pereira J, Correia C, Monteiro C, Araújo M, Brüggemann W, and Santos C. 2016. Physiological mechanisms to cope with Cr(VI) toxicity in lettuce: can lettuce be used in Cr phytoremediation? *Environ Sci Pollut Res Int* 23(15): 15627–15637.

Dinakar N, Nagajyothi PC, Suresh S, Udaykiran Y, and Damodharam T. 2008. Phytotoxicity of cadmium on protein, proline and antioxidant enzyme activities in growing *Arachis hypogaea* L. seedlings. *J Environ Sci (China)* 20(2): 199–206.

Dixon HB. 1996. The biochemical action of arsonic acids especially as phosphate analogues. *Adv Inorg Chem* 44: 191–227.

Dobrikova AG, Apostolova EL, Hanč A, Yotsova E, Borisova P, Sperdouli I, Adamakis IS, and Moustakas M. 2021. Cadmium toxicity in *Salvia sclarea* L.: An integrative response of element uptake, oxidative stress markers, leaf structure and photosynthesis. *Ecotoxicol Environ Saf* 209: 111851.

Dong Q, Fang J, Huang F, and Cai K. 2019. Silicon amendment reduces soil Cd availability and Cd uptake of two *Pennisetum* species. *Int J Environ Res Public Health* 16(9): 1624.

Dorneles AOS, Pereira AS, Rossato LV, Possebom G, Sasso VM, Bernardy K, Sandri RD, Nicoloso FT, Ferreira PA, and Tabaldi LA. 2016. Silicon reduces aluminum content in tissues and ameliorates its toxic effects on potato plant growth. *Cienc Rural* 46(3): 506–512.

Dresler S, Hawrylak-Nowak B, Kováčik J, Woźniak M, Gałązka A, Staniak M, Wójciak M, and Sowa I. 2021. Organic nitrogen modulates not only cadmium toxicity but also microbial activity in plants. *J Hazard Mater* 402: 123887.

Droux M. 2004. Sulfur assimilation and the role of sulfur in plant metabolism: a survey. *Photosynth Res* 79(3): 331–348.

Dutta S, Mitra M, Agarwal P, Mahapatra K, De S, Sett U, and Roy S. 2018. Oxidative and genotoxic damages in plants in response to heavy metal stress and maintenance of genome stability. *Plant Signal Behav* 13(8): e1460048.

Ernst WHO. 1969. Zur Physiologie der Schwermetallpflanzen –subzelluliire Speicherungsorte des Zinks. *Ber Deutsch Bot Ges* 82: 161–164.

Fariduddin Q, Yusuf M, Hayat S, and Ahmad A. 2009. Effect of 28-homobrassinolide on antioxidant capacity and photosynthesis in *Brassica juncea* plants exposed to different levels of copper. *Environ Exp Bot* 66: 418–424.

Farzadfar S, Zarinkamar F, Modarres-Sanavy SA, and Hojati M. 2013. Exogenously applied calcium alleviates cadmium toxicity in *Matricaria chamomilla* L. plants. *Environ Sci Pollut Res Int* 20(3): 1413–1422.

Foyer C and Noctor G. 2005. Oxidant and antioxidant signaling in plants: a re-evaluation of the concept of oxidative stress in a physiological context. *Plant Cell Environ* 28: 1056–71.

Gajewska E and Skłodowska M. 2009. Nickel-induced changes in nitrogen metabolism in wheat shoots. *J Plant Physiol* 166(10): 1034–1044.

Gallego SM, Pena LB, Barcia RA, Azpilicueta CE, Iannone MF, Rosales EP, Zawoznik MS, Groppa MD, and Benavides MP. 2012. Unravelling cadmium toxicity and tolerance in plants: insight into regulatory mechanisms. *Environ Exp Bot* 83: 33–46.

Gangwar S, Singh VP, and Maurya JN. 2011a. Responses of *Pisum sativum* L. to exogenous indole acetic acid application under manganese toxicity. *Bull Environ Contam Toxicol* 86(6): 605–609.

Gangwar S, Singh VP, Garg SK, Prasad SM, and Maurya JN. 2011b. Kinetin supplementation modifies chromium (VI) induced alterations in growth and ammonium assimilation in pea seedlings. *Biol Trace Elem Res* 144(1-3): 1327–1343.

Genchi G, Sinicropi MS, Lauria G, Carocci A, and Catalano A. 2020. The effects of cadmium toxicity. *Int J Environ Res Public Health* 17(11): 3782.

Gjorgieva D, Kadifkova Panovska T, Ruskovska T, Bačeva K, and Stafilov T. 2013. Influence of heavy metal stress on antioxidant status and DNA damage in Urtica dioica. *BioMed Res Int* 2013.

Godwill Azeh Engwa, Ferdinand Paschaline Udoka, Nwalo Friday Nweke and Unachukwu Marian N. June 19th 2019. Mechanism and health effects of heavy metal toxicity in humans, poisoning in the modern world - new tricks for an old dog? *Ozgur Karcioglu and Banu Arslan, IntechOpen*, DOI: 10.5772/intechopen.82511.

Godwin HA. 2001. The biological chemistry of lead. *Curr Opin Chem Biol* 5(2): 223–227.

Gong Q, Wang L, Dai TW, Kang Q, Zhou JY, and Li ZH. 2019. Effects of copper treatment on mineral nutrient absorption and cell ultrastructure of spinach seedlings. [铜处理对菠菜幼苗矿质营养吸收和细胞超微结构的影响. *Ying yong sheng tai xue bao*]= *J Appl Ecol* 30(3): 941–950.

Gratão PL, Polle A, Lea PJ, and Azevedo RA. 2005. Making the life of heavy metal-stressed plants a little easier. *Funct Plant Biol* 32: 481–494.

Grill E, Winnacker E-L, and Zenk MH. 1985. Phytochelatins: the principal heavy-metal complexing peptides of higher plants. *Science* 230: 674–676.

Guangqiu Q, Chongling Y, and Haoliang L. 2007. Influence of heavy metals on the carbohydrate and phenolics in mangrove, *Aegiceras corniculatum* L., Seedlings. *Bull Environ Contam Toxicol* 78: 440–444.

Guo WD, Liang J, Yang XE, Chao YE, and Feng Y. 2009. Response of ATP sulfurylase and serine acetyltransferase towards cadmium in hyperaccumulator *Sedum alfredii* Hance. *J Zhejiang Univ Sci B* 10(4): 251–257.

Gupta P and Gadre R. 2005. Increase in NADH-glutamate dehydrogenase activity by mercury in excised bean leaf segments. *Indian J Exp Biol* 43(9): 824–8.

Haddad SA, Tabatabai MA, Abdel-Moneim AMA, and Loynachan TE. 2015. Inhibition of nodulation and nitrogen nutrition of leguminous crops by selected heavy metals. *Air Soil Water Res* 8: ASWR–S21098.

Halliwell B. 2006. Reactive species and antioxidants. Redox biology is a fundamental theme of aerobic life. *Plant Physiol* 141: 312–322.

Handa N, Kohli SK, Thukral AK, Bhardwaj R, Alyemeni MN, Wijaya L, and Ahmad P. 2018. Protective role of selenium against chromium stress involving metabolites and essential elements in *Brassica juncea* L. seedlings. *3 Biotech* 8(1): 66.

Hasan MDK, Cheng Y, Kanwar MK, Chu XY, Ahammed GJ, and Qi ZY. 2017. Responses of plant proteins to heavy metal stress—a review. *Front Plant Sci* 8: 1492.

Hell R and Wirtz M. 2011. Molecular biology, biochemistry and cellular physiology of cysteine metabolism in *Arabidopsis thaliana*. *The Arabidopsis Book* 9: e0154.

Horio T, Higashi T, and Okunuki K. 1955. Copper resistance of *Mycobacterium tuberculosis avium*. II. The influence of copper ion on the respiration of the parent cells and copper-resistant cells. *J Biochem, Tokyo* 42: 491–498.

Hossain Z and Komatsu S. 2013. Contribution of proteomic studies towards understanding plant heavy metal stress response. *Front Plant Sci* 3: 310.

Howarth JR, Domínguez-Solís JR, Gutiérrez-Alcalá G, Wray JL, Romero LC, and Gotor C. 2003. The serine acetyltransferase gene family in *Arabidopsis thaliana* and the regulation of its expression by cadmium. *Plant Mol Biol* 51(4): 589–598.

Hu B, Liang D, Liu J, Lei L, and Yu D. 2014. Transformation of heavy metal fractions on soil urease and nitrate reductase activities in copper and selenium co-contaminated soil. *Ecotoxicol Environ Saf* 110: 41–48.

Imran M, Hussain S, Rana MS, Saleem MH, Rasul F, Ali KH, Potcho MP, Pan S, Duan M, and Tang X. 2021. Molybdenum improves 2-acetyl-1-pyrroline, grain quality traits and yield attributes in fragrant rice through efficient nitrogen assimilation under cadmium toxicity. *Ecotoxicol Environ Saf* 211: 111911.

Irfan M, Ahmad A, and Hayat S. 2014. Effect of cadmium on the growth and antioxidant enzymes in two varieties of *Brassica juncea*. *Saudi J Biol Sci* 21(2): 125–131.

Jiang XJ, Luo YM, Liu Q, Liu SL, and Zhao QG. 2004. Effects of cadmium on nutrient uptake and translocation by Indian Mustard. *Environ Geochem Health* 26(2): 319–324.

Kamiya T, Islam R, Duan G, Uraguchi S, and Fujiwara T. 2013. Phosphate deficiency signaling pathway is a target of arsenate and phosphate transporter OsPT1 is involved in As accumulation in shoots of rice. *Soil Sci Plant Nutr* 59(4): 580–590.

Kaya C, Ashraf M, Alyemeni MN, and Ahmad P. 2020. The role of nitrate reductase in brassinosteroid-induced endogenous nitric oxide generation to improve cadmium stress tolerance of pepper plants by upregulating the ascorbate-glutathione cycle. *Ecotoxicol Environ Saf* 196: 110483.

Keunen E, Remans T, Bohler S, Vangronsveld J, and Cuypers A. 2011. Metal-induced oxidative stress and plant mitochondria. *Int J Mol Sci* 12: 6894–6918.

Keunen ELS, Peshev D, Vangronsveld J, Ende WAD, and Cuypers A. 2013. Plant sugars are crucial players in the oxidative challenge during abiotic stress: extending the traditional concept. *Plant Cell Environ* 36: 1242–1255.

Kim CG, Power SA, and Bell JNB. 2003. Effects of cadmium and soil type on mineral nutrition and carbon partitioning in seedlings of *Pinus sylvestris*. *Water Air Soil Pollut* 145: 253–266.

Kolahi M, Mohajel Kazemi E, Yazdi M, and Goldson-Barnaby A. 2020. Oxidative stress induced by cadmium in lettuce (*Lactuca sativa* Linn.): Oxidative stress indicators and prediction of their genes. *Plant Physiol Biochem* 146: 71–89.

Kopittke PM, Gianoncelli A, Kourousias G, Green K, and McKenna BA. 2017. Alleviation of Al toxicity by Si is associated with the formation of Al-Si complexes in root tissues of *Sorghum*. *Front Plant Sci* 8: 2189.

Lamhamdi M, Galiou OEI, Bakrim A, Nóvoa-Muñoz JC, Arias-Estévez M, Aarab A, and Lafont R. 2013. Effect of lead stress on mineral content and growth of wheat (*Triticum aestivum*) and spinach (*Spinacia oleracea*) seedlings. *Saudi J Biol Sci* 20(1): 29–36.

Lang M, Hao M, Fan Q, Wang W, Mo S, Zhao W, and Zhou J. 2011. Functional characterization of BjCET3 and BjCET4, two new cation-efflux transporters from *Brassica juncea* L. *J Exp Bot* 62(13): 4467–80.

Li C, Liu Y, Tian J, Zhu Y, and Fan J. 2020. Changes in sucrose metabolism in maize varieties with different cadmium sensitivities under cadmium stress. *PloS one* 15(12): e0243835.

Lin A, Zhang X, Zhu YG, and Zhao FJ. 2008. Arsenate-induced toxicity: Effects on antioxidative enzymes and DNA damage in *Vicia faba*. *Environ Toxicol Chem* 27(2): 413–419.

Lindstrom K, and Mousavi SA. 2019. Effectiveness of nitrogen fixation in rhizobia. *Microb Biotechnol* 13: 1314–1335.

Llamas A, Chamizo-Ampudia A, Tejada-Jimenez M, Galvan A, and Fernandez E. 2017. The molybdenum cofactor enzyme mARC: Moonlighting or promiscuous enzyme? *BioFactors (Oxford, England)* 43(4): 486–494.

Lu C, Zhang L, Tang Z, Huang XY, Ma JF, and Zhao FJ. 2019. Producing cadmium-free Indica rice by overexpressing OsHMA3. *Environ Int* 126: 619–626.

Łukowski A and Dec D. 2018. Influence of Zn, Cd, and Cu fractions on enzymatic activity of arable soils. *Environ Monit Assess* 190(5): 278.

Luo JS and Zhang Z. 2019. Proteomic changes in the xylem sap of *Brassica napus* under cadmium stress and functional validation. *BMC Plant Biol* 19(1): 280.

Maathuis FJ and Diatloff E. 2013. Roles and functions of plant mineral nutrients. *Methods Mol Biol (Clifton, N.J.)* 953: 1–21.

Mahmud K and Burslem D. 2020. Contrasting growth responses to aluminium addition among populations of the aluminium accumulator *Melastoma malabathricum*. *AoB PLANTS* 12(5): plaa049.

Manara A. 2012. Plant responses to heavy metal toxicity. pp. 27–53. *In*: Furini A (ed.). *Plants and Heavy Metals*. Springer Briefs in Molecular Science. Springer, Dordrecht.

Marcato-Romain CE, Pinelli E, Pourrut B, Silvestre J, and Guiresse M. 2009. Assessment of the genotoxicity of Cu and Zn in raw and anaerobically digested slurry with the *Vicia faba* micronucleus test. *Mutat Res Genet Toxicol Environ Mutagen* 672(2): 113–118.

Mateos-Naranjo E, Andrades-Moreno L, Cambrollé J, and Perez-Martin A. 2013. Assessing the effect of copper on growth, copper accumulation and physiological responses of grazing species *Atriplex halimus*: ecotoxicological implications. *Ecotoxicol Environ Saf* 90: 136–142.

Meharg AA. 1993. The role of the plasmalemma in metal tolerance in angiosperms. *Physiol Planta* 88: 191–198.

Mellado M, Contreras RA, González A, Dennett G, and Moenne A. 2012. Copper-induced synthesis of ascorbate, glutathione and phytochelatins in the marine alga *Ulva compressa* (Chlorophyta). *Plant Physiol Biochem* 51: 102–108.

Mishra J, Singh R, and Arora NK. 2017. Alleviation of heavy metal stress in plants and remediation of soil by rhizosphere microorganisms. *Front Microbiol* 8: 1706.

Mittler R. 2006. Abiotic stress, the field environment and stress combination. *Trends Plant Sci* 11: 15–9.

Młodzińska E, and Zboińska M. 2016. Phosphate uptake and allocation - a closer look at *Arabidopsis thaliana* L. and *Oryza sativa* L. *Front Plant Sci* 7: 1198.

Monnet F, Vaillant N, Vernay P, Coudret A, Sallanon H, and Hitmi A. 2001. Relationship between PSII activity, CO2 fixation, and Zn, Mn and Mg contents of *Lolium perenne* under zinc stress. *J Plant Physiol* 158(9): 1137–1144.

Morales ME, Derbes RS, Ade CM, Ortego JC, Stark J, Deininger PL, and Roy-Engel AM. 2016. Heavy metal exposure influences double strand break DNA repair outcomes. *PloS one* 11(3): e0151367.

Moura DJ, Péres VF, Jacques RA, and Saffi J. 2012. Heavy metal toxicity: oxidative stress parameters and DNA repair. pp. 187–205. *In*: *Metal Toxicity in Plants: Perception, Signaling and Remediation*. Springer, Berlin, Heidelberg.

Murayama T. 1961. Studies on the metabolic pattern of yeast with reference to its copper resistance. *Memoirs Ehime Univ Sect II*, B4: 43–66.

Muszyńska E and Labudda M. 2020. Effects of lead, cadmium and zinc on protein changes in Silene vulgaris shoots cultured *in vitro*. *Ecotoxicol Environ Saf* 204: 111086.

Muthuchelian K, Bertamini M, and Nedunchezhian N. 2001. Triacontanol can protect *Erythrina variegata* from cadmium toxicity. *J Plant Physiol* 158(11): 1487–1490.

Mysliwa-Kurdziel B, Prasad MNV, and Strzalka K. 2004. Photosynthesis in heavy metal stressed plants. pp. 146–181. *In*: Prasad MNV (ed.). *Heavy Metal Stress in Plants: From Biomolecules to Ecosystems*. Narosa Publishing House, New Delhi India.

Nada E, Ferjani BA, Ali R, Bechir BR, Imed M, and Makki B. 2007. Cadmium-induced growth inhibition and alteration of biochemical parameters in almond seedlings grown in solution culture. *Acta Physiol Plant* 29(1): 57–62.

Navarrete A, González A, Gómez M, Contreras RA, Díaz P, Lobos G, Brown MT, Sáez CA, and Moenne A. 2019. Copper excess detoxification is mediated by a coordinated and complementary induction of glutathione, phytochelatins and metallothioneins in the green seaweed *Ulva compressa*. *Plant Physiol Biochem* 135: 423–431.

Nedjimi B, and Daoud Y. 2009. Cadmium accumulation in *Atriplex halimus* subsp. *schweinfurthii* and its influence on growth, proline, root hydraulic conductivity and nutrient uptake. *Flora-Morphology, Distribution, Functional Ecology of Plants* 204(4): 316–324.

Nie Z, Wang Y, and Li S. 2009. Heavy metal-induced DNA damage in *Arabidopsis thaliana* Protoplasts Measured by Single-cell Gel Electrophoresis. *Chin J Bot* 44(01): 117–123.

Nishida S, Aisu A, and Mizuno T. 2012. Induction of IRT1 by the nickel-induced iron-deficient response in *Arabidopsis*. *Plant Signal Behav* 7(3): 329–331.

Nishida S, Kato A, Tsuzuki C, Yoshida J, and Mizuno T. 2015. Induction of nickel accumulation in response to zinc deficiency in *Arabidopsis thaliana*. *Int J Mol Sci* 16(5): 9420–9430.

Nouairi I, Jalali K, Essid S, Zribi K, and Mhadhbi H. 2019. Alleviation of cadmium-induced genotoxicity and cytotoxicity by calcium chloride in faba bean (*Vicia faba* L. var. minor) roots. *Physiol Mol Biol Plants* 25(4): 921–931.

Ouariti O, Boussama N, Zarrouk M, Cherif A, and Ghorbal MH. 1997. Cadmium- and copper-induced changes in tomato membrane lipids. *Phytochemistry* 45(7): 1343–1350.

Pal SC. 1996. Effect of heavy metals on Legume-Rhizobium symbiosis. *In*: Rahman M, Podder AK, Van Hove C, Begum ZNT, Heulin T, and Hartmann A (eds.). *Biological Nitrogen Fixation Associated with Rice Production*. Developments in Plant and Soil Sciences, vol 70. Springer, Dordrecht.

Pandey N, Pathak GC, Pandey DK, and Pandey R. 2009. Heavy metals, Co, Ni, Cu, Zn and Cd, produce oxidative damage and evoke differential antioxidant responses in spinach. *Braz J Plant Physiol* 21(2): 103–111.

Pego JV, Kortstee AJ, Huijser C, and Smeekens SCM. 2000. Photosynthesis, sugars and the regulation of gene expression. *J Exp Bot* 51: 407–16.

Peralta-Videa JR, Lopez ML, Narayan M, Saupe G, and Gardea-Torresdey J. 2009. The biochemistry of environmental heavy metal uptake by plants: implications for the food chain. *Int J Biochem Cell Biol* 41(8-9): 1665–1677.

Perchlik M and Tegeder M. 2018. Leaf amino acid supply affects photosynthetic and plant nitrogen use efficiency under nitrogen stress. *Plant Physiol* 178(1): 174–188.

Pfeffer PE, Tu SI, Gerasimowicz WV, and Cavanaugh JR. 1986. *In vivo* ^{31}P NMR studies of corn root tissue and its uptake of toxic metals. *Plant Physiol* 80: 77–84.

Priyanka N, Geetha N, Manish T, Sahi SV, and Venkatachalam P. 2021. Zinc oxide nanocatalyst mediates cadmium and lead toxicity tolerance mechanism by differential regulation of photosynthetic machinery and antioxidant enzymes level in cotton seedlings. *Toxicol Rep* 8: 295–302.

Rizvi A, Zaidi A, Ameen F, Ahmed B, AlKahtani MD, and Khan MS. 2020. Heavy metal induced stress on wheat: phytotoxicity and microbiological management. *RSC Adv* 10(63): 38379–38403.

Rizzardo C, Tomasi N, Monte R, Varanini Z, Nocito FF, Cesco S, and Pinton R. 2012. Cadmium inhibits the induction of high-affinity nitrate uptake in maize (*Zea mays* L.) roots. *Planta* 236(6): 1701–1712.

Rosa M, Prado C, Podazza G, Interdonato R, González JA, Hilal M, and Prado FE. 2009. Soluble sugars—Metabolism, sensing and abiotic stress A complex network in the life of plants. *Plant Signal Behav* 4(5): 388–393.

Ruan YL. 2014. Sucrose metabolism: gateway to diverse carbon use and sugar signaling. *Annu Rev Plant Biol* 65: 33–67.

Saad RB, Hsouna AB, Saibi W, Hamed KB, Brini F, and Ghneim-Herrera T. 2018. A stress-associated protein, LmSAP, from the halophyte *Lobularia maritima* provides tolerance to heavy metals in tobacco through increased ROS scavenging and metal detoxification processes. *J Plant Physiol* 231: 234–243.

Sangwan P, Kumar V, and Joshi UN. 2014. Effect of chromium (VI) toxicity on enzymes of nitrogen metabolism in clusterbean (*Cyamopsis tetragonoloba* L.). *Enzyme Res* 2014: 784036.

Sarkar U, Tahura S, Das U, Mintu MRA, and Kabir AH. 2020. Mitigation of chromium toxicity in wheat (*Triticum aestivum* L.) through silicon. *Gesunde Pflanzen* 72: 237–244.

Schiavon M, Pilon-Smits EA, Wirtz M, Hell R, and Malagoli M. 2008. Interactions between chromium and sulfur metabolism in *Brassica juncea*. *J Environ Qual* 37(4): 1536–1545.

Schmitt M, Watanabe T, and Jansen S. 2016. The effects of aluminium on plant growth in a temperate and deciduous aluminium accumulating species. *AoB PLANTS* 8: plw065.

Schützendübel A, and Polle A. 2002. Plant responses to abiotic stresses: heavy metal-induced oxidative stress and protection by mycorrhization. *J Exp Bot* 53(372): 1351–1365.

Seneviratne M, Rajakaruna N, Rizwan M, Madawala H, Ok YS, and Vithanage M. 2019. Heavy metal-induced oxidative stress on seed germination and seedling development: a critical review. *Environ Geochem Health* 41(4): 1813–1831.

Serre N, Sarthou M, Gigarel O, Figuet S, Corso M, Choulet J, Rofidal V, Alban C, Santoni V, Bourguignon J, Verbruggen N, and Ravanel S. 2020. Protein lysine methylation contributes to modulating the response of sensitive and tolerant *Arabidopsis* species to cadmium stress. *Plant Cell Environ* 43(3): 760–774.

Shahid M, Pourrut B, Dumat C, Nadeem M, Aslam M, and Pinelli E. 2014. Heavy-metal-induced reactive oxygen species: phytotoxicity and physicochemical changes in plants. *Rev Environ Contam T* 232: 1–44.

Shahid MA, Balal RM, Khan N, Zotarelli L, Liu GD, Sarkhosh A, Fernández-Zapata JC, Martínez Nicolás JJ, and Garcia-Sanchez F. 2019. Selenium impedes cadmium and arsenic toxicity in potato by modulating carbohydrate and nitrogen metabolism. *Ecotoxicol Environ Saf* 180: 588–599.

Sharma P and Dubey RS. 2005. Lead toxicity in plants. *Braz J Plant Physiol* 17(1): 35–52.

Sharma SS and Dietz KJ. 2006. The significance of amino acids and amino acid-derived molecules in plant responses and adaptation to heavy metal stress. *J Exp Bot* 57(4): 711–726.

Shu X, Zhang Q, and Wang W. 2014. Lead induced changes in growth and micronutrient uptake of *Jatropha curcas* L. *Bull Environ Contam Toxicol* 93(5): 611–617.

Siddhu G and Ali Khan MA. 2012. Effects of cadmium on growth and metabolism of *Phaseolus mungo*. *J Environ Biol* 33(2): 173–179.

Singh CB, Verma SK, and Singh SP. 1987. Impact of heavy metals on glutamine synthetase and nitrogenase activity in Nostoc calcicola. *The Journal of General and Applied Microbiology* 33(1): 87–91.

Singh P, Singh I, and Shah K. 2019. Reduced activity of nitrate reductase under heavy metal cadmium stress in rice: An *in silico* Answer. *Front Plant Sci* 9: 1948.

Sreekanth TVM, Nagajyothi PC, Lee KD, and Prasad TNVKV. 2013. Occurrence, physiological responses and toxicity of nickel in plants. *Int J Sci Environ Technol* 10(5): 1129–1140.

Srivastava RK, Pandey P, Rajpoot R, Rani A, and Dubey R. 2014. Cadmium and lead interactive effects on oxidative stress and antioxidative responses in rice seedlings. *Protoplasma* 251(5): 1047–1065.

Stein O, and Granot D. 2019. An Overview of Sucrose Synthases in Plants. *Front Plant Sci* 10: 95.

Stitt M and Hurry V. 2002. A plant for all seasons: alterations in photosynthetic carbon metabolism during cold acclimation in Arabidopsis. *Curr Opin Plant Biol* 5(3): 199–206.

Stratton GW, Huber AL, and Corke CT. 1979. Effect of mercuric ion on the growth, photosynthesis, and nitrogenase activity of *Anabaena inaequalis*. *Applied and Environmental Microbiology* 38(3): 537–543.

Sundaramoorthy P, Chidambaram A, Ganesh KS, Unnikannan P, and Baskaran L. 2010. Chromium stress in paddy: (i) nutrient status of paddy under chromium stress; (ii) phytoremediation of chromium by aquatic and terrestrial weeds. *C R Biol* 333(8): 597–607.

Tejada-Jimenez M, Chamizo-Ampudia A, Calatrava V, Galvan A, Fernandez E, and Llamas A. 2018. From the eukaryotic molybdenum cofactor biosynthesis to the moonlighting enzyme mARC. *Molecules (Basel, Switzerland)* 23(12): 3287.

Tiwari KK, Singh NK, and Rai UN. 2013. Chromium phytotoxicity in radish (*Raphanus sativus*): effects on metabolism and nutrient uptake. *Bull Environ Contam Toxicol* 91(3): 339–344.

Venkatachalam P, Jayaraj M, Manikandan R, Geetha N, Rene ER, Sharma NC, and Sahi SV. 2017. Zinc oxide nanoparticles (ZnONPs) alleviate heavy metal-induced toxicity in *Leucaena leucocephala* seedlings: A physiochemical analysis. *Plant Physiol Biochem* 110: 59–69.

Vogeli-Lange R, and Wagner GJ. 1990. Subcellular localization of cadmium-binding peptides in tobacco leaves. Implications of a transport function for cadmium-binding peptides. *Plant Physiol* 92: 1086–1093.

Wagner SC. 2011. Biological nitrogen fixation. *Nature Education Knowledge* 3(10): 15.

Wang X, Shi M, Hao P, Zheng W, and Cao F. 2017. Alleviation of cadmium toxicity by potassium supplementation involves various physiological and biochemical features in *Nicotiana tabacum* L. *Acta Physiol Plant* 39(6): 132.

Wang S, Wang L, Zhou Q, and Huang X. 2013. Combined effect and mechanism of acidity and lead ion on soybean biomass. *Biol Trace Elem Res* 156(1-3): 298–307.

Wingler A, Lea PJ, Quick WP, and Leegood RC. 2000. Photorespiration: metabolic pathways and their role in stress protection. *Philos Trans R Soc Lond B Biol Sci* 355(1402): 1517–1529.

Wu L, Bradshaw AD, and Thurman DA. 1975. Potential for evolution of heavy metal tolerance in plants. III. The rapid evolution of copper tolerance in *Agrostis stolonifera*. United Kingdom: N. p., Web. doi: 10.1038/hdy.1975.21.

Wu Z, Ren H, McGrath SP, Wu P, and Zhao FJ. 2011. Investigating the contribution of the phosphate transport pathway to arsenic accumulation in rice. *Plant Physiol* 157(1): 498–508.

Xiong ZT, Zhao F, and Li MJ. 2006. Lead toxicity in *Brassica pekinensis* Rupr.: effect on nitrate assimilation and growth. *Environ Toxicol* 21(2): 147–153.

Zhang L, Yang Q, Wang S, Li W, Jiang S, and Liu Y. 2017. Influence of silicon treatment on antimony uptake and translocation in rice genotypes with different radial oxygen loss. *Ecotoxicol Environ Saf* 144: 572–577.

Zhu ZJ, Sun GW, Fang XZ, Qian QQ, and Yang XE. 2004. Genotypic differences in effects of cadmium exposure on plant growth and contents of cadmium and elements in 14 cultivars of bai cai. *Int J Environ Res Public Health B* 39(4): 675–687.

8

Photosynthetic Response of Plants Against Heavy Metals

Sujata Rathi,[1,]* *Neha Mittal*[2] and *Deepak Kumar*[3,]*

ABSTRACT

Photosynthesis is a physico-biochemical process that traps the solar energy and carbon from the atmosphere and reduces it to fulfill the energy needs of most of the organisms existing on earth. This process may be regarded as the main driving force to evolve various forms of life on earth. But this vital process is greatly affected by the increased concentration of heavy metals in the environment. Although, heavy metals such as Cu, Zn, Mn, Fe, Mo, Ni, and Co, serve as micronutrient for various physiological and growth promoting activities such as photosynthesis, protein synthesis, fat metabolism and acts as cofactors in many enzyme activities. Heavy metal exposure in increased concentration greatly influenced the photosynthetic process either directly or indirectly. Various experimental studies revealed the adverse effect of heavy metal on anatomical features of the plants involved in photosynthesis such as deformed chloroplast, poorly developed leaves, and abnormal stomata structure. Heavy metals also impose an adverse effect on the biochemistry of photosynthesis such as photosystem I (PSI) and photosystem II (PSII). Photosystem II, an oxygen evolving process of photosynthesis, is responsible for the maintenance of aerobic life on earth. It is noticeable that PSII is more sensitive to heavy metal in comparison to PSI. Most of the heavy metal studies revealed that the oxygen evolving system that plays a crucial role in generating electrons and hydrogen ions to maintain electrochemical gradient for generation of assimilatory power becomes the target site for the binding of various heavy metals. Heavy metals interfere with various enzymatic activities

[1] Department of Botany, Multanimal Modi College, Modinagar.
[2] Department of Botany, C.C.S. University, Meerut.
[3] Graphic Era Deemed (to be) University, Dehradun.
* Corresponding authors: srathi84@gmail.com; dtxyz007@gmail.com

involved in dark phase synthetic process of photosynthesis such as ALA synthase (precursor of tetrapyrrole), RuBPcase and PEP-carboxylase. Heavy metal ions alter the functioning of photosynthetic enzymes by interacting with the SH groups.

1. Introduction

Photosynthesis is a multistep reductive synthetic process that meets the energy and carbon demand of plants for the synthesis of organic molecules needed for growth and development. In this phenomenon, solar energy is harvested and transformed into chemical energy through a series of enzymatic reactions. Heavy metal concentration is increasing in the environment beyond the permissible limit according to various environmental-protection agencies. Elevated concentration of heavy metal affects adversely plant structural and physiological aspects. Heavy metals such as Cu, Mn, Zn, Co, Fe, Ni, and Mo, serving as micronutrients for plants play a very crucial role in various physiological functions of plants such as protein synthesis, photosynthesis, fat metabolism, etc. Most of the time heavy metal acts as a cofactor, required for the fully active enzyme. The oxygen evolving complex associated with PSII is consisting of Mn. Some heavy metals such as Al, Pd, As, Cr, and Cd, although not required for plant growth and development, are detrimental even in very low concentrations (Ernst et al., 2008; Garzón et al., 2011; Hayat et al., 2012; Shahid et al., 2012; Chong-qing et al., 2013). Zn is needed for the enzymatic activity involved in membrane stability, hormone balance, and reproduction (Marschner, 1995; Barker and Pilbeam, 2007; Briat et al., 2007; Williams and Pittman, 2010; Prasad, 2012; Ricachenevsky et al., 2013). Accumulation of Cr in higher concentrations in soil inhibits the uptake of macronutrients like the uptake of K, P, Fe, and Mg (Kabata-Pendias and Pendias, 2001). Heavy metals may pose and adverse effect on plants by changing the ion absorption strategy of a root, denaturing of various proteins, replacing groups from the specific receptor site of membrane, and by generating free radicals, ROS (reactive oxygen species) (Sharma and Dietz, 2009; DalCorso et al., 2013a). In the defense system of plants, ROS behaves as signaling molecules for the regulation of cell division and differentiation, root hair growth, and stomatal dynamics (Foreman et al., 2003; Kwak et al., 2006; Tsukagoshi et al., 2010). Generally, plants defend themselves from the adverse impact of heavy metals by avoiding metal entry into the root either by immobilization of metal by associated rhizobacteria or forming a complex with root exudates (Zengin and Munzuroglu, 2005). If a plant fails to prevent the uptake of heavy metal ions, then plants use another strategy of producing osmoproctectants to nullify the impact of taken heavy metal. Taken up heavy metals are sequestered and trapped inside vacuoles, form intracellular complex, and undergo chelation (Patra et al., 2004; Dalvi and Bhalerao, 2013).

2. Heavy metals impact on the anatomy of photosynthetic tissue

Leaf anatomical features play a crucial role in foliar photosynthetic capacity (Poorter et al., 2009a; Scafaro et al., 2011; Terashima et al., 2011). Cd toxicity caused a reduction in leaf size and the thickness of lamina and a decrease in intercellular spaces in mesophyll cells (Tran et al., 2013). Significant reduction in the photosynthetic area was reported due to the reduction in the size of mesophyll cells and destruction of palisade and spongy parenchyma in heavy metal treated plants or plants at contaminated sites (Srighar et al., 2005; Zhao et al., 2000). Plants used coping strategies such as increasing thickness of abaxial and adaxial sclerenchyma and pericycle due to the allocation of heavy metals in the cell walls thus minimizing the damaging impact of heavy metal on the photosynthetic region (Marcelo PedrosaGomes et al., 2011). This alternative strategy to deposit heavy metals in the non-photosynthetic area was also reported in Cd treated *Salixviminalis* to minimize Cd concentration in chlorophyll parenchyma (Olivier André et al., 2006). The extent to which photosynthesis is inhibited or reduced in response to heavy metal treatment also depends on the position of leaves as seen in wheat leaves. A substantial relationship exists between photosynthesis and transpiration as the stomata control both the processes due to its involvement in both the loss of water vapors as well gaseous exchange. Various experimental studies revealed the adverse effect of various heavy metals on stomata such as a decrease in the number of stomata per unit area of the leaf, reduction in the frequency of opened stomata per unit of leaf area, and anomalous development of the stomatal complex. Exposure to higher concentrations of Cd, Cu, and Zn leads to a decrease in the number of stomata per unit leaf area in the case of *Sorghum bicolor*, *Phaseolus vulgaris*, and in *Beta vulgaris* (Kasim, 2007; Sagardoy et al., 2010). Treatment of heavy metal also resulted in the reduction of the frequency of opened stomata per unit of leaf area (Greger and Johansson, 1992). Heavy metals cause more or less similar effects in both monocotyledon and dicotyledons plants. The frequencies of epidermal cells and stomata per unit of leaf area, guard cell indices and pore area, pore area indices, and total pore area were observed to decrease in most of the treated plants in barley and safflower as compared to untreated control (Mittal and Srivastava, 2014). In the case of *Helianthusannus* and *Betavulgaris*, treatment with heavy metal Cd resulted in reduced guard cell size due to water deficit in leaves thus lowers the rate of transpiration (Greger and Johansson, 1992; Kastori et al., 1992). In the condition of water deficit, water is driven at the expense of energy in active transport of solutes by root (Javot and Maurel, 2002; Steudle, 2000). Pb deposits were reported in the cuticle of guard cells and subsidiary cells in the treated plants expected to be responsible for the reduction in the number of opened stomata (Woźny et al., 1995). Various types of stomatal anomalies were also reported on exposure to Cd and Cr^{6+} in barley. Most of these anomalies are defined as the presence of common subsidiary cell, presence of contiguous stomata with common subsidiary cell, stomata with obliquely placed subsidiary cells, stomata with the unequally enlarged subsidiary cells as shown in Fig. 1 (Mittal and Srivastava, 2014). Abnormal shaped stomata resulted due to the interaction of heavy metals with microtubules that affected the mode of cell division and organization of cytoplasmic organelle during the formation of stomata. This interaction interfered with microtubule dynamicity (polymerization and depolymerization), resulted in poorly differentiated stomata with irregular wall thickening in as treated *Vignaradiata*. Abnormally thickened wall affects the stomata closing and opening thus influencing the stomatal conductance (Barceló et al., 1988; Bazihizina et al., 2015). Under the influence of heavy metals stomata failed to respond to external stimuli such as ABA and temperature signals (Sagardoy et al., 2010). Small apertured stomata with increased stomatal density were also utilized

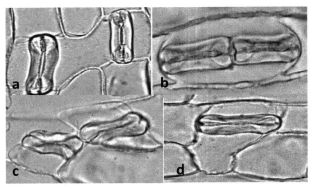

Figure 1. Stomatal Anomalies: (a) stomata with one common subsidiary cell, (b) stomata with both subsidiary cells in common, (c) obliquely placed stomata, (d) stomata with one enlarged subsidiary cell.

as an adaptation strategy to nullify the heavy metal toxicity in plants as this ensured optimum CO_2 supply with minimum water loss in transpiration (Melo et al., 2007).

Chloroplast, a complex intracellular organelle, has an intramembranous system called thylakoids. Heavy metals also disrupt the thylakoid membrane in the chloroplast. The intramembranous system of chloroplast plays a crucial role in photosynthesis as proteins and lipids of photosystems remain embedded in these membranes. Heavy metals cause lipid peroxidation and protein carbonylation leading to inhibition of electron flow by disrupting the intramembranous system (Gallego et al., 1996; Devi and Prasad, 2005; Tripathi et al., 2006). Cd toxicity leads to abnormal chloroplast with inflated thylakoids (Najeeb et al., 2011). Increased activity of lipoxygenase causes peroxidation of fatty acids and lipid content of chloroplast membrane resulting in a distortion in the shape of chloroplast (Remans et al., 2010). In the case of wheat, excessive copper ruptured chloroplast membrane, thylakoids and osmiophilic granules increased in number (Zhong et al., 2017). Significant decreases in the number and size of chloroplast was also noticed in various plant studies like *Picrisdivariticata*, *Hordeumvulgare*, and *Brassica* sp. (Xin et al., 2010; Wang et al., 2011; Elhiti et al., 2012).

3. Heavy metals impact on the chemistry of photosynthesis

Several investigations have revealed that photosynthetic activity is suppressed with a high concentration of heavy metals. Heavy metals differ in their site of action to influence the chemistry of photosynthesis. Their mode of action can be observed as follows:

3.1 Stomatal conductance

Heavy metals affect both the opening and closing movement of the stomatal thus affecting the rate of transpiration and net photosynthesis. Heavy metals affect stomatal movements in two ways through ion channels and water channels. Cd treatment in excised silver maple leaves showed decreased CO_2 conductance (Lamoreaux and

Chaney, 1978; Bazzaz et al., 1974a,b; Huang et al., 1974; Bazzaz et al., 1975; Malik, 1992a,b; Sheoran et al., 1990). The studies have proved that sub millimolar concentration of heavy metals significantly affect the stomatal movements. Studies also suggested the role of cytosolic Ca ion concentration in regulating the opening and closing of water channels thus affecting stomatal movements (Maurel et al., 1995; Johansson et al., 1996, 1998). Ca concentration in guard cell changes with the treatment of $LaCl_3$ that blocks the calcium ion channel. This disturbance in the concentration of Ca ions in the cytosol directly affects the concentration of rest ions such as K^+, Cl^+ and $malate^{2-}$ and thus opening and closing movements of the stomatal. Studies with mercury treatments also demonstrated the decreased water flow through water channels (Sun et al., 2001; Barone et al., 1997; Lu et al., 1999; Biela et al., 1999; Zhang et al., 1999; Clarkson et al., 2000; Yang et al., 2002; Krupa et al., 1999). The experiments have shown that heavy metals strikingly affected stomatal movements at sub millimolar concentrations, probably in different ways. $LaCl_3$, a Ca^{+2} channel blocker, apparently affected the changes in the cytosolic Ca^{+2} concentrations in guard cells, thus indirectly influencing the activities of other ion channels, such as K^+, Cl^-, and $malate^{2-}$ channels, and finally, stomatal opening or closing (Cuypers et al., 2001). The inhibition of water flow through water channels by lower concentrations of mercurial reagents has been tested in many experiments (Vassilev et al., 2011; Huang et al., 2002). Treatment of broad bean leaves with the combination of $HgCl_2$ and $LaCl_3$ resulted in complete inhibition of stomatal movements. However, changes in stomatal aperture varied with different concentrations of $HgCl_2$, $PbCl_2$ or $ZnCl_2$ and inhibited stomatal opening in the light and closing in darkness (Zhang et al., 1999). $HgCl_2$ inhibits water channels whereas $LaCl_3$ inhibits ion channels. $HgCl_2$ inhibits aquaporins in guard cells in *Viciafaba* leaves (Biela et al., 1999; Zhang et al., 1999). Some studies suggested the possibilities of membrane depolarization in the presence of $HgCl_2$ can influence stomatal movement (Biela et al., 1999; Zhang et al., 1999).

3.2 Chlorophyll

The amount of chlorophyll, the key molecule of photosynthesis, is significantly affected by environmental conditions. Chlorophyll a, b and carotenoids together constitute the main components of photosystems. Chlorophyll a is responsible for the evolution of O_2 whereas chlorophyll b absorbs blue light. Substances such as heavy metals directly affect the photosynthetic function by affecting the synthesis and decomposition of chlorophyll molecules. Ouzounidou (1997) demonstrated that the impact of various heavy metals varied on chloroplast structure and was found in the order of Pb > Cd > Hg (Ouzounidou et al., 1997). Heavy metals cause Mg substitution in chlorophyll rings with different magnitude in the order of Hg^{+2}, Cu^{+2}, Cd^{+2} > Zn^{+2} > Ni^{+2} > Pb^{+2} (Kupper et al., 1996). Cd interferes with biosynthesis of chlorophyll by substitution of Mg element in chlorophyll molecule (Joshi et al., 2004; McGrath et al., 2001; Franco et al., 1999). Mg component is inserted into protoporphyrin IX ring by Mg-chetalase which is inhibited by Cd^{+2} leading to accumulation of protoporphyrinogen IX without further conversion into the active form (Csatorday et al., 1984).

Cu dysfunctions the chlorophyll present in the reaction center and antenna molecules by removing Mg from the chlorophyll (Kupper et al., 2003). Mn^{+2} also caused the accumulation of protoporphyrin IX and Mg-protoporphyrin IX monoethyl ester in dark-grown barley plants. *Brassicajuncea* L. showed a reduction in chlorophyll and carotenoid content under the toxicity of Cd and Pb. Pb was reported to be more detrimental in comparison to Cd (John et al., 2009). In the case of *Pisumsativum*, Cu affects the photosynthetic process at higher concentrations (Hattab et al., 2009). Energy migration from the antenna complexes to the reaction center is disturbed on metal exposure leading to increased chlorophyll fluorescence. Noticeable changes occurred in fluorescence kinetics of Chlorophyll a upon exposure to heavy metal. This chlorophyll fluorescence parameter has been regarded as an important information tool to measure the effect of varying degrees exerted by various kinds of heavy metal on photosynthetic systems (Clijsters and Van Assche, 1985; Stiborova et al., 1986; Joshi and Mohanty, 2004).

3.3 Enzymes

Some heavy metal ions are essential for the growth and development of plants as they are required for many enzymatic activities. They are required in very low concentrations however at higher concentrations they affect the plant growth adversely by affecting enzymes involved in photosynthetic activity. Furthermore, the time duration to which plants are exposed to heavy metals also had a major impact on plant growth and development (Sheoran et al., 1990). Heavy metals also inhibit the key enzyme protochlorophyllide reductase responsible for the conversion of protochlorophyll to chlorophyll (De Filippis and Pallaghy, 1994). Cd could form covalent interaction with side groups of proteins such as protochlorophyllide reductase and plastocyanin disrupting both structure and functional role of these enzymes. Heavy metal ions alter the functioning of photosynthetic enzymes by interacting with the SH groups. Cadmium interacted with the thiol group of ALA synthase thus inhibiting the synthesis of the ALA (delta-aminolevulinic acid), a precursor of tetrapyrrole for the synthesis of the chlorophyll molecule, in dark-grown bean seedling (Padmaja et al., 1990). ALA is converted into porphobilinogen by ALA dehydratase (ALAD), a metal sensitive regulatory enzyme. The toxicity of metals Pb, Se and Hg on ALAD was investigated in mung bean and bajra and suggested that these metals interacted with the SH group at the active site of ALAD causing increased activity of ALAD in a treated plant as compared to control (Prasad and Prasad, 1987). Cu, Zn, and Hg combine with the SH group of NADP oxido reductase of PSI thereby altering the activity of the enzyme (Lucero et al., 1976; De Filippis et al., 1981; Clijsters and Van Assche, 1985). Heavy metals such as Zn and Cd induce disturbances in photosynthesis by affecting activities of the Calvin cycle enzyme. In dark-grown saplings Cu ions targets Chl molecules present in light harvesting complex II (Clijsters and Van Assche, 1985; Drzewiecka-Matuszek et al., 2005; Nonomura et al., 1997; Petrovic et al., 2006; Kupper et al., 2006). In various investigations, heavy metals bind with the key enzymes of CO_2 fixation such as RuBPcase and PEP-carboxylase and modify the catalytic activity (Eicchan et al., 1969; Ernst, 1980; Clijsters and Van Assche, 1985; Stilborova and Leblora, 1985;

Malik, 1989; Sheoran et al., 1990). Heavy metals such as Zn substitute metals from the metalloproteins resulting in decreased RuBPcase/oxygenase ratio (Wilder and Henkel, 1979). *In vitro* studies showed the substitution of Mg^{+2} by Co^{+2} or Mn^{+2} and Ni caused decreased RuBPcase/oxygenase activity ratio (Wilder and Henkel, 1979). This changed ratio could be manifested with increased CO_2 compensation. The activity of most of the enzymes of the Calvin cycle such as RuBPcase, 3-PGA kinase, aldolase and NADP-glyceraldehyde phosphate dehydrogenase were seen to be reduced to a higher extent with the application of 10 mM Cd^{+2}. Burzynski and Zurek et al., 2007 reported that treatment of cucumber cotyledons with Cu and Cd caused inhibition of PGK (Phosphoglyceric acid kinase) and GAPDH (Glyceraldehyde 3-phosphate dehydrogenase). Similar results were also noticed in the pigeon pea and maize leaves (Sheoran et al., 1990; Stilborova et al., 1986).

3.4 Photosystem I and Photosystem II

Photosynthetic carbon fixation is carried out with the aid of metal ions such as Mn, Mg, Cu, Fe, and Ca. These metal ions regulate both structural and functional components of photosynthesis. Assimilatory power (NADPH + ATP) for carbon fixation in photosynthesis is generated by electron flow through PSI and PSII in a light reaction. Electrons are derived from the oxidation of water molecules to generate oxygen. This electron flow through the photosystems generates a proton gradient across the thylakoid membrane that drives the formation of ATP. Various investigations revealed that PSII is more sensitive to heavy metal exposure in comparison to PSI. It may be due to the more complexity of PSII as an oxygen evolving system (which play a crucial role to generate electrons as well as hydrogen ion and maintain an electrochemical gradient) preferred as a target site for the binding of various heavy metals (Fodor, 2013). PSII, a multi-subunit protein complex consisting of 20 subunits, present in the appressed part of grana lamellae, becomes active at a wavelength shorter than 680 causing the release of electrons and molecular oxygen. It involves the reduction of plastoquinone by the electrons ejected from the oxidation of water into molecular oxygen (Andersson and Styring, 1991; Vermaas et al., 1993). Both reducing and oxidizing sides of PSII is directly or indirectly affected by heavy metal. Interaction of Pb^{+2}, Hg^{+2}, and Cd^{+2} with light harvesting complex (LHC II) was observed by using Fourier Transform Infrared (FTIR) in spinach (Ahmed and Tajmir-Riahi, 1993). Cd^{+2} deformed the light harvesting protein complex Chla/b in radish cotyledons (Krupa et al., 1988). Different target sites of various heavy metals have been shown in Fig. 2.

Studies carried out on isolated chloroplast of Maize showed that Cd^{+2} toxicity occurred between primary electron acceptor and $NADP^+$ in PSI (Siedlecka and Baszynski, 1993). Cd affected the light phase by inducing Fe deficiency. Site of action for Hg^{+2} and Cu^{+2} was found at ferredoxin that interrupts electron flow from ferredoxin to $NADP^+$ as seen in *Chlorella vulgaris* (Šeršeň and Kráľova, 1996; Šeršeň et al., 2013). Hg interacted with plastocyanin to interrupt PSI (Mysliwa-Kurdziel et al., 2004). Zn and Cd-induced effects in the light-dependent photosynthetic processes have been studied in both *in vitro* and *in vivo* conditions (Van Assche et al., 1986; Vassilev et al., 1999; Kalaji et al., 2007; Sagardoy et al., 2009). In *in vitro*

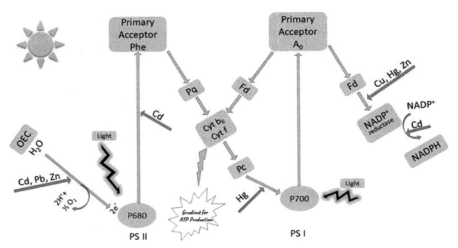

Figure 2. Photosystem I and II showing target sites for different heavy metals.

studies, it was established that both metals can significantly decrease the activities of PSII and, to a lesser extent, also of PSI as well as the rate of photosynthetic electron transport (Krupa, 1999; Vassilev et al., 2004). *In vitro* studies with isolated chloroplast in the presence of various heavy metals such as Pb, Hg, Cd, Zn, Cu, Ni, and Co indicated that the toxic impact of heavy metals is more pronounced on PSII in comparison to PSI and dark carbon fixing enzymatic reactions (Hampp et al., 1976; Tripathy and Mohanty, 1980; Krupa et al., 1987; Mohanty et al., 1989; Bazzaz and Govindjee, 1974; Li and Miles, 1975; Tripathy and Mohanty, 1980; Tripathy et al., 1981, 1983). Zn and Cd mediated inhibition of PSII have been studied in *Euglenagracilis* by tracing the artificial electron transport dye (De Filippis et al., 1981). Pb affects PSII more as compared to PSI (Joshi and Mohanty, 2004).

Zn interacts with the donor side of PSII (Prasad and Stezalka, 1999). It was observed that Fe in excess can reduce the photosynthesis by 40% in *Nicotiana plumbaginifolia* as it takes part in oxidation reduction reaction causing oxidative stress resulting in increased activity of oxidative enzymes (Kampfenkel et al., 1995; Keunen et al., 2011). Zn competes with Mn and substitutes it in water splitting in PSII. Pb also competes with Ca and Cl for binding site in the water splitting complex and causing conformational changes in extrinsic polypeptides of the thylakoid membrane (Rashid et al., 1994). Cd can also bind to both oxidizing and reducing sites of PSII (Sigfridsson et al., 2004). Chlorophyll fluorescence analysis associated with JIP test provides better information on the structure and functioning of photosynthetic apparatus with the energy fluxes between the complexes of photosystem II (Strasser et al., 1978, 1981). In Spinach, Cu inhibited the flow of electrons in both the photosystems PSI and PSII by disrupting the thylakoid membrane (Capsi et al., 1999). Cd binds the Ca^{+2} cofactor of the oxygen evolving center thus inhibiting oxygen generation by the oxygen evolving center (Faller et al., 2005). Heavy metals interfere with oxygen evolution by quenching fluorescence (Hsu et al., 1988; Arellano et al., 1995). Most of the time, oxygen evolving activity could be restored by washing the isolated chloroplast but this is not seen with the heavy metal ions Ni. Even the exogenous application of electron donor $MnCl_2$, benzidine, or NH_2OH could not restore the PSII activity in the chloroplast (Tripathy et al., 1981, 1983).

References

Ahmed A, and Tajmir-Riahi HA. 1993. Interaction of toxic metal ions Cd^{+2}, Hg^{+2} and Pb^{+2} with light harvesting proteins of chloroplast thylakoid membranes. An FITR spectroscopic study. *J Inorg Biochem* 50: 235–243.

Andersson B, and Styring S. 1991. Photosystem II: molecular organization, function, and acclimation. *Current Topic in Bioenergetics* 16: 1–81.

André O, Vollenweider P, and Günthardt-Goerg MS. 2006. Foliage response to heavy metal contamination in Sycamore Maple (*Acer pseudoplatanus* L.). *For Snow Landsc Res* 80, 3: 275–288.

Arellano JB, Lazaro JJ, Lopez-Gorge J, and Baron M. 1995. The donor side of Photosystem II as the copper-inhibitory binding site. *Photosynthetic Research* 50: 698–701.

Barceló J, Vazquez MD, and Poschendrieuder CH. 1988. Structural and ultrastructural disorders in cadmium treated bush bean plants (*Phaseolus vulgaris* L.). *New Phytol* 108: 31–49.

Barker AV, and Pilbeam DJ. 2007. Handbook of Plant Nutrition. Boca Raton, FL: Taylor and Francis.

Barone LM, Shih C, and Wasserman BP. 1997. Mercury-induced conformational changes and identification of conserved surface loops in plasma membrane aquaporins from higher plants: topology of PMIP31 from *Beta vulgaris* L. *J Biol Chem* 272: 30672–30677.

Bazihizina N, Colzi I, Giorni E, Mancuso S, and Gonnelli C. 2015. Photosynthesizing on metal excess: copper differently induced changes in various photosynthetic parameters in copper tolerant and sensitive *Silene paradoxa* L. populations. Plant Science. *Amsterdam* 232(2): 67–76.

Bazzaz FA, Rolfe GL, and Carlson RW. 1974a. Effect of cadmium on photosynthesis and transpiration of excised leaves of corn and sunflower. *Plant Physiol* 32: 373–376.

Bazzaz FA, Rolfe GL, and Carlson RW. 1974b. Differing sensitivity of corn and soybean photosynthesis and transpiration to lead contamination. *J Environ Qual* 3: 156–158.

Bazzaz FA, Carlson RW, and Rolfe GL. 1975. Inhibition of corn and sunflower photosynthesis by lead. *Physiol Plant* 34: 326–329.

Bazzaz MB and Govindjee. 1974. Effects of cadmium nitrate on spectral characteristics and light reactions of chloroplaste. *Environ Sci Lett* 6: 1–12.

Biela A, Grote K, Otto B, Hoth S, Hedrich R, and Kaldenhoff R. 1999. The nicotiana tabacum plasma membrane aquaporin NtAQP1 is mercury-insensitive and permeable for Glycerol. *Plant J* 18: 565–570.

Bria, JF, Curie C, and Gaymard F. 2007. Iron utilization and metabolism in plants. *Curr Opin Plant Biol* 10: 276–282.

Burzynski M, and Zurek A. 2007. Effects of copper and cadmium on photosynthesis in cucumber cotyledons. *Photosynthetica* 45: 239–244.

Capsi V, Droppa M, Horváth G, Malkin S, Marder JB, and Raskin VI. 1999. The effect of copper on chlorophyll organization during greening of barley leaves. *Photosynth Res* 62: 165–174.

Chong-Qing Jiao and Hong-Zhao Zhu. 2013. Resonance suppression and electromagnetic shielding effectiveness improvement of an apertured rectangular cavity by using wall losses. *Chinese Physics B* 22: 084101.

Clarkson DT, Carvajal M, Henzler T, Waterhouse RN, Smyth AJ, Cooke DT, and Steudle E. 2000. Root hydraulic conductance: diurnal aquaporin expression and the effects of nutrient stress. *J Exp Bot* 51: 61–70.

Clijsters H and Van Assche F. 1985. Inhibition of photosynthesis by heavy metals. *Photosynth Res* 7: 31–40.

Csatorday K, Gombos Z, and Szalontai C. 1984. Mn^{+2} and Co^{+2} toxicity in chlorophyll biosynthesis. *Proc Natl Acad Sci USA* 81: 476–478.

Cuypers A, Vangronsve J, and Clijsters H. 2001. The redox status of plant cells (AsA and GSH) is sensitive to zinc imposed oxidative stress in roots and primary leaves of *Phaseolus vulgaris*. *Plant Physiol Biochem* 39: 657–664.

Dal Corso G, Manara A, and Furini A. 2013a. An overview of heavy metal challenge in plants: from roots to shoots. *Metallomics* 5(9): 1117–1132.

Dalvi A, and Bhalerao SA. 2013. Response of plants towards heavy metal toxicity: an overview of avoidance, tolerance and uptake mechanism. *Annals of Plant Sciences* 2(9): 362–368.

De Filippis LF, and Pallaghy CK. 1994. Heavy metals: Sources and biological effects. pp. 31–77. *In*: Rai LC, Gaur JP, and Soeder CJ (eds.). *Algae and Water Pollution*. E. Schweizerbart'sche, Verlagsbuchhandlung, Stuttgart.

De Filippis LF, Hampp R, and Ziegler H. 1981a. The effects of sublethal concentrations of zinc, cadmium and mercury on Euglena. Growth and pigments. *Z Pflanzenphysiol* 101: 37–47.

De Filippis LF, Hampp R, and Ziegler H. 1981. The effect of sublethal concentrations of zinc, cadmium and mercury on euglena. Adenylates and energy charge. *Zeitschrift für Pflanzenphysiologie* 103(1): 1–7.

Devi SR and Prasad MNV. 2005. Antioxidant capacity of *Brassica juncea* plants exposed to elevated levels of copper. *Russian Journal of Plant Physiology* 52: 246–273.

Drzewiecka-Matuszek A, Skalna A, Karocki A, Stochel G, and Fiedor L. 2005. Effects of heavy central metal on the ground and excited states of chlorophyll. *J Biol Inorg Chem* 10: 453–462.

Eicchan GL, Clark P, and Tarian E. 1969. The interaction of metal ion with polynucleotides and related compounds. *J Biol Chem* 244: 937–942.

Elhiti M, Yang C, Chan A, Durnin DC, Belmonte MF, Ayele BT, Tahir M, and Stasolla C. 2012. Altered seed oil and glucosinolate levels in transgenic plants overexpressing the *Brassica napus* Shootmeristemless gene. *J Exp Bot* 63: 4447–4461.

Ernst WHO. 1980 Biochemical aspects of cadmium in plants. pp. 639–653. *In*: Nriagu JO (ed.). *Cadmium in the Environment Part I*, 10 hn Wiley & Sons, New York.

Ernst WHO, Krauss GJ, Verkleij JAC, and Wesenberg D. 2008. Interaction of heavy metal with the sulphur metabolism in angiosperms from an ecological point of view. *Plant Cell Environ* 31: 123–143.

Faller P, Kienzler K, and Krieger-Liszkay A. 2005. Mechanism of Cd+2 toxicity: Cd+2 inhibits photoactivation of Photosystem II by competitive binding to the essential Ca+2 site. *Biochim Biophy Acta Bioenerg* 1706: 158–164.

Fodor E. 2013. The RNA polymerase of influenza a virus: mechanisms of viral transcription and replication. *Acta Virol* 57(2): 113–22.

Foreman J, Demidchik V, Bothwell JHF, Mylona P, Miedema H, Torres MA, Linstead P, Costa S, Brownlee C, Jones JDG, Davies JM, and Dolan L. 2003. Reactive oxygen species produced by NADPH oxidase regulate plant cell growth. *Nature* 422(6930): 442–446.

Franco E, Alessandrelli S, Masojidek J, Margonelli A, and Giardi MT. 1999. Modulation of D1 protein turnover under cadmium and heat stresses monitored by [35S] methionine incorporation. *Plant Sci* 144: 53–61.

Gallego SM, Benavides MP, and Tomar ML. 1996. Effect of heavy metal ion excess on sunflower leaves: evidence for involvment of oxidative stress. *Plant Science* 121: 151–159.

Garzón MB, Alía R, Matthew Robson T, and Zavala MA. 2011. Intra-specific variability and plasticity influence potential tree species distributions under climate change. *Global Ecology and Biogeography* 20: 766–778.

Gomes MP, Nogueira MOG, Castro EM, and Soares ÂM. 2011. Ecophysiological and anatomical changes due to uptake and accumulation of heavy metal in *Brachiaria decumbens*. *Scientia Agricola* 68(5): 566–573.

Greger M, and Johansson M. 1992. Cadmium effects on leaf transpiration of sugar beet (*Beta vulgaris*) *Physiologia Plantarum* 86: 465–473.

Hampp R, Beulich K, and Zeigler H. 1976. Effects of zinc and cadmium on photosynthetic CO2 fixation and Hill activity of isolated spinach chloroplasts. *Z Pflanzenphysiol* 77: 336–344.

Hattab S, Dridi B, Chouba L, Kheder MB, and Bousetta H. 2009. Photosynthesis and growth responses of *Pisum sativum* L. under heavy metal stress. *J Env Sci* 2: 1552–1556.

Hayat S, Alyemeni MN, and Hasan SA. 2012. Foliar spray of brassinosteroid enhances yield and quality of *Solanum lycopersicum* under cadmium stress. *Saudi Journal of Biological Sciences* 19: 325–335.

Hsu BD, and Lee JY. 1988. Toxic effect of copper on Photosystem II of spinach chloroplasts. *Plant Physiology* 87: 116–119.

Huang CY, Bazzaz FA, and Vanderhoeff LN. 1974. The inhibition of soybean metabolism by cadmium and lead. *Plant Physiol* 54: 122–124.

Huang RF, Zhu MJ, Kang Y, Chen J, and Wang XC. 2002. Identification of plasma membrane aquaporin in guard cells of *Vicia faba* and its role in stomatal movement. *Acta bot Sinica* 44: 42–48.

Javot H and Maurel C. 2002. The role of aquaporins in root water uptake. *Ann Bot* 90(3): 301–313.

Johansson I, Larsson C, Ek B, and Kjellbom P. 1996. The major integral proteins of spinach leaf plasma membrane are putative aquaporins and are phosphorylated in response to Ca^{+2} and apoplastic Water Potential. *Plant Cell* 8: 1181–1191.

Johansson I, Karlsson M, Shukla VK, Chrispeels MJ, Larsson C, and Kjellbom P. 1998. Water transport activity of the plasma membrane aquaporin PM28A is regulated by phosphorylation. *Plant Cell* 10: 451–459.

John R, Ahmad P, Gadgil K, and Sharma S. 2009. Heavy metal toxicity: effect on plant growth, biochemical parameters and metal accumulation by *Brassica juncea* L. *Int J Plant Prod* 3: 65–75.

Joshi MK and Mohanty P. 2004. Chlorophyll a fluorescence as a probe of heavy metal ion toxicity in plants. pp. 637–661. *In*: Papageorgiou GC (ed.). *Chlorophyll a Fluorescence: A Signature of Photosynthesis, Advances in Photosynthesis and Respiration.* Springer, Dordrecht.

Kabata-Pendias A, and Pendias H. 2001. Trace Elements in Soils and Plants. CRC Press, Washington, D.C.

Kalaji HM, and Loboda T. 2007. Photosystem II of barley seedlings under cadmium and lead stress. *Soil Environ* 53: 511–516.

Kampfenkel K, Montagu MV, and Inze D. 1995. Effects of iron excess on *Nicotiana plumbaginifolia* plants (Implications to oxidative stress). *Plant Physiol* 107: 725–735.

Kasim WA. 2007. Physiological consequences of structural and ultra-structural changes induced by Zn stress in *Phaseolus vulgaris*. I. Growth and photosynthetic apparatus. *Int J Bot* 3(1): 15–22.

Kastori R, Petrovic M, and Petrovic N. 1992. Effect of excess lead, cadmium, copper, and zinc on water relations in sunflower. *Journal of Plant Nutrition* 15(11): 2427–2439.

Keunen E, Remans T, Bohler S, Vangronsveld J, and Cuypers A. 2011. Metal-induced oxidative stress and plant mitochondria. *Int J Mol Sci* 12: 6894–6918.

Krupa Z, Skórzynska E, Maksymiec W, and Baszynski T. 1987. Effect of cadmium treatment on the photosynthetic apparatus and its photochemical activities in greening radish seedlings. *Photosynthetica* 21: 156–164.

Krupa Z, Skorzynska E, Maksymiec W, and Baszynski T. 1988. Effect of cadmium treatment on the photosynthetic apparatus and its photochemical activities in greening radish seedlings. *Photosynthetica* 21: 156–164.

Krupa Z. 1999. Cadmium against higher plant photosynthesis—a variety of effects and where do they possibly come from? *Z Naturforsch* 54c: 723–729.

Kupper H, Kupper FC, and Spiller M. 1996. Environmental relevance of heavy metal-substituted chlorophylls using the example of water planta. *J Exp Bot* 47: 259–266.

Kupper H, Setlik I, Setlikova E, Ferimazova N, Spiller M, and Kupper FC. 2003. Copper-induced inhibition of photosynthesis: Limiting steps of *in vivo* copper chlorophyll formation in *Scenedesmus quadricauda. Functional Plant Biology* 30: 1187–1196.

Kupper H, Kupper FC, and Spiller M. 2006. [Heavy metal]-chlorophylls formed *in vivo* during heavy metal stress and degradation products formed during digestion, extraction and storage of plant material. pp. 67–77. *In*: Grimm B, Porra RJ, Rüdiger W, and Scheer H (eds.). *Chlorophylls and Bacteriochlorophylls: Biochemistry, Biophysics, Functions and Applications.* Springer, Amsterdam.

Kwak JM, Nguyen V, and Schroeder JI. 2006. The role of reactive oxygen species in hormonal responses. *Plant Physiol* 141(2): 323–329.

Lamoreaux RJ, and Chaney WR. 1978. The effect of cadmium on net photosynthesis, transpiration, and dark respiration of excised silver maple leaves. *Physiologia Plantarum* 43(3): 231–236.

Li EH and Miles CD. 1975. Effects of cadmium on photoreaction II of chloroplasts. *Plant Sci Lett* 5: 33–40.

Lu Z and Neumann PM. 1999. Water stress inhibits hydraulic conductance and leaf growth in rice seedlings but not the transport of water via mercury-sensitive water channels in the root. *Plant Physiol* 120: 143–151.

Lucero HA, Andreo CS, and Vallejos RH. 1976. Sulphydryl groups in photosynthetic energy conservation III. Inhibition of photophosphorylation in spinach chloroplasts by CdCI. *Plant Sci Lett* 6: 309–313.

Malik D. 1989. Effect of cadmium on photosynthetic efficiency of wheat (*Triticum aestivum* L.) Ph. D. Thesis, Haryana Agricultural University, Hisar.

Malik D, Sheoran IS, and Singh R. 1992a. Lipid composition of thylakoid membranes of cadmium treated wheat seedlings. *Indian J Biochem Biophys* 29: 350–354.

Malik D, Sheoran IS, and Singh R. 1992b. Carbon metabolism in leaves of cadmium treated wheat seedlings. *Plant Physiol Biochem* 30: 223–229.

Marschner H. 1995. Mineral nutrition of higher plants. Second edition. London: Academic Press, 889.

Maurel C, Kado RT, Guern J, and Chrispeels MJ. 1995. Phosphorylation regulates the water channel activity of the seed-specific Aquaporin alpha-TIP. *EMBO J* 14: 3028–3035.

McGrath S, Lombi E, and Zhao FJ. 2001. What's new about cadmium hyperaccumulation? *New Phytol* 149: 2–3.

Melo HC, Castro EM, Soares AM, Melo LA, and Alves JD. 2007. Anatomical and physiological alterations in *Setaria anceps* Stapf ex Massey and *Paspalum paniculatum* L. under water deficit conditions. Hoehnea 34: 145–153 (in Portuguese, with abstract in English).

Mittal N, and Srivastava AK. 2014. Analyses for Multiple Tolerance for Cadmium and Hexavalent Chromium Combination in Barley and Safflower. Ph.D. thesis, C. C. S. University, Meerut, India.

Mohanty N, Vass I, and Demeter S. 1989. Impairment of photosystem 2 activity at the level of secondary quinone electron acceptor in chloroplasts treated with cobalt, nickel and zinc ions. *Physiol Plant* 76: 386–390.

Mysliwa-Kurdzielt B, Prasad MNV, and Slrzatka K. 2004. Photosynthesis in heavy metal stressed plants. *Heavy Metal Stress in Plants*, 146–181.

Najeeb U, Jilani G, Ali S, Sarwar M, Xu L, and Zhou W. 2011. Insights into cadmium induced physiological and ultra-structural disorders in *Juncus effusus* L., and its remediation through exogenous citric acid. *Journal of Hazardous Materials* 186: 565–574.

Nonomura Y, Igarashi S, Yoshioka N, and Inoue H. 1997. Spectroscopic properties of chlorophylls and their derivatives: Influence of molecular structure on the electronic state. *Chem Phys* 220: 155–166.

Ouzounidou G. 1997. Sites of copper in the photosynthetic apparatus of maize leaves: kinetic analysis of chlorophyll fluorescence, oxygen evolution, absorption changes and thermal dissipation as monitored by photoacoustic signals. *Aust J Plant Physiol* 24: 81–90.

Padmaja K, Prasad DDK, and Prasad ARK. 1990. Inhibition of chlorophyll synthesis in *Phaseolus vulgaris* L. seedlings by cadmium acetate. *Photosynthetica* 24: 399–405.

Patra M, Bhowmik N, Bandopadhyay B, and Sharma A. 2004. Comparison of mercury, lead and arsenic with respect to genotoxic effects on plant systems and the development of genetic tolerance. *Environmental and Experimental Botany* 52(3): 199–223.

Petrovic J, Nikolic G, and Markovic D. 2006. *In vitro* complexes of copper and zinc with chlorophyll. *J Serb Chem Soc* 71: 501–512.

Poorter H, Niinemets U, Poorter L, Wright IJ, and Villar R. 2009a. Causes and consequences of variation in leaf mass per area (LMA): a meta-analysis. *New Phytologist* 182: 565–588.

Prasad AS. 2012. Discovery of human zinc deficiency: 50 years later. *J Trace Elem Med Biol* 26: 66–69.

Prasad DDK, and Prasad ARK. 1987. Altered delta-aminolevulinic acid metabolism by lead and mercury in germinating seedlings of bajra (*Pennisetum typhoideum*). *J Plant Physiol* 127: 241–249.

Prasad MNV, and Strzalka K. 1999. Impact of heavy metals on photosynthesis. pp. 117–138. *In*: Prasad MNV J, and Hagemeyer J (eds.). *Heavy Metal Stress in Plants: From Molecules to Ecosystems*. Springer, Berlin.

Rashid A, Camin EL, and Ekramoddoulah AKM. 1994. Molecular mechanism of action of Pb^{+2} and Zn^{+2} on water oxidizing complex of photosystem II. *FEBS Lett* 350: 296–298.

Remans K, Pauwels K, Ulsen PV, Buts L, Cornelis P, Tommassen J, Savvides SN, Decanniere K, and Gelder PV. 2010. Hydrophobic surface patches on LolA of *Pseudomonas aeruginosa* are essential for lipoprotein binding. *J Mol Biol* 401: 921–930.

Ricachenevsky FK, Menguer PK, Sperotto RA, Williams LE, and Fett JP. 2013. Roles of plant metal tolerance proteins (MTP) in metal storage and potential use in biofortification strategies. *Plant Physiology* 4: 00144.

Sagardoy R, Morales F, López-Millán AF, Abadía A, and Abadía J. 2009. Effects of zinc toxicity on sugar beet (*Beta vulgaris* L.) plants grown in hydroponic. *Plant Biol* 11: 339–350.

Sagardoy R, Vázquez S, Florez-Sarasa ID, Albacete A, Ribas-Carbó M, Flexas J, Abadía J, and Morales F. 2010. Stomatal and mesophyll conductances to CO2 are the main limitations to photosynthesis in sugar beet (*Beta vulgaris*) plants grown with excess zinc. *New Phytol* 187(1): 145–158.

Scafaro AP, Caemmerer SV, Evans JR, and Atwell BJ. 2011. Temperature response of mesophyll conductance in cultivated and wild Oryza species with contrasting mesophyll cell wall thickness. *Plant, Cell and Environment* 34: 1999–2008.

Šeršeň F, and Kráľova K. 1996. Concentration-dependent inhibitory and stimulating effects of amphiphilic ammonium salts upon photosynthetic activity of spinach chloroplasts. *Gen Physiol Biophys* 15: 27–36.

Šeršeň F, and Kráľova K. 2013. EPR spectroscopy—a valuable tool to study photosynthesizing organisms exposed to abiotic stresses. *Photosynthesis*, 247–283.

Shahid M, and Fatma Hussain F. 2012. Chemical composition and mineral contents of *Zingiber officinale* and *Alpinia allughas* (*Zingiberaceae*) Rhizomes. *International Journal of Chemical and Biochemical Sciences* 2: 101–104.

Shahid M, Pinelli E, and Dumat C. 2012. Review of Pb availability and toxicity to plants in relation with metal speciation; role of synthetic and natural organic ligands. *J Hazard Mater* 219: 1–12.

Sharma SS and Dietz KJ. 2009. The relationship between metal toxicity and cellular redox balance. *Trends in Plant Science* 14(1): 43–50.

Sheoran IS, Singal HR, and Singh R. 1990. Effect of cadmium and nickel on photosynthesis and the enzymes of the photosynthetic carbon reduction cycle in pigeon pea (*Cajanus cajan*). *Photosynth Res* 23: 345–351.

Siedlecka A, and Baszynski T. 1993. Inhibition of electron flow around photosystem I in chloroplasts of Cd-treated maize plants is due to Cd-induced iron deficiency. *Physiol Plant* 87: 199–202.

Sigfridsson KGV, Bernát G, Mamedov F, and Styring S. 2004. Molecular interference of Cd^{+2} with photosystem II. *Biochim Biophys Acta* 1659: 19–31.

Srighar BBM, Diehl SV, Han FX, Monts DL, and Su Y. 2005. Anatomical changes due to uptake and accumulation of Zn and Cd in Indian mustard (*Brassica juncea*). *Environmental and Experimental Botany* 54: 131–141.

Steudle E. 2000. Water uptake by Roots: Effects of Water Deficit 51(350): 1531–1542.

Stiborova M, Doubravova M, Brezinova A, and Friedrich A. 1986. Effect of heavy metal ions on growth and biochemical characteristics of photosynthesis of barley (*Hordeum vulgare* L.). *Photosynthetica* 20: 418–425.

Stiborová M, Doubravová M, and Leblová S. 1986. A comparative study of the effect of heavy metal ions on ribulose-1,5-bisphosphate carboxylase and phosphoenolpyruvate carboxylase. *Biochem Physiol Pflanzen* 181(6): 373–379.

Stilborova M, and Leblora S. 1985. Heavy metal inactivation of maize PEP-carboxylase isoenzyme. *Photosynthetica* 19: 500–503.

Strasser RJ. 1978. The grouping model of plant photosynthesis. pp. 513–524. *In*: Akoyunoglou G (ed.). *Chloroplast Development*. Amsterdam: Elsevier.

Strasser RJ. 1981. The grouping model of plant photosynthesis: heterogeneity of photosynthetic units in thylakoids photosynthesis III. pp. 727–737. *In*: Akoyunoglou G (ed.). *Structure and Molecular Organisation of the Photosynthetic Apparatus*. Philadelphia: Balaban Int. Sci. Services.

Sun MH, Xu W, Zhu YF, Su W, and Tang ZC. 2001. A simple method for *in situ* hybridization to RNA in guard cells of *Vicia faba* L.: the expression of aquaporins in guard cells. *Plant Mol Biol Rep* 19: 129–135.

Terashima I, Hanba YT, Danny Tholen, and Niinemets U. 2010. Leaf functional anatomy in relation to photosynthesis. *American Society of Plant Biologists* DOI: https://doi.org/10.1104/pp.110.165472.

Tran L, Nunan L, Redman RM, Mohney LL, Pantoja CR, Fitzsimmons K, and Lightner DV. 2013. Determination of the infectious nature of the agent of acute hepatopancreatic necrosis syndrome affecting penaeid shrimp. *Dis Aquat Organ* 105: 45–55.

Tripathi BN, Mehta SK, Amar A, and Guar JP. 2006. Oxidative stress in *Scenedesmus* sp. during short- and long-term exposure to Cu^{+2} and Zn^{+2}. *Chemosphere* 62: 538–544.

Tripathy BC, Bhatia B, and Mohanty P. 1981. Inactivation of chloroplast photosynthetic electron transport activity by Ni'. *Biochim Biophys Acta* 638: 217–224.

Tripathy BC, Bhatia B, and Mohanty P. 1983. Cobalt ions inhibit electron transport activity of photosystem II without affecting photosystem I. *Biochim Biophys Acta* 722: 88–93.

Tripathy BC and Mohanthy P. 1980. Zinc inhibited electron transport of photosynthesis in isolated barley chloroplasts. *Plant Physiol* 66: 1174.

Tsukagoshi H, Busch W, and Benfey PN. 2010. Transcriptional regulation of ROS controls transition from proliferation to differentiation in the root. *Cell* 12; 143(4): 606–16.

Van Assche FV and Clijsters H. 1986. Inhibition of photosynthesis by treatment of *Phaseolus vulgaris* with toxic concentration of zinc: Effects on electron transport and photophosphorylation. *Physiol Plant* 66: 717–721.

Vassilev A and Manolov P. 1999. Chlorophyll fluorescence of barley (*H. vulgare* L.) seedlings grown in excess of Cd. *Bulg J Plant Physiol* 25: 67–76.

Vassilev A, Lidon FC, Matos MD, Ramalho JC, and Bareiro MG. 2004. Shoot cadmium accumulation and photosynthetic performance of barley at high Cd treatments. *J Plant Nutr* 27: 773–793.

Vassilev A, Nikolova A, Koleva L, and Lidon F. 2011. Effects of excess Zn on growth and photosynthetic performance of young bean plants. *J Phytol* 3: 58–62.

Vermaas WFJ, Styring S, Schroder WP, and Andersson B. 1993. Photosynthetic water oxidation: the protein framework. *Photsynthetic Reseach* 38: 249–63.

Wang W, Wang S, Ma X, and Gong J. 2011. Recent advances in catalytichydrogenation of carbon dioxide. *Chemical Society Review*, 7.

Wilder GF, and Henkel J. 1979. The effect of divalent ions on the activity of Mif+ -depleted ribulose-1,5-bisphosphate oxygenase. *Planta* 146: 223–228.

Williams LE, and Pittman JK. 2010. Dissecting pathways involved in manganese homeostasis and stress in higher plants. pp. 95–117. *In*: Hell, R and Mendal RR (eds.). *Cell Biology of Metals and Nutrients, Plant Cell Monographs*, Vol. 17 (Berlin, Heidelberg: Springer-Verlag).

Woźny A, Schneider J, and Gwóźdź EA. 1995. The effects of lead and kinetin on greening barley leaves. *Biologia Plantarum* 37(4): 541–552.

Xin L, Hong-ying H, Ke G, and Ying-xue S. 2010. Effects of different nitrogen and phosphorus concentrations on the growth, nutrient uptake, and lipid accumulation of a freshwater microalga *Scenedesmus* sp. *Bioresource Technology* 101(14): 5494–5500.

Yang HM, Li Y, and Wang GX. 2002. Functions and roles of the channels in broad bean stomatal movements. *Acta Phytoecol Sinica* 26: 656–660.

Zengin FK, and Munzuroglu O. 2005. Effects of some heavy metals on content of chlorophyll, proline and some antioxidant chemicals in bean (*Phaseolus vulgaris* L.) seedlings. *Acta Biologica Cracoviensia Series Botanica* 47: 157–164.

Zhang WH, and Tyerman SD. 1999. Inhibition of water channels by HgCl2 in intact wheat root cells. *Plant Physiol* 120: 849–857.

Zhao FJ, Lombi E, Breedon T, and McGrath SP. 2000. Zinc hyperaccumulation and cellular distribution in *Arabidopsis halleri*. *Plant Cell and Environmental* 23: 507–514.

Zhong L, Lai CY, Shi LD, Wang KD, Yu-Jie Dai YJ, Liu YW, Ma F, Rittmann BE, Zheng P, and Zhao HP. 2017. Nitrate effects on chromate reduction in a methane-based biofilm. *Water Research* 115: 130–137.

9

A Mechanistic Overview of Heavy Metal Detoxification in Plants

Kavita Ghosal,[1] *Dwaipayan Sinha*[2,]* and *Satendra Pal Singh*[3]

ABSTRACT

Heavy metals (HMs) have been contaminating the earth much before the emergence of humans through natural geological activities. In the 17th and 18th centuries, the industrial revolution, followed by the rapid development of industries across the globe has fuelled the anthropogenic cause of heavy metal (HM) pollution. Presently HM pollution is a menace and is affecting the lives of all the biotic components of the ecosystem. Plants, being static are far more prone to HM pollution as compared to animals. They have thus devised several adaptive strategies to counter the pollution of HMs. There are several strategies for countering the metal contamination issue by the plant. Some plants tend to avoid metal contamination by attenuating the entry in their body through some protective morphological adaptations while others effectively compartmentalize the uptaken HMs in the intracellular organelles so that there is minimal toxicity response. Other strategies involve some biological chelators which make the HMs inactive both from a movement and chemical point of view. The compartmentalization involves the role of many proteins which act as carriers of the HMs. In addition to it, under a high concentration of HMs, plants also activate their signal transduction pathways and oxidative stress responses which are also unique in their aspects involving the tuning of micro-RNAs and long non-coding RNAs' regulation in Metallo-stress management. All of these processes result in the detoxification of HMs so that the deleterious effects of HMs are minimized. This

[1] Department of Botany, P.D. Women's College, Jalpaiguri, West Bengal-735101.
[2] Department of Botany, Government General Degree College, Mohanpur, Paschim Medinipur, West Bengal-721436.
[3] Y.D. (PG.) College, Lakhimpur Kheri, Uttar Pradesh-262701.
* Corresponding author: dwaipayansinha@hotmail.com
  ORCID iD: https://orcid.org/0000-0001-7870-8998

chapter is an attempt to illustrate various mechanisms involved in the detoxification process of HMs by the plants. Efforts have also been taken to delve into the molecular mechanisms of detoxification for a clearer representation of the entire process.

1. Introduction

Heavy metals (HMs) are present in the lithosphere and have a density of at least five times more than that of water (H_2O). At present, anthropogenic activities have led to the unplanned release of HMs from their geographical resources, leading to pollution across soil, river, oceans, and the atmosphere (Narendrula-Kotha et al., 2019), and all biological entities are getting affected for decades. An exponential surge of human intervention in numerous agricultural, industrial, technological processes and domestic appliances is causing human exposure to HMs to a higher degree (Bradl et al., 2002). Reported sources of HMs are agricultural, geogenic, pharmaceutical, domestic and industrial discharges, and atmospheric sources (He et al., 2005). Plants, mining, smelters, and other metal-based manufacturing operations are the key regions eliciting HM contamination in nature.

Selected HMs such as iron (Fe), molybdenum (Mo), manganese (Mn), zinc (Zn), and copper (Cu) act as a micronutrient and are required in the physiological process of plants (Rahman et al., 2019). However, most of the HMs cause toxicity to plants once they exceed the threshold levels (Fryzova et al., 2017) and show reduced growth and less productivity in context to fruiting and seeding. They exhibit challenged vitality in respect to the capacity of seed germination and sustainability in HM-infused soil. Toxicity is induced by elements like chromium (Cr), arsenic (As), lead (Pb), cadmium (Cd), antimony (Sb), mercury (Hg) aluminium (Al), and nickel (Ni). Most of them have extensively been reviewed to inspect their adverse impacts on plant growth when present above their verge of limits in the environment.

For millions of years, plants have learned to tolerate HMs in their habitat through a complex mechanistic approach of cellular biology called adaptation in the path of evolution. Different plants have a wide but variable range of tolerance and it differs from plant to plant. In the detoxification process, some plants employ an extracellular mechanism to fling them away from cells or may deposit out of the cell cytosol. Whereas, in intracellular mechanisms, they modulate their genes to produce some metal-chelator proteins which securely sequester metals via interacting and binding with HMs and store them in vacuoles, Golgi-bodies, or in endoplasmic reticulum protecting themselves from the deleterious effect of HMs (Talebi et al., 2019; De Caroli et al., 2020; Yan et al., 2020; Jogawat et al., 2021). Non-coding RNAs play some significant modulation in target gene's expression mainly in reducing or inhibiting at transcription and post-transcriptional modifications level. Metal transporters act on collecting HM ions and taking them back to vacuoles (Angulo-Bejarano et al., 2021), while non-coding RNA at higher HM levels inhibit their respective genes for translation and activation (Gao et al., 2019). Histone modification and methylation at DNA and chromosomal levels are the sophisticated strategies to initiate transcription (Gulli et al., 2018). Modifying at promoter region or surrounding the area by methylation shows resilience in uptaking HM ions into the cell. Heavy metal-mediated generation of free radicals jeopardizes the ion homeostasis of a plant cell by disrupting plasma membrane (PM) (Keyster et al., 2020), lipid peroxidation (Kapoor et al., 2019), and DNA breakage (Gechev and Petrov, 2020). Beyond threshold level, they start instilling in most of the metabolic pathway-specific substrates, oxidizing them rendering collapse or slow shutdown of the plant system.

For that reason, multidimensional research is adopted to find the way out to get rid of the uprising threat in cultivated and wild plant species. Here, this chapter lets readers understand the effects of HM stress in plants at several levels affecting all overproduction and vitality. Researches also shed light on some immediate induction to rescue mutilation by cell-signaling or utilizing the plant's inherent properties to face the HM stress in plants. There are some mechanisms like avoidance, creating defensive barriers collaborating with neighbouring microorganisms, or external applications like introducing chemicals, physical or some transgenic breeds into the challenging environment to counteract stress conditions.

2. Rhozospheric detoxification of heavy metals

Plants colonizing the HM contaminated sites can be of resistant types through the exhibition of adaptive mechanisms towards the stressed condition. On the other hand, some plants are avoiders and prevent the entry of metals into their body. Resistance towards HMs by plants can be achieved either by the process of avoidance or in conjugation with tolerance (Mehes-Smith et al., 2012). Avoidance is the adaptive strategy and enables the plant to limit HM uptake and restrict the onward movement within the plant. It forms the initial defensive barrier that acts outside the cell and involves several mechanisms including root adsorption (Parrotta et al., 2015), metal ion precipitation (Yan et al., 2020), and exclusion of metals (Wei et al., 2005). Upon exposure to HMs, the plants initially attempt to obstruct the same either through root sorption or through modification of metal ions (Yan et al., 2020). In this case, several organic compounds act as HM chelators thereby reducing the bio-accessibility and subsiding the toxicity (Dalvi and Bhalerao et al., 2013; Moe, 2013; Chen et al., 2017). Studies state that HMs induce the secretion of organic acid by the roots. It is reported that in *Amaranthus hypochondriacus*, exposure to Al resulted in oxalate and citrate secretion (Fan et al., 2016). Another report states that citric acid is instrumental in the immobilization of Pb in the soil (Kim et al., 2013). In another experiment, it was found that treatment of ectomycorrhiza infected or non-infected scots pine seedling with Pb and Cd results in increased secretion of oxalate and other low molecular weight organic acids, which immobilize metals under stressed conditions (Johansson et al., 2008). In another experiment, it was found that treatment of *Medicago polymorpha, Poa annua,* and *Malva sylvestris* with Cd, Cu, and Zn induced oxalic, malic, and citric acid production by the roots (Montiel-Rhozas et al., 2016). Similar reports were also obtained when maize plants were administered with Cd and Cu (Dresler et al., 2014). Metabolomic analysis of root exudates of *Sedum alfredii* under Cd stress has been performed. Analysis of exudates with gas chromatography-mass spectrometry (GC-MS) revealed the existence of compounds responsible for stabilization (erythritol, trehalose, d-pinitol, naphthalene) and mobilization (oxalic acid, tetradecanoic acid, threonic acid, glycine, and phosphoric acid) of Cd (Luo et al., 2014). From a mechanistic point of view, the root exudates induce rhizospheric pH alteration resulting in the precipitation of HMs. This limits the bioavailability and lessens the toxicity of the HMs (Dalvi and Bhalerao, 2013). Organic acids as mentioned before are instrumental in HM detoxification. The organic acid can decrease the uptake of HMs. This is because the presence of organic acids in the region surrounding the roots results in the reduction of the bioavailable ionized form of metals along with increased competition with protons for adsorption sites of the root cell walls. The organic acid also forms chelate complexes with HMs, which are less bioavailable, and thus increase more tolerance towards HMs (Osmolovskaya et al., 2018). The detoxification of HMs by organic acids can also be interpreted by comparing the stability constants with organic acid-metal complexes with metal biological ligands. For example, it is observed that Al-organic acid complexes have higher stability constant than Al-ATP complexes and thus the organic acids can precipitate the metal thereby preventing it from binding with ATP (Wu et al., 2018). The mechanism of metal exclusion involves an exclusion barrier between the shoot system and root system thereby limiting the entry of HMs from the soil only to the roots. In this case, upward translocation of HMs is restricted for protecting the aerial parts (Yan et al., 2020). Arbuscular mycorrhizal fungi also restrict the movement of HMs within the plant. Most of the arbuscular mycorrhizal fungi immobilize HMs in the cortical region of the root by attaching with them and preventing upward translocation in the aerial parts (Dhalaria et al., 2020). AMFs reduce metal stress by immobilizing them in their structure. They also precipitate and chelate the HMs in the rhizosphere, sequester in the vacuoles, and induce antioxidant defense mechanisms in the plant (Mishra et al., 2019).

3. Detoxification of heavy metals in the cell wall

Under a high concentration of rhizospheric HM, the cell wall (CW) also forms a layer of fortification. The root CW is equipped with cation binding sites that facilitate the binding of HMs (Szatanik-Kloc et al., 2017). The pectic constituent is largely responsible for the binding of metal cations (Schiewer and Iqbal, 2010; Yu et al., 2020). Cadmium has been reported to bind with pectin components and hemicellulose in the root CW of *Oryza sativa* (Wang et al., 2020). Other reports also state that hemicellulose is also responsible for the HM adsorption of *Arabidopsis* (Yang et al., 2011) and Cd by the roots of *Sedum alfredii* (Guo et al., 2019). The unique architecture of the CW provides a unique site for adsorption and consequent immobilization of HMs (Spain et al., 2021). The metals generally bind to carboxyl groups of polygalacturonic acid molecules (Meychik et al., 2014), history, and sulfhydryl moieties derived from proteins (Lešková et al., 2020). In addition, the callose layers are also impermeable to HMs and prevent their internalization (De Caroli et al., 2020). Mucilage also acts as a protective barrier and prevents entry of HMs into the root (Sharma and Dubey, 2005). The presence of silica in the CW of *Silene vulgaris* also accounts for HM tolerance (Jain et al., 2018).

The CW can provide a defensive structure in the HM avoidance mechanism. For the plants that are HM tolerant, the detoxification mechanism happens in the region of the plant body that is inside the cell. In this case, the CWs of the plants act as a region to perceive and respond to HM stress. The wall-associated kinases (WAKs) are the receptors present on the CW (Kohorn and Kohors, 2012) and are associated with the perception of biotic (Amsbury, 2020) and abiotic disturbances (Zhang et al., 2020) outside the cell and consequently transduce them into the interior of the cell (Delteil et al., 2016). They are the only known proteins that link the CW and PM and mediate rapid signal transduction response (Li et al., 2020). During HM stress, the transporters of the PM are instrumental in the uptake of ions and consequent detoxification through several processes. Table 1 represents the different types of transporters associated with the transportation of HMs.

Table 1: Illustration of selected transporters involved in the uptake of heavy metals.

S. No.	Transporter name	Nature of Transporter	References
1.	Proton Pumps (ATPase)	ATPase (H^+-ATPase) coupled to ATP hydrolysis for transportation of proton out of the cell	Duby and Boutry (2009)
2.	Natural Resistance-Associated Macrophage Proteins (NRAMPS)	Proton/metal transporters involved in importing protons and divalent metals	Cellier (2012)
3.	Cu transporter family of proteins	Transportation of Cu	Wang et al. (2018)

4. Retention of heavy metals in plant roots

After uptake, it is often observed that the HMs are retained in the root. In this way, the root system through its unique physiological machinery often restricts the upward movement of HMs and thus contributes to the detoxification process. In an experiment, it was observed that 57Co and 109Cd were strongly retained in the cluster roots of *Lupinus albus* cv. *Amiga* (Page et al., 2006). Another experiment reported that 57Co was also preserved in wheat plants instead of getting loaded in xylem (Page and Feller, 2005). Similar reports were also obtained for *Solanum nigrum* in which 57Co was retained in the roots (Wei et al., 2014). According to a new study, Cd accumulates in the roots of *Cicer arietinum*. It was further observed that bioaccumulation coefficients (BAC) and bio-concentration factors (BCF) were higher than the translocation factor (TF) indicating lesser translocation of the HMs through the conducting system (Ullah et al., 2020). According to another recent study, that 69% of Cr is accumulated in rhizomes whereas 18% were distributed in the root system indicating more accumulation in the basal region of the plant (Ranieri et al., 2020). Cr build-up was also shown to be higher in the roots in comparison to shoots of *Pennisetum sinese* (Chen et al., 2020). In *Sesbania grandiflora* and it was revealed that roots accumulated more Pb than shoots (Malar et al., 2014). The retention of the HMs in the root system is due to the presence of phytochelatins (PCs) (Nguyen et al., 2017) which chelates the HMs thereby inhibiting upward translocation in the aerial parts. One report indicates that treatment of Cr stressed rice plants with silicon (Si) reduced the metal accumulation in the shoots as compared to the roots. Furthermore, phytochelatin synthase (OsPCS1) was strongly induced when rice plants treated with Cr are also supplemented with Si indicating the possibility of binding of PCs with the HMs ultimately paving the way for vacuolar sequestration in the roots (Huda et al., 2017). Cr translocation from roots to shoots is also induced by calcium (Ca) in rice through the engagement of PCs (Mukta et al., 2019). There are also reports that *OsMT* genes (for metallothionein synthesis) are up-regulated in rice plants during Cr stressed conditions and may be instrumental in the chelation of HMs in roots (Yu et al., 2019). Similar reports were also obtained in Cd-stressed rice supplemented with Si. The addition of Si to the soil resulted in a decrease in the concentration of Cd in the shoots. In addition to it, Cd transporters namely *OsHMA2* and *OsNRAMP5* had down-regulated expression while the concentrations of PCs, glutathione (GSH), and cysteine were up-regulated along with expression of *OsPCS1* in the rice-roots exposed to higher Cd level. These results indicated that Si supplementation not only resulted in down-regulation of Cd transporters but also up-regulated synthesis of PCs that are possibly involved in sequestration of the metal in the vacuole of root cells (Bari et al., 2020). The PIt was observed that Si supplementation resulted in an increase of Cd content in the roots (Rahman et al., 2017). Further, the levels of GSH1 transcript (PC precursor) and MT_A (metallothionein) were largely expressed in roots under the influence of Si indicating that the chelating agents are possibly involved in vacuolar compartmentalization of Cd in the root cells (Rahman et al., 2017). In As stressed wheat plants, Si supplementation induced accumulation of the metalloid more in the roots than in the shoots. In addin, the levels of TaPCS1 and TaMT1 (metallothionein synthase) was increased upon treatment with Si suggesting possible synthesis of chelating molecules for sequestration of As in roots and limiting to aboveground parts (Hossain et al., 2018). Reports also state that transgenic *Ziziphus jujuba* expressing the ZjMT gene store more Cd ions in roots thereby minimizing the toxicity of the HM (Yang et al., 2015). In sugarcane, the metallothionein gene ScMT2-1-3 is associated with the detoxification of Cu and is up-regulated in roots and buds during the Cu stressed condition (Guo et al., 2013).

5. Translocation of heavy metals

The transport of HMs from roots to above-ground parts is an important aspect of the detoxification process. The metal hyper-accumulating plants translocate high concentrations of HMs through the symplastic pathway involving the xylem (Chaudhary et al., 2018; Kajala et al., 2019; Balafrej et al., 2020). For symplastic movement, the HMs are required to cross the PM of the cells and this process is aided by the high negative resting potential of the PM which directs the inward movement of the metal ions due to the generation of an electrochemical gradient (Thakur et al., 2016). In addition, metal ions trapped within the vacuoles may be transferred to the stele and ultimately to the xylem through the root symplastic pathway (Yan et al., 2020). Within the xylem, the metals are carried by various transporter proteins. Table 2 represents selected metal transporters related to the translocation of HMs in the vascular bundle.

6. Intracellular mechanism of heavy metal detoxification

Once the HMs have made their way from the roots to the aerial sections, they are sequestered and stored (Luo et al., 2016). In aerial parts of the plants, the intracellular

Table 2: Transporters involved in movement and translocation of heavy metals in plants

Transporters	Nature/properties	Specific transporters/proteins	Representative plant	Metal ions transported/Loaded	Functions	Reference
Heavy metal ATPase (HMA)	P-type ATPase	AtHMA2	*Arabidopsis thaliana*	Cd	Xylem loading	Fan et al. (2018a)
		AtHMA4	*Arabidopsis thaliana*	Cd	Xylem loading	Ceasar et al. (2020)
		AtHMA5	*Arabidopsis thaliana*	Cu	Translocation from roots to shoots, Cu detoxification in roots	Huang et al. (2016)
		AhHMA4	*Arabidopsis halleri*	Zn	Xylem loading	Nouet et al. (2015)
		OsHMA5	*Oryza sativa*	Cu	Loading and transport of Cu	Printz et al. (2016)
Multidrug And Toxic Compound Extrusion (MATE) Family of Efflux Proteins	Multidrug efflux transporters	Ferric reductase defective 3 (FRD3)	*Arabidopsis thaliana*	Fe	citrate-dependent translocation of iron	Wu et al. (2018); Zhang et al. (2019)
Oligopeptide Transporters Family (OPT)	PM-bound proteins are involved in the long-distance transportation of peptides and/or metals (Gomolplitinant and Saier, 2011)	Yellow stripe 1 protein (HvYS1)	*Hordeum vulgare*	Iron	Transportation of Iron-Phytosiderophores in the root cells	Murata et al. (2006)
		OsYSL15	*Oryza sativa*	Fe	Uptake of Iron in roots	Ionue et al. (2009)

Table 2 contd. ....

...*Table 2 contd.*

Transporters	Nature/properties	Specific transporters/ proteins	Representative plant	Metal ions transported/ Loaded	Functions	Reference
Natural resistance-associated macrophage protein (NRAMP)	Integral membrane protein acting as metal ion transporters	OsNRAMP3	*Oryza sativa*	Mn	Distribution of Mn.	Yang et al. (2013)
		OsNRAMP5	*Oryza sativa*	Mn	Translocation of Mn in the shoots	Yang et al. (2014)
		TpNRAMP3	*Triticum polonicum*	Cd, Co	Transportation of Co, Mn and Cd	Peng et al. (2018)
Zn-regulated, Fe-regulated transporter-like proteins (ZIP) family of proteins	Metal transporter family identified in plants (Guerinot, 2000)	AtZIP1	*Arabidopsis thaliana*	Mn	The vacuolar transporter is responsible for the remobilization of Mn from the vacuole to the cytoplasm	Milner et al. (2013)
		AtZIP2	*Arabidopsis thaliana*	Mn	Uptake of Mn by root stellar cells	
PCs	Chelators of HMs	AtABCC1 AtABCC2 AtABCC3	*Arabidopsis thaliana*	Cd, Hg	Chelation of PCs with Cd and consequent transportation and vacuolar sequestration	Park et al. (2012); Brunetti et al. (2015)

organelles offer the most suitable site for the storage of HMs (Zhang et al., 2018; Khatiwada et al., 2020). The approach of concentration of HMs in the cellular organelles comprises the intracellular detoxification mechanism of the plants. In this section, the involvement of various transporters in the movement and sequestration of HMs would be discussed.

The CPx type of ATPases aids in the mobility of Pb, Cu, and Cd across the PM by using ATP-P1B type form of energy. The ATPases are also responsible for the metabolic activities in plants in the detoxification processes (Jain et al., 2018). Multiple Drug Resistance (MDR) transporter homologs can be found in higher plants and can be distributed into three separate groups namely the Major Facilitator Superfamily (MFS), ATP-Binding Cassette (ABC) superfamily, and the Multidrug and Toxic Compound Extrusion (MATE) family. Primary detoxification is carried by MDR transporters, which involves removing toxic elements from the cytosol and storing them in the vacuole (Remy and Duque, 2014). ABC transporters are divided into 8 subfamilies in plants and are distinguished alphabetically as A, B, C, D, E, F, G, and I whose major function is the detoxification process (Lane et al., 2016). All but two of the 53 Arabidopsis standard ABC transporter members can be classified into three groups: the ABCB subfamily's P-glycoproteins (PGP) or multidrug resistance (MDR), the MRP/ABCC subfamily's multidrug resistance-associated protein (MRP), and the ABCG subfamily's pleiotropic drug resistance (PDR) (Sánchez-Fernández et al., 2001). Table 3 represents various types of transporters involved in metal detoxification.

7. Phytochelatins and their role in metal tolerance and detoxification

Phytochelatins are a class of peptides that bind to HM and are made from GSH. (Chaudhary et al., 2018; Cobbett, 2000). They actively participate in the detoxification of HMs by the plants (Pal and Rai, 2010). PCs are a combination of 3 amino acids namely L-cysteine (Cys), L-glutamate (Glu), and glycine (Gly) (Inouhe, 2005), and are ultimately synthesized from GSH (Wünschmann et al., 2007). PC oligomers have a general structure of (g-Glu-Cys)n-Gly, with 'n' generally spans from 2–5 (11) in some organisms (Cobbett, 2000). The biosynthesis of PCs shares a common pathway with GSH biosynthesis. In the initial step, the enzyme g-glutamylcysteine synthase catalyzes the ATP-dependent ligation of L-glutamate and L-cysteine to produce γ-glutamylcysteine which forms the first rate-limiting step of the biosynthesis of GSH (Misra and Griffith, 1998). In the next step, glycine is added to the molecule γ-glutamylcysteine by glutathione synthetase to form GSH. This reaction is also coupled to the hydrolysis of ATP (Forman et al., 2009). Phytochelatin molecules are made of GSH by phytochelatin synthase (PCS) (Rea et al., 2012) (Fig. 1). The HMs bind to high-affinity thiol groups of PCs and form complexes (Dennis et al., 2019). Several studies report the synthesis of PCs in response to HM stress. Table 4 represents the role of selected metals in the synthesis of PCs.

The gene responsible for the synthesis of phytochelatin synthase (PCS) is reported to get induced by several metal ions. A report states that the manifestation of *Arabidopsis* phytochelatin synthase 2 (AtPCS2) gene significantly increased

Table 3: Selected transporter representative responsible for heavy metal detoxification.

S. No.	Name of the transporter	Location	Plant species	Metal detoxified	References
ABC Transporters/Heavy metal ATPase (HMA)					
1.	AtABCC1 AtABCC2	Vacuolar membrane	*Arabidopsis thaliana*	Cd and Hg sequestration in the vacuole	Park et al. (2012)
2.	AtABCC3	Vacuolar membrane	*Arabidopsis thaliana*	Sequestration of Cd in vacuole	Brunetti et al. (2015)
3.	OsABCC1	Vacuolar membrane	*Oryza sativa*	As is sequestered in the vacuoles companion cells	Song et al. (2014)
4.	AtHMA3	Vacuolar membrane	*Arabidopsis thaliana*	Sequestration of Cd in vacuoles	Morel et al. (2009)
5.	OsHMA3	Vacuolar membrane	*Oryza sativa*	Cd sequestration in the root cell vacuoles	Sasaki et al. (2014)
				Zn sequestration	Cai et al. (2019)
6.	TcHMA3	Vacuolar membrane	*Thlaspi caerulescens*	Cd sequestration	Ueno et al. (2011)
NRAMP transporters					
7.	AtNRAMP3 and AtNRAMP4	Vacuolar membrane	*Arabidopsis thaliana*	Mobilization of vacuolar Fe	Lanquar et al. (2005)
8.	AtNRAMP2	Trans-Golgi network	*Arabidopsis thaliana*	Mn influx transport activity	Gao et al. (2018)
9.	AtNRAMP6	Endomembrane compartment	*Arabidopsis thaliana*	Transportation of Cd	Cailliatte et al. (2006)
10.	NcNRAMP1	Plasma membrane	*Noccaea caerulescens*	Transportation of Cd	Milner et al. (2014)
11.	OsNRAMP1	Plasma membrane	*Oryza sativa*	Uptake of Cd in the cells	Takahashi et al. (2011)
Cation Diffusion Facilitator (CDF)/MTP					
12.	AtMTP1	Vacuolar membrane	*Arabidopsis thaliana*	Sequestration of Zn in the vacuole	Kobae et al. (2004)
13.	AtMTP3	Vacuolar membrane	*Arabidopsis thaliana*	Sequestration of Zn in the vacuole	Arrivault et al. (2006)
14.	AtMTP8	Vacuolar membrane	*Arabidopsis thaliana*	Sequestration of Mn in vacuole	Chu et al. (2017)
15.	AtMTP11	Golgi network	*Arabidopsis thaliana*	Mn sequestration	Peiter et al. (2007)
16.	TgMTP1	Vacuolar membrane	*Thlaspi goesingense*	Zn sequestration	Gustin et al. (2009)
17.	OsMTP1	Vacuolar membrane	*Oryza sativa*	n sequestration	Menguer et al. (2013)
18.	OsMTP8.1	Vacuolar membrane	*Oryza sativa*	Sequestration of Mn in vacuole	Chen et al. (2013)
19.	MTP8.2	Vacuolar membrane	*Oryza sativa*	Sequestration of Mn in vacuole	Takemoto et al. (2017)
20.	OsMTP11	Golgi network	*Oryza sativa*	Compartmentalization of Mn	Tsunemitsu et al. (2018)

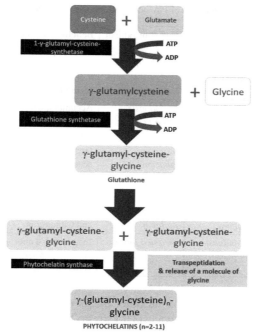

Figure 1: Biosynthetic pathway of phytochelatins.

Table 4: Role of selected metals in the generation of phytochelatins.

S. No.	Name of the metal	Name of the plants studied	Reference
1.	Hg	*Hydrilla verticillata, Vallisneria spiralis*	Gupta et al. (1998)
2.		*Brassica napus*	Iglesia-Turiño et al. (2006)
3.	Cd	*Bacopa monnieri*	Mishra et al. (2006)
4.		*Brassica juncea*	Seth et al. (2008)
5.		*Cuscuta reflexa*	Srivastava et al. (2004)
6.	Cd and Mn	*Phytolacca americana*	Gao et al. (2013)
7.	Cu	*Colobanthus squitensis*	Contreras et al. (2018)
8.	Zn	*Arabidopsis thaliana*	Kühnlenz et al. (2016)
9.	As	*Wolffia globosa*	Zhang et al. (2012)
10.		*Nasturtium officinale*	Namdjoyan et al. (2016)

upon treatment with 100 and 200 mM sodium chloride (NaCl) (Kim et al., 2019). Treatment of *Morus alba* with 200 μM Zn^{2+}/30 or 100 μM Cd^{2+} expressed PCS genes (Fan et al., 2018b). It was also noticed that Zn supplementation results in the synthesis of PCS mRNA in marine diatom *Nitzschiapalea* (Nguyen-Deroche et al., 2012). *Salvinia minima* is reported to respond to Pb stress by increasing PCs concentration (Estrella-Gómez et al., 2009). In *Brassica juncea*, exposure to Cd increased transcript expression of PCS (Shanmugaraj et al., 2013). The heterologous

Table 5: Heterologous expression of Phytochelatin synthase to confer heavy metal tolerance.

S. No.	Name of gene	Source organism	Target organism	Beneficial effect	References
1.	CdPCS1	*Ceratophyllum demersum*	*Oryza sativa*	Sequestration of As in roots thereby lessening accumulation in grains.	Shri et al. (2014)
2.			*Arabidopsis thaliana*	Accumulation of heavy metalloids in aerial parts.	Shukla et al. (2013)
3.	CdPCS1	*Ceratophyllum demersum*	*Nicotiana tabacum*	Increase in PC As accumulation.	Shukla et al. (2012)
4.	PaPCS	*Phragmites australis*	*Festuca arundinacea*	Higher content of PCs and Cd in roots.	Zhao et al. (2014)
5.	NnPCS1	*Nelumbo nucifera*	*Arabidopsis thaliana*	Higher content of PCS and Cd in roots.	Liu et al. (2012)
6.	AtPCS1	*Arabidopsis thaliana*	*Brassica juncea*	Increased Cd and As tolerance and concentration of PCs.	Gasic and Korban (2007)

expression of the PCS gene to some other plants of agricultural importance also confers forbearance to the plants against HM stress. Table 5 illustrates the expression of the PCS gene in bringing about HM tolerance in selected plants.

8. Role of metallothioneins in heavy metal tolerance and detoxification

Metallothioneins (MTs) are cytoplasmic metal-binding proteins with low molecular weight and high cysteine content, and detected in all kingdoms and are related to metal allocation and homeostasis (Hassinen et al., 2011; Rono et al., 2021). They were first isolated by Margoshes and Vallee in the form of low-molecular-weight protein having affinity to Cd from the kidney of the horse (Takahashi et al., 2012). In plants, the metallothioneins are categorized into four subfamilies depending upon the arrangement of cysteine residues (Guo et al., 2008). They are p1 (Type/Class 1), p2 (Type/Class 2), p3 (Type/Class 3), and pec (Type/Class 4). *Arabidopsis thaliana*'s MT1a and MT1c, as well as *Cicer arietinum*'s MT1, belong to the p1 subfamily. The p2 subfamily includes *A. thaliana*'s MT2a and MT2b, as well as *C. arietinum*'s MT2. The p3 subfamily comprises MT3 of *A. thaliana* and *Musa acuminata*. MT4a and MT4b of *Arabidopsis thaliana*, as well as Ec-1 of wheat, are members of the pec subfamily (Hassinen et al., 2011). Plant MTs contain anywhere from 45 to 87 amino acids (Freisinger, 2009). The number of Cys residues varies between 10 and 17 and aromatic amino acids vary from 0 (pec) to many (p1, p2, and p3), with a minimal amount of histidine (His) (Hassinen et al., 2011). The types 1, 2, and 3 metallothioneins are distinguished by their cysteine-rich domains interrupted by a linker consisting of around forty amino acids and a linker of about 15 amino acids, and they vary in the pattern in which cysteines are arranged in cysteine-rich domains. The type 4 metallothioneins have a link of around 15 amino acids (Zúñiga et al., 2019). It is reported that metallothionein forms two distinct metal-thiolate clusters, in which each group, 12 metal coordinates to four cysteinyl thiols forming a tetrahedral geometry (Ngu and Stillman, 2009). The metal binding capacities of metallothioneins vary from metal to metal. For example, Four Zn (II) or Cd (II) ions can bind to MT1 of *Cicer arietinum* (Schicht et al., 2009). Characterization of *Quercus suber* MT2 metallothionein revealed the coordination with 3.5 Zn (II) or 5.6 Cd (II) ions to the metal-thiolate cluster (Domènech et al., 2008). It is also reported that the metallothionein MT E(c)-1 from *Triticum aestivum* can bind to six Zn (II) ions in two metal-binding domains (Loebus et al., 2011). *Musa acuminata* MT3 can bind with three or four divalent d^{10} metal ions (Cabral et al., 2018). Metal homeostasis is aided by plant metallothioneins (Ferraz et al., 2012). Since they are readily complex with HMs, they have the potential to sequester HMs and render them ineffective and thus helping in metal tolerance. A recent study reported that states that plants species namely *Bacharissalicifolia, Tessariaintegrifolia, Chenopodium murale, Eleocharis montevidensis* which grow along with the Cu mining areas of Peru resort to the production of type I and type II metallothioneins (Llerena et al., 2021). It was observed in another investigation that heterologous expression of metallothionein gene (OsMT1e-P) from *Oryza sativa* in *Nicotiana tabacum* resulted in abiotic stress tolerance that included both salinity and HMs (Kumar et al., 2012). Another set of work heterologous expression of rgMT gene (rice) in *Arabidopsis thaliana* enabled the plant to better adapt to HM and peroxide stress (Jin et al., 2014). SsMT2 is a type 2 metallothionein gene found in the halotolerant plant, *Suaeda salsa*, when expressed in *Arabidopsis thaliana* resulted in improved metal stress resistance (Jin et al., 2017). A type 4 metallothionein gene (CsMT4) from *Cucumis sativus* has also been characterized and it was inferred that metal stress causes the gene to get activated and induces Cd tolerance (Duan et al., 2019).

9. Heat shock protein and heavy metal stress

Heat shock proteins (HSPs) are the proteins that play a role in the folding of proteins and maturation and are activated by heat shock or other stressors (Wu et al., 2017). The HSPs are related to the heat shock response (HSR) pathway. It is an emergency pathway and is generally triggered by elevated temperatures, HMs, and changes in pH. This pathway results in the launch of heat shock transcription factors (HSF1) and consequent expression of HSPs, molecular chaperones that help proteins that have been unfolded or misfolded to revert to their original state (Steurer et al., 2018). One of the most common cell chaperones is the HSP70 protein family (Radons et al., 2016). HSF1 is instrumental in the HSR (Barna et al., 2018). Upon activation, HSF1 undergoes trimerization and binds to DNA (Heat shock responsive genes), concentrates in the nuclear stress granules, and is phosphorylated in multiple locations, which are related to its transcriptional activity (Hietakangas et al., 2003). These responsive components are called heat shock elements (HSEs) and are characterized by the presence of 3–6 alternating, inverted pentameric repeats (nTTCnnGAAnnTTCn) (Verghese et al., 2012). HMs are reported to induce heat shock response (Kim et al., 2014). It is reported that HMs activate the HSF1 of the heat shock pathway which then communicates with the HSE of target genes (Shinkai et al., 2017). HSPs can thus act as a suitable biomarker for HM stress.

10. Heavy metal-induced oxidative stress in plants

Oxidative stress is an occurrence initiated by a disproportion between the production and accretion of ROS in tissues and the capacity of a living system to detoxify these reactive products. "ROS" usually exists in many forms like superoxide anion, hydroxyl, peroxyl, and alkoxyl radicals, hydrogen peroxide (H_2O_2), singlet oxygen ($^1O_2$), and organic hydroperoxide (ROOH) (Circu and Aw et al., 2010; Corpas et al., 2011). ROS are ephemeral, weak, and chemically vastly reactive molecules (Sharma et al., 2012; Pizzino et al., 2017).

HMs are one of the most significant abiotic stressors that wreak havoc on agricultural and economically important crops' productivity and nutritional value (Bhat et al., 2019). The HM accumulation is quite alarming for the cultivators and agricultural markets that have a direct impact on the country's economy. Among the variety of HMs present in the soil here, only poignant metals are to be talked about and discussed. Like Pb, Cd, Cr, As, Hg, Cu, Ni, Zn, Fe. Some of which are required as micronutrients (Cu, Zn, Fe) but when they exceed the limit of tolerance capacity they manifest their negative appearance in plants' cycle. Therefore, when oxidative stress is introduced during HM accumulation physiological metabolism faces a series of catastrophes. It may start from seed germination to seed production after fertilization.

Catalytic enzymes such as proteases, alpha-amylases, and acid phosphatases are distinguished to assist in seed germination up to seedling growth utilizing nutrient mobilization in the endosperm. Starch is restricted in the presence of HMs, and nutritional supplies become scanty. Moreover, under HM stress, a decrease in proteolytic enzyme activity (Hasan et al., 2017) and an increase in protein and amino acid content (Shah et al., 1998; Seneviratne et al., 2019). The amino acid, Proline is vital for cellular metabolism. Several reports have confirmed a boost in proline content in higher plants during oxidative stress (Georgiadou et al., 2018; Sytar et al., 2019; Zhao et al., 2021).

Heat shock protein (HSP) acts as a reducer rather than a repairing photosystem-II when HM disrupts the enzyme equilibrium, causing photosystem II to oxidize (PSII) by ROS and leading to disruption of the electron transport pathway and mineral metabolism ted (Seneviratne et al., 2019). Not only interfering with chloroplasts and mitochondria the consequence of the interaction between HM and ROS dismantles the cellular membrane and alters the redox status of a cell (Seneviratne et al., 2019). Biological macromolecule depreciation, lipid peroxidation, DNA-strand cleavage, and ion leakage are observed when the duration of exposure exceeds in Cd stressed plants in the additional presence of Cr and As. The stage of plant development also determines the vulnerability versus tolerance or repairing the system in HM stress (Ghori et al., 2019).

Algae counter to HMs by stimulation of several antioxidants, containing diverse enzymes such as catalase (CAT), superoxide dismutase (SOD), glutathione peroxidase (GPx), ascorbate peroxidase, and the low-molecular-weight-compounds-production namely GSH and carotenoids (Pinto et al., 2003). At the acute level of metal pollutants, damage to algal cells appears because ROS levels surpass the capability of the cell to manage so they can transmit them onto organisms of other trophic levels such as crustaceans, and fishes, and mollusks. We can now realize the results of biomagnifications and the circumlocutory deposition of HM in our system.

Temporary (24 h) responses of *Cladonia arbuscula* subsp. *Mitis* (previously known as *Cladina*) and *Cladonia furcata* to CuII or CrIII excesses (10 or 100 μM) were evaluated. *Cladina* collected much Cr and Cu at higher metal concentrations but together species exposed exhaustion of K and/or Ca amount raising ROS development (fluorescence microscopy uncovering of overall H_2O_2 and ROS) and down nitric oxide (NO) signal, with Cu presenting a more damaging effect on lipid peroxidation (BODIPY 581/591 C11 staining reagent). A chiefly different impression of Cr and Cu was detected at the level of the antioxidative metabolite, signifying various means of metal-stimulated ROS elimination and/or metal chelation. Cu strongly exhausted stimulated phytochelatin 2 (PC2) and GSH content while the ascorbic acid stock was diminished by Cu and accelerated by Cr. Following experiment with GSH biosynthetic inhibitor (buthionine sulfoximine, BSO) disclosed that 48 h of exposure is required to exhaust GSH and BSO-induced depletion of GSH and PC2 quantities under Cr or Cu excess promoted ROS but drained NO. This information proposes close affairs between thiols, NO, and the appearance of oxidative stress (ROS production) under metallic stress also in lichens (Kováčik et al., 2018).

According to a recent study, melatonin which ubiquitously exists in most of the living organisms acting as a free radical scavenger showed a significant response under oxidative stress augmented by HMs in soil. A significant reduction in malondialdehyde (MDA) and oxygen-free radicals (OFR) in *Exophiala pisciphila* was noted under Zn, Cd and Pb stresses in comparison to control. Pre-treatment of *E. pisciphila* with exogenic melatonin considerably improved SOD activity under Zn and Pb stresses reduced the HM contents. Melatonin production was increased by Cu, Cd, and Zn after 2 d, and melatonin biosynthetic enzyme genes like serotonin N-acetyltransferase (*EpSNAT1*) and *E. pisciphila* tryptophan decarboxylase (*EpTDC1*), were upregulated in RNA levels. EpTDC1 and N-acetylserotonin O-methyltransferase (EpASMT1) over-expression in *Escherichia coli* and *Arabidopsis thaliana* heightened its HM-stimulated stress forbearance. The over-expression of EpTDC1 and EpASMT1 decreased the accumulation of Cd in *Arabidopsis thaliana* roots (Yu et al., 2021).

11. Signal transduction pathways

Hyperaccumulator plants always acquire the center of attention by scientists on how they tolerate, sequester, or exude the HM ions. There are reports linked to the variation of gene transcription of metal chelators or transporters that aid HM elimination or detoxification (Arbelaez et al., 2017; Gulli et al., 2018; Zhang et al., 2019). A reversible epigenetic mechanism may control these genes, specifically on plants having hyperaccumulator properties that can thrive in soils without showing any toxic symptoms.

Rice has several strategies for dealing with Al stress. Like Multi-antimicrobial extrusion protein–MATE transporters, OsFRDL2 and OsFRDL4 participate in the transportation of OA (Famoso et al., 2010; Delhaize et al., 2012; Yokosho et al., 2016). For Al^{+3} sequestration and translocation in the vacuole discrete Al transporters namely bacterial-type ABC and Nramp transporters are involved (Huang et al., 2009; Xia et al., 2010; Li et al., 2014). 69 probable candidate genes linked to Al sufferance, recognized in a set of 150 rice landraces via a joint GWAS-transcriptomic method along with the report of 48 QTLs on chromosomes 1,3,9, and 12 (Zhang et al., 2019).

Organic Acid (OA) efflux is a ubiquitous mechanism exploited by plants to exude AL+3 ions by forming stable compounds with citric and or malic acid in their rhizospheres, found in wheat, sorghum, barley. Discoveries identify associates of two transporters families, the OA/HC transport channel (MATE and Al-activated malate transporter (ALMT), which are responsible for releasing malate and citrate from root cells into the rhizosphere, respectively in the presence of Al (Kochian et al., 2015).

Another layer of control strategy employed by plants showing pinnacle importance is epigenetic mechanisms in handling HM. In gene expression regulation, three epigenetic techniques are involved.: (i) DNA methylation (changes at genomic level), (ii) histone modifications (chromatin modifications), and (iii) Small RNA modifications (RNA directed DNA Methylation-RdDM route) (Sudan et al., 2018; Chang et al., 2020). Among those mentioned, DNA methylation is more recognized and in a more multifaceted way, it is manifested mostly in cytosine residues in the CHG, CHH, and CG sequence context (H maybe A, C, or T) (Bender, 2004; He et al., 2010). Typical DNA methylation profiling of *Oryza sativa* L. (cultivated rice), has displayed that repetitive sequences and transposable elements are the genome's most heavily methylated DNA regions (He et al., 2010; Yan et al., 2010; Li et al., 2012). Gene methylation is most common in the context of CG, while methylation of transposon follows in each of the three outlined situations (He et al., 2010; Yan et al., 2010; Li et al., 2012).

In plants, the methylome is chiefly checked and sustained at the time of replication of DNA and division of cells by DNA methyltransferases. DNA methyltransferases are divided into three categories, like (i) DNA methyltransferases (METs), are in charge of CG methylation; (ii) the plant-exclusive enzymes chromo methyltransferases (CMTs), that are documented to determine CHH and CHG methylation; and (iii) the domain rearranged methyltransferases (DRMs), that are deals with non-CG methylation monitoring and *de novo* methylation in all 3 cases: CG, CHG, and CHH (Lanciano et al., 2017). On the flip side, DNA demethylation is achieved by DNA glycosylases such as the Repressor of Silencing 1 (ROS1) and the Demeter (DME) enzyme (Lanciano et al., 2017). Histone modifications along with DNA methylation is another revertible additional feature found in HM stressed phenotypes (Mirouze et al., 2011). Genomic sequences whose changes in their methylation pattern are preserved across generations, without modifying the obtained methylated pattern, are recognized as epialleles (Kalisz et al., 2004). These epialleles randomly occur over genomes and are not specific only to stressed response plants but transmitted through progenies (Verhoeven et al., 2010). To suppress gene expression transposons pose themselves onto or near the stress-response gene thus silencing through methylating them or near that gene of interest (Saze et al., 2007; Galindo-González et al., 2018).

Henceforth scientists are in surge for the general epigenetic mechanisms that will be sustained through progenies. Proof from earlier reports proposes that DNA methylation possibly will instrumental HM-stress adaptation by at least a couple of mechanisms (Aina et al., 2004; Choi et al., 2007; Arif et al., 2016). The first refers to methylation's protective impact against HM-stimulated DNA damage by single-strand breaks or transposition having multiple copies (Bender, 1998). For instance, Gulli et al., 2018, discovered that in Ni stressed *Noccaea caerulescens* were there was hypermethylation at the genomic level in comparison to Ni sensitive *Arabidopsis thaliana*. The reasons lie in the differential expression of the genes namely MET1, DRM2, and HDA8 that are held for DNA methylation and histone modification in the plants *N. caerulescens* and *A. thaliana*. Similarly, comparing methylation levels in HM sensitive to varieties of clover and moderately tolerant hemp revealed that HM tolerance is related to an elevated level of methylation of hemp roots in comparison to clover. Similarly, hypermethylation to counter radiation genotoxic impact was evidenced by a report (Volkova et al., 2018) in ionizing radiation-tolerant *Pinus sylvestris* revealed loci with substantial hypermethylation than non-accustomed plants. Another type of mechanism shows that the regulation is not restricted not only to the promoter regions but also to their coding regions (Choi et al., 2007). In the period in-between, exon/intron methylation happens generally in the CG context and its function remains indistinct. Methylation of the genome has been linked to upregulation in transcriptional activity and thus protects genes from uncharacteristic transcription caused by cryptic promoters (Feng et al., 2016). Activation of transcription can be initiated by histone acetylation around the promoter region of genes (Finnegan, 2001). Degree of gene expression stimulated by HM stress positively correlating with the methylation variation in Cd dose-response in plants. Genes enciphering for HM transporters are also subjected to methylation in DNA (Oono et al., 2016). The relevance of the events was also supported by the Cd-specific stress-induced *Oryza sativa* ssp. *japonica* cv *Nipponbare* which showed unique differentially methylated areas.

Similarly, other studies have indicated the inheritance and consistency of methylation alterations caused by HM stress (Ou et al., 2012). The progeny of *Arabidopsis thaliana* has developed forbearance to HMs. More lately, a report unveiled that HM stress causes unique methylation changes. The methylation differs at the Tos17 retrotransposon and revealed trans generational bequest across 3 generations (Cong et al., 2019). As a result, the evidence suggests that epigenetic methods have an impact on HM stress adaption between successive plant generations.

When HMs heads to oxidative stress by mounting up ROS, loss of PM integrity, increasing lipid peroxidation, oxidation of protein, and DNA damage (Cortés-Eslava et al., 2018; Pandey et al., 2019; Agarwal and Khan, 2020). Fascinatingly, the supplement of sodium hydrogen sulphide (NaHS) lessened the detrimental effects of HMs including Cr, Cd, Zn, Cu, etc. (Santisree et al., 2019).

An elevated level of hydrogen sulfide (H_2S) was found in Cd stress-induced rice seedlings. Furthermore, these expression of the H_2S-producing genes DES1 and DCS1 was also elevated during Cd reception (Ahmad, 2016). In foxtail millet seedlings, exposure to Cr also increased the expression of H_2S-related genes such as DCD2, DES, and LCD (Fang et al., 2014). Scientists have testified improvement of toxicity induced by As through NaHS treatment in pea by counteracting the high levels of cysteine, harmful effects on proteins, lipids, and membranes (Singh et al., 2015). Furthermore, NaHS reduced Cr-stimulated H_2O_2 and MDA overproduction in wheat seedlings (Zhang et al., 2010a,b). Seed germination is improved when NaHS is applied to Cr-exposed wheat seeds along with improvement in esterase, amylase, and antioxidant enzymes (Zhang et al., 2010a,b). These consequences prove a vital role for H_2S in balancing the HMs (Cr and As) stimulated toxicity in wheat and pea (Singh et al., 2015). A report of NaHS addition in Cu-induced stress in wheat has also been found (Zhang et al., 2008). NaHS boosts activities of esterase and amylase while subsiding MDA and H_2O_2 content in Cu stressed wheat. Hydrogen sulphide, on the other hand, upregulates the translation of metallothioneins (MTs) to protect against Zn poisoning. It additionally boosts the expression of antioxidant enzymes counterbalancing the expression and genes related to Zn-homoeostasis such as Zn and Fe-regulated transporter (Liu et al., 2016). There have been contrasting reports of the concentration of NO and its act upon plants in stress. Where the addition of SNP the donor of NO reduces oxidative damage by Al-induced stress, found in *Cassia tora*, wheat, soybean, and kidney bean plants (Wang et al., 2019); donor (NaHS) of hydrogen sulfide (H_2S) enhanced the activity of the antioxidant enzyme and separated Al in the vacuole. Besides, a decline in NO content perhaps be participating in the H_2S-facilitated mitigation of Al toxicity in roots of rice. The H_2S noticeably decreased NO amounts in rice roots under Al stress. The application of SNP notably raised the content of Al in roots and enhanced the blockage of root extension. This effect was retracted by the remedy of NO scavenger, pointing to the crosstalk between H_2S and NO in minimizing the toxicity of Al in rice (Zhu et al., 2018).

12. Emerging aspects of non-coding RNAs in heavy metal stress

In HM stress, presently one of the most cultivated areas is to identify epigenetic regulation to control the damage of plants via non-coding RNAs (Wei et al., 2017). These are of two kinds-micro-RNA and Long non-coding RNA (lnc RNA) of transcripts that don't appear for the next phase, i.e., translation of a specific protein.

A microRNA (miRNA) is a tiny single-stranded non-coding RNA molecule (comprising around 22 nucleotides) that acts in RNA silencing and post-transcriptional control of genes (Lu and Rothenberg, 2018) expression, located in plants, animals, and some viruses. A long non-coding RNA is a long single-stranded RNA molecule ($\geq$ 200 nucleotides) (Bhat et al., 2016) functionally similar to microRNA in tuning target gene's expression at the transcriptional and post-transcriptional level.

Instead, they tune to change the expression of the assigned target genes that get up-regulated during HM stress and also other abiotic stresses. To identify the microRNAs' role in Cd-mediated stress, maize seedlings were treated with $CdCl_2$ (200 mg/L). 2.5 H_2O over different exposure times in a recent exploration (Gao et al., 2019). To confirm Cd stress, enzymatic assays of superoxide dismutase and peroxidase were assayed. Employing quantitative real-time PCR (qRT-PCR) techniques in the manifestation of six candidate miRNAs and target genes were corroborated. Differential expression of the miRNAs and their corresponding target genes over a different period of Cd stress occurred in seedlings suggested that candidate miRNAs such as *Zma-miR166d*, *Zma-miR156b*, *Zma-miR393b*, and *Zma-miR171b* may negatively adjust their corresponding target genes in maize during stress, which likely encompasses transcriptional and post-transcriptional control systems (Gao et al., 2019).

A few years back, a study in *Oryza sativa* (rice) plants in response to Cd stress interpreted the part of miRNA166. Plant growth, development, and reaction to environmental challenges, HM stress are all known to be influenced by miRNA166. In Cd stressed condition, the down-regulation of miRNA166 was found in the tissues of rice seedlings' leaves and roots. In rice, overexpression of miR166 increased Cd toxicity, along with the decrease in oxidative stress. Furthermore, overexpression of miR166 inhibited Cd transport from roots to shoots and Cd accretion in grains. The class-III homeodomain-Leu zipper (HD-Zip) family proteins encrypting genes are triggered by miR166 in plants. *HOMEODOMAIN CONTAINING PROTEIN4* (*OsHB4*) gene in rice, which translates into an HD-Zip protein, was down-regulated by overexpression of miR166 but up-regulated by Cd treatment in transgenic rice plants. The excess expression of *OsHB4* amplified sensitivity to Cd and Cd accretion in transgenic rice plants. On the contrary, silencing *OsHB4* by RNA intervention boosted Cd acceptance in transgenic rice plants. Thus in rice, Cd accumulation and tolerance can be manipulated by miR166 through its objective gene, *OsHB4* (Ding et al., 2018).

In another study performed on ramie (*Boehmeria nivea*) that is economically famous for its natural fiber production was tested for phytoremediation through targeting specific miRNAs to reduce Cd stress in plants. Four small RNAs libraries were created from the ramie leaves and roots of Cd stressed plants as experimental and non-stressed plants as control sets. 73 novel miRNAs were recognized using RNA sequencing. A set of miRNAs displayed differential expression discovered through Genome-wide expression analysis. *In silico* target prediction detected 426 potential miRNA targets which were validated real-time quantification by q-PCR. Through such potential experiments with convincing outcomes, this study let us identify six randomly selected genes whose expression was reciprocally correlated with the expression of the corresponding miRNA in Cd stress plants. Thus the study not only enriches the numbers of miRNAs that regulate Cd stress also overlays the route to survive from Cd stress (Chen et al., 2018).

Like microRNAs, lnc-RNAs also play an influential role to modulate their corresponding genes in curbing abiotic stresses like HM stress faced by HM susceptible plants (Chen et al., 2018; Jha et al., 2020). Cd stress in economically crucial crop plants rice poses a great menace to the cultivator and agronomy on a large scale (He et al., 2015; Chen et al., 2018). Here another study showed lnc RNAs are differentially expressed in two unlike conditions (Cd stress vs control condition) through Deep sequencing (Chen et al., 2018). In the control and Cd stress groups, 144 differentially expressed lncRNAs from 143 lncRNA genes were categorized. The 144 lncRNAs are composed of 1 intronic lncRNA, 23 antisense-lncRNAs, and 120 over-long intergenic non-coding RNAs (linc RNAs) in which, 75 lncRNAs were down-regulated and 69 lncRNAs were up-regulated in this study. Besides, 386 matched lncRNA-mRNA pairs were unearthed for 120 divergently expressed lncRNAs and 362 differentially expressed genes in cis. Target gene-related route investigations displayed considerable variations in methionine and cysteine modulating genes in the GO (Gene Ontology) analysis. For the genes in trans, a total of 28,276 interactive relationships for 144 lncRNAs and divergently expressed protein-coding genes were spotted. Phenylpropanoids and phenylalanine like secondary metabolites, and photosynthesis pathway-associated genes, were identified through the pathway analysis program were significantly changed by Cd stress. Therefore, this study could provide valuable information to alter Cd-stress in rice plants in the future.

13. Modulation of genes during heavy metal stress

It's been million years of practice of altering or modulating genes' expression in a living system. It can happen at any level starting from the start of transcription to RNA processing and modification after translation. Since one gene control another gene's expression it's better to say in a gene network modulation happening at any time to alleviate any stress plants/animals face.

In context to HM stress to plants, here is presented some fresh and impactful research that would help us to understand a wide spectacle of modulations of genes' reactions occurring in cultivated and among wild variety to take them out of the stress they learn to face, tolerate or to refuse to accept any stress.

In the case of one of the most popular staple crops, *Oryza sativa* shows some significant genes' modulations that's been discovered through thorough implications of genome, transcriptome analysis, and real-time approach to check differential expression at the various stages of development along with changes of HM varieties and their grades of concentrations they are facing in their entire life cycle. In rice, (Cong et al., 2019) the crucial role in the uptake and translocation of HMs in plants has been executed by the P1B subfamily of the HM-transporting P-type ATPases (HMAs) including OsHMA1-OsHMA9. When HM treatment is acted upon, some locus-specific alterations of HMA genes are seen along with a low copy of several genes and transposons occur during transgenerational inheritance. The state of change in the next progeny is evidenced but the viability of the change through progenies was verified by checking on a Tos17 transposon for bisulfite sequencing and studying its methylation state across three generations.

Some homologous proteins destined for detoxification have been discovered recently (Li et al., 2020). 46 HMPs in *Oryza sativa* and 55 HMPs in *Arabidopsis thaliana* are named corresponding to their chromosomal locations as *OsHMP* 1–46 rice and *AtHMP* 1–55 in Arabidopsis. HMPs possibly be controlled by different transcription factors that were revealed in a cis-element analysis. An expression profile study revealed that Out of 46, only eight OsHMPs were imminent in rice. Among them, the manifestation of *OsHMP37* was superior to the rest seven genes where as *OsHMP28* was exclusive for the roots. A higher level of constitutive transcripts was found for the nine *AtHMPs* in *Arabidopsis*. These selected OsHMPs showed a broad spectrum in respect to their quantitative revelation under different HM stresses. Exceptionally *OsHMP09*, *OsHMP18*, and *OsHMP22* only displayed a general higher expression for all tissues during stress. On the other side, most of the preferred *AtHMP*s had nearly varying levels of expression in different tissues under diverse HM stresses. In their leaves and roots higher expression of the *AtHMP20*, *AtHMP23*, *AtHMP25*, *AtHMP31*, *AtHMP35*, *AtHMP46*. These discoveries of homology of the HMPs' advancement in monocots and dicots brought light to functionally illustrate HMPs in the future (Li et al., 2020).

In proteomics study, several metal-responsive gene families have been identified to date. Among them, cytochrome-P450 (CYPs) members encode enzymes that perform the detoxification of exogenous molecules (Rai et al., 2015). A study discovered a CYP-like protein encoded by the *Os08g01480* locus in rice that assists plants in fighting against HM and other stresses. A transgenic-lines were developed through cDNA and promoter isolation from rice to develop them in *Arabidopsis* to decipher the functional characterization of the CYP-like gene. Heterologous expression of *Os08g01480* in *Arabidopsis* offered substantial tolerance towards abiotic stresses. To fight against stress, a modulation in auxin metabolism was found to take place with the assistance of the Os08g01480 gene that was revealed by *in silico* analysis. Transgenic lines exhibiting reporter gene under the switch of *Os08g01480* promoter exhibited divergent promoter activity in unalike plant-tissues in abiotic stresses. These studies pointed to that degree of expression of *Os08g01480* might be modifying the response of plants according to the different stages of development, tissues, and environmental stresses like HM (Rai et al., 2015).

In practice, Si has shown promising results in curbing HM stress by increasing plants' tolerance levels. One of these studies was performed to verify whether Si treatment to rice plants can enhance the tolerance level in Cd and Cu-induced stress. Experiments have revealed something interesting modulation and tuning of genes of interest in this context (Kim et al., 2014). In context to the increase in growth and biomass with reduced toxicity for the above two HMs, Si application not only showed success but also enhanced the root function and structure in comparison to the non-treated plants that experienced severed damage to roots. The Cu/Cd concentration was considerably lesser in rice plants in presence of Si along with a decrease in peroxidation of lipids, desaturation of fatty acid in plants. The external application of Si changed the game inside resulting in a decrease in uptake of metals in roots tuned the signaling of phytohormones engaged in responses to stress and host weaponry, like Jasmonic acid, abscisic acid, and salicylic acid. The less concentration was of HM positively correlated with the mRNA level of the enzymes coding HM transporters (OsHMA3 and OsHMA2) in Si-treated rice plants. Whereas, the genes liable for Si transport (sLSi2 and OsLSi1), exhibited a substantial up-regulation of mRNA translation in Si-treated rice plants. Thus, the study encourages the application of Si in Cd/Cu stress management in rice plants through changes in the root morphology.

In Cd stressed soil how tomato seedlings responded were recorded and altered genes' expression reflecting real-time proteome profiles presented something fascinating witnessed by a group of scientists (Khanna et al., 2019). Plant growth-promoting rhizobacteria (PGPR) are known for cessation of HMs and moderating their transportation in plants via rain, complex configuration, and adsorption. To check on their efficiency in the alleviation of Cd stress Tomato seedlings, *Burkholderia gladioli*, and *Pseudomonas aeruginosa* were taken into account. The growth features like photosynthetic pigments, metal tolerance index, metal-abstracting, and the substances of metal-chelating complexes in microbes inoculated Cd treated tomato seedlings. Differential expression of the genes for the metal transporters was profiled for quantitative analysis. Outcomes revealed the improvement in the aforementioned characteristics of growth after the inoculations of *P. aeruginosa* and *B. gladioli* when Cd-treated seedlings of reduced growth were inoculated. Here the length of shoot-root ratio, fresh weight, and photosynthetic pigments were monitored and assessed. In Cd-stressed seedlings, metal transporter genes' expression was differentially up-regulated for many folds which significantly diminished after the inoculation of the referred PGPRs at the molecular level. Thus the application of PGPRs declined the toxicity of HM for healthier growth of tomato seedlings.

On account of HM stress tolerance, a very novel study was performed on the trees of *Phoenix dactylifera* (Palm) (Chaâbene et al., 2021). The leaves showing significant accumulation of Cd, Cu, and Cr drew the attention of the researchers. Based on this discovery, the cellular means that justify these metal stores were explored in controlled states. The next four months of subjection to Cr, Cu, or Cd low TF values were found for saplings indulged with Cr and Cu but high bioaccumulation and translocation factor (TF > 1) have been displayed for date palm saplings exposed to Cd. In Cu and Cd exposed roots accretion of oxidants and antioxidants activating enzymes were noticed. Whereas, when plants were exposed at low metal concentrations, the secondary metabolites, such as flavonoids and polyphenols, were increased initially and dropped thereafter. A positive correlation was found with increased expression of a transporter gene Pdmate5 known for the secondary metabolites transportation and flavonoids accumulation. Another transporter gene acted completely to metal-inclusion, chiefly *Pdhma2* besides *Pdnramp6* and *Pdabcc* as new HM stress-sensitive candidate genes. Expression profiles of genes coding proteins as metal chelators were also inspected. *Pdmt3* and *Pdpcs1* revealed a robust stimulation in plants subjected to Cr. These variations of the manifestation of some molecular and biochemical based-markers in date palm aided to recognize more the capability of the plant to endure metals. They could be beneficial in evaluating HM pollutions in adulterated soils and may advance the accumulation capability of other plants.

The cross-talk between calcineurin B-like protein (CBL) and CBL-interacting protein kinase (CIPK) are instrumental in signaling cascades during stress (Pan et al., 2018). Bioinformatics and molecular approaches were employed in which 12 hsSPIK genes from wild barley have been isolated to characterize their role in response to HM toxicity (Pan et al., 2018). The results presented that several HsCIPKs were transcriptionally controlled by HM noxiousness and other abiotic stresses. When the transgenic plants (*Oryza sativa* cv *Nipponbare*) of rice built up for the over-expression of multiple HsCIPKs, individually they revealed developed root growth tolerance to HM toxicity, drought, and salt stresses. These outcomes indicate that HsCIPKs are concerned with the consequence of HM toxicities and other abiotic stresses. Thus HsCIPKs of wild barley of the Tibetan Plateau hold wide-ranging applications in genetically engineered rice with endurance to HM toxicities.

An escapism mechanism to avoid tolerance of roots of *Arabidopsis thaliana* was observed in the research held in Korea (Khare et al., 2017). Plants restructure their root construction to keep growth away from unsuitable regions of the rhizosphere.

In a test grounded on chimeric repressor gene-silencing skill, the *Arabidopsis thaliana* GeBP-LIKE 4 (GPL4) transcription factor was identified as an inhibitor of root development that is stimulated fasting root tips in reaction to Cd. The theory that GPL4 performs in the root refrainment of Cd was verified through examining root propagation in a split-medium, in which 50% of the medium covered toxic concentrations of Cd. On a comparative account, the WT plants displayed root escaping by preventing root development on the Cd wing but growing biomass of root on the control wing. By distinction, GPL4-suppression lineages displayed almost alike root growth in the Cd and control sides and escalated further Cd in the shoots than acted in the WT. GPL4 silencing furthermore modified the root evasion of toxic concentrations of other essential metals, influenced the expression of several genes linked to oxidative stress, and reliably reduced ROS concentrations. This study opined that GPL4 prevents the progress of roots subjected to metal toxicity by modifying ROS concentrations, thereby letting roots establish fresh zones of the rhizosphere.

14. Future prospects and conclusion

Heavy metal pollution has posed a serious threat to civilization. By and large, HM pollution is affecting the human race directly by causing physiological damage or through plants. The plants, being immobile are more prone to HM toxicity. However, they have devised several strategies to cope up with the HM pollution. Some plants altogether tend to avoid the HM stress by limiting the access of HMs in their body. Whereas, the others tend to sequester HM in their cellular organelles thereby reducing their detrimental effects. The plants which uptake and sequester the HMs in their organelle have improvised several physiological and genetic machinery for the purpose. They are equipped with a series of HM transporters that acts in the uptake of HMs, loading the same in the conducting tissues and finally unloading or sequestering them in the organelles. These plants also possess several groups of organic compounds which also aid in the uptake and transportation of HMs. These compounds include organic acids, PCs, metallothioneins, amnio acids whose primary action is to chelate the HMs and depending upon the ambient condition, either restrict their movement or transfer them to the conduction stream for final sequestration in the above-ground part. As per the rule of a biological system, the synthesis of every organic compound (including the transporters) is related to the course of gene expression. Many studies have publicized that HM stress has activated specific genes required to cope up with the stress. Thus, it becomes extremely relevant to delve into the mechanistic aspects of these genes and study the mechanisms of their action. In nature, there are lots of plants that grow in metal-contaminated sites. Thus, exploration of these plants and further looking into their genetic structure can help us in unraveling some such wonder genes which aid in HM tolerance. The third world countries suffer from acute H_2O scarcity and thus resort to irrigation with treated or untreated waste H_2O, the direct contamination of which is the accumulation of HMs in the crops and consequent transmission of the same in the food chain through sequential consumption. Moreover, in addition to HM accumulation, the majority of the plants suffer from low yield and reduced mass as a result of stress. This poses a threat to the ever-increasing demand for food by an exponentially growing population. The appropriate remedy to this problem is possibly the adoption of genetic engineering techniques through which HM stress-resistant genes may be incorporated in the genomes of susceptible cultivars. This either gives them tolerance towards HM-stress or restrict accumulations of HMs in the underground parts thereby optimizing the yield under stressed condition. The non-coding RNA has also emerged as a very effective candidate to confer HM tolerance in plants. In a different approach, some of these plants can be functioned to emend the metal-polluted soil for phytoremediation purposes. Being a cost-effective and green approach it has already established itself as a very efficient strategy for metal decontamination. The HM tolerance in plants can be further fortified through the use of pg PGPR which has proved itself to be a candidate of multifarious benefit. They also possess an HM resistance gene which can be engineered into the plant to evaluate its efficacy. In a nutshell, it can be concluded that nature has devised intricate mechanisms to counter HM stress in several organisms. We must make a successful exploration of the candidates for potent HM resistance and express the gene in susceptible varieties. The future of HM detoxification lies in the genetic level and serious investigations are required to achieve further success for the gross profit of manhood.

References

Agarwal S, and Khan S. 2020. Heavy metal phytotoxicity: DNA damage. *In*: Faisal M, Saquib Q, Alatar AA, and Al-Khedhairy AA (eds.). *Cellular and Molecular Phytotoxicity of Heavy Metals*. Nanotechnology in the Life Sciences. Springer, Cham.

Ahmad P. 2016. Plant Metal Interaction: Emerging Remediation Techniques. Elsevier, Netherlands.

Aina R, Sgorbati S, Santagostino A, Labra M, Ghiani A, and Citterio S. 2004. Specific hypomethylation of DNA is induced by heavy metals in white clover and industrial hemp. *Physiol Plant* 121: 472–480.

Amsbury S. 2020. Sensing attack: the role of wall-associated kinases in plant pathogen responses. *Plant Physiol* 183(4): 1420–1421.

Angulo-Bejarano PI, Puente-Rivera J, and Cruz-Ortega R. 2021. Metal and metalloid toxicity in plants: an overview on molecular aspects. *Plants (Basel)* 10(4): 635.

Arbelaez JD, Maron LG, Jobe TO, Piñeros MA, Famoso AN, Rebelo AR, and McCouch SR. 2017. Aluminum Resistance Transcription Factor 1 (ART1) contributes to natural variation in aluminum resistance in diverse genetic backgrounds of rice (*O. sativa*). *Plant Direct* 1: 4.

Arif N, Yadav V, Singh S, Singh S, Ahmad P, Mishra RK et al. 2016. Influence of high and low levels of plant-beneficial heavy metal ions on plant growth and development. *Front Environ Sci* 4: 69.

Arrivault S, Senger T, and Krämer U. 2006. The Arabidopsis metal tolerance protein AtMTP3 maintains metal homeostasis by mediating Zn exclusion from the shoot under Fe deficiency and Zn oversupply. *Plant J* 46(5): 861–79.

Balafrej H, Bogusz D, Triqui ZA, Guedira A, Bendaou N, Smouni A, and Fahr M. 2020. Zinc Hyperaccumulation in Plants: A Review. *Plants (Basel)* 9(5): 562.

Bari MA, Prity SA, Das U, Akther MS, Sajib SA, Reza MA, and Kabir AH. 2020. Silicon induces phytochelatin and ROS scavengers facilitating cadmium detoxification in rice. *Plant Biol (Stuttg)* 22(3): 472–479.

Barna J, Csermely P, and Vellai T. 2018. Roles of heat shock factor 1 beyond the heat shock response. *Cell Mol Life Sci* 75(16): 2897–2916.

Bender J. 1998. Cytosine methylation of repeated sequences in eukaryotes: The role of DNA pairing. *Trends Biochem Sci* 23: 252–256.

Bender J. 2004. DNA-Methylation and epigenetics. *Annu Rev Plant Biol* 55: 41–68.

Bhat JA, Shivaraj SM, Singh P, Navadagi DB, Tripathi DK, Dash PK, Solanke AU, Sonah H, and Deshmukh R. 2019. Role of silicon in mitigation of heavy metal stresses in crop plants. *Plants (Basel)* 8(3): 71.

Bhat SA, Ahmad SM, Mumtaz PT, Malik AA, Dar MA, Urwat U, Shah RA, and Ganai NA. 2016. Long non-coding RNAs: Mechanism of action and functional utility. *Noncoding RNA Res* 1(1): 43–50.

Bradl H (ed.). 2002. Heavy Metals in the Environment: Origin, Interaction and Remediation Volume 6. London: Academic Press.

Brunetti P, Zanella L, De Paolis A, Di Litta D, Cecchetti V, Falasca G, Barbieri M, Altamura MM, Costantino P, and Cardarelli M. 2015. Cadmium-inducible expression of the ABC-type transporter AtABCC3 increases phytochelatin-mediated cadmium tolerance in Arabidopsis. *J Exp Bot* 66(13): 3815–29.

Cabral ACS, Jakovleska J, Deb A, Penner-Hahn JE, Pecoraro VL, and Freisinger E. 2018. Further insights into the metal ion binding abilities and the metalation pathway of a plant metallothionein from *Musa acuminata*. *J Biol Inorg Chem* 23(1): 91–107.

Cai H, Huang S, Che J, Yamaji N, and Ma JF. 2019. The tonoplast-localized transporter OsHMA3 plays an important role in maintaining Zn homeostasis in rice. *J Exp Bot* 70(10): 2717–2725.

Cailliatte R, Lapeyre B, Briat JF, Mari S, and Curie C. 2009. The NRAMP6 metal transporter contributes to cadmium toxicity. *Biochem J* 422(2): 217–28.

Ceasar SA, Lekeux G, Motte P, Xiao Z, Galleni M, and Hanikenne M. 2020. di-Cysteine residues of the *Arabidopsis thaliana* HMA4 C-terminus are only partially required for cadmium transport. *Front Plant Sci* 11: 560.

Cellier MF. 2012. Nramp: from sequence to structure and mechanism of divalent metal import. *Curr Top Membr* 69: 249–93.

Chaâbene Z, Rorat A, Kriaa W, Rekik I, Mejdoub H, Vandenbulcke F, and Elleuch A. 2021. In-site and ex-site date palm exposure to heavy metals involved infra-individual biomarkers upregulation. *Plants* 10: 137.

Chang YN, Zhu C, Jiang J, Zhang H, Zhu JK, and Duan CG. 2020. Epigenetic regulation in plant abiotic stress responses. *J Integrat Plant Biol* 62: 563–580.

Chaudhary K, Agarwal S, and Khan S. 2018. Role of Phytochelatins (PCs), Metallothioneins (MTs), and Heavy Metal ATPase (HMA) genes in heavy metal tolerance. pp. 39–60. *In*: Prasad R (eds.). *Mycoremediation and Environmental Sustainability*. Fungal Biology. Springer, Cham.

Chen K, Yu Y, Sun K, Xiong HYC, Chen P, Zhu A et al. 2018. The miRNAome of ramie (*Boehmeria nivea* L.): identification, expression, and potential roles of novel microRNAs in regulation of cadmium stress response. *BMC Plant Biology* 18(1): 369.

Chen L, Shi S, Jiang N, Khanzada H, Wassan GM, Zhu C, Peng X et al. 2018. Genome-wide analysis of long non-coding RNAs affecting roots development at an early stage in the rice response to cadmium stress. *BMC Genomics* 19(1): 460.

Chen X, Tong J, Su Y, and Xiao L. 2020. *Pennisetum sinese*: A potential phytoremediation plant for chromium deletion from soil. *Sustainability* 12(9): 3651.

Chen YT, Wang Y, and Yeh KC. 2017. Role of root exudates in metal acquisition and tolerance. *Curr Opin Plant Biol* 39: 66–72.

Chen Z, Fujii Y, Yamaji N, Masuda S, Takemoto Y, Kamiya T, Yusuyin Y, Iwasaki K, Kato S, Maeshima M, Ma JF, and Ueno D. 2013. Mn tolerance in rice is mediated by MTP8.1, a member of the cation diffusion facilitator family. *J Exp Bot* 64(14): 4375–87.

Choi CS and Sano H. 2007. Abiotic-stress induces demethylation and transcriptional activation of a gene encoding a glycerophosphodiesterase-like protein in tobacco plants. *Mol Genet Genomics* 277: 589–600.

Chu HH, Car S, Socha AL, Hindt MN, Punshon T, and Guerinot ML. 2017. The Arabidopsis MTP8 transporter determines the localization of manganese and iron in seeds. *Sci Rep* 7(1): 11024.

Circu ML and Aw TY. 2010. Reactive oxygen species, cellular redox systems, and apoptosis. *Free Radic Biol Med* 48(6): 749–62. doi: 10.1016/j.freeradbiomed.2009.12.022.

Cobbett CS. 2000. Phytochelatins and their roles in heavy metal detoxification. *Plant Physiol* 123(3): 825–32.

Cong W, Miao Y, Xu L, Zhang Y, Yuan C, Wang J et al. 2019. Transgenerational memory of gene expression changes induced by heavy metal stress in rice (*Oryza sativa* L.). *BMC Plant Biol* 19: 1–14.

Contreras RA, Pizarro M, Köhler H, Sáez CA, and Zúñiga GE. 2018. Copper stress induces antioxidant responses and accumulation of sugars and phytochelatins in Antarctic Colobanthus quitensis (Kunth) Bartl. *Biol Res* 51(1): 48.

Corpas FJ, Leterrier M, Valderrama R, Airaki M, Chaki M, Palma JM, and Barroso JB. 2011. Nitric oxide imbalance provokes a nitrosative response in plants under abiotic stress. *Plant Science* 181: 604–611.

Cortés-Eslavaet J, Gómez-Arroyo S, Risueño MC, and Testillano PS. 2018. The effects of organophosphorus insecticides and heavy metals on DNA damage and program med cell death in two plant models. *Environ Pollut* 240: 77–86.

Dalvi AA and Bhalerao SA. 2013. Response of plants towards heavy metal toxicity: an overview of avoidance, tolerance and uptake mechanism. *Ann Plant Sci* 2(9): 362–8.

De Caroli M, Furini A, DalCorso G, Rojas M, and Di Sansebastiano GP. 2020. Endomembrane reorganization induced by heavy metals. *Plants (Basel)* 9(4): 482.

Delhaize E, Ma JF, and Ryan PR. 2012. Transcriptional regulation of aluminium tolerance genes. *Trends Plant Sci* 17: 341–348.

Delteil A, Gobbato E, Cayrol B, Estevan J, Michel-Romiti C, Dievart A, Kroj T, and Morel JB. 2016. Several wall-associated kinases participate positively and negatively in basal defense against rice blast fungus. *BMC Plant Biol* 16(1): 1–0.

Dennis KK, Uppal K, Liu KH, Ma C, Liang B, Go YM, and Jones DP. 2019. Phytochelatin database: a resource for phytochelatin complexes of nutritional and environmental metals. *Database (Oxford)* 2019: baz083.

Dhalaria R, Kumar D, Kumar H, Nepovimova E, Kuča K, Torequl Islam M, and Verma R. 2020. *Arbuscular mycorrhizal* fungi as potential agents in ameliorating heavy metal stress in plants. *Agronomy* 10(6): 815.

Ding Y, Gong S, Wang Y, Wang F, Bao H, Sun J, Cai C, Yi K, Chen Z, Zhu C. 2018. MicroRNA166 modulates cadmium tolerance and accumulation in rice. *Plant Physiol* 177(4): 1691–1703.

Domènech-Casal J, Tinti A, and Torreggiani A. 2008. Research progress on metallothioneins: insights into structure, metal binding properties and molecular function by spectroscopic investigations. *Biopolym Res Trends* 2008: 11–48.

Dresler S, Hanaka A, Bednarek W, and Maksymiec W. 2014. Accumulation of low-molecular-weight organic acids in roots and leaf segments of *Zea mays* plants treated with cadmium and copper. *Acta Physiol Plant* 36(6): 1565–75.

Duan L, Yu J, Xu L, Tian P, Hu X, Song X, and Pan Y. 2019. Functional characterization of a type 4 metallothionein gene (CsMT4) in cucumber. *Hortic Plant J* 5(3): 120–8.

Duby G and Boutry M. 2009. The plant plasma membrane proton pump ATPase: a highly regulated P-type ATPase with multiple physiological roles. *Pflugers Arch* 457(3): 645–55.

Estrella-Gómez N, Mendoza-Cózatl D, Moreno-Sánchez R, González-Mendoza D, Zapata-Pérez O, Martínez-Hernández A, and Santamaría JM. 2009. The Pb-hyperaccumulator aquatic fern Salvinia minima Baker, responds to Pb(2+) by increasing phytochelatins via changes in SmPCS expression and in phytochelatin synthase activity. *Aquat Toxicol* 91(4): 320–8.

Famoso AN, Clark RT, Shaff JE, Craft E, McCouch SR and Kochian LV. 2010. Development of a novel aluminum tolerance phenotyping platform used for comparisons of cereal aluminum tolerance and investigations into rice aluminum tolerance mechanisms. *Plant Physiol* 153: 1678–1691.

Fan W, Xu JM, Lou HQ, Xiao C, Chen WW, and Yang JL. 2016. Physiological and molecular analysis of aluminium-induced organic acid anion secretion from grain amaranth (*Amaranthus hypochondriacus* L.) Roots. *Int J Mol Sci* 17(5): 608.

Fan W, Liu C, Cao B, Qin M, Long D, Xiang Z, and Zhao A. 2018a. Genome-wide identification and characterization of four gene families putatively involved in cadmium uptake, translocation and sequestration in mulberry. *Front Plant Sci* 9: 879.

Fan W, Guo Q, Liu C, Liu X, Zhang M, Long D, Xiang Z, and Zhao A. 2018b. Two mulberry phytochelatin synthase genes confer zinc/cadmium tolerance and accumulation in transgenic Arabidopsis and tobacco. *Gene* 645: 95–104.

Fang H, Jing T, Liu Z, Zhang L, Jin Z, and Pei Y. 2014. Hydrogen sulfide interacts with calcium signaling to enhance the chromium tolerance in *Setaria italica*. *Cell Calcium* 56: 472–48.

Feng SJ, Liu XS, Tao H, Tan SK, Chu SS, Oono Y et al. 2016. Variation of DNA methylation patterns associated with gene expression in rice (*Oryza sativa*) exposed to cadmium. *Plant Cell Environ* 39: 2629–2649.

Ferraz P, Fidalgo F, Almeida A, and Teixeira J. 2012. Phytostabilization of nickel by the zinc and cadmium hyperaccumulator *Solanum nigrum* L. Are metallothioneins involved? *Plant Physiol Biochem* 57: 254–60.

Finnegan EJ. 2001. Is plant gene expression regulated globally? *Trends Genet* 17: 361–365.

Forman HJ, Zhang H, and Rinna A. 2009. Glutathione: overview of its protective roles, measurement, and biosynthesis. *Mol Aspects Med* 30(1-2): 1–12.

Freisinger E. 2009. Metallothioneins in plants. *Met Ions Life Sci* 5: 107–153.

Fryzova R, Pohanka M, Martinkova P, Cihlarova H, Brtnicky M, Hladky J, and Kynicky J. 2017. Oxidative stress and heavy metals in plants. *Residue Rev* 245: 129–156.

Galindo-González L, Sarmiento F, and Quimbaya MA. 2018. Shaping plant adaptability, genome structure and gene expression through transposable element epigenetic control: Focus on methylation. *Agronomy* 8: 180.

Gao H, Xie W, Yang C, Xu J, Li J, Wang H, Chen X, and Huang CF. 2018. NRAMP2, a trans-Golgi network-localized manganese transporter, is required for Arabidopsis root growth under manganese deficiency. *New Phytol* 217(1): 179–193.

Gao J, Luo M, Peng H, Chen F, and Li W. 2019. Characterization of cadmium-responsive MicroRNAs and their target genes in maize (*Zea mays*) roots. *BMC Mol Biol* 20(1): 14.

Gao L, Peng K, Xia Y, Wang G, Niu L, Lian C, and Shen Z. 2013. Cadmium and manganese accumulation in *Phytolacca americana* L. and the roles of non-protein thiols and organic acids. *Int J Phytoremediation* 15(4): 307–19.

Gasic K and Korban SS. 2007. Transgenic Indian mustard (*Brassica juncea*) plants expressing an *Arabidopsis phytochelatin* synthase (AtPCS1) exhibit enhanced As and Cd tolerance. *Plant Mol Biol* 64(4): 361–9.

Gechev T and Petrov V. 2020. Reactive oxygen species and abiotic stress in plants. *Int J Mol Sci* 21(20): 7433.

Georgiadou EC, Kowalska E, Patla K, Kulbat K, Smolińska B, Leszczyńska J, and Fotopoulos V. 2018. Influence of heavy metals (Ni, Cu, and Zn) on nitro-oxidative stress responses, proteome regulation and allergen production in basil (*Ocimum basilicum* L.) Plants. *Front Plant Sci* 9: 862.

Ghori NH, Ghori T, Hayat MQ, Imadi SR, Gul A, Altay V, and Ozturk M. 2019. Heavy metal stress and responses in plants. *Int J Environ Sci Technol* 16: 1807–1828.

Gomolplitinant KM and Saier MH Jr. 2011. Evolution of the oligopeptide transporter family. *J Membr Biol* 240(2): 89–110.

Guerinot ML. 2000. The ZIP family of metal transporters. *Biochim Biophys Acta* 1465(1-2): 190–8.

Gulli M, Marchi L, Fragni R, Buschini A, and Visioli G. 2018. Epigenetic modifications preserve the hyperaccumulator *Noccaeacaerulescens* from Ni geno-toxicity. *Environ Mol Mutagenesis* 59: 464–475.

Guo J, Xu L, Su Y, Wang H, Gao S, Xu J, and Que Y. 2013. ScMT2-1-3, a metallothionein gene of sugarcane, plays an important role in the regulation of heavy metal tolerance/accumulation. *Biomed Res Int* 2013: 904769.

Guo WJ, Meetam M, and Goldsbrough PB. 2008. Examining the specific contributions of individual Arabidopsis metallothioneins to copper distribution and metal tolerance. *Plant Physiol* 146(4): 1697–706.

Guo X, Liu Y, Zhang R, Luo J, Song Y, Li J, Wu K, Peng L, Liu Y, Du Y, and Liang Y. 2019. Hemicellulose modification promotes cadmium hyperaccumulation by decreasing its retention on roots in *Sedum alfredii*. *Plant Soil* 5: 1–5.

Gupta M, Tripathi RD, Rai UN, and Chandra P. 1998. Role of glutathione and phytochelatin in *Hydrilla verticillata* (If) royle and *Valusneria spiraus* L. under mercury stress. *Chemosphere* 37(4): 785–800.

Gustin JL, Loureiro ME, Kim D, Na G, Tikhonova M, and Salt DE. 2009. MTP1-dependent Zn sequestration into shoot vacuoles suggests dual roles in Zn tolerance and accumulation in Zn-hyperaccumulating plants. *Plant J* 57(6): 1116–27.

Hasan MK, Cheng Y, Kanwar MK, Chu XY, Ahammed GJ, and Qi ZY. 2017. Responses of plant proteins to heavy metal stress-a review. *Front Plant Sci* 8: 1492.

Hassinen VH, Tervahauta AI, Schat H, and Kärenlampi SO. 2011. Plant metallothioneins—metal chelators with ROS scavenging activity? *Plant Biol (Stuttg)* 13(2): 225–232.

He F, Liu Q, Zheng L, Cui Y, Shen Z, and Zheng L. 2015. RNA-Seq analysis of rice roots reveals the involvement of post-transcriptional regulation in response to cadmium stress. *Front Plant Sci* 6: 1136.

He G, Zhu X, Elling AA, Chen L, Wang X, Guo L et al. 2010. Global epigenetic and transcriptional trends among two rice subspecies and their reciprocal hybrids. *Plant Cell* 22: 17–33.

He ZL, Yang XE, and Stoffella PJ. 2005. Trace elements in agroecosystems and impacts on the environment. *J Trace Elem Med Biol* 19(2-3): 125–140.

Hietakangas V, Ahlskog JK, Jakobsson AM, Hellesuo M, Sahlberg NM, Holmberg CI, Mikhailov A, Palvimo JJ, Pirkkala L, and Sistonen L. 2003. Phosphorylation of serine 303 is a prerequisite for the stress-inducible SUMO modification of heat shock factor 1. *Mol Cell Biol* 23(8): 2953–68.

Hossain MM, Khatun MA, Haque MN, Bari MA, Alam MF, Mandal A, and Kabir AH. 2018. Silicon alleviates arsenic-induced toxicity in wheat through vacuolar sequestration and ROS scavenging. *Int J Phytoremediation* 20(8): 796–804.

Huang CF, Yamaji N, Mitani N, Yano M, Nagamura Y, and Ma JF. 2009. A bacterial-type ABC transporter is involved in aluminum tolerance in rice. *Plant Cell* (2): 655–667.

Huang XY, Deng F, Yamaji N, Pinson SR, Fujii-Kashino M, Danku J, Douglas A, Guerinot ML, Salt DE, and Ma JF. 2016. A heavy metal P-type ATPase OsHMA4 prevents copper accumulation in rice grain. *Nat Commun* 7(1): 1–3.

Huda AK, Haque MA, Zaman R, Swaraz AM, and Kabir AH. 2017. Silicon ameliorates chromium toxicity through phytochelatin-mediated vacuolar sequestration in the roots of *Oryza sativa* (L.). *Int J Phytoremediation* 19(3): 246–253.

Iglesia-Turiño S, Febrero A, Jauregui O, Caldelas C, Araus JL, and Bort J. 2006. Detection and quantification of unbound phytochelatin 2 in plant extracts of *Brassica napus* grown with different levels of mercury. *Plant Physiol* 142(2): 742–9.

Inoue H, Kobayashi T, Nozoye T, Takahashi M, Kakei Y, Suzuki K, Nakazono M, Nakanishi H, Mori S, and Nishizawa NK. 2009. Rice OsYSL15 is an iron-regulated iron(III)-deoxymugineic acid transporter expressed in the roots and is essential for iron uptake in early growth of the seedlings. *J Biol Chem* 284(6): 3470–9.

Inouhe M. 2005. Phytochelatins. *Braz J Plant Physiol* 17(1): 65–78.

Jain S, Muneer S, Guerriero G, Liu S, Vishwakarma K, Chauhan DK, Dubey NK, Tripathi DK, and Sharma S. 2018. Tracing the role of plant proteins in the response to metal toxicity: a comprehensive review. *Plant Signal Behav* 13(9): e1507401.

Jha UC, Nayyar H, Jha R, Khurshid M, Zhou M, Mantri N, and Siddique KHM. 2020. Long non-coding RNAs: emerging players regulating plant abiotic stress response and adaptation. *BMC Plant Biol* 20(1): 466.

Jin S, Sun D, Wang J, Li Y, Wang X, and Liu S. 2014. Expression of the rgMT gene, encoding for a rice metallothionein-like protein in *Saccharomyces cerevisiae* and *Arabidopsis thaliana*. *J Genet* 93(3): 709–18.

Jin S, Xu C, Li G, Sun D, Li Y, Wang X, and Liu S. 2017. Functional characterization of a type 2 metallothionein gene, SsMT2, from alkaline-tolerant *Suaeda salsa*. *Sci Rep* 7(1): 17914.

Jogawat A, Yadav B, and Narayan OP. 2021. Metal transporters in organelles and their roles in heavy metal transportation and sequestration mechanisms in plants. *Physiol Plant* doi: https://doi.org/10.1111/ppl.13370.

Johansson EM, Fransson PM, Finlay RD, and van Hees PA. 2008. Quantitative analysis of root and ectomycorrhizal exudates as a response to Pb, Cd and As stress. *Plant Soil* 313(1): 39–54.

Kajala K, Walker KL, Mitchell GS, Krämer U, Cherry SR, and Brady SM. 2019. Real-time whole-plant dynamics of heavy metal transport in *Arabidopsis halleri* and *Arabidopsis thaliana* by gamma-ray imaging. *Plant Direct* 3(4): e00131.

Kalisz S and Purugganan MD. 2004. Epialleles via DNA methylation: Consequences for plant evolution. *Trends Ecol Evol* 19: 309–314.

Kapoor D, Singh MP, Kaur S, Bhardwaj R, Zheng B, and Sharma A. 2019. Modulation of the functional components of growth, photosynthesis, and anti-oxidant stress markers in cadmium exposed *Brassica juncea* L. *Plants (Basel)* 8(8): 260.

Keyster M, Niekerk LA, Basson G, Carelse M, Bakare O, Ludidi N, Klein A, Mekuto L, and Gokul A. 2020. Decoding heavy metal stress signalling in plants: towards improved food security and safety. *Plants (Basel)* 9(12): 1781.

Khanna K, Jamwal VL, Gandhi SG, Ohri P, and Bhardwaj R. 2019. Metal resistant PGPR lowered Cd uptake and expression of metal transporter genes with improved growth and photosynthetic pigments in *Lycopersicon esculentum* under metal toxicity. *Sci Rep* 9(1): 5855.

Khare D, Mitsuda N, Lee S, Song WY, Hwang D, Ohme-Takagi M, Martinoia E, Lee Y, and Hwang JU. 2017. Root avoidance of toxic metals requires the GeBP-LIKE 4 transcription factor in *Arabidopsis thaliana*. *New Phytol* 213(3): 1257–1273.

Khatiwada B, Hasan MT, Sun A, Kamath KS, Mirzaei M, Sunna A, and Nevalainen H. 2020. Probing the role of the chloroplasts in heavy metal tolerance and accumulation in *Euglena gracilis*. *Microorganisms* 8(1): 115.

Kim BM, Rhee JS, Jeong CB, Seo JS, Park GS, Lee YM, and Lee JS. 2014. Heavy metals induce oxidative stress and trigger oxidative stress-mediated heat shock protein (hsp) modulation in the intertidal copepod *Tigriopus japonicus*. *Comp Biochem Physiol C Toxicol Pharmacol* 166: 65–74.

Kim JO, Lee YW, and Chung J. 2013. The role of organic acids in the mobilization of heavy metals from soil. *KSCE J Civ Eng* 17(7): 1596–602.

Kim YH, Khan A, Kim DH, Lee, SY, Kim KM, Waqas M et al. 2014. Silicon mitigates heavy metal stress by regulating P-type heavy metal ATPases, *Oryza sativa* low silicon genes, and endogenous phytohormones. *BMC Plant Biology* 14(1): 13.

Kim YO, Kang H, and Ahn SJ. 2019. Overexpression of phytochelatin synthase AtPCS2 enhances salt tolerance in *Arabidopsis thaliana*. *J Plant Physiol* 240: 153011.

Kobae Y, Uemura T, Sato MH, Ohnishi M, Mimura T, Nakagawa T, and Maeshima M. 2004. Zinc transporter of *Arabidopsis thaliana* AtMTP1 is localized to vacuolar membranes and implicated in zinc homeostasis. *Plant Cell Physiol* 45(12): 1749–58.

Kochian LV, Piñeros MA, Liu J, and Magalhaes JV. 2015. Plant adaptation to acid soils: the molecular basis for crop aluminum resistance. *Annu Rev Plant Biol* 66: 571–598.

Kohorn BD and Kohorn SL. 2012. The cell wall-associated kinases, WAKs, as pectin receptors. *Front Plant Sci* 3: 88.

Kováčik J, Dresler S, Peterková V, and Babula P. 2018. Metal-induced oxidative stress in terrestrial macrolichens. *Chemosphere* 203: 402–409.

Kühnlenz T, Hofmann C, Uraguchi S, Schmidt H, Schempp S, Weber M, Lahner B, Salt DE, and Clemens S. 2016. Phytochelatin synthesis promotes leaf Zn accumulation of *Arabidopsis thaliana* plants grown in soil with adequate Zn supply and is essential for survival on Zn-contaminated soil. *Plant Cell Physiol* 57(11): 2342–52.

Kumar G, Kushwaha HR, Panjabi-Sabharwal V, Kumari S, Joshi R, Karan R, Mittal S, Pareek SL, and Pareek A. 2012. Clustered metallothionein genes are co-regulated in rice and ectopic expression of OsMT1e-P confers multiple abiotic stress tolerance in tobacco via ROS scavenging. *BMC Plant Biol* 12: 107.

Lanciano S and Mirouze M. 2017. DNA methylation in rice and relevance for breeding. *Epigenomes* 1: 10.

Lane TS, Rempe CS, Davitt J, Staton ME, Peng Y, Soltis DE, Melkonian M, Deyholos M, Leebens-Mack JH, Chase M, Rothfels CJ, Stevenson D, Graham SW, Yu J, Liu T, Pires JC, Edger PP, Zhang Y, Xie Y, Zhu Y, Carpenter E, Wong GK, and Stewart CN Jr. 2016. Diversity of ABC transporter genes across the plant kingdom and their potential utility in biotechnology. *BMC Biotechnol* 16(1): 47.

Lanquar V, Lelièvre F, Bolte S, Hamès C, Alcon C, Neumann D, Vansuyt G, Curie C, Schröder A, Krämer U, Barbier-Brygoo H, and Thomine S. 2005. Mobilization of vacuolar iron by AtNRAMP3 and AtNRAMP4 is essential for seed germination on low iron. *EMBO J* 24(23): 4041–51.

Lešková A, Zvarík M, Araya T, and Giehl RF. 2020. Nickel toxicity targets cell wall-related processes and PIN2-mediated auxin transport to inhibit root elongation and gravitropic responses in Arabidopsis. *Plant Cell Physiol* 61(3): 519–35.

Li J, Zhang M, Sun J, Mao X, Wang J, Liu H, Zheng H, Li X, Zhao H, and Zou D. 2020. Heavy metal stress-associated proteins in rice and *Arabidopsis*: Genome-Wide Identification, Phylogenetics, Duplication, and Expression Profiles Analysis. *Front Genet* 11: 477.

Li JY, Liu J, Dong D, Jia X, McCouch SR, and Kochian LV. 2014. Natural variation underlies alterations in Nrampaluminum transporter (NRAT1) expression and function that play a key role in rice aluminum tolerance. *Proc Natl Acad Sci USA* 111: 6503–6508.

Li X, Zhu J, Hu F, Ge S, Ye M, Xiang H et al. 2012. Single-base resolution maps of cultivated and wild rice methylomes and regulatory roles of DNA methylation in plant gene expression. *BMC Genomics* 13: 1–15.

Li Q, Hu A, Qi J, Dou W, Qin X, Zou X, Xu L, Chen S, and He Y. 2020. CsWAKL08, a pathogen-induced wall-associated receptor-like kinase in sweet orange, confers resistance to citrus bacterial canker via ROS control and JA signaling. *Hortic Res* 7(1): 1–5.

Liu X, Chen J, Wang GH, Wang WH, Shen ZJ, Luo MR, Gao GF, Simon M, Ghoto K, and Zheng HL. 2016. Hydrogen sulfide alleviates zinc toxicity by reducing zinc uptake and regulating genes expression of antioxidative enzymes and metallothioneins in roots of the cadmium/zinc hyperaccumulator *Solanum nigrum* L. *Plant Soil* 400: 177–192.

Liu Z, Gu C, Chen F, Yang D, Wu K, Chen S, Jiang J, and Zhang Z. 2012. Heterologous expression of a Nelumbo nuciferaphytochelatin synthase gene enhances cadmium tolerance in *Arabidopsis thaliana*. *Appl Biochem Biotechnol* 166(3): 722–34.

Llerena JPP, Coasaca RL, Rodriguez HOL, Llerena SÁP, Valencia YD, and Mazzafera P. 2021. Metallothionein production is a common tolerance mechanism in four species growing in polluted Cu mining areas in Peru. *Ecotoxicol Environ Saf* 212: 112009.

Loebus J, Peroza EA, Blüthgen N, Fox T, Meyer-Klaucke W, Zerbe O, and Freisinger E. 2011. Protein and metal cluster structure of the wheat metallothionein domain γ-E(c)-1: the second part of the puzzle. *J Biol Inorg Chem* 16(5): 683–94.

Lu TX and Rothenberg ME. 2018. MicroRNA. *J Allergy Clin Immunol* 141(4): 1202–1207.

Luo Q, Sun L, Hu X, and Zhou R. 2014. The variation of root exudates from the hyperaccumulator *Sedum alfredii* under cadmium stress: metabonomics analysis. *PloS one* 9(12): e115581.

Luo ZB, He J, Polle A, and Rennenberg H. 2016. Heavy metal accumulation and signal transduction in herbaceous and woody plants: Paving the way for enhancing phytoremediation efficiency. *Biotechnol Adv* 34(6): 1131–1148.

Malar S, Manikandan R, Favas PJ, VikramSahi S, and Venkatachalam P. 2014. Effect of lead on phytotoxicity, growth, biochemical alterations and its role on genomic template stability in *Sesbania grandiflora*: a potential plant for phytoremediation. *Ecotoxicol Environ Saf* 108: 249–57.

Mehes-Smith M, Nkongolo K, and Cholewa E. 2013. Coping mechanisms of plants to metal contaminated soil. *Environmental Change and Sustainability* 54: 53–90.

Menguer PK, Farthing E, Peaston KA, Ricachenevsky FK, Fett JP, and Williams LE. 2013. Functional analysis of the rice vacuolar zinc transporter OsMTP1. *J Exp Bot* 64(10): 2871–83.

Meychik N, Nikolaeva Y, Kushunina M, and Yermakov I. 2014. Are the carboxyl groups of pectin polymers the only metal-binding sites in plant cell walls? *Plant Soil* 381(1): 25–34.

Milner MJ, Mitani-Ueno N, Yamaji N, Yokosho K, Craft E, Fei Z, Ebbs S, Clemencia Zambrano M, Ma JF, and Kochian LV. 2014. Root and shoot transcriptome analysis of two ecotypes of *Noccaea caerulescens* uncovers the role of NcNramp1 in Cd hyperaccumulation. *Plant J* 78(3): 398–410.

Milner MJ, Seamon J, Craft E, and Kochian LV. 2013. Transport properties of members of the ZIP family in plants and their role in Zn and Mn homeostasis. *J Exp Bot* 64(1): 369–81.

Mirouze M and Paszkowski J. 2011. Epigenetic contribution to stress adaptation in plants. *Curr Opin Plant Biol* 14: 267–274.

Mishra A, Bhattacharya A, and Mishra N. 2019. Mycorrhizal symbiosis: An effective tool for metal bioremediation. pp. 113–128. *In*: Singh JS (ed.). *New and Future Developments in Microbial Biotechnology and Bioengineering*. Elsevier: Amsterdam, The Netherlands.

Mishra S, Srivastava S, Tripathi RD, Govindarajan R, Kuriakose SV, and Prasad MN. 2006. Phytochelatin synthesis and response of antioxidants during cadmium stress in *Bacopa monnieri* L. *Plant Physiol Biochem* 44(1): 25–37.

Misra I and Griffith OW. 1998. Expression and purification of human gamma-glutamylcysteine synthetase. *Protein Expr Purif* 13(2): 268–76.

Moe LA. 2013. Amino acids in the rhizosphere: from plants to microbes. *Am J Bot* 100(9): 1692–705.

Montiel-Rozas MM, Madejón E, and Madejón P. 2016. Effect of heavy metals and organic matter on root exudates (low molecular weight organic acids) of herbaceous species: An assessment in sand and soil conditions under different levels of contamination. *Environ Pollut* 216: 273–281.

Morel M, Crouzet J, Gravot A, Auroy P, Leonhardt N, Vavasseur A, and Richaud P. 2009. AtHMA3, a P1B-ATPase allowing Cd/Zn/Co/Pb vacuolar storage in Arabidopsis. *Plant Physiol* 149(2): 894–904.

Mukta RH, Khatun MR, and Nazmul Huda AK. 2019. Calcium induces phytochelatin accumulation to cope with chromium toxicity in rice (*Oryza sativa* L.). *J Plant Interact* 14(1): 295–302.

Murata Y, Ma JF, Yamaji N, Ueno D, Nomoto K, and Iwashita T. 2006. A specific transporter for iron (III)-phytosiderophore in barley roots. *Plant J May* 46(4): 563–72. doi: 10.1111/j.1365-313X.2006.02714.x. Erratum in: *Plant J* 61(1): 188.

Namdjoyan SH and Kermanian H. 2016. Phytochelatin synthesis and responses of antioxidants during arsenic stress in *Nasturtium officinale*. *Russ J Plant Physiol* 63(6): 739–48.

Narendrula-Kotha R, Theriault G, Mehes-Smith M, Kalubi K, and Nkongolo K. 2019. Metal toxicity and resistance in plants and microorganisms in terrestrial ecosystems. *Residue Rev* 249: 1–27.

Ngu TT, and Stillman MJ. 2009. Metal-binding mechanisms in metallothioneins. *Dalton Trans* (28): 5425–33.

Nguyen XV, Le-Ho KH, and Papenbrock J. 2017. Phytochelatin 2 accumulates in roots of the seagrass Enhalusacoroides collected from sediment highly contaminated with lead. *Biometals* 30(2): 249–260.

Nguyen-DerocheTle N, Caruso A, Le TT, Bui TV, Schoefs B, Tremblin G, and Morant-Manceau A. 2012. Zinc affects differently growth, photosynthesis, antioxidant enzyme activities and phytochelatin synthase expression of four marine diatoms. *Sci World J* 2012: 982957.

Nouet C, Charlier JB, Carnol M, Bosman B, Farnir F, Motte P, and Hanikenne M. 2015. Functional analysis of the three HMA4 copies of the metal hyperaccumulator *Arabidopsis halleri*. *J Exp Bot* 66(19): 5783–95.

Oono Y, Yazawa T, Kanamori H, Sasaki H, Mori S, Handa H et al. 2016. Genome-wide transcriptome analysis of cadmium stress in rice. *BioMed Res Int* 2016: 9739505.

Osmolovskaya N, Dung VV, and Kuchaeva L. 2018. The role of organic acids in heavy metal tolerance in plants. *Biol Commun* 63(1).

Ou X, Zhang Y, Xu C, Lin X, Zang Q, Zhuang T et al. 2012. Transgenerational inheritance of modified DNA methylation patterns and enhanced tolerance induced by heavy metal stress in rice (*Oryza sativa* L.). *PLoS One* 7: 41143.

Page V and Feller U. 2005. Selective transport of zinc, manganese, nickel, cobalt and cadmium in the root system and transfer to the leaves in young wheat plants. *Ann Bot* 96(3): 425–34.

Page V, Weisskopf L, and Feller U. 2006. Heavy metals in white lupin: uptake, root-to-shoot transfer and redistribution within the plant. *New Phytol* 171(2): 329–41.

Pal R and Rai JP. 2010. Phytochelatins: peptides involved in heavy metal detoxification. *Appl Biochem Biotechnol* 160(3): 945–63.

Pan W, Shen J, Zheng Z, Yan X, Shou J, Wang W, Jiang L, and Pan J. 2018. Overexpression of the *Tibetan Plateau* annual wild barley (*Hordeum spontaneum*) HsCIPKs enhances rice tolerance to heavy metal toxicities and other abiotic stresses. *Rice (N Y)* 11(1): 51.

Pandey AK, Gautam A, and Dubey RS. 2019. Transport and detoxification of metalloids in plants in relation to plantmetalloid tolerance. *Plant Gene* 17: 100171.

Park J, Song WY, Ko D, Eom Y, Hansen TH, Schiller M, Lee TG, Martinoia E, and Lee Y. 2012. The phytochelatin transporters AtABCC1 and AtABCC2 mediate tolerance to cadmium and mercury. *Plant J* 69(2): 278–88.

Parrotta L, Guerriero G, Sergeant K, Cai G, and Hausman JF. 2015. Target or barrier? The cell wall of early- and later-diverging plants vs cadmium toxicity: differences in the response mechanisms. *Front Plant Sci* 6: 133.

Peiter E, Montanini B, Gobert A, Pedas P, Husted S, Maathuis FJ, Blaudez D, Chalot M, and Sanders D. 2007. A secretory pathway-localized cation diffusion facilitator confers plant manganese tolerance. *Proc Natl Acad Sci USA* 104(20): 8532–7.

Peng F, Wang C, Cheng Y, Kang H, Fan X, Sha L, Zhang H, Zeng J, Zhou Y, and Wang Y. 2018. Cloning and characterization of TpNRAMP3, a metal transporter from polish wheat (*Triticum polonicum* L.). *Front Plant Sci* 9: 1354.

Pinto E, Sigaud-Kutner TCS, Leita⁓o MAS, Okamoto OK, Morse D, and Colepicolo P. 2003. Heavy metal–induced oxidative stress in algae. *Departamento J Phycol* 39: 1008–1018.

Pizzino G, Irrera N, Cucinotta M, Pallio G, Mannino F, Arcoraci V, Squadrito F, Altavilla D, and Bitto A. 2017. Oxidative Stress: harms and benefits for human health. *Oxid Med Cell Longev* 2017: 8416763.

Printz B, Lutts S, Hausman JF, and Sergeant K. 2016. Copper trafficking in plants and its implication on cell wall dynamics. *Front Plant Sci* 7: 601.

Radons J. 2016. The human HSP70 family of chaperones: where do we stand? *Cell Stress Chaperones* 21(3): 379–404.

Rahman MF, Ghosal A, Alam MF, and Kabir AH. 2017. Remediation of cadmium toxicity in field peas (*Pisum sativum* L.) through exogenous silicon. *Ecotoxicol Environ Saf* 135: 165–172.

Rahman Z and Singh VP. 2019. The Relative Impact of Toxic Heavy Metals (THMs) (Arsenic (As), Cadmium (Cd), Chromium (Cr) (VI), Mercury (Hg), and Lead (Pb)) On the Total Environment: An Overview. *Environ Monit Assess* 191: 1–21.

Rai A, Singh R, Shirke PA, Tripathi RD, Trivedi PK, and Chakrabarty D. 2015. Expression of rice CYP450-Like Gene (Os08g01480) in arabidopsis modulates regulatory network leading to heavy metal and other abiotic stress tolerance. *PLoS One* 10(9): e0138574.

Ranieri E, Tursi A, Giuliano S, Spagnolo V, Ranieri AC, and Petrella A. 2020. Phytoextraction from chromium-contaminated soil using moso bamboo in mediterranean conditions. *Water Air Soil Pollut* 231(8): 1–2.

Rea PA. 2012. Phytochelatin synthase: of a protease a peptide polymerase made. *Physiol Plant* 145(1): 154–64.

Remy E and Duque P. 2014. Beyond cellular detoxification: a plethora of physiological roles for MDR transporter homologs in plants. *Front Physiol* 5: 201.

Rono JK, Le Wang L, Wu XC, Cao HW, Zhao YN, Khan IU, and Yang ZM. 2021. Identification of a new function of metallothionein-like gene OsMT1e for cadmium detoxification and potential phytoremediation. *Chemosphere* 265: 129136.

Sánchez-Fernández R, Davies TG, Coleman JO, and Rea PA. 2001. The *Arabidopsis thaliana* ABC protein superfamily, a complete inventory. *J Biol Chem* 276(32): 30231–44.

Santisree P, Adimulam SS, Sharma K, Bhatnagar-Mathur P, and Sharma KK. 2019. Insights into the nitric oxide mediated stress tolerance in plants. pp. 385–406. *In*: Khan MIR (ed.). *Plant Signaling Molecules*. Woodhead Publishing.

Sasaki A, Yamaji N, and Ma JF. 2014. Overexpression of OsHMA3 enhances Cd tolerance and expression of Zn transporter genes in rice. *J Exp Bot* 65(20): 6013–21.

Saze H and Kakutani T. 2007. Heritable epigenetic mutation of a transposon flanked *Arabidopsis* gene due to lack of the chromatin-remodeling factor DDM1. *EMBO J* 26: 3641–3652.

Schicht O and Freisinger E. 2009. Spectroscopic characterization of *Cicer arietinum* metallothionein 1. *Inorganica Chimica Acta* 362(3): 714–24.

Schiewer S and Iqbal M. 2010. The role of pectin in Cd binding by orange peel biosorbents: a comparison of peels, depectinated peels and pectic acid. *J Hazard Mater* 177(1-3): 899–907.

Seneviratne M, Rajakaruna N, Rizwan M, Madawala HMSP, OK YS, and Vithanage M. 2019. Heavy metal-induced oxidative stress on seed germination and seedling development: a critical review. *Environ Geochem Health* 41(4): 1813–183.

Seth CS, Kumar Chaturvedi P, and Misra V. 2008. The role of phytochelatins and antioxidants in tolerance to Cd accumulation in *Brassica juncea* L. *Ecotoxicol Environ Saf* 71(1): 76–85.

Shah K and Dubey R. 1998. Cadmium elevates level of protein, amino acids and alters activity of proteolytic enzymes in germinating rice seeds. *Acta Physiol. Plant* 20(2): 189–196.

Shanmugaraj BM, Chandra HM, Srinivasan B, and Ramalingam S. 2013. Cadmium induced physio-biochemical and molecular response in *Brassica juncea*. *Int J Phytoremediation* 15(3): 206–18.

Sharma P and Dubey RS. 2005. Lead toxicity in plants. *Braz J Plant Physiol* 17(1): 35–52.

Sharma P, Jha AB, Dubey RS, and Pessarakli M. 2012. Reactive oxygen species, oxidative damage, and antioxidative defense mechanism in plants under stressful conditions. *Journal of Botany*, 1–26.

Shinkai Y, Masuda A, Akiyama M, Xian M, and Kumagai Y. 2017. Cadmium-mediated activation of the HSP90/HSF1 pathway regulated by reactive Persulfides/Polysulfides. *Toxicol Sci* 156(2): 412–421.

Shri M, Dave R, Diwedi S, Shukla D, Kesari R, Tripathi RD, Trivedi PK, and Chakrabarty D. 2014. Heterologous expression of *Ceratophyllum demersumphytochelatin* synthase, CdPCS1, in rice leads to lower arsenic accumulation in grain. *Sci Rep* 4: 5784.

Shukla D, Kesari R, Mishra S, Dwivedi S, Tripathi RD, Nath P, and Trivedi PK. 2012. Expression of phytochelatin synthase from aquatic macrophyte *Ceratophyllum demersum* L. enhances cadmium and arsenic accumulation in tobacco. *Plant Cell Rep* 31(9): 1687–99.

Shukla D, Kesari R, Tiwari M, Dwivedi S, Tripathi RD, Nath P, and Trivedi PK. 2013. Expression of *Ceratophyllum demersumphytochelatin* synthase, CdPCS1, in *Escherichia coli* and *Arabidopsis* enhances heavy metal(loid)s accumulation. *Protoplasma* 250(6): 1263–72.

Singh VP, Singh S, Kumar J, and Prasad SM. 2015. Hydrogen sulfide alleviates toxic effects of arsenate in pea seedlings through up-regulation of the ascorbate–glutathione cycle: possible involvement of nitric oxide. *J Plant Physiol* 181: 20–29.

Song WY, Yamaki T, Yamaji N, Ko D, Jung KH, Fujii-Kashino M, An G, Martinoia E, Lee Y, and Ma JF. 2014. A rice ABC transporter, OsABCC1, reduces arsenic accumulation in the grain. *Proc Natl Acad Sci USA* 111(44): 15699–704.

Spain O, Plöhn M, and Funk C. 2021. The cell wall of green microalgae and its role in heavy metal removal. *Physiol Plant* 2021.

Srivastava S, Tripathi RD, and Dwivedi UN. 2004. Synthesis of phytochelatins and modulation of antioxidants in response to cadmium stress in Cuscutareflexa—an angiospermic parasite. *J Plant Physiol* 161(6): 665–74.

Steurer C, Eder N, Kerschbaum S, Wegrostek C, Gabriel S, Pardo N, Ortner V, Czerny T, and Riegel E. 2018. HSF1 mediated stress response of heavy metals. *PLoS One* 2018 Dec 19; 13(12): e0209077.

Sudan J, Raina M, and Singh R. 2018. Plant epigenetic mechanisms: role in abiotic stress and their generational heritability. *3 Biotech* 8(172): 1–12.

Sytar O, Kumari P, Yadav S, Brestic M, and Rastogi A. 2019. Phytohormone priming: regulator for heavy metal stress in plants. *J Plant Growth Regul* 38(2): 739–52.

Szatanik-Kloc A, Szerement J, Cybulska J, and Jozefaciuk G. 2017. Input of different kinds of soluble pectin to cation binding properties of roots cell walls. *Plant Physiol Biochem* 120: 194–201.

Takahashi R, Ishimaru Y, Senoura T, Shimo H, Ishikawa S, Arao T, Nakanishi H, and Nishizawa NK. 2011. The OsNRAMP1 iron transporter is involved in Cd accumulation in rice. *J Exp Bot* 62(14): 4843–50.

Takahashi S. 2012. Molecular functions of metallothionein and its role in hematological malignancies. *J Hematol Oncol* 5: 41.

Takemoto Y, Tsunemitsu Y, Fujii-Kashino M, Mitani-Ueno N, Yamaji N, Ma JF, Kato SI, Iwasaki K, and Ueno D. 2017. The tonoplast-localized transporter MTP8.2 contributes to manganese detoxification in the shoots and roots of *Oryza sativa* L. *Plant Cell Physiol* 58(9): 1573–1582.

Talebi M, Tabatabaei BES, and Akbarzadeh H. 2019. Hyperaccumulation of Cu, Zn, Ni, and Cd in *Azolla* species inducing expression of methallothionein and phytochelatin synthase genes. *Chemosphere* 230: 488–497.

Thakur S, Singh L, Ab Wahid Z, Siddiqui MF, Atnaw SM, and Din MF. 2016. Plant-driven removal of heavy metals from soil: uptake, translocation, tolerance mechanism, challenges, and future perspectives. *Environ Monit Assess* 188(4): 206.

Tsunemitsu Y, Genga M, Okada T, Yamaji N, Ma JF, Miyazaki A, Kato SI, Iwasaki K, and Ueno D. 2018. A member of cation diffusion facilitator family, MTP11, is required for manganese tolerance and high fertility in rice. *Planta* 248(1): 231–241.

Ueno D, Milner MJ, Yamaji N, Yokosho K, Koyama E, Clemencia Zambrano M, Kaskie M, Ebbs S, Kochian LV, and Ma JF. 2011. Elevated expression of TcHMA3 plays a key role in the extreme Cd tolerance in a Cd-hyperaccumulating ecotype of Thlaspicaerulescens. *Plant J* 66(5): 852–62.

Ullah S, Khan J, Hayat K, AbdelfattahElateeq A, Salam U, Yu B, Ma Y, Wang H, and Tang ZH. 2020. Comparative study of growth, cadmium accumulation and tolerance of three chickpea (*Cicer arietinum* L.) cultivars. *Plants* 9(3): 310.

Verghese J, Abrams J, Wang Y, and Morano KA. 2012. Biology of the heat shock response and protein chaperones: budding yeast (*Saccharomyces cerevisiae*) as a model system. *Microbiol Mol Biol Rev* 76(2): 115–58.

Verhoeven KJF, Jansen JJ, Dijk PJ, and Biere A. 2010. Stress induced DNA methylation changes and their heritability in asexual dandelions. *N Phytol* 185: 1108–1118.

Volkova PY, Geras'kin SA, Horemans N, Makarenko ES, Saenen E, Duarte GT et al. 2018. Chronic radiation exposure as an ecological factor: Hypermethylation and genetic differentiation in irradiated Scots pine populations. *Environ Poll* 232: 105–112.

Wang H, Du H, Li H, Huang Y, Ding J, Liu C, Wang N, Lan H, and Zhang S. 2018. Identification and functional characterization of the ZmCOPT copper transporter family in maize. *PLoS One* 13(7): e0199081.

Wang H, Ji F, Zhang Y, Hou J, Liu W, Huang J, and Liang W. 2019. Interactions between hydrogen sulphide and nitric oxide regulate two soybean citrate transporters during the alleviation of aluminium toxicity. *Plant Cell Environ* 42(8): 2340–2356.

Wang L, Li R, Yan X, Liang X, Sun Y, and Xu Y. 2020. Pivotal role for root cell wall polysaccharides in cultivar-dependent cadmium accumulation in *Brassica chinensis* L. *Ecotoxicol Environ Saf* 194: 110369.

Wei JW, Huang K, Yang C, and Kang CS. 2017. Non-coding RNAs as regulators in epigenetics (Review). *Oncol Rep* 37(1): 3–9.

Wei S, Anders I, and Feller U. 2014. Selective uptake, distribution, and redistribution of (109)Cd, (57)Co, (65)Zn, (63)Ni, and (134)Cs via xylem and phloem in the heavy metal hyperaccumulator *Solanum nigrum* L. *Environ Sci Pollut Res Int* 21(12): 7624–30.

Wei S, Zhou Q, and Wang X. 2005. Identification of weed plants excluding the uptake of heavy metals. *Environ Int* 31(6): 829–34.

Wu J, Liu T, Rios Z, Mei Q, Lin X, and Cao S. 2017. Heat shock proteins and cancer. *Trends Pharmacol Sci* 38(3): 226–256.

Wu L, Kobayashi Y, Wasaki J, and Koyama H. 2018. Organic acid excretion from roots: a plant mechanism for enhancing phosphorus acquisition, enhancing aluminum tolerance, and recruiting beneficial rhizobacteria. *Soil Sci Plant Nutr* 64(6): 697–704.

Wu TY, Gruissem W, and Bhullar NK. 2018. Facilitated citrate-dependent iron translocation increases rice endosperm iron and zinc concentrations. *Plant Sci* 270: 13–22.

Wünschmann J, Beck A, Meyer L, Letzel T, Grill E, and Lendzian KJ. 2007. Phytochelatins are synthesized by two vacuolar serine carboxypeptidases in *Saccharomyces cerevisiae*. *FEBS Lett* 581(8): 1681–7.

Xia J, Yamaji N, Kasai T, and Ma JF. 2010. Plasma membrane-localized transporter for aluminum in rice. *Proc Natl Acad Sci USA* 107(43): 18381–18385.

Yan A, Wang Y, Tan SN, Yusof ML, Ghosh S, and Chen Z. 2020. Phytoremediation: a promising approach for revegetation of heavy metal-polluted land. *Front Plant Sci* 11: 359.

Yan H, Kikuchi S, Neumann P, Zhang W, Wu Y, Chen F et al. 2010. Genome-wide mapping of cytosine methylation revealed dynamic DNA methylation patterns associated with genes and centromeres in rice. *Plant J* 63: 353–365.

Yang JL, Zhu XF, Peng YX, Zheng C, Li GX, Liu Y, Shi YZ, and Zheng SJ. 2011. Cell wall hemicellulose contributes significantly to aluminum adsorption and root growth in *Arabidopsis*. *Plant Physiol* 155(4): 1885–1892.

Yang M, Zhang W, Dong H, Zhang Y, Lv K, Wang D, and Lian X. 2013. OsNRAMP3 is a vascular bundles-specific manganese transporter that is responsible for manganese distribution in rice. *PLoS One* 8(12): e83990.

Yang M, Zhang Y, Zhang L, Hu J, Zhang X, Lu K, Dong H, Wang D, Zhao FJ, Huang CF, and Lian X. 2014. OsNRAMP5 contributes to manganese translocation and distribution in rice shoots. *J Exp Bot* 65(17): 4849–61.

Yang M, Zhang F, Wang F, Dong Z, Cao Q, and Chen M. 2015. Characterization of a type 1 metallothionein gene from the stresses-tolerant plant *Ziziphus jujuba*. *Int J Mol Sci* 16(8): 16750–62.

Yokosho K, Yamaji N, Fujii-Kashino M, and Ma JF. 2016. Functional analysis of a mate gene osfrdl2 revealed its involvement in al-induced secretion of citrate, but a lower contribution to all tolerance in Rice. *Plant Cell Physiol* 57: 976–985.

Yu H, Guo J, Li Q, Zhang X, Huang H, Huang F, Yang A, and Li T. 2020. Characteristics of cadmium immobilization in the cell wall of root in a cadmium-safe rice line (*Oryza sativa* L.). *Chemosphere* 241: 125095.

Yu XZ, Lin YJ, and Zhang Q. 2019. Metallothioneins enhance chromium detoxification through scavenging ROS and stimulating metal chelation in *Oryza sativa*. *Chemosphere* 220: 300–313.

Yu Y, Teng Z, Mou Z et al. 2021. Melatonin confers heavy metal-induced tolerance by alleviating oxidative stress and reducing the heavy metal accumulation in *Exophialapisciphila*, a dark septateendophyte (DSE). *BMC Microbiol* 21: 40.

Zhang B, Li P, Su T, Li P, Xin X, Wang W, Zhao X, Yu Y, Zhang D, Yu S, and Zhang F. 2020. Comprehensive Analysis of wall-associated kinase genes and their expression under abiotic and biotic stress in chinese cabbage (*Brassica rapa* ssp. pekinensis). *J Plant Growth Regul* 39: 72–86.

Zhang H, Hu LY, Hu KD, He YD, Wang SH, and Luo JP. 2008. Hydrogen sulfide promotes wheat seed germination and alleviates oxidative damage against copper stress. *J Integr Plant Biol* 50(12): 1518–29.

Zhang H, Hu LY, Li P, Hu KD, Jiang CX, and Luo JP. 2010a. Hydrogen sulfide alleviated chromium toxicity in wheat. *Biol Plant* 54: 743–747.

Zhang H, Wang MF, Hua LY, Wang SH, Hua KD, Bao LJ et al. 2010b. Hydrogen sulfide promotes wheat seed germination under osmotic stress. *Russ J Plant Physiol* 57: 532–539.

Zhang J, Martinoia E, and Lee Y. 2018. Vacuolar transporters for cadmium and arsenic in plants and their applications in phytoremediation and crop development. *Plant Cell Physiol* 59(7): 1317–1325.

Zhang P, Zhong K, Zhong Z, and Tong H. 2019. Mining candidate gene for rice aluminum tolerance through genome wide association study and transcriptomic analysis. *BMC Plant Biol* 19: 490.

Zhang X, Uroic MK, Xie WY, Zhu YG, Chen BD, McGrath SP, Feldmann J, and Zhao FJ. 2012. Phytochelatins play a key role in arsenic accumulation and tolerance in the aquatic macrophyte Wolffiaglobosa. *Environ Pollut* 165: 18–24.

Zhang X, Zhang D, Sun W, and Wang T. 2019. The adaptive mechanism of plants to iron deficiency via iron uptake, transport, and homeostasis. *Int J Mol Sci* 20(10): 2424.

Zhao C, Xu J, Li Q, Li S, Wang P, and Xiang F. 2014. Cloning and characterization of a *Phragmites australis phytochelatin* synthase (PaPCS) and achieving Cd tolerance in tall fescue. *PLoS One* 9(8): e103771.

Zhao H, Guan J, Liang Q, Zhang X, Hu H, and Zhang J. 2021. Effects of cadmium stress on growth and physiological characteristics of sassafras seedlings. *Sci Rep* 11(1): 9913.

Zhu CQ, Zhang JH, Sun LM, Zhu LF, Abliz B, Hu WJ, Zhong C, Bai ZG, Sajid H, Cao XC, and Jin QY. 2018. Hydrogen sulfide alleviates aluminum toxicity via decreasing apoplast and symplast Al contents in rice. *Front Plant Sci* 9: 294.

Zúñiga A, Laporte D, González A, Gómez M, Sáez CA, and Moenne A. 2019. Isolation and characterization of copper- and zinc- binding metallothioneins from the marine alga ulvacompressa (Chlorophyta). *Int J Mol Sci* 21(1): 153.

10

Amelioration of Heavy Metal Toxicity by Natural and Synthetic Hormones

Nidhi Verma,[1,2] *Jitendra Kumar*[1,3],* and *Sheo Mohan Prasad*[1],*

ABSTRACT

Heavy metals are known as important natural and non-biodegradable constituents of the Earth's crust. Due to anthropogenic activities, these heavy metals can accumulate and persist for an indefinite time in the ecosystem. The high concentration of heavy metals in an ecosystem leads to a major yield loss in plants. They can interrupt plant metabolic activities by enhancing ROS accumulation and reducing the level of antioxidant enzymes. The removal of these heavy metals has become an essential concern of scientific interest. In recent years, researches marked that plant hormones have the potential to provide metal stress tolerance to plants. To sustain life, plant hormones can regulate all the physiological and biochemical parameters of the plant and ultimately enhance the growth responses against various heavy metal stresses. These specific plant hormones can boost internal antioxidant machinery thereby releasing different toxic oxidative radicles which are vigorously formed during heavy metal stresses. Additionally, they can also promote genetic modifications and programmed cell death to cope with adverse situations by modulating different signaling pathways that ultimately provide a better crop yield to the growing population. Interaction of plant growth hormones with other signaling molecules showed their defensive-signalling

---

[1] Ranjan Plant Physiology and Biochemistry Laboratory, Department of Botany, University of Allahabad, Prayagraj, U.P. 211002 (INDIA).
Email: nidhi111verma@gmail.com
[2] B K Birla Institute of Higher Education (Shree Krishnarpana Charity Trust) Pilani, Rajasthan- 333031 (INDIA).
[3] Institute of Engineering and Technology, Dr. ShakuntlaMisra National Rehabilitation University, Mohaan Road, Lucknow, U.P. 226017 (INDIA).
* Corresponding authors: jitendradhuria@gmail.com; profsmprasad@gmail.com

strategies to cope up with stressful environments. In the present study, we have gathered all the current information about the role of natural and synthetic plant growth hormones like auxins, gibberellins, cytokinins, abscisic acid, ethylene, jasmonic acid, salicylic acid, brassinosteroids, and 5-aminolevulinic acid in defense mechanisms, signalling pathways, and lessening of heavy metal toxicity.

1. Introduction

During fast development and urbanization, heavy metals are directly dumped into the life saving rivers and fields from a variety of sources and unbalanced anthropogenic activities. They accumulate in crop plants and later on, they mingle into the food chain, thereby metal toxicity increases the risk and poses concerns for human health (Sytar et al., 2019). The accumulation of heavy metals above threshold levels poses a worldwide environmental threat. Heavy metal contamination can easily damage the plants' metabolism by interfering with their internal defensive mechanisms (by lowering antioxidant enzymes thereby enhancing ROS accumulation) (Choudhury et al., 2017). Different physiological attributes in plants like growth, reproduction, the permeability of the cell membrane, and the synthesis of many biochemical compounds are negatively affected by heavy metal stress. All these negative impacts can directly harm the genetic efficacy of plants thereby overall health of plant get compromised (Bücker-Neto et al., 2017). Thus, the priority is to find eco-friendly and economically affordable medications to tackle this problem. Scientists nowadays work in collaboration to develop potential strategies to combat the adverse impacts of heavy metal stress in plants. In addition to different strategies, exogenous application of phytohormones, i.e., auxins, gibberellins, cytokinins, abscisic acid, ethylene, jasmonic acid, salicylic acid, brassinosteroids, and 5-aminolevulinic acid is considered a cost effective and promising technique. These phytohormones when applied to heavy metal stressed plants they lead to enhance the level of the antioxidant defense system, minimize the levels of ROS and lipid peroxidation and also improve the rate of photosynthesis. On the other hand, Sharma et al. (2019) has been reported that phytohormones play an important role in accumulating osmolytes during various abiotic stresses. Accumulation of these osmolytes under abiotic stress conditions potentially confer the tolerance to cells without interfering with the cellular machinery of the plant. Glycine betaine, polyamines, polyols, sugars, and proline are the efficient osmolytes having the ability to scavenge ROS inside the plant body to provide abiotic stress tolerance. Moreover, phytohormones can also interact with different signaling molecules (like NO and H_2O_2) to enhance the plant's internal defence strategies to fight against various stresses (Verma et al., 2020). Overall, information about the multidisciplinary role of phytohormones during heavy metal stresses tolerance in plants has been collected and demonstrated in this study.

2. Amelioration of heavy metal toxicity by natural and synthetic auxins

Auxins, the bioactive phytohormones play a crucial role in numerous essential aspects of plant development and growth at the molecular, biochemical and physiological levels not only under normal conditions but also during stress conditions (Kazan, 2013). Previous research by Finet and Jaillais (2012) showed that indole-3-acetic acid (IAA), phenylacetic acid (PAA) and indole-3-butyric acid (IBA) which are considered as natural auxins can potentially regulate cell division and its growth, root growth, and development, formation of leaf, apical dominance and biosynthesis of ethylene as well as vascular tissues differentiation process and fruit development. Similarly, 1-naphtha-leneacetic acid (NAA), a synthetic auxin shows similar physiological responses during plant growth and development (Imin et al., 2005). The protective role of IAA was also noticed against Cr(VI) and Mn toxicity in pea seedlings where exogenously applied IAA improves the metabolic processes by enhancing the overall antioxidant defense system thereby balancing its redox homeostasis (Gangwar et al., 2011a,b). The study of Srivastava et al. (2013) noticed a positive potential of exogenously supplied IAA in As stress alleviation by expressing miRNAs in *Brassica juncea*. Earlier it has been well stated that the distribution of IAA is governed by PINFORMED1 (PIN1) protein during the heavy metal stress conditions. Thus As stress can cause toxicity in *Brassica juncea* by altering the endogenous levels of auxins in plant cells (i.e., indole-3-acetic acid, indole-3-butyric acid, and naphthalene acetic acid) which ultimately caused a greater hindrance in the expressions of almost 69 miRNAs belonging to 18 plant miRNA families. However, improved growth in As-stressed *Brassica juncea* has been noticed when IAA was supplied exogenously (Table 1) by enhancing the expressions of miR167, miR319, and miR854 which shows a defensive role of IAA in providing metal stress tolerance (Gangwar et al., 2014). Likewise, Li et al. (2015) found that distribution of IAA was hampered during boron (B) deficiency which leads to down-regulation of PIN1 protein thereby limiting the root elongation in *Arabidopsis thaliana*. In *Arabidopsis*, an increased level of NO can potentially reduce the PIN1-dependent acropetal auxin transport inside the root apical meristem, thereby inhibiting the polar auxin transport. Finally, this act indicates that NO acts as a downstream regulator molecule of auxin (Fernandez-Marcos et al., 2011, 2012), and a similar signaling mechanism was observed by Chen et al. (2010) during iron deficiency (Fig. 1). On the other hand, studies also demonstrated that the exogenous supplementation of IAA can also empower the phytoextraction of heavy metals from the plant system and enable them able to survive under stressful situations (Fa¨ssler et al., 2010; Hadi et al., 2010). The study of Krishnamurthy and Rathinasabapathi (2013) has also depicted that the distribution and active pool of auxins play an important role in providing metal stress tolerance in plants. Previously, it has been revealed that the mutants of auxin transporter, i.e., aux1, pin1, and pin2 showed more sensitive behavior to As(III) stress rather than the wild type varieties, thus inhibitors of auxin transport significantly reduced the plant tolerance against As(III) stress in the wild variety by enhancing the endogenous levels of H_2O_2. Another interesting phenomenon was noticed regarding the interactive behavior of nitric oxide (NO) in auxin signal transduction during the Cu-stress (Peto et al., 2011). Similarly, Verma et al. (2020) also demonstrated the positive interactions of NO with IAA in defensive strategies to cope-up with the stressful environment.

3. Amelioration of heavy metal toxicity by gibberellins (GAs)

Gibberellins (GAs), a large family belonging to the tetracyclic diterpenoid are known to be associated with various vital processes of plants such as seed germination, stem and hypocotyl elongation, leaf expansion, initiation of floral organs along with their development, and fruit development. The involvement of GAs was also noticed in source-to-sink formation during phloem loading and unloading (Gangwar et al., 2014). The GA mediated alleviation of heavy metal stresses like Cd toxicity alleviation in *Arabidopsis thaliana* (Zhu et al., 2012), Ni in wheat seedlings (Siddiqui et al., 2011), and Cr stress in pea seedlings (Gangwar et al., 2011a) (Table 1). It is also reported that the level of some phytoestrogens such as daidzein and genistein are also enhanced significantly during salt stress conditions which unambiguously indicated the protective behavior of GA (Hamayun et al., 2010). The involvement of GA in photosynthesis improvement and also in the regulation of ionic distribution and hormonal homeostasis was also proved by Iqbal and Ashraf (2013).

The beneficiary role of GA in counteracting the adverse impacts of various stressors is also more strongly proven by transcriptomic analysis. As Sun et al. (2013) reported that the GA is potentially involved in sustaining the abiotic stress tolerance to *Arabidopsis* by up-regulating the *GAST*1 gene (identified as a GA-stimulated transcript gene) which leads to modulating the endogenous accumulation of ROS (Fig. 1). Similarly, Liang et al. (2013) suggested that the GA enhances the expression of the OsAOP gene which encodes a tonoplast intrinsic protein (TIP) in rice plants which indicates the essential role of the OsAOP gene in defensive responses against various stresses in rice. Previously, it was well recorded that the sulfur-containing compounds are also known for their crucial role in plant protection against various abiotic and biotic stresses. The GA signaling considerably enhances the activity of adenosine 50-phosphosulfate reductase, which is a necessary key enzyme of sulfate assimilation under stressful situations while other hormones are unable to show such kind of effects (Koprivova et al., 2008). Thus, the overall concept is enough to understand the potential role of GA signaling to protect the plant metabolism under stress conditions.

4. Amelioration of heavy metal toxicity by cytokinins (CKs)

Exogenous supplementation of some plant hormones is potentially able to boost-up the plant-tolerance by increasing their capacity of detoxification along with the physiological and biochemical functioning of plants (Hasan et al., 2011; Ahammed et al., 2013). Kinetin (KN), a synthetic cytokinin plays multiple roles in inducing plant tolerance against different abiotic stresses like pesticides, salinity, and heavy metals (Aldesuquy et al., 2014; Bashri et al., 2018; Tiwari et al., 2020). Eser and Aydemir (2016) also found that exogenous application of KN can provide tolerance against boron stress in wheat seedlings by boosting their antioxidant defence system and thereby regulating oxidative stress (Fig. 1). Likewise, Wang et al. (2015) have also stated that the exogenously applied KN enhances the production of ROS scavenger

Table 1: Hormone mediated responses during various heavy metal stress situations and their interactions with other signaling molecules.

Hormones involved	Plants	Hormone mediated effects	Targeted metal	Involvement of signaling molecule or other hormones	References
Auxins	*Bassicajuncea*	modulate the expression of *miR167*, *miR319*, and *miR854*	As	-	Srivastava et al. (2013)
	Oryza sativa	Improve the level of stress indicators (Chlorophyll, protein, MDA) and modulators (cysteine, proline)	As	Selenium (Se)	Pandey and Gupta (2015)
Gibberellins (GAs)	*Arabidopsis. thaliana*	Suppress the iron-regulated transporter 1 (*IRT1*), a transporter involved in Cd uptake	Cd	NO	Zhu et al. (2012)
	Wheat seedlings	enhanced antioxidant potential	Ni	Ca	Siddiqui et al. (2011)
	Pea seedlings	regulating oxidative stress and the antioxidant system thereby enhance overall growth and ammonium assimilation	Cr	-	Gangwar et al. (2011a)
Cytokinins (CKs)	*Arabidopsis. thaliana*	Enhance the accumulation of thiol compounds like Phytochelatins	As	-	Mohan et al. (2016)
	Tobacco plant	Enhance the expression of an isopentenyltransferase (ipt) gene by enhancing metallothionein gene (*MT-L2*)	Cu	-	Thomas et al. (2005)
	Norway spruce	Enhance and ETS activity and improve overall root growth	Pb	-	Shukla et al. (2017)

Abscisic Acid (ABA)	*Bassicajuncea*	Interaction with miR159	As	-	Srivastava et al. (2013)
	Atractylodesmacrocephala	Enhance antioxidant enzyme activities and reduced the oxidative stress	Pb	-	Wang et al. (2013)
	Populus x Canescens	down-regulation of genes involved in Zn up-take viz., yellow stripe-like family protein 2 (YSL2) and plant cadmium resistance protein 2 (PCR2)	Zn	GA and SA	Shi et al. (2015)
Ethylene	*Bassicajuncea*	Enhance the rate of photosynthesis by changing the activity of photosystem II	Ni and Zn		Khan and Khan (2014)
	wheat plant	improve proline and glutathione production	Cd	Se and S	Khan et al. (2015)
Jasmonic Acid (JA)	*Cajanuscajan*	Enhance the accumulation of osmolytes like proline, glutathione, carotenoids as well as antioxidant enzymes	Cu	-	Poonam et al. (2013)
	Pea plant	induction of pathogenesis-related and heat shock proteins	Cd	ethylene, ROS, NO and Ca	Rodriguez-Serrano et al. (2009)
	Phaseoluscoccineus	Regulation of oxidative stress mechanism	Cu	-	Hanaka et al. (2016)
Brassinosteroids (BRs)	*Oryza sativa*	Strengthen the antioxidant defence system by up-regulating the expressions of superoxide dismutase (Mn-SOD, Cu/Zn-SOD), catalase (Cat A, Cat B), ascorbate peroxidase (APX) and glutathione reductase (GR)	Cr	-	Sharma et al. (2016)
	Raphanus sativa	Enhance the rate of photosynthesis and reduce the accumulation of ROS	Zn	-	Ramakrishna and Rao (2015)
	Bassicajuncea	improve proline metabolism and antioxidants	Cu	Se	Yusuf et al. (2016)

Table 1 contd. ...

...Table 1 contd.

Hormones involved	Plants	Hormone mediated effects	Targeted metal	Involvement of signaling molecule or other hormones	References
Salicylic acid (SA)	Maize seedlings	Up-regulation of antioxidants	Cr	-	Singh et al. (2016)
	Phaseolusaureus and *Vicia sativa*	Enhancement of antioxidant enzyme pool in both apoplastic and symplastic compartments and thereby decrease the endogenous accumulation of H_2O_2	Cd		Zhang et al. (2011)
	Cannabis sativa	Improvement of photosynthetic efficiency and enhancement of antioxidant activities	Cd	-	Shi et al. (2009)
	Oryza sativa	Change the expression of As transporter genes and enhance antioxidant activities	As	-	Singh et al. (2015)
5-aminolevulinic acid (ALA)	*Helianthus annuus*	Improve photosynthetic rate and gaseous exchange attributes	Cr	-	Farid et al. (2019)
	Oilseed rape	Reduce the level of oxidative stress biomarkers and boost antioxidant machinery	Cd	-	Ali et al. (2013)

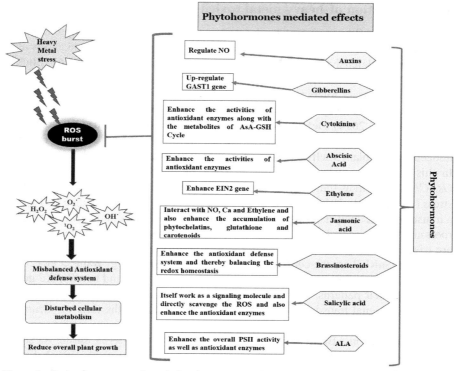

Figure 1: Systemic representation of phytohormones mediated effects and alterations of heavy metal induced damages in plant system.

antioxidants in arsenic-stressed maize seedlings as compared to those treated with arsenic alone. Mohan et al. (2016) described that CKs enhance the accumulation of thiol compounds like phytochelatins in *Arabidopsis thaliana* to provide resistivity against As stress (Table 1). The ascorbate biosynthetic pathway is considered a novel process of metal detoxification by balancing redox homeostasis in the plant system. Singh et al. (2018) have described that the exogenous KN lessens the Cd-mediated damages on PS II photochemistry by effectively improving the components of the ascorbate-glutathione (AsA-GSH) cycle and also by inducing a probable mechanism to reduce the endogenous Cd accumulation in tomato seedlings thereby improving their overall growth performance. Likewise, Singh and Prasad (2014) have also described that KN is efficiently able to improve the photosynthetic apparatus structurally and functionally to equip the Cd stress tolerance in *Solanum melongena*. But the mechanism of KN in metal stress removal is not so linear. Several recent studies have reported the synergistic interaction between KN and other signaling molecules during abiotic stress situations (Shao et al., 2010; Verma et al., 2020). Tiwari et al. (2020) noticed the interactive behavior of KN and NO in the alleviation of cypermethrin toxicity in the economically important paddy field cyanobacterium *Nostoc muscorum*.

5. Amelioration of heavy metal toxicity by abscisic acid (ABA)

Abscisic acid (ABA) is a carotenoid derivative and multifunctional phytohormone that plays an essential role during different stages of a plant's life cycle and is also known for its defensive responses against various environmental stresses (Shukla et al., 2017). This hormone has been linked with its positive potential for providing environmental stress tolerance as well as its signaling pathway and its interaction with the MAPK pathway is a crucial regulator of abiotic stress management in plants (Danquah et al., 2014). Defense responses of ABA signaling have been more clearly demonstrated by transcriptomic analysis in As-stressed rice plants where it has been noticed that the plant's self-defense system is activated to cope-up with this situation. In this response, strong expressions of ABA biosynthetic genes *OsNCED2* and *OsNCED3* as well as four other ABA signaling genes were also activated against As stress (Huang et al., 2012). On the other hand, Lin et al. (2013) have also performed a transcriptomic analysis using a whole-genome array in vanadium (V)-stressed rice roots and it is found that genes that are involved in signaling and biosynthesis of ABA are triggered by this metal. The three main protein components PYL/PYR/RCAR, PP2C, and SnRK2 are the complex regulators of ABA signaling. Wang et al. (2014) successfully identified and monitored nine *PYL*, three *PP2C*, and two *SnRK2* genes along with their patterns which are strongly involved in ABA signal transduction. These ABA signal transducing genes were identified during the time of cucumber seed germination which is exposed to Cu and Zn stress.

An increase in ABA synthesis and its interaction withmiR159 during As stress has been well recognized in *B. juncea* (Table 1) (Srivastava et al., 2013). Under arsenate and arsenite stress marked alterations in ABA metabolism genes were successfully detected after transcriptomic analysis in rice (Shukla et al., 2017). The positive role of exogenously supplied ABA in providing metal stress tolerance has also been identified in earlier studies of Wang et al. (2013) and Shi et al. (2015). The exogenously supplied ABA significantly enhanced the activities and production of ROS scavenging antioxidant enzymes and thereby reduced the endogenous level of oxidative stress biomarkers (Fig. 1) in the Pb-stressed *Atractylodes macrocephala* plant (Wang et al., 2013). Moreover, exogenous ABA is also able to enhance the levels of GA and salicylic acid (SA) along with the foliar ascorbate in Zn-stressed *Populus* x *Canescen*. Exogenous application of ABA remarkably suppressed the yellow stripe-like family protein 2 (YSL2) and plant cadmium resistance protein 2 (PCR2) which are involved in Zn uptake and accumulation in plants (Shi et al., 2015).

6. Amelioration of heavy metal toxicity by ethylene

Ethylene, a gaseous plant hormone is known for its potential involvement in many plant growth and developmental processes. Studies have also reported its positive involvement in several abiotic stress regulations (Khan et al., 2015). There are five receptors of ethylene which are ERT1, ETR2, ERS1, ERS2, and EIN4, and all are positioned on the endoplasmic reticulum membrane (Shan et al., 2012). The potential role of ethylene in metal stress regulation has also been described by transcriptomic analysis of Cr-excessed rice roots. Thus in this study, it was found that ACS1, ACS2,

ACO4, and ACO5 genes which are the four ethylene biosynthesis-related genes are expressed significantly during Cr stress which shows the positive involvement of ethylene in the Cr signaling process in rice (Steffens, 2014; Trinh et al., 2014). Similarly, the study of Schellingen et al. (2014) also described that Cd enhanced the biosynthesis of ethylene by expressing ACS2 and ACS6 genes in *Arabidopsis thaliana*. The *Arabidopsis* EIN2 gene is an essential constituent of the ethylene signaling pathway which participates as an ethylene transducer and also acts as a promoter of various stress responses. In the study of Cao et al. (2009), the increased transcript level of EIN2 was found in Pb-stressed *Arabidopsis* seedlings which indicated the potential role of the EIN2 gene in metal stress tolerance acquisition (Fig. 1). Moreover, several studies found that endogenous level of ethylene is enhanced markedly during Cd (Asgher et al., 2014) Ni and Zn stress (Khan and Khan 2014) in *Brassica juncea* (Table 1). Overall, future researches still need to focus on the probable mechanism of ethylene in metal stress alleviation in plants.

7. Amelioration of heavy metal toxicity by jasmonate (JA)

Jasmonates are oxylipin signalling molecules and the oxylipins are the family of oxygenated products of fatty acids. Among variousoxylipins, jasmonic acid (JA) and its methyl ester, i.e., methyl jasmonate (MeJA) are the best known oxylipins that are formed enzymatically and thereafter accumulated inside the plant cells during various stress conditions particularly during biotic stresses such as wounding, pathogen infection and during some abiotic stress situations like heavy metal and salt stress, etc. (Block et al., 2005; Azooz et al., 2015). Moreover, some of the biologically active oxylipins can also be produced through non–enzymatic reactions mainly via the action of reactive oxygen species (ROS), and they can also get accumulate inside the plant cells in response to the stress situation. In addition, the study of Wasternack and Hause (2013) has also reported that the JA considerably enhanced the auxin biosynthesis, in response to this auxin attenuates JA signaling. The role of JA in metal stress alleviation is demonstrated by several studies. The JA considerably enhanced the accumulation of phytochelatins, glutathione, and carotenoids to cope-up with the toxic impacts of Cu and Cd in *Arabidopsis thaliana* (Shukla et al., 2017). A synergistic interaction among the JA, ethylene, ROS, NO, and Ca was reported in pea plants (Table 1) in response to Cd. These phytohormones potentially induce pathogenesis-related and heat shock proteins (Fig. 1) (Rodriguez-Serrano et al., 2009). JA also provides resistance against Ni toxicity by regulating biomembrane peroxidation and boosting the antioxidant mechanism in higher plants (Sirhindi et al., 2015). Overall the above facts show the active participation of JA in the acquisition of metal stress tolerance in plants by interacting with some other signaling molecules and phytohormones as well.

8. Amelioration of heavy metal toxicity by brassinosteroids (BRs)

Brassinosteroids are categorized as plant steroids that participate in several plant growth and developmental processes. Its physiological role has been identified in releasing several abiotic stresses (Bücker-Neto et al., 2017). Cao et al. (2005) reported that the BRs were also involved in balancing antioxidant enzyme activities during heavy metal-induced phytotoxicity. The positive involvement of BRs in metal stress tolerance is more strongly proven when the role of exogenously applied 24epiBL has been estimated in the amelioration of Ni-stressed *Brassica juncea* by boosting the activities of antioxidant enzymes (Kanwar et al., 2013). Similarly, exogenously applied 28 homoBL has been found to enhance the activities of SOD, POD, and CAT enzymes to protect the wheat plant against Ni stress (Bücker-Neto et al., 2017). Moreover, Ahammed et al. (2013) have also observed that exogenously applied 24-epibrassinolidecan enhance the activities of APX and GR in *Solanum Lycopersicum* seedlings to provide resistance against Cd stress (Table 1). Foliar application of homoBL and normal exogenous application of BRs improves Cd and Cu tolerance in *Brassica juncea* and *Raphanus sativus* by boosting the activities of antioxidant enzymes (CAT, POD, and SOD) and thereby decreasing the accumulation of endogenous ROS species (Hayat et al., 2007; Kapoor et al., 2014). Thus, the above facts show that the BRs provide resistivity against heavy metal stress by enhancing the antioxidant defense system of the plants and thereby regulating the cellular redox homeostasis (Fig. 1).

9. Amelioration of heavy metal toxicity by salicylic acid (SA) and its exogenous supply

Salicylic acid is an essential phytohormone that is phenolic, and it is most commonly known for its important function in acquiring plant stress tolerance by modulating the activities of some antioxidative enzymes (Shukla et al., 2017). The SA is also known to modulate several physiological and biochemical factors involved during plant stress tolerance by regulating several stress-mediated signaling pathways and their response mechanisms. Previous studies reported that the exogenously applied SA can potentially alleviate various abiotic stresses as mentioned by Senaratna et al. (2000) for water stress bean and tomato plants, by Azooz et al. (2011) for salt stress in broad bean seedlings, and by Ahmad et al. (2011) for heavy metal stress in mustard plants. Exogenous application of SA can stimulate certain signalling pathways regarding their plant defense actions to minimize the Cd-induced oxidative stress (Belkadhi et al., 2016). Moreover, SA also considered itself as a signaling molecule and its exogenous application in the priming of maize seedlings can potentially reduce the endogenous accumulation of Cr by up-regulating antioxidant activities (Singh et al., 2016). During metal stress conditions, excessive ROS like hydroxyl radicals can directly be scavenged by exogenous SA (Fig. 1). The externally applied SA in Cd-stressed in *Phaseolus aureus* and *Vicia sativa* has markedly boosted the antioxidative enzyme pool of both apoplastic and symplastic compartments which resulted in a significant decline in H_2O_2 accumulation thereby providing the resistance against Cd stress (Table 1) (Zhang et al., 2011). Further, exogenous SA can also enhance the rate of photosynthesis and antioxidant system in *Cannabis sativa* to reduce the Cd accumulation inside the cells. Exogenous application of SA is potentially able to reverse all the negative impacts of As in rice plants by up-regulating the antioxidant enzymes and expressions of As transporter genes (Singh et al., 2015; Shukla et al., 2017).

10. Amelioration of heavy metal-stressed crop plants by 5-aminolevulinic acid (ALA)

Plant growth regulators are frequently used to enhance the stress tolerance capacity in plants by modulating their metabolism along with various physiological and biochemical activities. 5-aminolevulinicacid (ALA) is one of the most important phytohormones as well as plant growth regulators. It is also known as a necessary precursor for the biosynthesis of tetrapyrroles such as protochlorophyllide (which can convert into chlorophyll when exposed to light), heme, cytochrome, and vitamin B_{12} (Senge et al., 2014). The potential of ALA in heavy metal stress alleviation in various plant species has also been reported.

ALA improves the overall photosynthetic performance of the plant by enhancing pigments contents and PSII activity, boosting the activity of the antioxidant defense system, and also enhancing the phenolic contents (Fig. 1) which ultimately resulted in enhanced tolerance rate of the plants against abiotic stress conditions (Aksakal et al., 2017; Anwar et al., 2018; Gupta and Prasad, 2021). The study of Ali et al. (2014) noticed that exogenously applied ALA considerably enhance the uptake of several micro and macro-nutrients and also declined the production of ROS by increasing the activities of enzymatic antioxidants in the leaves and roots of *Brassica napus* plant to provide resistivity against Pb stress. Previously, Ali et al. (2013) have also noticed that the total glutathione ratio is also improved by exogenous ALA in seedlings of oilseed rape under Cd stress (Table 1). The study of Tian et al. (2014) has noticed that the exogenous application of ALA impressively recovered the Pb-induced destructions in mesophyll cells and root tips of the *Brassica napus* plant. After microscopic analysis, a well-developed nucleus in mesophyll cells and several starch granules were found in chloroplasts after exogenous supplementation of ALA in Pb-stressed *Brassica napus* plants. It shows the positive emphasis of ALA in equipping Pb stress tolerance to the plants. Thus, considering the above facts it is clear that the ALA can overcome several abiotic stresses by modulating several morphological, physiological, and biochemical processes. There might be a possibility of the involvement of several signaling molecules like NO, H_2O_2, and H_2S in the mechanistic defensive pathway of ALA during the heavy metal stress conditions which is still to be explored in further studies.

11. Conclusion

Plant hormones trigger various physiological, biochemical, and molecular mediators in plants growing under heavy metal stress conditions. They maintain the cellular redox balance by enhancing the efficiency of ROS scavenging enzymes and this process results in an effective reduction of oxidative damage in plant cells. The multidimentional interactions of phytohormones with other signaling molecules and characterization of key genes involved in phytohormone-mediated metal stress tolerance will give a new direction to develop stress-resistant crop varieties.

References

Ahammed GJ, Choudhary SP, Chen S, Xia X, Shi K, Zhou Y, and Yu J. 2013. Role of brassinosteroids in alleviation of phenanthrene-cadmium co-contamination-induced photosynthetic inhibition and oxidative stress in tomato. *J Exp Bot* 64: 199–213.

Ahmad P, Nabi G, and Ashraf M. 2011. Cadmium-induced oxidative damage in mustard [*Brassica juncea* (L.) Czern. & Coss.] plants can be alleviated by salicylic acid. *South Afr J Bot* 77: 36–44. doi: 10.1016/j.sajb.2010. 05.003.

Aksakal O, Algur OF, Aksakal FI, and Aysin F. 2017. Exogenous 5-aminolevulinic acid alleviates the detrimental effects of UV-B stress on lettuce (*Lactuca sativa* L.) seedlings. *Acta Physiol Plantarum* 39: 55. https://doi.org/10.1007/s11738-0172347-3.

Aldesuquy H, Baka Z, and Mickky B. 2014. Kinetin and spermine mediated induction of salt tolerance in wheat plants: Leaf area, photosynthesis and chloroplast ultrastructure of flag leaf at ear emergence. *Egy J Basic App Sci* 1: 77–87.

Ali B, Huang CR, Qi ZY, Ali S, DaudAminolevulinic MK, Geng XX, Liu HB, and Zhou WJ. 2013. 5-acid ameliorates and cadmium-induced morphological, biochemical ultrastructural changes in seedlings of oilseed rape. *Environ Sci Pollut Res* 20: 7256–7267.

Ali B, Xu X, Gill RA, Yang S, Ali S, Tahir M, and Zhou W. 2014. Promotive role of 5-aminolevulinic antioxidative acid on mineral nutrients and defense system under lead toxicity in *Brassica napus*. *Industrial Crops Products* 52: 617–626.

Anwar A, Yan Y, Liu Y, Li Y, and Yu X. 2018. 5-aminolevulinic acid improves nutrient uptake and endogenous hormone accumulation, enhancing low-temperature stress tolerance in cucumbers. *Int J Molecul Sci* 19: 3379. https://doi.org/10.3390/ijms19113379.

Asgher M, Khan NA, Khan MIR, Fatma M, and Masood A. 2014. Ethylene production is associated with alleviation of cadmium induced oxidative stress by sulfur in mustard types differing in ethylene sensitivity. *Ecotox Environ Saf* 106: 54–61.

Azooz MM, Youssef AM, and Ahmad P. 2011. Evaluation of salicylic acid (SA) application on growth, osmotic solutes and antioxidant enzyme activities on broad bean seedlings grown under diluted seawater. *Int J Plant Physiol Biochem* 3: 253–264.

Azooz MM, Metwally A, and Abou–Elhamd MF. 2015. Jasmonate–induced tolerance of Hassawi okra seedlings to salinity in brackish water. *Acta Physiol Plant* 37: 77.

Bashri G, Singh M, Mishra RK, Kumar J, Singh VP, and Prasad SM. 2018. Kinetin regulates UV-B-induced damage to growth, photosystem II Photochemistry and nitrogen metabolism in tomato seedlings. *J Plant Growth Regul* 37: 233–245.

Belkadhi A, Djebali W, Hediji H, and Chaibi W. 2016. Cellular and signalling mechanisms supporting Cd-tolerance in salicylic acid treated seedlings. *Plant Sci Today* 3: 41–47.

Block A, Schmelz E, Jones JB, and Klee HJ. 2005. Coronatine and salicylic acid: the battle between *Arabidopsis* and *Pseudomonas* for phytohormone control. *Mol Plant Pathol* 6: 79–83.

Bücker-Neto L, Paiva ALS, Machado RD, Arenhart RA, and Margis-Pinheiro M. 2017. Interactions between plant hormones and heavy metals responses. *Genetics and Molecular Biology* 40: 373–386.

Cao S, Xu Q, Cao Y, Qian K, An K, Zhu Y, Binzeng H, Zhao H, and Kuai B. 2005. Loss-of-function mutations in DET2 gene lead to an enhanced resistance to oxidative stress in Arabidopsis. *Physiol Plant* 123: 57–66.

Cao S, Chen Z, Liu G, Jiang L, Yuan H, Ren G, Bian X, Jian H, and Ma X. 2009. The Arabidopsis *Ethylene-Insensitive 2* gene is required for lead resistance. *Plant Physiol Biochem* 47: 308312.

Chen J, Xiao Q, Wu F, Dong X, He J, Pei Z, Zheng H, and Näsholm T. 2010. Nitric oxide enhances salt secretion and Na$^+$ sequestration in a mangrove plant, Avicennia marina, through increasing the expression of H$^{(+)}$-ATPase and Na$^{(+)}$/H$^{(+)}$ antiporter under high salinity. *Tree Physiol* 30: 1570–1585.

Choudhury FK, Rivero RM, Blumwald E, and Mittler R. 2017. Reactive oxygen species, abiotic stress and stress combination. *Plant J* 90: 856–867.

Danquah A, de Zelicourt A, Colcombet J, and Hirt H. 2014. The role of ABA and MAPK signaling pathways in plant abiotic stress responses. *Biotechnol Adv* 32: 40–52.

Eser A and Aydemir T. 2016. The effect of kinetin on wheat seedlings exposed to boron. *Plant Physiol Biochem* 108: 158–164.

Farid M, Ali S, Saeed R, Rizwan M, Bukhari SAH, Abbasi GH, Hussain A, Ali B, Zamir MSI, and Ahmad I. 2019. Combined application of citric acid and 5-aminolevulinic acid improved biomass, photosynthesis and gas exchange attributes of sunflower (*Helianthus annuus* L.) grown on chromium contaminated soil. *International J Phytoremed* 1522–6514. https://doi.org/10.1080/15226514.2018.1556595.

Fa¨ssler E, Evangelou MW, Robinson BH, and Schulin R. 2010. Effects of indole-3-acetic acid (IAA) on sunflower growth and heavy metal uptake in combination with ethylene diamine disuccinic acid (EDDS). *Chemosph* 80: 901–907.

Fernández-Marcos M, Sanz L, Lewis DR, Muday GK, and Lorenzo O. 2011. Nitric oxide causes root apical meristem defects and growth inhibition while reducing PIN-FORMED 1 (PIN1)-dependent acropetalauxin transport. *Proc Proc Natl Acad Sci USA* 108: 18506–18511.

Fernández-Marcos M, Sanz L, and Lorenzo O. 2012. Nitric oxide: an emerging regulator of cell elongation during primary root growth. *Plant Signal Behav* 7: 196–200.

Finet C and Jaillais Y. 2012. Auxology: when auxin meets plant evo–devo. *Dev Biol* 369: 19–31. doi: 10.1016/j.ydbio.2012.05.039.

Gangwar S, Singh VP, Srivastava PK, and Maurya JN. 2011a. Modification of chromium (VI) phytotoxicity by exogenous gibberellic acid application in *Pisumsativum* (L.) seedlings. *Acta Physiol Plant* 33: 1385–1397.

Gangwar S, Singh VP, Prasad SM, and Maurya JN. 2011b. Differential responses of pea seedlings to indole acetic acid under manganese toxicity. *Acta Physiol Plant* 33: 451–462.

Gangwar S, Singh VP, Tripathi DK, Chauhan DK, Prasad SM, and Maurya JN. 2014. Plant responses to metal stress: the emerging role of plant growth hormones in toxicity alleviation. pp. 215–248. *In*: Ahmad P, and Rasool S (eds.). *Emerging Technologies and Management of Crop Stress Tolerance*. Academic Press.

Gupta D and Prasad SM. 2021. Priming with 5-aminolevulinic acid (ALA) attenuates UV-B induced damaging effects in two varieties of *Cajanuscajan* L. seedlings by regulating photosynthetic and antioxidant systems. *South African J Bot* 138: 129–140.

Hadi F, Bano A, and Fuller MP. 2010. The improved phytoextraction of lead (Pb) and the growth of maize (*Zea mays* L.): the role of plant growth regulators (GA3 and IAA) and EDTA alone and in combinations. *Chemosph* 80: 457–462.

Hamayun M, Khan SA, Khan AL, Shin JH, Ahmad B, and Shin DH. 2010. Exogenous gibberellic acid reprograms soybean to higher growth and salt stress tolerance. *J Agric Food Chem* 58: 7226–7232.

Hanaka A, Wojcik M, Dreslar S, Mroczek-Zdyrska M, and Maksymiec W. 2016. Does methyl jasmonate modify the oxidative stress response in Phaseoluscoccineus treated with copper? *Ecotox Environ Saf* 124: 480–488.

Hasan SA, Hayat S, and Ahmad A. 2011. Brassinosteroids protect photosynthetic machinery against the cadmium induced oxidative stress in two tomato cultivars. *Chemosph* 84: 1446–1451.

Hayat S, Ali B, Hasan SA, and Ahmad A. 2007. Brassinosteroid enhanced the level of antioxidants under cadmium stress in *Brassica juncea*. *Environ Exp Bot* 60: 33–41.

Huang TL, Nguyen QTT, Fu SF, Lin CY, Chen YC, and Huang HJ. 2012. Transcriptomic changes and signalling pathways induced by arsenic stress in rice roots. *Plant Mol Biol* 80: 587608.

Imin N, Nizamidin M, Daniher D, Nolan KE, Rose RJ, and Rolfe BG. 2005. Proteomic analysis of somatic embryogenesis in Medicagotruncatula explants culture grown under 6-benzylaminopurine and 1-naphthaleneacetic acid treatments. *Plant Physiol* 137: 1250–1260. doi:10.1104/pp.104.055277

Iqbal M and Ashraf M. 2013. Gibberellic acid mediated induction of salt tolerance in wheat plants: growth, ionic partitioning, photosynthesis, yield and hormonal homeostasis. *Environ Exp Bot* 86: 76–85.

Kanwar MK, Bhardwaj R, Chowdhary SP, Arora P, Sharma P, and Kumar S. 2013. Isolation and characterization of 24-Epibrassinolide from *Brassica juncea* L. and its effects on growth, Ni ion uptake, antioxidant defence of *Brassica* plants and *in vitro* cytotoxicity. *Acta Physiol Plant* 35: 1351–1362.

Kapoor D, Rattan A, Gautam V, Kapoor N, Bhardwaj R, Kapoor D, Rattan A, Gautam V, and Kapoor N. 2014. 24-Epibrassinolide mediated changes in photosynthetic pigments and antioxidativedefence system of radish seedlings under cadmium and mercury stress. *Physiol Biochem* 10: 110–121.

Kazan K. 2013. Auxin and the integration of environmental signals into plant root development. *Ann Bot* 112: 1655–1665.

Khan MIR and Khan NA. 2014. Ethylene reverses photosynthetic inhibition by nickel and zinc in mustard through changes in PS II activity, photosynthetic nitrogen use efficiency, and antioxidant metabolism. *Protoplasma* 251: 1007–1019.

Khan MIR, Nazir F, Asgher M, Per TS, and Khan NA. 2015. Selenium and sulfur influence ethylene formation and alleviate cadmium-induced oxidative stress by improving proline and glutathione production in wheat. *J Plant Physiol* 173: 9–18.

Koprivova A, North KA, and Kopriva S. 2008. Complex signaling network in regulation of adenosine 50-phosphosulfate reductase by salt stress in *Arabidopsis* roots. *Plant Physiol* 146: 1408–1420.

Krishnamurthy A and Rathinasabapathi B. 2013. Auxin and its transport play a role in plant tolerance to arsenite-induced oxidative stress in *Arabidopsis thaliana*. *Plant Cell Environ* 36: 1838–1849.

Li K, Kamiya T, and Fujiwara T. 2015. Differential roles of PIN1 and PIN2 in root meristem maintenance under low-B conditions in *Arabidopsis thaliana*. *Plant Cell Physiol* 56: 1205–1214.

Liang WH, Li L, Zhang F, Liu YX, Li MM, and Shi HH. 2013. Effects of abiotic stress, light, phytochromes and phytohormones on the expression of OsAQP, a rice aquaporin gene. *Plant Growth Regul* 69: 21–27.

Lin CY, Trinh NN, Lin CW, and Huang HJ. 2013. Transcriptome analysis of phytohormone, transporters and signaling pathways in response to vanadium stress in rice roots. *Plant Physiol Biochem* 66: 98–104.

Mohan TC, Castrillo G, Navarro C, Zarco-Fernandez S, Ramireddy E, Mateo C, Zanarreno AM, Paz-Ares J, Munoz R, Garcia-Mina JM, Hernandez LE, Schmulling T, and Leyva A. 2016. Cytokinin determines thiol-mediated arsenic tolerance and accumulation. *Plant Physiol* 171: 1418–1426.

Pandey C and Gupta M. 2015. Selenium and auxin mitigates arsenic stress in rice by combining the role of stress indicators, modulators and genotoxicity assay. *J Hazard Mater* 287: 384–391.

Peto A, Lehotai N, Lozano-Juste J, Leo´n J, Tari I, and Erdei L. 2011. Involvement of nitric oxide and auxin in signal transduction of copper-induced morphological responses in *Arabidopsis* seedlings. *Ann Bot* 108: 449–457.

Poonam S, Kaur H, and Geetika S. 2013. Effect of jasmonic acid on photosynthetic pigments and stress akers in *Cajanuscajan* (L.) *Mil* sp. seedlings under copper stress. *Am J Plant Sci* 4: 817–823.

Ramakrishna B and Rao SSR. 2015. Foliar application of brassinosteroids alleviates adverse effects of zinc toxicity in radish (*Raphanus sativus* L.) plants. *Protoplasma* 252: 665–677.

Rodriguez-Serrano M, Romero-Puertas MC, Pazmino DM, Testillano PS, Risueno MC, del Rio LA, and Sandalio LM. 2009. Cellular responses of pea plants to cadmium toxicity: Cross talk between ROS, Nitric oxide and calcium. *Plant Physiol* 150: 229–243.

Schellingen K, Straeten D Van Der, Vandenbussche F, Prinsen E, and Remans T. 2014. Cadmium-induced ethylene production and responses in *Arabidopsis thaliana* rely on ACS2 and ACS6 gene expression. *BMC Plant Biol* 14: 214.

Senaratna T, Touchell D, Bunns E, and Dixon K. 2000. Acetyl salicylic acid (aspirin) and salicylic acid induce multiple stress tolerance in bean and tomato plants. *Plant Growth Regul* 30: 157–161. doi: 10.1023/A:1006386800974.

Senge MO, Ryan AA, Letchford KA, MacGowan SA, and Mielke T. 2014. Chlorophylls, symmetry, chirality, and photosynthesis. *Symmetry* 6: 781–843. https://doi.org/ 10.3390/sym60 30781.

Shan X, Yan J, and Xie D. 2012. Comparison of phytohormonessignalling mechanisms. *Curr Opin Plant Biol* 15: 84–91.

Shao R, Wang K, and Shangguan Z. 2010. Cytokinin-induced photosynthetic adaptability of *Zea mays* L. to drought stress associated with nitric oxide signal: probed by ESR spectroscopy and fast OJIP fluorescence rise. *J Plant Physiol* 167: 472–479.

Sharma A, Shahzad B, Kumar V, Kohli SK, Sidhu GPS, Bali AS, Handa N, Kapoor D, Bhardwaj R, and Zheng B. 2019. Phytohormones regulate accumulation of osmolytes under abiotic stress. *Biomolecules* 9: 285.

Sharma P, Kumar A, and Bhardwaj R. 2016. Plant steroidal Hormone epibrassinolide regulate-Heavy metal stress tolerance in *Oryza sativa* L. by modulating antioxidant defense expression. *Environ Exp Bot* 122: 1–9.

Shi GR, Cai QS, Liu QQ, and Wu L. 2009. Salicylic acid-mediated alleviation of cadmium-toxicity in hemp plants in relation to cadmium uptake, photosynthesis and antioxidant enzymes. *Acta Physiol Plant* 31: 969–977.

Shi WG, Li H, Liu TX, Polle A, Peng CH, and Luo ZB. 2015. Exogenous abscisic acid alleviates zinc uptake and accumulation in *Populus* × *canescens* exposed to excess zinc. *Plant Cell Environ* 38: 207–223.

Shukla A, Srivastava S, and Suprasanna P. 2017. Genomics of metal stress-mediated signaling and plant adaptive responses in reference to phytohormones. *Current Genom* 18: 512–522.

Siddiqui MH, Al-Whahibi MH, and Basalah MO. 2011. Interactive effect of calcium and gibberellins on nickel tolerance in relation to antioxidant system in *Triticumaestivum* L. *Protoplasma* 248: 503–511.

Singh AP, Dixit G, Mishra S, Dwivedi S, Tiwari M, Mallick S, Pandey V, Trivedi PK, Chakrabarty D, and Tripathi RD. 2015. Salicylic acid modulates arsenic toxicity by reducing its root to shoot translocation in rice (*Oryza sativa* L.). *Front Plant Sci* 6: 340.

Singh S and Prasad SM. 2014. Growth, photosynthesis and oxidative responses of *Solanum melongena* L. seedlings to cadmium stress: mechanism of toxicity amelioration by kinetin. *Sci Hortic* 176: 1–10.

Singh S, Singh A, Srivastava PK, and Prasad SM. 2018. Cadmium toxicity and its amelioration by kinetin in tomato seedlings vis-à-vis ascorbate-glutathione cycle. *J Photochem Photobiol B Biol* 178: 76–84.

Singh VP, Kumar J, Singh M, Singh S, Prasad SM, Dwivedi R, and Singh MPVVB. 2015. Role of salicylic acid-seed priming in the regulation of Cr(VI) and UV-B toxicity in maize seedlings. *Plant Growth Regul* 78: 79–91.

Singh VP, Kumar J, Singh M, Singh S, Prasad SM, Dwivedi R, and Singh MPVVB. 2016. Role of salicylic acid-seed priming in the regulation of Cr(VI) and UV-B toxicity in maize seedlings. *Plant Growth Regul* 78: 79–91.

Sirhindi G, Mir MA, Sharma P, Gill S, Singh KH, and Mushtaq R. 2015. Modulatory role of jasmonic acid on photosynthesis pigments, antioxidants and stress makers of *Glycine max* L. under nickel stress. *Physiol Mol Biol Plant* 21: 559–565.

Srivastava S, Srivastava AK, Suprasanna P, and D'Souza SF. 2013. Identification and profiling of arsenic stress induced microRNA in *Brassica juncea*. *J Exp Bot* 64: 303–315.

Steffens B. 2014. The role of ethylene and ROS in salinity, heavy metal, and flooding responses in rice. *Front Plant Sci* 5: 685.

Sun S, Wang H, Yu H, Zhong C, Zhang X, and Peng J. 2013. GASA14 regulates leaf expansion and abiotic stress resistance by modulating reactive oxygen species accumulation. *J Exp Bot* 64: 1637–1647.

Sytar O, Kumari P, Yadav S, Brestic M, and Rastogi A. 2019. Phytohormone priming: regulator for heavy metal stress in plants. *Plant Growth Regul* 38: 739–752.

Thomas JC, Perron M, LaRosa PC, and Smigocki AC. 2005. Cytokinin and the regulation of a tobacco metallothionein-like gene during copper stress. *Physiol Plant* 123: 262–271.

Tian T, Ali B, Qin Y, Malik Z, Gill RA, Ali S, and Zhou W. 2014. Alleviation of lead toxicity by 5-Aminolevulinic acid is related to elevated growth, photosynthesis, and suppressed ultrastructural damages in oilseed rape. *BioMed Res International* 530642: 1–11 http://dx.doi.org/10.1155/2014/530642.

Tiwari S, Verma N, Prasad SM, and Singh VP. 2020. Cytokinin alleviates cypermethrin toxicity in *Nostocmuscorum* by involving nitric oxide: Regulation of exopolysaccharides secretion, PS II photochemistry and reactive oxygen species homeostasis. *Chemosph* 259: 127356.

Trinh N, Huang T, Chi W, Fu S, and Chen C. 2014. Chromium stress response effect on signal transduction and expression of signaling genes in rice. *Physiol Plant* 150: 205–224.

Verma N, Tiwari S, Singh VP, and Prasad SM. 2020. Nitric oxide in plants: an ancient molecule with new tasks. *Plant Growth Regul* 90: 1–13.

Wang H, Dai B, Shu X, Wang H, and Ning P. 2015. Effect of kinetin on physiological and biochemical properties of maize seedlings under arsenic stress. *Adv Mater Sci Eng* 2015: 1–7.

Wang J, Chen J, and Pan K. 2013. Effect of exogenous abscisic acid on the level of antioxidants in *Atractylodes macrocephala* Koidz under lead stress. *Environ Sci Pollut Res* 20: 1441–1449.

Wang Y, Wang Y, Kai W, Zhao B, Chen P, Sun L, Ji K, Li Q, Dai S, and Sun Y. 2014. Transcriptional regulation of abscisic acid signal core components during cucumber seed germination and under Cu, Zn, NaCl and simulated acid rain stresses. *Plant Physiol Biochem* 76: 67–76.

Wasternack C and Hause B. 2013. Jasmonates: Biosynthesis, perception, signal transduction and action in plant stress response, growth and development. *Ann Bot* 111: 1021–1058.

Yusuf M, Khan TA, and Fariduddin Q. 2016. Interaction of epibrassinolide and selenium ameliorates the excess copper in *Brassicajuncea* through altered proline metabolism and antioxidants. *Ecotox Environ Saf* 129: 25–34.

Zhang F, Zhang H, Xia Y, Wang G, Xu L, and Shen Z. 2011. Exogenous application of salicylic acid alleviates Cd-toxicity and reduces hydrogen peroxide accumulation in root apoplasts of *Phaseolus aureus* and *Vicia sativa*. *Plant Cell Rep* 30: 1475–1483.

Zhu XF, Jiang T, Wang ZW, Lei GJ, Shi YZ, and Li GX. 2012. Gibberellic acid alleviates cadmium toxicity by reducing nitric oxide accumulation and expression of IRT1 in *Arabidopsis thaliana*. *J Hazard Mater* 239: 302–307.

11

Heavy Metal Sequestration in Plants

Rupinderpal Kaur,[1] *Bhekam Pal Singh*[1,2]
and *Yumnam Devashree*[1,*]

ABSTRACT

Heavy metal accumulation is increasing rapidly because of various natural as well as anthropogenic activities. Due to their non-biodegradable nature, they persist for a long time in the environment and enter the food chain via crop plants, then ultimately get accumulated in human beings through the process of biomagnifications and pose a threat to human health as well as the whole ecosystem. That is why the remedy of contaminated soil is of utmost significance. The phytoremediation technique is eco-friendly and cost-effective for the regeneration of polluted land. For efficient phytoremediation, it is a must to understand the mechanism of heavy metals tolerance, and accumulation in plants is a necessary condition. This chapter covers various mechanisms viz. uptake of metal, their detoxification, and translocation in detail.

1. Introduction

Heavy metals are metals with a density of over 5 gm/cm3. Among these, Hg, Pb, and Cd are included in non-essential heavy metals, and Zn and Cu are considered essential. In other words, "heavy metals" are metals having relatively high density and which show detrimental effects on living organisms even at very small concentrations (Morkunas et al., 2018). Heavy metals show undesirable effects on animal and plant health (Asharf et al., 2019). Heavy metals produce reactive oxygen species that cause damage to cell membranes and the peroxidation of lipids (Shahzad et al., 2018). Moreover, heavy metal also replaces essential metal in the cells that cause severe metabolic and physiological deformities in vegetation (Nwaichi and Dhankher, 2016). Some heavy metals like arsenic, cadmium, mercury, and lead are

[1] Department of Botany, School of Bioengineering and Biosciences, Lovely Professional University.
[2] Department of Botany, Govt. Degree (PG) College Bhaderwah, J&K, India.
* Corresponding author: devashreeyumnam@gmail.com

found in the environment as pollutants as they have deleterious effects on plant life and human physiology (Dhalaria et al., 2020). An optimum level of these heavy metals is already accumulated in the environment but their concentration is increasing at an alarming rate due to contamination of the environment by anthropogenic activities like mining, industry, agricultural production, fertilisers, and sewage discharge (Gupta et al., 2010).

Heavy metals cause a detrimental effect on ecosystems and living organisms (Grimm et al., 2008). When humans consume heavy metal contaminated foods, it enters the human body via food chains and shows adverse effects on human health (Liu et al., 2013). Heavy metals also reduce the growth and productivity of plants in various ways as given below:

(i) HMs interfere with the activity of essential nutrients in the soil. (ii) They show harmful effects on plant metabolism. (iii) They have negative effects on soil properties. (iv) They have a detrimental impact on soil microorganisms (Mittler, 2006; Miransari, 2016). (v) HMs damage the root tips of plants. (vi) They cause wilting or even death of plants (Gong and Tian, 2019). (vii) They produce free radical that damages the cells and disrupt ATP production (Reddy et al., 2005; Igiri et al., 2018).

Heavy metals degradation by biological actions is not possible. Therefore, heavy metal remedy is the need of today to save our environment as well as living organisms (Yan et al., 2020). Governments are also taking steps for the remedy of heavy metal contaminated soils at an accepted level (Asharf et al., 2019). The measures include soil replacement, soil leaching, chemical immobilization, etc. However, these methods have low efficiency, high cost, damage to local area, soil erosion secondary contaminations, etc. (Huang et al., 2015). Recent studies have been done on vegetation for remediation of contaminated soil, called phytoremediation (Nwaichi et al., 2016). So phytoremediation is a technique to remove pollutants from degraded soil, rendering it non-hazardous for growing plants (Teofilo et al., 2010).

2. Phytoremediation technology

Due to urbanization and industrialisation, the amount of heavy metals have been increasing during the last few decades (Ashraf et al., 2019). These heavy metals originate from natural and man-made sources like oil, gas industry, fertilizers, sewage sludge, mining and smelting, pesticide, fossil fuel burning, and electroplating (Rafique and Tariq, 2016).

Heavy metals, being non degradable can be persistent in the environment for a longer period (Suman et al., 2018). They are required for various biochemical and physiological processes of plants (Cempel and Nikel, 2006) but they are toxic when exceeding the required amount. Nonessential heavy metals cause toxicity with no role in plant life (Fasani et al., 2018) and reduce agricultural productivity by affecting biochemical and physiological functions (Clemens, 2006) and cause biomagnifications and therefore threat to living beings (Rehman et al., 2017).

Therefore there is a need to take up remedies to ward off heavy metals from reaching the environment and lessen soil contamination soil (Hasan et al., 2019). A large variety of remediation techniques has been made for the reclamation of soil contaminated by heavy metals. These techniques are mainly based on physicochemical and mechanical methods, e.g., soil incineration, soil washing, excavation, electric field application, solidification, etc. (DalCorso et al., 2019). Physicochemical approaches are costly, inefficient, cause deterioration of soil and also introduce secondary pollutants (DalCorso et al., 2019). Therefore, it is needed to develop efficient, cheap, and eco-friendly remedies for soil reclamation.

Phytoremediation is a plant based method to use plants for extraction and removal of pollutants and lower their concentration from the soil (Berti and Cunningham, 2000). Plants have the ability to absorb ionic compounds from the soil by extending their roots and establishing the rhizosphere to reclaim the polluted soil and its fertility (DalCorso et al., 2019). The main advantage of phytoremediation technology includes economic feasibility, eco-friendly, high applicability, preventing erosion and metal leaching, improving soil fertility (Jacob et al., 2018). During the last few years, the main focus has been to understand the mechanism of heavy metal tolerance so as to develop efficient methods of phytoremediation that includes a strategy to develop heavy-metal bioavailability, their accumulation, as well as tolerance, including applying genetic engineering for improving plant performance during process of phytoremediation has been achieved. There is a vast array of strategies of phytoremediation including phytoextraction, phytostabilization, phytovolatilization and phytofiltration (Marques et al., 2009).

3. Phytostabilization

This method uses plants with metal tolerance to immobilise heavy metals present below the ground and decrease their bio-availability, thereby preventing them from entering into the ecosystem as well as the food web (Marques et al., 2009). It involves precipitation reaction in the metal valence of rhizosphere and absorption in the cell wall of roots (Gerhardt et al., 2017). As the plant grows, it preserves the soil health in areas polluted with heavy metals as well as prevent heavy metal dispersion through the wind (Mench et al., 2010). Disposal of hazardous biomass is not necessary for this system (Wuana and Okieimen, 2011). Plant species chosen for this should possess a dense root system, produce large biomass, grow fast and are easy to maintain (Burges et al., 2018). By altering the pH of the soil, Organic and inorganic substitutes can be applied for altering metal specification, to reduce heavy-metal solubility (Burges et al., 2018). Microorganism can also improve heavy metal immobilization by adsorption of metals in their cell walls, producing chelators, and increasing the plant root system (Ma et al., 2011).

4. Phytoextraction

In this method, plants are used to take up, accumulate and translocate contaminants in all plant parts (Jacob et al., 2018). Recently it is the most approachable technique to reclaim heavy metals from polluted soil (Sarwar et al., 2017). The process involves the following steps: (a) movement of heavy metals in the rhizosphere, (b) compartmentalisation of heavy metal ions in plant cells (c) translocation of ions of heavy metals (d) metal uptake by roots (Ali et al., 2013). The effectiveness of the process depends upon the right selection of the plant with the following efficiencies:

(i) toxic effect tolerance, (ii) good extraction capacity, (iii) fast growth, (iv) extensive shoot/root system, (v) good adaptability, (vi) highly resistant to pathogens, pest, herbivore (Ali et al., 2013). Out of these, above groundmass and metal-accumulating capacity determines the phytoextraction potential of a plant species. So, there are mainly two strategies for the selection of plants (a) use of hyperaccumulator plants, (b) plants with high aboveground biomass (Ali et al., 2013). Hyperaccumulators include the plants can assuage the range of heavy metal without affecting plants even 100 times than non-hyper accumulators (van der Ent et al., 2013), and should have a high shoot/root ratio (Marques et al., 2009) and high shoot to soil ratio (McGrath and Zhao, 2003). More than 450 plant species are recognized as hyperaccumulators including trees, shrubs, and herbs (Suman et al., 2018; Dushenkov, 2003). *Sedum alfredii* is one of them, it can accumulate more than two elements (Yang et al., 2004).

5. Phytovolatilization

It includes the uptake of pollutants from soil and then converts them into volatile compounds with less toxicity to release them into the atmosphere (Mahar et al., 2016). For instance, Inorganic Selenium is absorbed to its organic form selenoamino acids selenocysteine and selenomethionine, which is then biomethylated less toxic form dimethylselenide (DMSe) and then evaporated in the air (Terry et al., 2000). Similarly, Hg exists as a Hg_2C. Then the roots take methyl-Hg and convert it into an ionic form that is non-toxic and evaporated in the air (Marques et al., 2009). This method does not remove the pollutants completely from the source and can be redeposited again (Vangronsveld et al., 2009).

6. Phytofiltration

This involves the use of plant parts including root, shoot, seedlings to remove pollutants that are called rhizofiltration, caulofiltration, and blastofiltration respectively (Mesjasz-Przybyłowicz et al., 2004). In the process of rhizofiltration, heavy metals get adsorbed in the plant root system that forms precipitates of heavy metals on the root system, and roots become saturated, and heavy metals are disposed when harvested (Javed et al., 2019). Plant species like Azolla, water hyacinth, *Brassica juncea*, duckweed, sunflower, etc., are some plants that are commonly used for rhizofiltration (Dhanwal et al., 2017).

7. Improvement in plant performance

The plants used for phytoremediation have limitations including slow growing speed (Sarwar et al., 2017), and lesser adaptability to a wide range of environmental conditions (Gerhardt et al., 2017). To overcome this, genetic engineering plant species are used for the improvement of growth and biomass productivity (DalCorso et al., 2019). For instance, a mutagen called ethyl methanesulfonate has been used to produce a sunflower mutant called "giant mutant" that increased extraction ability of heavy metals many fold, e.g., 7.5 times for Zinc and 8.2 times for Lead (Nehnevajova et al., 2007).

8. Genetic engineering

By this method, there is a possibility for modification of plants with the required characters for the phytoremediation method within a short period of time. It also becomes possible to transfer the required genes from a hyperaccumulator to sexually in-compatible plants (Marques et al., 2009). Heavy metals accumulation leads to oxidative stress, so its tolerance is based upon strengthening the oxidative stress mechanism of defense. Therefore, a common method to accelerate the heavy metals tolerance system is to increase the antioxidant potential of the plant by the overexpression of the genes involved in the uptake as well as translocation of heavy metals by use of genes of MTP, HMA, ZIP, and the MATE family that help to encode ion transporters to make metal translocation improvement (Das et al., 2016; Kozminska et al., 2018). But, this mechanism involving genes is time consuming, complicated, and laborious.

9. Use microbes for improving plant performance

The use of plant related microbes is another method to improve the performance of plants for the accumulation of heavy metals by root proliferation stimulation, promotion of plant growth to increase metal absorption and tolerance, and by producing biochemicals like phytohormones, enzymes, organic acids, and antibiotics (Fasani et al., 2018; Ma et al., 2011). Plant growth promoting rhizobacteria (PGPR) can synthesise the 1-aminocyclopropane-1-carboxylate (ACC) deaminase, which further degrades ACC, an ethylene precursor that promotes plant growth (Glick, 2014). These plants have high root and shoot systems with increased phytoremediation ability (Arshad et al., 2007). PGPR also produces bacterial auxin (IAA) that causes lateral root as well as root hair production to stimulate and facilitate the process of phytoremediation (DalCorso et al., 2019). Arbuscular mycorrhizal (AM) fungi can increase the absorptive area of roots to increase heavy metal accumulation in continuation with minerals and water uptake (Göhre and Paszkowski, 2006). It also can produce phytohormones, which help in promoting plant growth and thereby phytoremediation (Vamerali et al., 2010).

10. Progress in heavy metal phytoremediation of contaminated soils

10.1 Heavy metal speciation and factors

Total heavy metal concentration is used as an indication for the reflection of heavy-metal contamination in the environment and assesses environmental quality (Rao et al., 2008). The properties of heavy metals depend on their speciation, and species are sometimes accompanied by reactions, not with the concentration of heavy metals (Kang et al., 2017). The heavy metal specification is broadly categorized into organic, manganese, and iron oxides, exchangeable, and carbonate, and residual species (Pizarro et al., 2016). Out of these, exchangeable have a tendency to be easily transferred and absorbed by plants. Carbonate has a sensitivity to pH value as heavy metals for precipitates on carbonate and forms insoluble compounds that are sensitive to acids. Manganese and Iron oxides are stable with heavy metals, are easy to absorb due to their large surface area and good absorption power (Zhang et al., 2014).

10.2 Mechanism of phytoremediation

The presence of heavy metals is an important topic to reclaim heavy metal contaminated areas (Olaniran et al., 2013). The absorption of heavy metals depends upon the bio-availability of heavy metals in the land (Park et al., 2013). High bioavailability is directly proportional to absorptivity and toxicity in living organisms (Zhang et al., 2014). There is a direct relationship between bioavailability and the concentration of Lead, Cadmium, and Arsenic (Wang et al., 2011). However, some claim that bioavailability depends on the solid to liquid phase ratio and molecular form (Marcin et al., 2014; Reeder et al., 2006). For instance, the bioavailability power of Cr^{+3}, which largely get precipitated in the soil, is much lower than Cr^{+6} mainly in the case of chromate (Reeder et al., 2006). Moreover, Teofilo et al. (2010) mention that organic acid having low molecular weight improves the bioavailability and solubility of heavy metals).

10.3 Mechanism of the absorption of heavy metals by plants

The main pathways for heavy metals migration in the cell membrane are Apoplast absorption and infiltration (Li and Zu, 2011). The apoplast absorption is the radial migration of heavy metal between the cell wall and inter-cellular spaces through channels. Cd and Pb are radially transported through apoplast in maize roots (Seregin et al., 2004). Infiltration involves heavy metal infiltrating across cytoplasm and cells. Absorption mechanism to Cadmium in heavy metal sensitive and heavy metal tolerant barley revealed cadmium content is cytoplasm and cell wall of sensitive barley is much higher (Wu et al., 2005). Heavy metal concentration in indicator plants is known by the determination of heavy metals in leaves, where it is found much higher amounts (Memon and Schröder, 2009). *Pteris vittata* tend to accumulate Arsenic ions in leaves in the amount of 9677 mg/kg dry weight (Danh et al., 2014).

10.4 Mechanisms of plant tolerance to heavy metal stress

There is an array of mechanisms to detoxification of adverse effects of heavy metals in the plants for their proper growth and metabolism. These methods are divided into the following categories discussed below:

Absorption rejection

This method involves immobilisation of the heavy metals in the environment by rhizosphere organisms to reduce the absorption of the heavy metals into plants (Hashimoto et al., 2011). It has been observed that arbuscular mycorrhizal fungi (AMF) can enhance the adverse effect of Cadmium on plants growth (Vogel-Mikus et al., 2006). Other ways to reduce metal penetration include a metal chelating agent, immobilization, and accumulation of heavy metal inside a fungal structure like hyphae and vesicles (Miransari, 2011).

Combination-inactivation

This method involves inactivating heavy metals by binding to the cell walls by means of cellulose, pectin, and lignin. It has been known that a great array of heavy metals accumulates in cell walls with functional groups like –COOH, –SH and, –OH in plants (Mehes-Smith et al., 2013). It has been observed that –OH in pectin has a tendency to bind lead in the cell to control its toxicity (Jiang and Liu, 2010b; De et al., 2012). This reduction in toxicity is due to transportation through the cation-efflux transporter protein present in the plasmalemma (Huang and Xin, 2013). Heavy metals bind to the pectin, phenolic groups, amino acid, saccharides in the cell membrane and forms stable chelates (Yuan et al., 2007).

Compartmentalization

In this process, the absorbed heavy metal gets transported to some other position in the plant cell and gets separated from the active molecule to help control the cell damage. It occurs through the vacuole and epidermis (Li and Zu, 2016). It has been found that *Thlaspi goesingense*, a Nickel hyper-accumulator, can accumulate two times more Ni in the vacuole in comparison to a non-accumulator (Krämer et al., 2000). Similarly, *Pteris vittata*, an Arsenic accumulator, accumulates 61% Arsenic in its vacuoles (Chen et al., 2006). Similarly (Jiang and Wang, 2008), have reported increased Zn in vacuoles. Transgenic *Arabidopsis* under Cadmium and Arsenic stresses increases the toxicity of thiol chelating toxic metal (Guo et al., 2012). It has been noticed that hyperaccumulators avoid the adverse effects of heavy metals stress by accumulation into vacuoles (Leitenmaier and Küpper, 2013).

10.5 Phytoremediation assessment

Out of all methods of heavy metal detoxification, the absorption of the heavy metals by plants species is totally different (Marques et al., 2009). It can be estimated by the measurement of heavy metals in the soil (Zhuang et al., 2009; Murakami and Ae, 2009). when *Oryza sativa* and *Glycine max* were compared for their remediation potential in soil for Pb, Cu, and Zn, it has been found that *O. sativa* has a greater selection in Copper absorption, while *G. max* has a greater selection in Zinc absorption and this is because Zn is essential for plant enzymes as well as for photosynthesis, and lead and Cadmium cause toxicity to plants. By applying of EDTA and $(NH_4)_2SO_4$ in the soil, the absorption ability of *S. bicolor* to Cu, Pb, and Zn were compared, and the results have shown that Pb concentrations in *S. bicolor* leaves was higher as compared to others (Zhuang et al., 2009).

10.6 Selection principles of phytoremediation

Large concentrations of heavy metals result in stress in plants, by slowing down growth and even death of the plant. Based on appropriate reference, the following criteria are made: (1) the selected plant should have good adaptation to heavy metals stress (2) Heavy metals concentrations in a plant is higher than in the growing area (3) should have strong resistance to adverse climate as well as developed root systems (4) should have large biomass and fast growth (Koptsik, 2014). Moreover, it has been found that intercropping has a more promising effect than mono-planted plants (Jiang et al., 2010; An et al., 2011).

11. Development of new technological approach to mitigate metal stress

Remediation is necessary to protect and conserve the environment (Glick, 2010). Various biochemical, as well as physic-chemical techniques, are used for heavy metal elimination. Physicochemical approaches are rapid but have high costs and are technically complex. Moreover, they cause ill effects on physical, biological, and chemical properties of soil, which ultimately cause secondary pollution (Glick, 2010; Ali et al., 2013; Ullah et al., 2015). Bioremediation is the most effective, natural, environment friendly, and low cost way to remove toxic metals (Doble and Kumar, 2005). The use of plant-microbe for bioremediation is a beneficial technology with relation to global climate change and the use of fertilizer (Tiwari et al., 2016). To survive in metal toxic environments, microorganisms have developed various ways to develop resistance (Mustapha and Halimoon, 2015). Microbes enhance the bioavailability of metal in various ways, like acidification, chelation, and precipitation. Organic acids released by the plant roots lower the pH of the soil to help to metal ions removal (Mishra et al., 2017). Microbes have developed resistance factors like bioaccumulation, metal sorption, oxidation reduction, and efflux of toxic compounds from the cell (Mustapha and Halimoon, 2015).

11.1 Remediation of heavy metals by bacteria

Bacteria are used for the remediation of heavy metal contaminated soil (Chen et al., 2015). They cause alleviation of heavy metal toxicity by utilizing, mobilizing, uptake and, transformation mechanism (Hassan et al., 2017). Symbiotic and free living bacteria present in the soil environment alter plant growth as well as productivity by producing growth regulators and facilitating nutrient uptake in soil (Nadeem et al., 2014).

Various studies showed that microbes are elicitors for stress and metal tolerance (Tiwari et al., 2016, 2017b). They form complexes with metabolites and bacterial transporters to limit their availability (Ahemad, 2012). These microorganisms have some strategies for tolerance like, (a) biosorption in the cell wall (b) transport across membrane (c) precipitation (d) metal entrapment (e) metal detoxification (Zubair et al., 2016). Metal tolerating bacteria like *Bacillus, Methylobacterium* and *Pseudomonas*, tend to improve crop growth and productivity by decreasing metal toxicity (Sessitsch et al., 2013). Various studies have shown Cadmium resistant *Ochrobactrum*, and Arsenic and Lead resistant *Bacillus* species have characters for the promotion of growth in rice (Pandey et al., 2013). Several rhizobacteria are also used as hyperaccumulators for the uptake of heavy metals (Thijs et al., 2017). Various reports have shown that additives used with microbes are significantly more effective for reducing toxicity by heavy metals than without additives (Mishra et al., 2017). For example, an additive named as thiosulfate, when used with microbes increases the uptake of Mercury and Arsenic in *Brassica juncea* (Franchi et al., 2017). Apart from this, genetically engineered bacteria have greater power to remove toxic metals, but this practice is only at the laboratory level (Ullah et al., 2015; Gupta and Singh,

2017; Ashraf et al., 2017). Recently, great consideration is given to the elimination of toxic metals through genetically engineered microbes (Ullah et al., 2015). In this relation, genes for metal transporters, chelators, stress tolerance and biodegradative enzymes are an important consideration for the production of genetically engineered bacteria (Singh et al., 2011).

12. Evolution of plants in metal-contaminated soil

Metals toxicity causes oxidative damage to plants, which ultimately alter their phenotype (Kachout et al., 2009). In polluted areas, the microbial and plant community are intertwined with the soil properties (Schimel et al., 2007). It has been found that fast growing plants remove metals from the soil, and this has proved to be cost effective (McIntyre, 2003). Some plants themself translocate metal, e.g., *B. populifolia* (Gallagher et al., 2008a), sometimes the metal tolerant ability of plants is directly proportional to organisms present in its rhizosphere, which has caught the attention of researchers (Kuiper et al., 2004; Sessitsch et al., 2013). Understanding the ecology of soil is very necessary for metal removal (Bissett et al., 2013), which in turn is necessary for the proper application of mechanisms. Microbes like fungi and bacteria alter the soil environment by translocating, absorbing, and remediating contaminants from soil (Clarholm and Skyllberg, 2013). After four successive years it has been found that the micro and meso-fauna tend to increase the tendency to remove heavy metals by facilitating plant growth (Frey et al., 2006; Krumins, 2014) even in metal contamination (Neher, 2010). The composition of soil nematode population and other taxa is used for indication of soil health (Fiscus and Neher, 2002; Ellis et al., 2002). Enchytraeid populations are sensitive to metal contamination (Kapusta et al., 2011). Diverse soil fauna indicates primary productivity and plant community (De Deyn et al., 2003). Nitrogen fixation can be measured by enzyme activity and can help the plant from nutrient deficiency (Burns et al., 2013). Metal stress increases enzyme activity. It was found that in microbe communities near industrialized areas there is decreased soil biota diversity and enzyme action (He et al., 2010; Wang et al., 2007) and increased metal load, but an increase in the metal load in soil was correlated to an increase in enzymatic action (Pascual et al., 2004). Upon calculation of microbial composition and their functional activity, it was found that there is a drastic shift in microbes' population, and function were correlated with plant aging and composition (Zhang et al., 2007).

Some plants are more adapted for growing as well as reproducing in heavy metals stressed soil. Plants growing in non-contaminated soils also evolved more tolerant ecotypes to grow in toxic soil (Kruckeberg, 1967). In this way, plants growing in stress areas are genetically more different from noncontaminated areas (Assunção et al., 2003). Heavy metals contaminated tolerant ecotypes occur at an outlooked rate, and thus maintain a high degree of polymorphism in population. This process appears a less changing parameter in comparison to the non-tolerant population (Mengoni et al., 2000). Moreover, inter-population changes do not have any correlation with respect to the geography of the areas and heavy metals stress tolerance (Lefèbvre and Vernet, 1990). By using molecular markers, there is a possibility to state relationships between the genetic variation and heavy-metal tolerance. These molecular markers for metal tolerance are necessary to improve the phytoremediation technique as well as interpretation of molecular characterisation of stress tolerance (Mengoni et al., 2000).

12.1 Antioxidant defense system

Heavy-metal toxicity leads to the over productivity of reactive oxidation species that result in the peroxidation of lipids in cells s an important consequence. As a result of this, plants have remarkable defense systems including enzymatic as well as non-enzymatic antioxidants. A wide range of enzymatic antioxidants that include glutathione-*s*-transferase (GST), peroxidase (POD), superoxide dismutase (SOD), and catalase (CAT), can convert the high molecular weight of superoxide radicals into water and oxygen. On the other hand, non-enzymatic low molecular weight antioxidants like glutathione, ascorbic acid, and proline tend to direct the detoxification of ROS (Yadav et al., 2014). These antioxidants have a wide range of quenching activity that depends upon the localizations at different compartments of the cells. SOD is a group of metallo-enzymes that can facilitate the conversion of superoxide radicals (SOR, $O2 \bullet^-$) into hydrogen peroxide (H_2O_2) and a wide array of PODs can catalyze the breakdown of H_2O_2 (Gill and Tuteja, 2010). Similarly, various studies have revealed that during stress conditions proline acts like osmolyte and increases the antioxidant enzyme to lower the effects of oxidative stress in the cell (Ashraf and Foolad, 2007; Islam et al., 2009). Various studies show that when proline is applied exogenously, it reduces the phytotoxic effect caused by Se in *Phaseolus vulgaris* L. Seedling and improves plant growth (Aggarwa et al., 2011) while in some case it increases the damaging effect of Cadmium and increase the rate of photosynthesis also (Okuma et al., 2004; Hayat et al., 2013).

12.2 Cellular homeostasis

During stress conditions, proline accumulates in the cytosol (Ashraf and Foolad, 2007; Parida and Das, 2005). According to various studies (Hayat et al., 2013), when proline is applied exogenously, it enhances the inner proline level during heavy metal stress conditions that may help in maintaining the redox homeostasis potential intracellularly (Hoque at al., 2008), it also protects enzymes, the 3-dimensional structures of protein and organelles as well as reduces the effect of peroxidation of protein as well as lipid (Okuma et al., 2004; Islam et al., 2009; Paleg, 1981). Proline helps to enhance the tolerance ability of plants by chelating heavy metals in the cytoplasm and prevents the organelles from damage, it maintains the osmotic balance and homeostasis (Forago and Mullen, 1979; Costa and More l, 1994; Wu et al., 1998). In *Anacystis nidulans* it is also observed that during Cu stress, when proline is applied it showed the protective role of proline and minimized the efflux of K ions (Wu et al., 1995).

12.3 Role of genes in metal uptake and their transportation

Genes activate special enzymes to overcome the negative responses of heavy metal stress. It is observed that genetically modified tobacco is more resistance to methyl mercury (CH₃Hg⁺). The *merB* genes encoding the *MerB* enzyme, lead to the dissociation of the CH_3Hg^+ to Hg^{+2}, which is less toxic and accumulates in the form of Hg-polyP complex in the cells of the tobacco plant (Nagata et al., 2010). Apart from this, the over expression of the *AtPCS1* and *CePCS* gene also increase the detoxification rate of tobacco plant under Cd and As stress, by increasing the phytochelatins (PC) level (Clemens et al., 2006). Other researchers (Clemens et al., 2006), found that Cadmium ions entered into the plant cell by $Fe2+/Zn2+$ transporter of ZIP family. Furthermore, some orthologous genes are over-expressed in Cd and Zn hyper tolerant variants than in non-hyper accumulator metallophytes, for example *SvHMA4* and *SpHMA4* in *Silene vulgaris* and *S. paradoxa* (Arnetoli and Schat), and MT2b orthologous genes are over expressed in hyper-accumulators (Hasinen et al., 2009). Proline is found to enhance the tolerance ability of transgenic alga namely, *Chlamydomonas reinhardtii* under hyper dosage of Cd (Siripornadulsil et al., 2002) with the help of the gene encoding moth bean D1-pyrroline-5-carboxylate synthetase (P5CS). The transgenic alga produced 80% higher proline content and maximum growth in Cd stress conditions than the wild type (Siripornadulsil et al., 2002).

13. Arbuscular mycorrhiza fungi (AMF) in phytoemediation of heavy metals contaminated soil

The Flamentous fungi of genera *Trichoderma, Mucor, Penicillium*, and *Aspergillus* have the capacity for stress tolerance (Ezzouhri et al., 2009; Oladipo et al., 2017). Fungi cell wall has the property to bind metal due to different functional groups with negative charge viz. carboxylic, amine phosphate, in their cell walls (Ong et al., 2017). It has been found that interactions of *Aspergillus niger* var. *tubingensis* strain Ed8 with Chromium (VI) is a reduction and sorption process (Coreño-Alonso et al., 2014). It has been found that reduced Arsenic stress in chickpea occurs via the *Trichoderma* species (Tripathi et al., 2013, 2017). Arbuscular mycorrhizal fungi establish a link between soil and roots to increase the absorptive surface area and alleviate metal toxicity (Saxena et al., 2017; Meharg, 2003). AMF develop a symbiotic association with roots of plants to get food from the plants and assists the plant in the absorption of mineral nutrients, withstanding heavy metal stress and improving photosynthetic efficiency thus enabling the plant to exhibit normal growth and development even under metal stress conditions (Birhane et al., 2012; Chen et al., 2017; Mitra et al., 2019). The role of arbuscular mycorrhizae with heavy metal exposure varies by various factors like the fungal species, the metal type; topology of soil, and plant growth (Pawlowska and Charvat, 2004). This process involves various steps, (a) heavy metal bind cell wall and deposition in the AMF vacuole (b) metal compartmentalisation through siderophores (c) metal bind to metallo-thioneins inside the cells (d) metal transporters help in transport (Jan and Parray, 2016).

13.1 Role of AMF in the absorption of nutrients by plants

AMF facilitates absorption of mineral nutrients like P, N, S, Ca, Zn, K, etc., from heavy metal contaminated and nutrient deficient soil and thus promote plant growth under mineral deficient conditions. The fact that inoculation of plant roots with AMF promotes plant growth and enhances their tolerance for HMs has also been supported by studies carried out by Shakeel and Yaseen (2015). Thus the level of HM tolerance of plants having mycorhhizal associations depends upon the growth and biomass of mycorrhizal fungi (Yang et al., 2015). The plants having symbiotic association with arbuscular mycorrhizal fungi show a high growth rate, high biomass, well developed roots with increased nutrient absorption capacity, and stronger tolerance to heavy metals (Khade and Adholeya, 2007). Wei et al. (2012) had also reported that the higher the plant biomass, the stronger the metal tolerance of the plant (Wei et al., 2012). AMF also produce certain growth promoters like gibberellins, cyrtokinins, indole acetic acid, etc. (Karthikeyan et al., 2008).

Glomus caledonium is one of the most effective fungi used for the rehabilitation of HM contaminated soil through bioremediation (Liao et al., 2003). The plants used for phytoextraction of heavy metal from contaminated soil possess either high biomass, e.g., willow (Landberg and Greger, 1996), or low biomass with hyperaccumulating traits, e.g., *Arabidopsis* and *Thlaspi* species (Reeves, 2003). Insoluble phosphate present in the soil is hydrolysed and converted into a soluble form with an increased degree of absorption by the action of phosphatase enzyme released by AMF into the soil. Various organic acids like succinic acid, oxalic acid, propionic acid, etc., are also released by AMF into the soil which acts upon insoluble phosphate, and facilitate its absorption by plant root and then upward transport to aerial parts of the plant. Thus AMF are beneficial for the growth and survival and of plants under nutrient deficiency (Zhang et al., 2003).

13.2 Affect of AMF on soil pH and metal solubility

The solubility and availability of metals for plant roots, inter alia, depend upon the pH of the soil (Harter, 1983). The change in soil pH changes the degree of solubility and absorption of heavy metals in plants (Lin et al., 2013). It has been observed that AMF releases certain organic compounds in the soil which reduce the pH of the soil and thus influence the solubility and absorption of HMs (Ying, 2008). Decrease in soil pH facilitates phytoextraction of HM from the soil by increasing their solubility and bioavailability (Karaca, 2004). On the contrary, an increase in pH of soil reduces the availability of Cd and Zn for absorption by the roots of *Thlaspi caerulescens* (Wang et al., 2006).

Phytostabilisation is defined as a reduction in mobility and bioavailability of pollutants especially metals by using their immobilization thereby preventing their spread in the ecosystem and circulation in the food chain (Wong, 2003). It prevents the spread of toxic metals through leaching, rain, or wind (Salt et al., 1995). Phytostabilisation takes place by sequestering absorbed metal ions inside root tissue, adsorption of HMs on the cell walls of plant roots or precipitation of HMs (Gerhardt et al., 2017; Grinn et al., 2008). AMF also assist in the fixation of metals in the roots of plants growing in soil with high metal concentration and thus many toxic metals fixed and accumulated in the plant roots are prevented from their transport to the aerial parts thereby alleviating the toxic effect of heavy metals on plants (Gong and Tian, 2019).

AMF can accumulate HMs in their vesicles and mycelium and thus immobilize them by preventing their transport to the aerial part of plants (Vogel-Mikus et al.,

2006). They serve as filtration barriers against the transportation of HM to the aerial part of the plant (Aggarwal et al., 2017). The number of vesicles in mycorrhizal fungi increases with an increase in concentrations of HM in soil (Yang et al., 2015). The fungal mycelium growing on the plant roots in HM contaminated soil plays a vital role in immobilizing metals as the mycelium having a large surface area adsorbs more metal ions than the plant roots (Bodong, 2002). Metals are adsorbed on the surface of the fungal cell wall which acts as a barrier for metal ions regulating their absorption (Liu et al., 2008). Joner et al. (2000) also supported in their studies that the metal sorption capacity of the outer surface of extra radical mycelium of AMF is much higher than root cells and reported that the Cd sorption capacity of extra radical mycelium of AMF *Glomus mosseae* was 10 times higher than some other fungal biosorbents. Moreover, the adsorption of metal ions is also promoted by the negative charge of functional groups like carboxyl and hydroxyl groups as well as free amino acids present in the fungal cell walls which can bind with metal ions (Meier et al., 2012; Zhou, 1999).

When Pakchoi plants were inoculated with mycorrhizal fungi like *Glomus versiforme, Rhizophagus intraradices,* and *Funneliformis mosseae,* a significant decrease in the concentration of Pb has been observed in its shoots (Zhiping et al., 2016). Addition of mycorrhizal fungi to the roots of *Aster tripolium* results in an increase in tolerance of the plant to Cadmium due to retention of Cd in fungal mycelium (Carvalho et al., 2006).

13.3 *Role of AMF secretion in chelation of HM*

AMF assists plants to withstand heavy metal stress by immobilization of HM in the fungal body, precipitation of HM and their chelation in the soil, sequestration of HM in fungal vacuoles and through antioxidant mechanisms in plants (Mishra et al., 2019). AMF secrete glycoprotein glomalin and release it into the soil (Driver et al., 2005). Glomalin is generally secreted by the fungi that belong to group the Glomeromycota and its production depends upon the concentration of the metal ion in soil (Singh et al., 2012).

Presence of different chelating agents like metallothioneins (MTs), phytochelatins (PCs), amino acid, organic acid in AMF and plants assist the plants in resisting HM stress by chelating these substances (Clemens, 2006; Anjum et al., 2015). The formation of glycoprotein-metal complex immobilize HMs in the soil and thus reduces their solubility, bioavailability, and absorption by the plants. The glycoprotein-metal complexes move through the cytoplasm and reach cell vacuoles (Hall, 2002). Thus glomalin protein of AMF helps plants to reduce the toxic effects of metals and to resist metal stress (Wright and Upadhyaya, 1998). Glomalin also reduces soil erosion and helps in soil reconstruction by promoting the formation of much stable soil aggregates (Dhalaria et al., 2020; Hongxin, 2010).

AMF promotes the synthesis of thiol peptides phytochelatins which form complexes with heavy metals and thus provide metal tolerance to the plant (Garg and Chandel, 2012; Begum et al., 2019). These phytochelatins are synthesized by using glutathione as substrate and the reaction is mediated by the enzyme phytochelatin synthase when the concentration of HM increases in soil (Grill et al., 1989). Since

AMF facilitates the biosynthesis of phytochelatins, they promote the formation of phytochelatin-Cadmium complexes which are transferred to the vacuole thereby enabling the plant to resist metal stress (Jiang et al., 2016). Introduction of mycorrhiza in *Zea mays* changes Cd into an inert form by increasing phytochelatins and glutathione content and thus reduces mobility and toxicity of Cd (Zhang et al., 2019). It is found that the introduction of *F. mosseae* in *Nicotiana tabacum* increases the concentration of glutathione which reduces the level of As and Cd in this plant (Degola et al., 2015).

13.4 Role of AMF in the expression of regulatory gene

Inoculation of AMF in the plant roots induces expression of plant genes encoding proteins involved in heavy metals tolerance (Rivera-Becerril et al., 2005). Some genes of AMF are involved in resisting HM stress in plants and are induced when exposed to HMs. GmarMT1 gene has been reported from germinating spores of AM fungus *Gigaspora margarita* and the polypeptide encoded by this gene plays an important role in alleviating Cu and Cd stress (Lanfranco et al., 2002). The expression of GmarMT1 is upregulated by exposure to HMs. Another gene GintABC1 has been reported from the extra radial mycelium of AMF *Glomus intraradices* and it encodes polypeptide which helps in withstanding Cu and Cd stress (Gonzalez-Guerrero et al., 2010). RintZnT1 gene and RintPDX1 gene of *Rhizophagus intraradices* plays a vital role in sequestration of Zn in the fungal vacuoles and reduction of hydrogen peroxide and Cu respectively. In *Glomus intraradices*, some genes encode proteins involved in HM tolerance like a glutathione S-transferase, a Zinc transporter, metallothionein, a 90 kD heat shock protein. In this fungus, upregulation of several transcriptional factors involved in activation of glutathione S-transferase and Zn transporter has been reported on exposure to metal stress (Hildebrandt et al., 2007).

Metallithioneins are proteins that have competence for binding with metals and are produced in different organisms on exposure to high concentrations of metals like Zn and Cd (Kumar et al., 2005). Upregulation of metallothionein gene (BI451899) in the extraradical mycelium of *G. intraradices* upon exposure to Zn and Cu stress indicates that they are involved in the detoxification of these metals. Glutathione S-transferases catalyses conjugation between glutathione and some reactive electrophilic compounds and help in oxidative stress tolerance (Smith et al., 2004). Zn transporters assist in the uptake of HMs from outside the cell or their mobilization inside the cell (Gaither and Eide, 2001). In *G. intraradices*, the Zn transporter gene was upregulated by exposure to Cd and Cu stress (Hildebrandt et al., 2007).

14. Functional significance of metal ligands in hyperaccumulating plants

Contamination of the environment by heavy metals (HMs) is causing a serious threat to living beings as heavy metals are non-biodegradable, i.e., they cannot be degraded to non toxic forms and persist in the environment for several years (Jabeen et al., 2009). For example, Pb remains in the soil for about 150–500 years (Kumar et al., 1995).

Different plants have developed different mechanism for tolerating heavy metal stress (Verkleij and Schat, 1990). Some plants, called excluders, can resist HMs stress by restricting root uptake of HMs and also their translocation from root to shoot. Another category of plants, called metal hyperaccumulators, are those plant species that have the competence to absorb metals from the soil, translocating to the aerial parts where they are stored in much higher concentrations, about 100 to 1000 times higher than in non-hyperaccumulator species without suffering toxic effects (Baker and Brooks, 1989). The term hyperaccumulator was used for the first time by Brooks (Brooks et al., 1977) for such plant species which can accumulate Ni in their aerial parts in natural habitat in concentrations of more than 1 mg per gram of dry weight. Brooks (Brooks, 1998) also defined Zn hyperaccumulators as those plants which can accumulate Zn in their aerial parts in natural habitat in concentrations of more than 10 mg per gram of dry weight. *Thlaspi caerulescens* is a good example of a Zn hyperaccumulator (Baker and Reeves, 2000). Metal hyperaccumulators are receiving attention due to their potential in phytoremediation of heavy metal contaminated areas (McGrath, 1998).

Ligand is a large molecule attached to the central metal ion forming a ring structure called complex compound or chelate. A chemical compound that forms several bonds to a single metal ion is called a chelating agent or chelator. In other words, chelating agent may be defined as multidentate ligand, e.g., EDT, citrate, maleate, oxalate, succinate, phthalate, etc. (Jabeen et al., 2009).

Ligands play the following significant roles in hyperaccumulating plants.

(i) Metal uptake.

(ii) Translocation of metals in plants.

(iii) Metal detoxification and stress toleration.

14.1 Metal uptake

Metals in the soil solution are present not only as free ions but also as complexes with organic or inorganic ligands or in association with mineral colloids (Degryse et al., 2009). The presence of aqueous metal complexes in the soil enhances metal uptake by plant roots and thus the metal uptake depends not only on free ion activity but also on metal complexes (Degryse et al., 2006). Chelates are used to help the phytoextraction of various metals from the soil like Pb, Cu, Zn, Cd, etc., as they increase the bioavailability of the metal in soil by increasing the solubility of precipitated metals (Prasad, 2003). Plants secrete and release organic ligands like organic acids into the soil to increase solubility and absorption of metals (Knight et al., 1997) by acting as a source of H^+ as well as metal chelating anions (Li et al., 2010; Devevre et al., 1996). Natural biodegradable organic acids like citric and gallic acid secreted by the plants facilitate phytoextraction of Cd, Cu, Zn, and Ni from the soil without an increase in their leaching (Nascimento et al., 2006). However, some synthetic ligands like EDTA have also been utilized for increasing solubility of metals thereby enhancing the metal uptake through plant roots (Huang et al., 1997).

Low bioavailability of metal in the soil causes a decrease in their absorption by plant roots (Brun et al., 2001). However, this problem can be overcome by using chemical chelates which help to remove metals from soil (Chiu et al., 2005).

For example, iron plays a significant role in various metabolic processes in living organisms and hence sustenance of life, as it has the unique ability to catalyse oxidation-reduction reactions. Fe acts as a prosthetic group in several enzymes. But there is limited availability of Fe to the plant roots due to its insolubility in an inorganic state. To cope with the deficiency of iron, certain specialized systems are operating in plants for efficient iron uptake. For example, phytosiderophores, which are synthesized from nicotinamide, are ligands that have a high affinity for Fe. These are exclusively produced by the members of the Poaceae (grass family) and are released into the soil for chelation of Fe to form Fe–phytosiderophore complex and thus help in Fe uptake (Higuchi et al., 1999). Phytosiderophores have been found in the xylem sap of several plants, e.g., *Hordeum vulgare* (Shah et al., 2001). Maize YS1 is a membrane transporter that helps to uptake Fe–phytosiderophore from the soil by plant roots and its long distance transport through vascular tissues (Curie et al., 2009). However non-grass monocotyledons and dicotyledons increase the solubility of Fe in the soil by decreasing soil pH due to secretion and release of organic acids and protons into the soil (Curie and Briat, 2003).

In rhizosphere, microorganisms produce organic acids and H^+ which play a significant role in metal mobilization and absorption by plants (Crowley et al., 1991). Thus it can be concluded that ligands facilitate absorption of metal by hyperaccumulating plants in metal deficient soils.

14.2 Translocation of metals in plants

Various ligands are involved in the transportation of metals in plants. For example, nicotianamine (NA) is a significant ligand involved in the loading of Ni and Cu in the xylem and it also prevents precipitation of iron in the leaves. In the xylem sap nicotianamine forms chelates with Cu and thus helps in its transport from roots to aerial parts. NA also helps in phloem loading and unloading of Fe, Cu and Zn (Curie et al., 2009). The amount of Ni in the leaf vacuoles of hyperaccumulator *Thlaspi goesingense* is found about 2 times higher than the non-accumulator *T. arvense* when both plants were exposed to nontoxic concentrations of Ni. *Thlaspi caerulescens* is a polymetallic hyperaccumulator plant that can tolerate and hyperaccumulate Cd, Zn, and Ni (Kramer et al., 2000). The increased transcription of genes encoding chelator proteins may promote the translocation and localisation of the metal in the plants (Cherian and Oliveira, 2005). For example, there was 16-fold increase in Cd tolerance in cauliflower when an MT-gene (CUP1) was overexpressed (Hasegawa et al., 1997).

The genome of *T. caerulescens* contains a TcNAS gene which encodes for functional nicotianamine synthase (NAS) and its expression is confined to the leaves. The exposure of *T. caerulescens* to Ni results in the accumulation of nicotianamine (NA) in the roots most probably due to translocation of NA from leaves through phloem sap. In the roots, NA along with Ni moves to the xylem for its transport to aerial parts. A part of Ni binds with NA and is translocated in the form of stable Ni-NA chelates in the xylem sap. This circulation of Ni, NA, and Ni-NA chelates is lacking in non-hyperaccumulator species *T. ravens*. The presence of NA in the phloem and xylem sap, as revealed in different studies, indicates the role of NA plants for long-distance transport (Mari et al., 2006). Specific transport systems are needed for the transportation of Ni-NA complexes in the plant body across membranes. For example, in *Arabidiopsis thaliana*, YS1-like proteins (YSLs) act as metal-NA transporters (Curie et al., 2001).

In the root and leaf tissues of *T. caerulescens*, in addition to NA, there are some other Ni chelators like citrate, histidine, and malate which help in the transport of Ni (Ouerdane et al., 2006). In hyperaccumulator species, a major part of the leaf Ni was associated with the cell wall and the remaining Ni has an association with His and citrate ligands localized in the cytoplasm and vacuoles respectively. Kramer (2000) has also reported that in hyperaccumulator species, Ni was localised in the vacuole in form of Ni-organic acid. Ligands like free His participate in the transport of Ni across the cytoplasm for its loading into vacuoles in a similar manner as its loading in xylem in the *Alyssum* species (Kramer et al., 1996).

At cytoplasmic pH (pH about 7.5) His is a suitable ligand of Ni because of the high stability constant and Ni-His complex as well as protonation constant of His. However, in the vacuoles where pH values are comparatively smaller (pH about 5.5), protonation of imidazole nitrogen of this amino acid takes place. It causes a decrease in stability constants of Ni-His complexes which favours chelation of Ni by organic acids like citrate (Kramer et al., 2000). The hyperaccumulating *Sedum alfredii* shows a comparatively high degree of Zn and Cd translocation from soil to aerial parts than non-accumulating plant species (Long et al., 2002).

In *Alyssum lesbiacum*, histidine is found to be involved in the translocation of Ni in the xylem (Kramer et al., 1996). However, in *Thlaspi goesingense*, histidine does not play any role in hyperaccumulation and tolerance of Ni (Persans et al., 1999). Thus same ligands may play different roles in different plant species. In *Sebertia acuminate*, another ligand that binds with Ni in its latex is citrate and about 20% of the dry weight of its latex contains Ni (Jaffre et al., 1976; Schaumloffel et al., 2003).

14.3 Metal detoxification and stress tolerance

Plants can withstand HM stress through the inactivation and detoxification of HM ions in their cytoplasm by chelation or converting toxic ions into comparatively less toxic forms or by their sequestration (Jabeen et al., 2009). Phytochelatins are the chelators produced by all plant species which can detoxify certain metals like Cd, Ag and As and perhaps Hg and Cu by binding to them (Ha et al., 1999). Phytochelatin-metal complex formation probably also helps in the translocation of metals through xylem sap (Przemeck and Haase, 1991). Some low molecular weight chelators like citrate, malate, and histidine were also involved in the transportation of metal ions in plants (Von Wiren et al., 1999). Citrate (Sanger et al., 1998) and His (Kramer et al., 1996) were also involved in Ni detoxification in hyperaccumulating plants.

Metallothioneins are metal binding proteins with low molecular weight involved in metals chelation through thiols. In *Brassica juncea* the absorbed As (III or V) is converted to AS(III)-tris-thiolate complex by coordinating with the ligand glutathione or phytochelatin and stored as AS(III)-tris-thiolate Studies have shown that there are three sulphur ligands coordinated to the As and thus the complex has trigonal bipyramidal geometry (Pickering et al., 2000).

The results of different studies have revealed that vacuolar localization of metals in hyperaccumulator plants plays a significant role in metal detoxification (Kramer et al., 2000). Moreover, in *Saccharomyces cerevisiae* the accumulation of Ni in the vacuoles is essential for Ni tolerance (Ramsay and Gadd, 1997).

In plants, organic acids like citrate and malate are mainly stored in the vacuoles (Ryan and Walker-Simmons, 1983), where at the acidic pH citrate forms chelate with Ni (Dawson et al., 1986). Ni stored in vacuoles of hyperaccumulator plants in form of Ni-organic acid complexes plays a significant role in the Ni tolerance of such plants (Kramer et al., 2000). A high level of intracellular Histidine in yeast resulted in increased tolerance to the metal ions mainly Ni^{2+} (Joho et al., 1986). A much higher level of Fe in living organisms could have detrimental effects as Fe facilitates the formation of reactive oxygen species (ROS) like OH^-. Free radicals can generate oxidative stress and cause damage to the cell. To withstand oxidative stress caused due to higher concentration of Fe, plants store it in vacuoles complexed with organic acids or engulf it in ferretin—an Fe storage protein, located in plastids (Briat et al., 2006). Inside living cells, Fe exists mainly in the form of stable coordination complexes with inorganic phosphates or organic ligands.

The sequestration of Zn in *T. caerulescens* leaves preferably in vacuoles of the epidermal cell probably plays an important role in providing Zn resistance to this plant (Kupper et al., 1999).

In cytoplasm having neutral pH, Ni is expected to bind with such ligands which contain nitrogen, oxygen and, sulphur like, glutathione, free amino acids, amino acid moieties like Cys, His, and Gln in proteins and different nucleotides (Dawson et al., 1986). The binding of Ni to these ligands in large quantities in the cell may cause toxicity or even cell death. To cope with this problem, hyperaccumulator *T. goesingense* has devised an efficient mechanism for the storage of Ni in the vacuoles which helps the plant to tolerate Ni toxicity. However, in non-hyperaccumulator *T. arvense*, which is highly sensitive to Ni toxicity, such an efficient Ni storage mechanism is lacking (Kramer et al., 1997). Exposure of different species of hyperaccumulator plant *Alyssum* to Ni resulted in a significant increase in histidine level in xylem-sap (Kramer et al., 1996).

This result indicates that His – Ni complex formation may provide Ni tolerance in roots as well as transport of Ni from roots to aerial parts.

One positive effect of highly elevated Ni concentration in tissues of hyperaccumulator plants is the protection against pathogens and herbivores (Boyd, 1998). Recent studies in *Arabidiopsis thaliana* transgenic plants have shown that overexpresion of the TcNAS gene results in Ni resistance (Pianelli et al., 2005) and it supports the view that nicotianamine (NA) is involved in metal tolerance.

15. Conclusion

Heavy metal uptake by the plants, using the technique of phytoremediation is a flourishing method to detoxify the heavy metal contaminated areas. It contains some good characteristics in comparison to conventional technologies. Industrialisation has resulted in a significant escalation in the level of environmental pollution. The addition of some heavy metal pollutants like arsenic, cadmium, lead, mercury, etc., to the environment is also going on unabated due to anthropogenic activities. Such toxic metals enter the food chain and show detrimental effects on growth and development of living organisms including human beings. Keeping in view the threat posed by HMs to living organisms, there is a need to decontaminate HM contaminated soils. Recent studies have revealed that phytoremediation is appearing as a most promising, eco-friendly, and cost-effective technology to detoxify soil contaminants with the help of plant species. AMF can grow in metal contaminated soils and their interactions with hyperaccumulator plant roots facilitate phytoremediation by the increased absorptive surface of their hyphae, chelating heavy metals, metal sequestration, secretion of glomalin protein, inducing expression of specific genes involved in resisting metal stress, etc. Genetic engineering involving the production of transgenic plants for enhanced phytoremediation can also go a long way in decontaminating HM polluted soils. Ligands secreted by plants assist in phytoremediation as they increase metal solubility in the soil, act as an H$^+$ source and metal chelating ions thereby facilitating enhanced metal absorption.

Several factors are necessary to realize a serious concert of remedy. The main necessary factor is to use suitable plant species that can effectively uptake contaminants. Although the phytoremediation technique is the best alternative technique, but it also has some drawbacks. A prolonged investigation is necessary to minimise this drawback to apply this method effectively.

References

Aggarwal A, Singh J, and Singh AP. 2017. Arbuscular mycorrhizal fungi and its role in sequestration of heavy metals. *Trends Biosci* 10(21): 4068–4077.

Aggarwal M, Sharma S, Kaur N, Pathania D, Bhandhari K, Kaushal N, Kaur R, Singh K, Srivastava A, and Nayyar H. 2011. Exogenous proline application reduces phytotoxic effects of selenium by minimising oxidative stress and improves growth in bean (*Phaseolus vulgaris* L.) seedlings. *Biol Trace Elem Res* 140: 354–367.

Ahemad M. 2012. Implication of bacterial resistance against heavy metals in bioremediation: a review. *J Inst Integr Omics Appl Biotechnol* 3: 39–46.

Ali H, Khan E, and Sajad MA. 2013. Phytoremediation of heavy metals-Concepts and applications. *Chemosphere* 91: 869–881.

Alvarez-Vázquez LJ, Martínez A, Rodríguez C, Vázquez-Méndez ME, and Vilar MA. 2019. Mathematical analysis and optimal control of heavy metals phytoremediation techniques. *Applied Mathematical Modelling* 73: 387–400.

An L, Pan Y, Wang Z, and Zhu C. 2011. Heavy metal absorption status of five plant species in monoculture and intercropping. *Plant & Soil* 345(1-2): 237–245.

Anjum NA, Hasanuzzaman M, Hossain MA Thangavel P, Roychoudhury A, Gill SS, Rodrigo MAM, Adam V, Fujita M, Kizek R, Duarte AC, Pereira E, and Ahmad I. 2015. Jacks of metal/metalloid chelation trade in plants-an overview. *Front Plant Sci* 6: 10–17.

Arshad M, Saleem M, and Hussain S. 2007. Perspectives of bacterial ACC deaminase in phytoremediation. *Trends Biotechnol* 25: 356–362.

Ashraf M, and Foolad MR. 2007. Roles of glycine betaine and proline in improving plant abiotic stress resistance. *Environ Exp Bot* 59: 206–216.

Ashraf MA, Hussain I, Rasheed R, Iqbal M, Riaz M, and Arif MS. 2017. Advances in microbe-assisted reclamation of heavy metal contaminated soils over the last decade: A review. *J Environ Manage* 198: 132–143. doi: 10.1016/j. jenvman.2017.04.060.

Ashraf S, Ali Q, Zahir ZA, Ashraf S, and Asghar HN. 2019. Phytoremediation: environmentally sustainable way for reclamation of heavy metal polluted soils. *Ecotox Environ Safe* 174: 714–727.

Assunção AG, Bookum WM, Nelissen HJ, Vooijs R, Schat H, and Ernst WH. 2003. Differential metal-specific tolerance and accumulation patterns among *Thlaspi caerulescens* populations originating from different soil types. *New Phytol* 159: 411–419. doi: 10.1046/j.1469-8137.2003. 00819.x.

Baker AJM, and Brooks R. 1989. Terrestrial higher plants which hyperaccumulate metallic elements-a review of their distribution, ecology and phytochemeistry. *Biorecovery* 1: 81–126.

Baker AJ, and Walker PL. 1990. Ecophysiology of metal uptake by tolerant plants. *Heavy Metal Tolerance in Plants: Evolutionary Aspects* 2: 155–165.

Baker AJ, McGrath SP, Reeves RD, and Smith JA. 2000. Metal hyperaccumulator plants: a review of the ecology and physiology of a biochemical resource for phytoremediation of metal-polluted soils. *Phytoremediation of Contaminated Soil and Water. Boca Raton, Florida, USA, Lewis Publishers*, pp. 85–107.

Begum N, Qin C, Ahanger MA, Raza S, Khan MI, Ashraf M, Ahmed N, and Zhang L. 2019. Role of arbuscular mycorrhizal fungi in plant growth regulation: Implications in abiotic stress tolerance. *Front Plant Sci* 10: 1068.

Berti WR, and Cunningham SD. 2000. Phytostabilization of metals. pp. 71–88. *In*: Raskin I and Ensley BD (eds.). *Phytoremediation of Toxic Metals: Using Plants to Clean-up the Environment* (New York, NY: John Wiley & Sons, Inc.).

Bhunia B, Prasad UU, Oinam G. Mondal A, and Bandyopadhyay TK. 2018. Characterization, genetic regulation and production of cyanobacterial exopolysaccharides and its applicability for heavy metal removal. *Carbohydrate Polymers* 179: 228–243.

Birhane E, Sterck FJ, Fetene M, Bongers F, and Kuyper TW. 2012. Arbuscular mycorrhizal fungi enhance photosynthesis, water use efficiency, and growth of frankincense seedlings under pulsed water availability conditions. *Oecologia* 169: 895–904.

Bissett A, Brown MV, Siciliano SD, and Thrall PH. 2013. Microbial community responses to anthropogenically induced environmental change: towards a systems approach. *Ecology Letters* 16: 128e139.

Bodong C. 2002. Role of arbuscular mycorrhizae in alleviation of zinc and cadmium phytotoxicity[D]. Beijing: China Agricultural University.

Boyd RS. 1998. Hyperaccumulation as a plant defense strategy. pp. 181–201. *In*: Brooks RR (ed.). *Plants that Hyperaccumulate Heavy Metals*. Wallingford, UK: CAB International.

Briat JF, Cellier F, and Gaymard F. 2006. Ferritins and iron accumulation in plant tissues. *In*: Barton L, Abadia J (eds.). *Iron Nutrition in Plants and Rhizospheric Microorganisms*. Amsterdam: Kluwer Academic Publishers.

Brooks R, Lee J, Reeves R, and Jaffre T. 1977. Detection of nickeliferous rocks by analysis of herbarium specimens of indicators plants. *Journal of Geochemical Exploration* 7: 49–57.

Brooks RR. 1998. Geobotany and hyperaccumulators. pp. 55–94. *In*: Brooks RR (ed.). *Plants that Hyperaccumulate Heavy Metals*. CAB International, Wallingford, UK.

Brun LA, Maillet J, Hinsinger P, and Pepin M. 2001. Evaluation of copper availability to plants in copper-contaminated vineyard soils. *Environ Pollut* 111: 293–302.

Burges A, Alkorta I, Epelde L, and Garbisu C. 2018. From phytoremediation of soil contaminants to phytomanagement of ecosystem services in metal contaminated sites. *Int. J. Phytoremediat.* 20: 384–397.

Burns RG, DeForest JL, Marxsen J, Sinsabaugh RL, Stromberger ME, Wallenstein MD, Weintraub MN, and Zoppini A. 2013. Soil enzymes in a changing environment: current knowledge and future directions. *Soil Biology and Biochemistry* 58: 216e234.

Carvalho LM, Cacador I, and Martins-Loucao MA. 2006. Arbuscular mycorrhizal fungi enhance root cadmium and copper accumulation in the roots of the salt marsh plant *Aster tripolium* L. *Plant Soil* 285: 161–169.

Cempel M, and Nikel G. 2006. Nickel: a review of its sources and environmental toxicology. *Pol J Environ Stud* 15: 375–382.

Chen BD, Zhu YG, and Smith FA. 2006. Effects of *Arbuscular mycorrhizal* inoculation on uranium and arsenic accumulation by Chinese brake fern (*Pteris vittata* L.) from a uranium mining-impacted soil. *Chemosphere* 62(9): 1464–1473.

Chen M, Xu P, Zeng G, Yang C, Huang D, and Zhang J. 2015. Bioremediation of soils contaminated with polycyclic aromatic hydrocarbons, petroleum, pesticides, chlorophenols and heavy metals by composting: applications, microbes and future research needs. *Biotechnol Adv* 33: 745–755.

Chen S, Zhao H, Zou C, Li Y, Chen Y, Wang Z, Jiang Y, Liu A, Zhao P, Wang M, and Ahammed GJ. 2017. Combined inoculation with multiple arbuscular mycorrhizal fungi improves growth, nutrient uptake and photosynthesis in cucumber seedlings. *Front Microbiol* 8: 2516.

Cherian S, and Oliveira MM. 2005. Transgenic plants in phytoremediation: recent advances and new possibilities. *Environ Sci Technol* 39: 9377–9390.

Chiu KK, Ye ZH, and Wong MH. 2005. Enhanced uptake of As, Zn, and Cu by Vetiveria zizanioides and Zea mays using chelating agents. *Chemosphere* 60(10): 1365–1375.

Clarholm M, and Skyllberg U. 2013. Translocation of metals by trees and fungi regulates pH, soil organic matter turnover and nitrogen availability in acidic forest soils. *Soil Biology and Biochemistry* 63: 142e153.

Clemens S. 2006. Toxic metal accumulation, responses to exposure and mechanisms of tolerance in plants. *Biochimie* 88: 1707–1719.

Colzi I, Arnetoli M, Gallo A, Doumett S, Del Bubba M, Pignattelli S, Gabbrielli R, and Gonnelli C. 2012. Copper tolerance strategies involving the root cell wall pectins in *Silene paradoxa*, L. *Environmental & Experimental Botany* 78(1): 91–98.

Coreño-Alonso A, Solé A, Diestra E, Esteve I, Gutiérrez-Corona JF, López GR, Fernández FJ, and Tomasini A. 2014. Mechanisms of interaction of chromium with *Aspergillus niger* var. tubingensis strain Ed8. *Bioresour Technol* 158: 188–192.

Costa G, and Morel JL. 1994. Water relations, gas exchange and amino acid content in cadmium-treated lettuce. *Plant Physiol Biochem* 32: 561–570.

Crowley DE, Wang YC, Reid CPP, and Szansiszlo PJ. 1991. Mechanism of iron acquisition from siderophores by microorganisms and plants. *Plant Soil* 130: 179–198.

Curie C, Panaviene Z, Loulergue C, Dellaporta SL, Briat JF, and Walker EL. 2001. Maize yellow stripe1 encodes a membrane protein directly involved in Fe(III) uptake. *Nature* 409: 346–349.

Curie C, and Briat JF. 2003. Iron transport and signaling in plants. *Annual Review of Plant Biology* 54: 183–206.

Curie C, Cassin G, Couch D, Divol F, Higuchi K, Le Jean M, Misson J, Schikora A, Czernic P, and Mari S. 2009. Metal movement within the plant: contribution of nicotianamine and yellow stripe 1-like transporters. *Annals of Botany* 103(1): 1–11.

DalCorso G, Fasani E, Manara A, Visioli G, and Furini A. 2019. Heavy metal pollutions: state of the art and innovation in phytoremediation. *Int J Mol Sci* 20: 3412.

Danh LT, Truong P, Mammucari R, and Foster N. 2014. A critical review of the arsenic uptake mechanisms and phytoremediation potential of *Pteris vittata*. *International Journal of Phytoremediation* 16(5): 429–453.

Das N, Bhattacharya S, and Maiti MK. 2016. Enhanced cadmium accumulation and tolerance in transgenic tobacco overexpressing rice metal tolerance protein gene *OsMTP1* is promising for phytoremediation. *Plant Physiol Biochem* 105: 297–309.

Dawson RMC, Elliott DC, Elliott WH, and Jones KM (eds.). 1986. Data for Biochemical Research, Ed 3. Clarendon Press, Oxford, UK.

De Deyn GB, Raaijmakers CE, Zoomer HR, Berg MP, de Ruiter PC, Verhoef HA, Bezemer TM, and van der Putten WH. 2003. Soil invertebrate fauna enhances grassland succession and diversity. *Nature* 422: 711e713.

De Silva ND, Cholewa E, and Ryser P. 2012. Effects of combined drought and heavy metal stresses on xylem structure and hydraulic conductivity in red maple (*Acer rubrum* L.). *Journal of Experimental Botany* 63(16): 5957–5966.

Degola F, Fattorini L, Bona E, Sprimuto CT, Argese E, Berta G, and di Toppi LS. 2015. The symbiosis between *Nicotiana tabacum* and the endomycorrhizal fungus *Funneliformis mosseae* increases the plant glutathione level and decreases leaf cadmium and root arsenic contents. *Plant Physiol Biochem* 92: 11–18.

Degryse F, Smolders E, and Parker DR. 2006. Metal complexes increase uptake of Zn and Cu by plants: implications for uptake and deficiency studies in chelator-buffered solutions. *Plant and Soil* 289(1): 171–185.

Degryse F, Smolders E, and Parker DR. 2009. Partitioning of metals (Cd, Co, Cu, Ni, Pb, Zn) in soils: concepts, methodologies, prediction and applications-a review. *European Journal of Soil Science* 60(4): 590–612.

Devevre O, Garbaye J, and Botton B. 1996. Release of complexing organic acids by rhizosphere fungi as a factor in Norway Spruce yellowing in acidic soils. *Mycol Res* 100: 1367–1374.

Dhalaria R, Kumar D, Kumar H, Nepovimova E, Kuča K, Torequl Islam M, and Verma R. 2020. Arbuscular mycorrhizal fungi as potential agents in ameliorating heavy metal stress in plants. *Agronomy* 10(6): 815.

Dhanwal P, Kumar A, Dudeja S, Chhokar V, and Beniwal V. 2017. Recent advances in phytoremediation technology. pp. 227–241. *In*: Kumar R, Sharma AK, and Ahluwalia SS (eds.). *Advances in Environmental Biotechnology*. (Singapore: Springer).

Doble M, and Kumar A. 2005. Biotreatment of Industrial Effluents. Oxford: Butterworth-Heinemann, 19–38.

Driver JD, Holben WE, and Rillig MC. 2005. Characterization of glomalin as a hyphal wall component of arbuscular mycorrhizal fungi. *Soil Biol Biochem* 37: 101–106.

Dushenkov S. 2003. Trends in phytoremediation of radionuclides. *Plant Soil* 249: 167–175.

Ellis RJ, Best JG, Fry JC, Morgan P, Neish B, Trett MW, and Weightman AJ. 2002. Similarity of microbial and meiofaunal community analyses for mapping ecological effects of heavy-metal contamination in soil. *Fems Microbiology Ecology* 40: 113e122.

Ezzouhri L, Castro E, Moya M, Espinola F, and Lairini K. 2009. Heavy metal tolerance of filamentous fungi isolated from polluted sites in Tangier, Morocco. *Afr J Microbiol Res* 3: 35–48.

Farago ME, and Mullen WA. 1979. Plants which accumulate metals. Part IV. A possible copper-proline complex from the roots of *Armeria maritima*. *Chim Acta* 32: L93–L94.

Fasani E, Manara A, Martini F, Furini A, and DalCorso G. 2018. The potential of genetic engineering of plants for the remediation of soils contaminated with heavy metals. *Plant Cell Environ* 41: 1201–1232.

Fiscus DA, and Neher DA. 2002. Distinguishing sensitivity of free-living soil nematode genera to physical and chemical disturbances. *Ecological Applications* 12: 565e575.

Franchi E, Rolli E, Marasco R, Agazzi G, Borin S, Cosmina P, Pedron F, Rosellini I, Barbafieri M, and Petruzzelli G. 2017. Phytoremediation of a multi contaminated soil: mercury and arsenic phytoextraction assisted by mobilizing agent and plant growth promoting bacteria. *J Soils Sediments* 17: 1224–1236.

Frey B, Stemmer M, Widmer F, Luster J, and Sperisen C. 2006. Microbial activity and community structure of a soil after heavy metal contamination in a model forest ecosystem. *Soil Biology and Biochemistry* 38: 1745e1756.

Gaither LA, and Eide DJ. 2001. Eukaryotic zinc transporters and their regulation. *Biometals* 14: 251–270.

Gallagher F, Pechmann I, Bogden J, Grabosky J, and Weis P. 2008a. Soil metal concentrations and productivity of *Betula populifolia* (gray birch) as measured by field spectrometry and incremental annual growth in an abandoned urban Brownfield in New Jersey. *Environmental Pollution* 156: 699e706.

Garg N, and Chandel S. 2012. Role of arbuscular mycorrhizal (AM) fungi on growth, cadmium uptake, osmolyte and phytochelatin synthesis in *Cajanus cajan* (L.) Millsp. under NaCl and Cd stresses. *J. Plant Growth Regul* 31: 292–308.

Gerhardt KE, Huang X-D, Glick BR, and Greenberg BM. 2009. Phytoremediation and rhizoremediation of organic soil contaminants: potential and challenges. *Plant Sci* 176: 20–30.

Gerhardt KE, Gerwing PD, and Greenberg BM. 2017. Opinion: taking phytoremediation from proven technology to accepted practice. *Plant Sci* 256: 170–185.

Gill SS, and Tuteja N. 2010. Reactive oxygen species and antioxidant machinery in abiotic stress tolerance in crop plants. *Plant Physiol Biochem* 48: 909–930.

Ginn BR, Szymanowski JS, and Fein JB. 2008. Metal and proton binding onto the roots of *Fescue rubra*. *Chem Geol* 253: 130–135.

Glick BR. 2010. Using soil bacteria to facilitate phytoremediation. *Biotechnol Adv* 28: 367–374.

Göhre V, and Paszkowski U. 2006. Contribution of the arbuscular mycorrhizal symbiosis to heavy metal phytoremediation. *Planta* 223: 1115–1122.

Gong X, and Tian DQ. 2019. May. Study on the effect mechanism of Arbuscular Mycorrhiza on the absorption of heavy metal elements in soil by plants. *In*: *IOP Conference Series: Earth and Environmental Science* (Vol. 267, No. 5, p. 052064). IOP Publishing.

Gonzalez-Guerrero M, Benabdellah K, Valderas A, Azcon-Aguilar C, and Ferrol N. 2010. GintABC1 encodesa putative ABC transporter of the MRP subfamily induced by Cu, Cd, and oxidative stress in *Glomus intraradices*. *Mycorrhiza* 20: 137–146.

Grill E, Loffler S, Winnacker EL, and Zenk MH. 1989. Phytochelatins, the heavy-metal-binding peptides of plants, are synthesized from glutathione by a specific γ-glutamylcysteine dipeptidyl transpeptidase (phytochelatin synthase). *Proceedings of the National Academy of Sciences* 86(18): 6838–6842.

Grimm NB, Foster D, Groffman P, Grove JM, Hopkinson CS, Nadelhoffer KJ, Pataki DE, and Peters DP. 2008. The changing landscape: ecosystem responses to urbanization and pollution across climatic and societal gradients. *Frontiers in Ecology and the Environment* 6(5): 264–272.

Guo J, Xu W, and Ma M. 2012. The assembly of metals chelation by thiols and vacuolar compartmentalization conferred increased tolerance to and accumulation of cadmium and arsenic in transgenic *Arabidopsis thaliana*. *Journal of Hazardous Materials* 199-200(51): 309–313.

Gupta N, Khan DK, and Santra SC. 2010. Determination of public health hazard potential of wastewater reuse in crop production. *World Rev Sci Technol Sustain Dev* 7: 328–340.

Gupta S, and Singh D. 2017. Role of genetically modified microorganisms in heavy metal bioremediation. pp. 197–214. *In*: Kumar R, Sharma A, and Ahluwalia S *Advances in Environmental Biotechnology*. (Singapore: Springer).

Ha SB, Smith AP, Howden R, Dietrich WM, Bugg S, O'Connell MJ, Goldsbrough PB, and Cobbett CS. 1999. Phytochelatin synthase genes from *Arabidopsis* and the yeast *Schizosaccharomyces pombe*. *Plant Cell* 11: 1153–1163.

Hall JÁ. 2002. Cellular mechanisms for heavy metal detoxification and tolerance. *Journal of Experimental botany* 53(366): 1–11.

Hamzah A, Hapsari RI, and Wisnubroto EI. 2016. Phytoremediation of Cadmium-contaminated agricultural land using indigenous plants. *Int J Environ Agric Res* 2: 8–14.

Harter RD. 1983. Effect of soil pH on adsorption of lead, copper, zinc, and nickel. *Soil Science Society of America Journal* 47(1): 47–51.

Hasan MM, Uddin MN, Ara-Sharmeen FI, Alharby H, Alzahrani Y, Hakeem KR, and Zhang L. 2019. Assisting phytoremediation of heavy metals using chemical amendments. *Plants* 8: 295.

Hasegawa I, Terada E, Sunairi M, Wakita H, Shinmachi F, Noguchi A, Nakajima M, and Yazaki J. 1997. Genetic improvement of heavy metal tolerance in plants by transfer of the yeast metallothionein gene (CUP1). pp. 391–395. *In*: *Plant Nutrition for Sustainable Food Production and Environment*. Springer, Dordrecht.

Hashimoto Y, Takaoka M, and Shiota K. 2011. Enhanced transformation of lead speciation in rhizosphere soils using phosphorus amendments and phytostabilization: An X-ray absorption fine structure spectroscopy investigation. *Journal of Environment Quality* 40(3): 696–703.

Hassan TU, Bano A, and Naz I. 2017. Alleviation of heavy metals toxicity by the application of plant growth promoting rhizobacteria and effects on wheat grown in saline sodic field. *Int J Phytoremediation* 19: 522–529.

Hassinen VH, Tuomainen M, Peräniemi S, Schat H, Kärenlampi SO, and Tervahauta A. 2009. Metallothioneins 2 and 3 contribute to the metal-adapted phenotype but are not directly linked to Zn accumulation in the metal hyperaccumulator, Thlaspi caerulescens. *J Exp Bot* 60: 187–196.

Hayat S, Hayat Q, Alyemeni MN, and Ahmad A. 2013. Proline enhances antioxidative enzyme activity, photosynthesis and yield of *Cicer arietinum* L. exposed to cadmium stress. *Acta Bot Croat* 2: 323–335.

He LY, Zhang YF, Ma HY, Su LN, Chen ZJ, Wang QY, Qian M, and Sheng XF. 2010. Characterization of copper-resistant bacteria and assessment of bacterial communities in rhizosphere soils of copper-tolerant plants. *Applied Soil Ecology* 44: 49e55.

Higuchi K, Suzuki K, Nakanishi H, Yamaguchi H, Nishizawa NK, and Mori S. 1999. Cloning of nicotianamine synthase genes, novel genes involved in the biosynthesis of phytosiderophores. *Plant Physiology* 119(2): 471–480.

Hildebrandt U, Regvar M, and Bothe H. 2007. Arbuscular mycorrhiza and heavy metal tolerance. *Phytochemistry* 38: 139–146.

Hongxin W. 2010. Application of arbuscular mycorrhizae in phytoremediation of heavy metal contamination soils [J]. *Soil and Fertilizer Sciences in China* (5): 1–5.

Hoque MA, Banu MNA, Nakamura Y, Shimoishi Y, and Murata Y. 2008. Proline and glycinebetaine enhance antioxidant defense and methylglyoxal detoxification systems and reduce NaCl-induced damage in cultured tobacco cells. *J Plant Physiol* 165: 813–824.

Hrynkiewicz K, and Baum C. 2014. Application of microorganisms in bioremediation of environment from heavy metals. pp. 215–227. *In*: Malik A, Grohmann E, and Akhtar R (eds.). *Environmental Deterioration and Human Health*. (Dordrecht: Springer).

Huang BF, and Xin JL. 2013. Mechanism of heavy metal accumulation in plants: a review. *Acta Prataculturae Sinica* 22(1): 300–307 (In Chinese).

Huang B, Xin J, Dai H, Liu A, Zhou W, Yi Y, and Liao K. 2015. Root morphological responses of three hot pepper cultivars to Cd exposure and their correlations with Cd accumulation. *Environmental Science and Pollution Research* 22(2): 1151–1159.

Huang JW, Chen JJ, Berti WR, and Cunningham SD. 1997. Phytoremediation of lead-contaminated soils: role of synthetic chelates in lead phytoremediation. *Environ Sci Technol* 31: 800–805.

Igiri BE, Okoduwa SIR, Idoko GO, Akabuogu EP, Adeyi AO, and Ejiogu IK. 2018. Toxicity and bioremediation of heavy metals contaminated ecosystem from tannery wastewater: a review. *J Toxicol* 1–16.

Islam E, Khan MT, and Irem S. 2015. Biochemical mechanisms of signalling: perspectives in plant under arsenic stress. *Ecotoxicol Environ Saf* 114: 126–133.

Islam MM, Hoque MA, Okuma E, Banu MN, Shimoishi Y, Nakamura Y, and Murata Y. 2009. Exogenous proline and glycine betaine increase antioxidant enzyme activities and confer tolerance to cadmium stress in cultured tobacco cells. *J Plant Physiol* 166: 1587–1597.

Jabeen R, Ahmad A, and Iqbal M. 2009. Phytoremediation of heavy metals: physiological and molecular mechanisms. *The Botanical Review* 75(4): 339–364.

Jacob JM, Karthik C, Saratale RG, Kumar SS, Prabakar D, Kadirvelu K, and Pugazhendhi A. 2018. Biological approaches to tackle heavy metal pollution: a survey of literature. *J Environ Manage* 217: 56–70.

Jaffre T, Brooks RR, Lee KJ, and Reeves RD. 1976. *Sebertia acuminata*: a hyperaccumulator of nickel from New Caledonia. *Science* 193: 579–580.

Jan S, and Parray JA (eds.). 2016. Use of mycorrhiza as metal tolerance strategy in plants. pp. 57–68. *In*: *Approaches to Heavy Metal Tolerance in Plants* (Singapore: Springer).

Javed MT, Tanwir K, Akram MS, Shahid M, Niazi NK, and Lindberg S. 2019. Chapter 20—Phytoremediation of cadmium-polluted water/sediment by aquatic macrophytes: role of plant-induced pH changes. pp. 495–529. *In*: Hasanuzzaman M, Prasad MNV, and Fujita M (eds.). *Cadmium Toxicity and Tolerance in Plants*. (London: Academic Press).

Jiang CA, Wu QT, and Sterckeman T. 2010a. Co-planting can phytoextract similar amounts of cadmium and zinc to monocropping from contaminated soils. *Ecological Engineering* 36(4): 391–395.

Jiang W, and Liu D. 2010b. Pb-induced cellular defense system in the root meristematic cells of *Allium sativum*, L. *Bmc Plant Biology* 10(1): 1–8.

Jiang QY, Zhuo F, Long SH, Zhao HD, Yang DJ, Ye ZH, Li SS, and Jing YX. 2016. Can arbuscular mycorrhizal fungi reduce Cd uptake and alleviate Cd toxicity of *Lonicera japonica* grown in Cd-added soils? *Sci Rep* 6: 1–9.

Jiang X, and Wang C. 2008. Zinc distribution and zinc-binding forms in *Phragmites australis*, under zinc pollution. *Journal of Plant Physiology* 165(7): 697–704.

Joho M, Ishikawa Y, Kunikane M, Inouhe M, Tohoyama H, and Murayama T. 1992. The subcellular distribution of nickel in Ni-sensitive and Ni-resistant strains of *Saccharomyces cerevisiae*. *Microbios* 71: 149–159.

Kang X, Song J, Yuan H, Duan L, Li X, Li N, Liang X, and Qu B. 2017. Speciation of heavy metals in different grain sizes of Jiaozhou Bay sediments: Bioavailability, ecological risk assessment and source analysis on a centennial timescale. *Ecotoxicology and Environmental Safety* (143): 296–306.

Kapusta P, Szarek-Łukaszewska G, and Stefanowicz AM. 2011. Direct and indirect effects of metal contamination on soil biota in a ZnePb post-mining and smelting area (S Poland). *Environmental Pollution* 159: 1516e1522.

Karaca A. 2004. Effect of organic wastes on the extractability of cadmium, copper, nickel, and zinc in soil. *Geoderma* 122(2-4): 297–303.

Karthikeyan B, Jaleel CA, Changxing Z, Joe MM, Srimannarayan J, and Deiveekasundaram M. 2008. The effect of AM fungi and phosphorous level on the biomass yield and ajmalicine production in *Catharanthus roseus. EurAsian Journal of BioSciences* 2(1): 26–33.

Khade SW, and Adholeya A. 2007. Feasible bioremediation through arbuscular mycorrhizal fungi imparting heavy metal tolerance: a retrospective [J]. *Bioremediation Journal* 11(1): 33–43.

Knight B, Zhao FJ, McGrath SP, and Shen ZG. 1997. Zinc and cadmium uptake by the hyperaccumulator *Thlaspi caerulescens* in contaminated soils and its effects on the concentration and chemical speciation of metals in soil solution. *Plant Soil* 197: 71–78.

Koptsik GN. 2014. Problems and prospects concerning the phytoremediation of heavy metal polluted soils: A review. *Eurasian Soil Science* 4(9): 923–939.

Koźmińska A, Wiszniewska A, Hanus-Fajerska E, and Muszyńska E. 2018. Recent strategies of increasing metal tolerance and phytoremediation potential using genetic transformation of plants. *Plant Biotechnol Rep* 12: 1–14.

Kramer U, Cotter-Howells JD, Charnock JM, Baker AJM, and Smith JAC. 1996. Free histidine as a metal chelator in plants that accumulate nickel. *Nature* 379: 635–638.

Kramer U, Smith RD, Wenzel WW, Raskin I, and Salt DE. 1997. The role of metal transport and tolerance in nickel hyperaccumulation by *Thlaspi goesingense* (Halacsy). *Plant Physiol* 115: 1641–1650.

Kramer U, Pickering IJ, Prince RC, Raskin I, and Salt DE. 2000. Subcellular localization and speciation of nickel in hyperaccumulator and non-accumulator *Thlaspi* species. *Plant Physiology* 122(4): 1343–1354.

Kruckeberg AR. 1967. Ecotypic response to ultramafic soils by some plant species of northwestern United States. *Brittonia* 19: 133–151. doi: 10.2307/2805271.

Krumins JA. 2014. The positive effects of trophic interactions in soils. *In*: Dighton J, and Krumins JA (eds.). *Interactions in Soil: Promoting Plant Growth.* Springer, Dordrecht, The Netherlands.

Kui Z, and Lin L. 2009. The status of heavy metal pollution and its control in the Pb and Zn mining districts of China[J]. *Journal of Anhui Agri-Sci* 37(30): 14837–14838.

Kuiper I, Lagendijk EL, Bloemberg GV, and Lugtenberg BJ. 2004. Rhizoremediation: a beneficial plant-microbe interaction. *Molecular Plant-Microbe Interactions* 17: 6e15.

Kumar KS, Dayananda S, and Subramanyam C. 2005. Copper alone, but not oxidative stress, induces copper-metallothionein gene in *Neurospora crassa*. FEMS Microbiol Lett 242: 45–50.

Kumar PN, Dushenkov V, Motto H, and Raskin I. 1995. Phytoextraction: the use of plants to remove heavy metals from soils. *Environmental Science & Technology* 29(5): 1232–1238.

Kupper H, Zhao FJ, and McGrath SP. 1999. Cellular compartmentation of zinc in leaves of the hyperaccumulator *Thlaspi caerulescens. Plant Physiology* 119(1): 305–312.

Landberg T, and Greger M. 1996. Differences in uptake and tolerance to heavy metals in Salix from unpolluted and polluted areas. *Applied Geochem* 11(1-2): 175–180.

Lanfranco L, Bolchi A, Ros EC, Ottonello S, and Bonfante P. 2002. Differential expression of a metallothioneingene during the presymbiotic versus the symbiotic phase of an arbuscular mycorrhizal fungus. *Plant Physiol* 130: 58–67.

Lefèbvre C, and Vernet P. 1990. Micro-evolutionary processes on contaminated deposits. pp. 286–297. *In*: Shaw AJ (ed.). *Heavy Metal Tolerance in Plants: Evolutionary Aspects.* (BocaRaton, FL: CRC Press).

Leitenmaier B, and Küpper H. 2013. Compartmentation and complexation of metals in hyperaccumulator plants. *Frontiers in Plant Science* 374(4): 1–13.

Li WC, Ye ZH, and Wong MH. 2010. Metal mobilization and production of short-chain organic acids by rhizosphere bacteria associated with a Cd/Zn hyperaccumulating plant, *Sedum alfredii. Plant and Soil* 326(1): 453–467.

Li Y, and Zu YQ. 2016. Heavy metal pollution ecology and ecological remediation. Beijing, Science Press. (In Chinese)

Liao JP, Lin XG, Cao ZH, Shi YQ, and Wong MH. 2003. Interactions between arbuscular mycorrhizae and heavy metals under sand culture experiment. *Chemosphere* 50(6): 847–853.

Lin S, Sun X, Wang X, Li YY, Luo QY, Sun L, and Jin L. 2013. Mechanism of plant tolerance to heavy metals enhanced by arbuscular mycorrhizal fungi [J]. *Pratacultural Science* 30(3): 365–374.

Liu X, Song Q, Tang Y, Li W, Xu J, Wu J, Wang F, and Brookes PC. 2013. Human health risk assessment of heavy metals in soil–vegetable system: A multi-medium analysis. *Sci Total Environ* 463: 530–540.

Liu YG, Feng BY, Fan T, Pan C, and Peng LJ. 2008. Study on the biosorption of heavy metals by fungi [J]. *Journal of Hunan University (Natural Sciences)* 35(1): 71–74.

Long, XX, Yang XE, Ye ZQ, Ni WZ, and Shi WY. 2002. Differences of uptake and accumulation of zinc in four species of Sedum. *Acta Bot Sin* 44: 152–157.

Ma Y, Prasad M, Rajkumar M, and Freitas H. 2011. Plant growth promoting rhizobacteria and endophytes accelerate phytoremediation of metalliferous soils. *Biotechnol Adv* 29: 248–258.

Mahar A, Wang P, Ali A, Awasthi MK, Lahori AH, Wang Q, Li R, and Zhang Z. 2016. Challenges and opportunities in the phytoremediation of heavy metals contaminated soils: a review. *Ecotox Environ Safe* 126: 111–121.

Pietrzykowski M, Socha J, and van Doorn NS. 2014. Linking heavy metal bioavailability (Cd, Cu, Zn and Pb) in Scots pine needles to soil properties in reclaimed mine areas. *Science of the Total Environment* 470-471: 501–510.

Mari S, Gendre D, Pianelli K, Ouerdane L, Lobinski R, Briat JF, Lebrun M, and Czernic P. 2006. Root-to-shoot long-distance circulation of nicotianamine and nicotianamine–nickel chelates in the metal hyperaccumulator *Thlaspi caerulescens*. *Journal of Experimental Botany* 57(15): 4111–4122.

Marques APGC, Rangel AOSS, and Castro PML. 2009. Remediation of heavy metal contaminated soils: Phytoremediation as a potentially promising clean-up technology. *Critical Reviews in Environmental Science & Technology* 39(8): 622–654.

McGrath SP. 1998. Phytoextyraction for soil remediation. pp. 261–287. *In*: Brooks RR (ed.). *Plants that Accumulate Heavy Metals*. CAB International, Wallingford, UK.

McGrath SP, and Zhao F-J. 2003. Phytoextraction of metals and metalloids from contaminated soils. *Curr Opin Biotechnol* 14: 277–282.

McIntyre T. 2003. Phytoremediation of Heavy Metals from Soils, Phytoremediation. Springer, pp. 97e123.

Meharg AA. 2003. The mechanistic basis of interactions between mycorrhizal associations and toxic metal cations. *Mycol Res* 107: 1253–1265.

Mehes-Smith M, Nkongolo K, and Cholewa E. 2013. Coping mechanisms of plants to metal contaminated soil. *Environmental Change & Sustainability* 53–90.

Meier S, Borie F, Bolan N, and Cornejo P. 2012. Phytoremediation of metal-polluted soils by arbuscular mycorrhizal fungi. *Critical Reviews in Environmental Science and Technology* 42(7): 741–775.

Memon AR, and Schröder P. 2009. Implications of metal accumulation mechanisms to phytoremediation. *Environmental Science & Pollution Research* 16(2): 162–175.

Mench M, Lepp N, Bert V, Schwitzguébel J-P, Gawronski SW, Schröder P, and Vangronsveld J. 2010. Successes and limitations of phytotechnologies at field scale: outcomes, assessment and outlook from COST Action 859. *J Soil Sediment* 10: 1039–1070.

Mengoni A, Gonnelli C, Galardi F, Gabbrielli R, and Bazzicalupo M. 2000. Genetic diversity and heavy metal tolerance in populations of *Sileneparadoxa* L. (Caryophyllaceae): arandom amplified polymorphic DNA analysis. *Mol Ecol* 9: 1319–1324.

Mesjasz-Przybyłowicz J, Nakonieczny M, Migula P, Augustyniak M, Tarnawska M, Reimold U, Koeberl C, Przybylowicz W, and Glowacka E. 2004. Uptake of cadmium, lead nickel and zinc from soil and water solutions by the nickel hyperaccumulator *Berkheya coddii*. *Acta Biol Cracoviensia Ser Bot* 46: 75–85.

Miransari M. 2011. Hyperaccumulators, arbuscular mycorrhizal fungi and stress of heavy metals. *Biotechnology Advances* (29): 645–653.

Miransari M. 2016. Soybean production and heavy metal stress. pp. 197–216. *In*: Miransari M (ed.). Abiotic and biotic stresses in soybean production. Academic Press: Cambridge, MA, USA; Elsevier: Cambridge, MA, USA.

Mishra A, Bhattacharya A, and Mishra N. 2019. Mycorrhizal symbiosis: An effective tool for metal bioremediation. pp. 113–128. *In*: Singh JS (ed.). *New and Future Developments in Microbial Biotechnology and Bioengineering*. Elsevier: Amsterdam, The Netherlands.

Mishra J, Singh R, and Arora NK. 2017. Alleviation of heavy metal stress in plants and remediation of soil by rhizosphere microorganisms. *Front Microbiol* 8: 1706.

Mitra D, Uniyal N, Panneerselvam P, Senapati A, and Ganeshamurthy AN. 2019. Role of mycorrhiza and its associated bacteria on plant growth promotion and nutrient management in sustainable agriculture. *Int J Life Sci Appl Sci* 1: 1–10.

Mittler R. 2006. Abiotic stress, the field environment and stress combination. *Trends Plant Sci* 11: 15–19.

Moons A. 2003. Osgstu3 and osgstu4, encoding tau class glutathione Stransferases, are heavy metal- and hypoxic stress-induced and differentially salt stress-responsive in rice roots. *FEBS Lett* 553: 427–432.

Morkunas I, Wozniak A, Mai VC, Rucinska-Sobkowiak R, and Jeandet P. 2018. The role of heavy metals in plant response to biotic stress. *Molecules* 23(9): 2320.

Murakami M, and Ae N. 2009. Potential for phytoextraction of copper, lead, and zinc by rice (*Oryza sativa* L.), soybean (*Glycine max* [L.] Merr.), and maize (*Zea mays* L.). *Journal of Hazardous Materials* 162(2-3): 1185–1192.

Mustapha MU, and Halimoon N. 2015. Screening and isolation of heavy metal tolerant bacteria in industrial effluent. *Procedia Environ Sci* 30: 33–37.

Nadeem SM, Ahmad M, Zahir ZA, Javaid A, and Ashraf M. 2014. The role of mycorrhizae and plant growth promoting rhizobacteria (PGPR) in improving crop productivity under stressful environments. *Biotechnol Adv* 32: 429–448.

Nagata T, Morita H, Akizawa T, and Pan-Hou H. 2010. Development of a transgenic tobacco plant for phytoremediation of methylmercury pollution. *Appl. Microbiol. Biotechnol.* 87: 781–786.

Nascimento CWA, Amarasiriwardena D, and Xing B. 2006. Comparison of natural organic acids and synthetic chelates at enhancing phytoextraction of metals from a multi-metal contaminated soil. *Environ Pollut* 140: 114–123.

Neher DA. 2010. Ecology of plant and free-living nematodes in natural and agricultural soil. pp. 371e394. *In*: VanAlfen NK, Bruening G, and Leach JE (eds.). *Annual Review of Phytopathology*, vol. 48. Annual Reviews, Palo Alto.

Nehnevajova E, Herzig R, Federer G, Erismann K-H, and Schwitzguébel J-P. 2007. Chemical mutagenesis—a promising technique to increase metal concentration and extraction in sunflowers. *Int J Phytoremediat* 9: 149–165.

Nwaichi EO, and Dhankher OP. 2016. Heavy metals contaminated environments and the road map with phytoremediation. *Journal of Environmental Protection* 7(1): 41–51.

Okuma E, Murakami Y, Shimoishi Y, Tada M, and Murata Y. 2004. Effects of exogenous application of proline and betaine on the growth of tobacco cultured cells under saline conditions. *Soil Sci Plant Nutr* 50: 1301–1305.

Oladipo OG, Awotoye OO, Olayinka A, Bezuidenhout CC, and Maboeta MS. 2017. Heavy metal tolerance traits of filamentous fungi isolated from gold and gemstone mining sites. *Braz J Microbiol* 49: 29–37.

Olaniran AO, Adhika B, and Balakrishna P. 2013. Bioavailability of heavy metals in soil: impact on microbial biodegradation of organic compounds and possible improvement strategies. *International Journal of Molecular Sciences* 14(5): 10197–10228.

Ong GH, Ho XH, Shamkeeva S, Fernando MS, Shimen A, and Wong LS. 2017. Biosorption study of potential fungi for copper remediation from *Peninsular Malaysia. Remediat J* 27: 59–63.

Ouerdane L, Mari S, Czernic P, Lebrun M, and Lobinski R. 2006. Speciation of non-covalent nickel species in plant tissue extracts by electrospray Q-TOF MS/MS after their isolation by 2D size exclusion– hydrophilic interaction LC (SEC–HILIC) monitored by ICP MS. *Journal of Analytical and Atomic Spectrometry* 21: 676–683.

Paleg LG, Doughlas TJ, Vandaal A, and Keech DB. 1981. Proline and betaine protect enzymes against heat inactivation. *Aust J Plant Physiol* 8: 107–114.

Pandey S, Ghosh PK, Ghosh S, De TK, and Maiti TK. 2013. Role of heavy metal resistant *Ochrobactrum* sp. and *Bacillus* spp. strains in bioremediation of a rice cultivar and their PGPR like activities. *J Microbiol* 51: 11–17.

Parida AK, and Das AB. 2005. Salt tolerance and salinity effects on plants: a review. *Ecotoxicol Environ Saf* 60: 324–349.

Park S, Kim KS, Kang D, Yoon H, and Sung K. 2013. Effects of humic acid on heavy metal uptake by herbaceous plants in soils simultaneously contaminated by petroleum hydrocarbons. *Environmental Earth Sciences* 68(8): 2375–2384.

Pascual I, Antolín MC, García C, Polo A, and Sánchez-Díaz M. 2004. Plant availability of heavy metals in a soil amended with a high dose of sewage sludge under drought conditions. *Biology and Fertility of Soils* 40: 291e299.

Pawlowska TE, and Charvat I. 2004. Heavy-metal stress and developmental patterns of arbuscular mycorrhizal fungi. *Appl Environ Microbiol* 70: 6643–6649.

Persans MW, Yan X, Patnoe JM, Kramer U, and Salt DE. 1999. Molecular dissection of the role of histidine in nickel hyperaccumulation in *Thlaspi goesingense* (Halacsy). *Plant Physiology* 121: 1117–1126.

Pianelli K, Mari S, Marques L, Lebrun M, and Czernic P. 2005. Nicotianamine over-accumulation confers resistance to nickel in *Arabidopsis thaliana*. *Transgenic Research* 14: 739–748.

Pickering IJ, Prince RC, George MJ, Smith RD, George GN, and Salt DE. 2000. Reduction and coordination of arsenic in Indian mustard. *Plant Physiology* 122(4): 1171–1178.

Pizarro I, Gomez M, Roman D, and Palacios AM. 2016. Bioavailability, bioaccesibility of heavy metal elements and speciation of as in contaminated areas of Chile. *Journal of Environmental Analytical Chemistry* 3: 175.

Prasad MNV. 2003. Metal hyperaccumulators in plants-Biodiversity prospecting for phytoremediation technology. *Electronic J Biotech* 6: 276–312.

Przemeck E, and Haase NU. 1991. On the bonding of manganese, copper and cadmium to peptides of the xylem sap of plant roots. *Water, Air, and Soil Pollution* 57(1): 569–577.

Rafique N, and Tariq SR. 2016. Distribution and source apportionment studies of heavy metals in soil of cotton/wheat fields. *Environ Monit Assess* 188: 309.

Rajkumar M, Vara Prasad MN, Freitas H, and Ae N. 2009. Biotechnological applications of serpentine bacteria for phytoremediation of heavy metals. *Critical Reviews in Biotechnology* 29(2): 120–130.

Rajkumar, M, Ae N, Prasad MNV, and Freitas H. 2010. Potential of siderophore-producing bacteria for improving heavy metal phytoextraction. *Trends Biotechnol* 28: 142–149.

Ramsay LM, and Gadd GM. 1997. Mutants of *Saccharomyces cerevisiae* defective in vacuolar function confirm a role for the vacuole in toxic metal ion detoxification. *FEMS Microbiol Lett* 152: 293–298.

Rao CRM, Sahuquillo A, and Sanchez JFL. 2008. A review of the different methods applied in environmental geochemistry for single and sequential extraction of trace elements in soils and related materials. *Water Air & Soil Pollution* 189(1-4): 291–333.

Rascio N, and Navari-Izzo F. 2011. Heavy metal hyperaccumulating plants: how and why do they do it? And what makes them so interesting? *Plant Sci* 180: 169–181.

Reddy AM, Kumar SG, Jyonthsnakumari G, Thimmanaik S, and Sudhakar C. 2005. Lead induced changes in antioxidant metabolism of horse gram (*Macrotyloma uniflorum (Lam.) Verdc.*) and Bengal gram (*Cicer arietinum* L.). *Chemosphere* 60: 97–104.

Reeder RJ, Schoonen MAA, and Lanzirotti A. 2006. Metal speciation and its role in bioaccessibility and bioavailability. *Reviews in Mineralogy & Geochemistry* 64(4): 3–59.

Reeves RD. 2003. Tropical hyperaccumulators of metals and their potential for phytoextraction. *Plant Soil* 249(1): 57–65.

ur Rehman MZ, Rizwan M, Ali S, Ok YS, Ishaque W, Nawaz MF, Akmal F, and Waqar M. 2017. Remediation of heavy metal contaminated soils by using *Solanum nigrum*: a review. *Ecotox Environ Safe* 143: 236–248.

Repetto O, Bestel-Corre G, Dumas-Gaudot E, Berta G, Gianinazzi-Pearson V, and Gianinazzi S. 2003. Targeted proteomics to identify cadmium-induced protein modifications in *Glomus mosseae*-inoculated pea roots. *New Phytol.* 157: 555–567.

Rivera-Becerril F, van Tuinen D, Martin-Laurent F, Metwally A, Dietz KJ, Gianinazzi S, Gianinazzi-and Pearson V. 2005. Molecular changes in *Pisum sativum* L. roots during arbuscular mycorrhiza buffering of cadmium stress. *Mycorrhiza* 16: 51–60.

Ryan CA, and Walker-Simmons M. 1983. Plant vacuoles. *Methods Enzymol* 96: 580–589.

Sagner S, Kneer R, Wanner G, Cosson JP, Deus-Neumann B, and Zenk MH. 1998. Hyperaccumulation, complexation and distribution of nickel in *Sebertia acuminata*. *Phytochemistry* 47: 339–347.

Salt DE, Blaylock M, Kumar NP, Dushenkov V, Ensley BD, Chet I, and Raskin I. 1995. Phytoremediation: A novel strategy for the removal of toxic metals from the environment using plants. *Biotechnology* 13: 468–474.

Sarwar N, Malhi SS, Zia MH, Naeem A, Bibi S, and Farid G. 2010. Role of mineral nutrition in minimizing cadmium accumulation by plants. *J Sci Food Agric* 90: 925–937.

Sarwar N, Imran M, Shaheen MR, Ishaque W, Kamran MA, Matloob A, Rehim A, and Hussain S. 2017. Phytoremediation strategies for soils contaminated with heavy metals: *Modifications and future perspectives, Chemosphere* 171: 710–721.

Saxena B, Shukla K, and Giri B. 2017. Arbuscular mycorrhizal fungi and tolerance of salt stress in plants. pp. 67–97. *In*: Wu QS (ed.). *Arbuscular Mycorrhizas and Stress Tolerance of Plants*. (Singapore: Springer).

Schaumloffel D, Ouerdane L, Boussiere B, and Lobinski R. 2003. Speciation analysis of nickel in the latex of a hyperaccumulating tree *Sebertia acuminata* by HPLC and CZE with ICP MS and electrospray MS–MS detection. *Journal of Analytical Atomic Spectrometry* 18: 120–127.

Schimel J, Balser TC, and Wallenstein M. 2007. Microbial stress-response physiology and its implications for ecosystem function. *Ecology* 88: 1386e1394.

Seregin IV, Shpigun LK, and Ivanov VB. 2004. Distribution and toxic effects of cadmium and lead on maize roots. *Russian Journal of Plant Physiology* 51(4): 525–533.

Sessitsch A, Kuffner M, Kidd P, Vangronsveld J, Wenzel WW, Fallmann K, and Puschenreiter M. 2013. The role of plant-associated bacteria in the mobilization and phytoextraction of trace elements in contaminated soils. *Soil Biol Biochem* 60: 182–194.

Shah A, Kamei S, and Kawai S. 2001. Metal micronutrients in xylem sap of iron-deficient barley as affected by plant-borne, microbial and synthetic metal chelators. *Soil Sci Plant Nutr* 47: 149–156.

Shahzad B, Tanveer M, Che Z, Rehman A, Cheema SA, Sharma A, Song H, ur Rehman S, and Zhaorong D. 2018. 2018. Role of 24- epibrassinolide (EBL) in mediating heavy metal and pesticide induced oxidative stress in plants: A review. *Ecotoxicology and Environmental Safety* 147: 935–944.

Shakeel M, and Yaseen T. 2015. An insight into phytoremediation of heavy metals from soil assisted by ancient fungi from glomeromycota-arbuscular mycorrhizal fungi. *Sci. Technol and Dev* 34(4): 215–220.

Singh PK, Singh M, and Tripathi BN. 2012. Glomalin: An arbuscular mycorrhizal fungal soil protein. *Protoplasma* 250: 663–669.

Singh S, and Prasad SM. 2014. Growth photosynthesis and oxidative responses of *Solanum melongena* L. seedlings to cadmium stress: Mechanism of toxicity amelioration by kinetin. *Sci Hortic* 176: 1–10.

Singh S, Parihar P, Singh R, Singh VP, and Prasad SM. 2016. Heavy metal tolerance in plants: role of transcriptomics, proteomics, metabolomics, and ionomics. *Front Plant Sci* 6: 1143.

Siripornadulsil S, Traina S, Verma DP, and Sayre RT. 2002. Molecular mechanisms of proline-mediated tolerance to toxic heavy metals in transgenic microalgae. *Plant Cell* 14: 2837–2847.

Smith AP, DeRidder BP, Guo WJ, Seeley EH, Regnier FE, and Goldsbrough PB. 2004. Proteomic analysis of *Arabidopsis* glutathione S-transferases from benoxacor- and copper-treated seedlings. *J Biol Chem* 279: 26098–26104.

Suman J, Uhlik O, Viktorova J, and Macek T. 2018. Phytoextraction of heavy metals: a promising tool for clean-up of polluted environment? *Front Plant Sci* 9: 1476.

Teofilo V, Marianna B, and Giuliano M. 2010. Field crops for phytoremediation of metal-contaminated land. a review. *Environmental Chemistry Letters* 8(1): 1–17.

Tepanosyan G, Sahakyan L, Belyaeva O, Maghakyan N, and Saghatelyan A. 2017. Human health risk assessment and riskiest heavy metal origin identification in urban soils of Yerevan, Armenia. *Chemosphere* 184: 1230–1240.

Terry N, Zayed AM, De Souza MP, and Tarun AS. 2000. Selenium in higher plants. *Annu Rev Plant Phys* 51: 401–432.

Thijs S, Langill T, and Vangronsveld J. 2017. Chapter two-the bacterial and fungal microbiota of hyperaccumulator plants: small organisms, large influence. *Adv Bot Res* 83: 43–86.

Tiwari S, Lata C, Chauhan PS, and Nautiyal CS. 2016. *Pseudomonas putida* attunes morphophysiological, biochemical and molecular responses in *Cicer arietinum* L. during drought stress and recovery. *Plant Physiol Biochem* 99: 108–117.

Tiwari S, Prasad V, Chauhan PS, and Lata C. 2017b. *Bacillus amyloliquefaciens* confers tolerance to various abiotic stresses and modulates plant response to phytohormones through osmoprotection and gene expression regulation in rice. *Front Plant Sci* 8: 1510.

Tripathi P, Singh PC, Mishra A, Srivastava S, Chauhan R, Awasthi S, Mishra S, Dwivedi S, Tripathi P, Kalra A, and Tripathi RD. 2017. Arsenic tolerant *Trichoderma* sp. reduces arsenic induced stress in chickpea (*Cicer arietinum*). *Environ Pollut* 223: 137–145.

Ullah A, Heng S, Munis MFH, Fahad S, and Yang X. 2015. Phytoremediation of heavy metals assisted by plant growth promoting (PGP) bacteria: a review. *Environ Exp Bot* 117: 28–40.

Vamerali T, Bandiera M, and Mosca G. 2010. Field crops for phytoremediation of metal-contaminated land. A review. *Environ Chem Lett* 8: 1–17.

van der Ent A, Baker AJ, Reeves RD, Pollard AJ, and Schat H. 2013. Hyperaccumulators of metal and metalloid trace elements: facts and fiction. *Plant Soil* 362: 319–334.

Vangronsveld J, Herzig R, Weyens N, Boulet J, Adriaensen K, Ruttens A, Thewys T, Vassilev A, Meers E, Nehnevajova E, and van der Lelie D. 2009. Phytoremediation of contaminated soils and groundwater: lessons from the field. *Environ Sci Pollut R* 16: 765–794.

Verkleij JAC, and Schat H. 1990. Mechanisms of metal tolerance in higher plants. *Heavy Metal Tolerance in Plants: Evolutionary Aspects* 1990: 179–194.

Vogel-Mikus K, Pongrac P, Kump P, Necemer M, and Regvar M. 2006. Colonisation of a Zn, Cd and Pb hyperaccumulator *Thlaspi praecox Wulfen* with indigenous arbuscular mycorrhizal fungal mixture induces changes in heavy metal and nutrient uptake. *Environmental Pollution* 139(2): 362–371.

Von Wiren N, Klair S, Bansal S, Briat JF, Khodr H, Shioiri T, Leigh RA, and Hider RC. 1999. Nicotinamide chelates both Fe III and Fe II. Implications for metal transport in plants. *Plant Physiol* 119: 1107–1114.

Wang AS, Angle JS, Chaney RL, Delorme TA, and Reeves RD. 2006. Soil pH effects on uptake of Cd and Zn by *Thlaspi caerulescens*. *Plant and Soil* 281(1): 325–337.

Wang D, Wei W, Liang DL, Wang SS, and Hu B. 2011. Transformation of copper and chromium in co-contaminated soil and its influence on bioavailability for pakchoi (*Brassica chinensis*). *Environmental Science* 32(10): 3113–3120. (In Chinese)

Wang YP, Shi JY, Wang H, Lin Q, Chen XC, and Chen YX. 2007. The influence of soil heavy metals pollution on soil microbial biomass, enzyme activity, and community composition near a copper smelter. *Ecotoxicology and Environmental Safety* 67: 75e81.

Wei S, Li Y, Zhan J, Wang S, and Zhu J. 2012. Tolerant mechanisms of *Rorippa globosa (Turcz.) Thell.* hyperaccumulating Cd explored from root morphology. *Bioresource Technology* 118: 455–459.

Wojas S, Clemens S, Hennig J, Skłodowska A, Kopera E, Schat H, Bal W, and Antosiewicz DM. 2008. Overexpression of phytochelatin synthase in tobacco: distinctive effects of AtPCS1 and CePCS genes on plant response to cadmium. *J Exp Bot* 59: 2205–2219.

Wojas S, Clemens S, Sklodowska A, and Antosiewicz DM. 2010. Arsenic response of AtPCS1- and CePCS-expressing plants—effects of external As(V) concentration on As-accumulation pattern and NPT metabolism. *J Plant Physiol* 167: 169–175.

Wong MH. 2003. Ecological restoration of mine degraded soils, with emphasis on metal contaminated soils. *Chemosphere* 50: 775–780.

Wright SF, and Upadhyaya AA. 1998. Survey of soils for aggregate stability and glomalin, a glycoprotein produced by hyphae of arbuscular mycorrhizal fungi. *Plant Soil* 198: 97–107.

Wu FB, Dong J, Qian QQ, and Zhang GP. 2005. Subcellular distribution and chemical form of Cd and Cd-Zn interaction in different barley genotypes. *Chemosphere* 60(10): 1437–1446.

Wu G, Kang H, Zhang X, Shao H, Chu L, and Ruan C. 2010. A critical review on the bio-removal of hazardous heavy metals from contaminated soils: issues, progress, eco-environmental concerns and opportunities. *J Hazard Mater* 174: 1–8.

Wu JT, Chang SC, and Chen KS. 1995. Enhancement of intracellular proline level in cells of *Anacystis nidulans* (Cyanobacteria) exposed to deleterious concentrations of copper. *J Phycol* 31: 376–379.

Wu JT, Hsieh MT, and Kow LC. 1998. Role of proline accumulation in response to toxic copper in *Chlorella* sp. (Chlorophyceae) cells. *J Phycol* 34: 113–117.

Wuana RA, and Okieimen FE. 2011. Heavy metals in contaminated soils: a review of sources, chemistry, risks and best available strategies for remediation. *Isrn Ecology* 2011: 402647.

Xu J, Yin H, and Li X. 2009. Protective effects of proline against cadmium toxicity in micropropagated hyperaccumulator, *Solanum nigrum* L. *Plant Cell Rep* 28: 325–333.

Yadav G, Srivastava PK, Singh VP, and Prasad SM. 2014. Light intensity alters the extent of arsenic toxicity in *Helianthus annuus* L. seedlings. *Biol Trace Elem Res* 158: 410–421.

Yadav SK. 2010. Heavy metals toxicity in plants: an overview on the role of glutathione and phytochelatins in heavy metal stress tolerance of plants. *S Afr J Bot* 76: 167–179. doi: 10.1016/j.sajb.2009.10.007.

Yan A, Wang Y, Tan SN, Yusof MLM, Ghosh S and Chen Z. 2020. Phytoremediation: a promising approach for revegetation of heavy metal-polluted land. *Frontiers in Plant Science* 11.

Yang XE, Long XX, Ye HB, He ZL, Calvert D, and Stoffella P. 2004. Cadmium tolerance and hyperaccumulation in a new Zn-hyperaccumulating plant species (*Sedum alfredii* Hance). *Plant Soil* 259: 181–189.

Yang Y, Han X, Liang Y, Ghosh A, Chen J, and Tang M. 2015. The Combined effects of arbuscular mycorrhizal fungi (AMF) and lead (Pb) stress on Pb accumulation, plant growth parameters, photosynthesis, and antioxidant Enzymes in *Robinia pseudoacacia* L. *PLoS ONE* 10(12): e0145726.

Ying W. 2008. Research progress of microbial diversity in heavy metal contaminated soil[J]. *Modern Agricultural Sciences and Technology* (17): 174–175.

Yuan H, Li Z, Ying J, and Wang E. 2007. Cadmium(II) removal by a hyperaccumulator fungus *Phoma* sp. F2 isolated from blende soil. *Current Microbiology* 55(3): 223–227.

Zhang CB, Huang LN, Shu WS, Qiu JW, Zhang JT, and Lan CY. 2007. Structural and functional diversity of a culturable bacterial community during the early stages of revegetation near a Pb/Zn smelter in Guangdong, PR China. *Ecological Engineering* 30: 16e26.

Zhang C, Yu ZG, Zeng GM, Jiang M, Yang ZZ, Cui F, Zhu MY, Shen LQ, and Hu L. 2014. Effects of sediment geochemical properties on heavy metal bioavailability. *Environment International* 73(4): 270–281.

Zhang XF, Hu ZH, Yan TX, Lu RR, Peng CL, Li SS, and Jing YX. 2019. Arbuscular mycorrhizal fungi alleviate Cd phytotoxicity by altering Cd subcellular distribution and chemical forms in *Zea mays*. *Ecotoxicol Environ Saf* 171: 352–360.

Zhang YF, Feng G and Li XL. 2003. The effect of arbuscular mycorrhizal fungi on the components and concentrations of organic acids in the exudates of mycorrhizal red clover. *Acta Ecol Sin* 23: 30–37.

Zhipeng WU, Weidong WU, Shenglu ZHOU, and Shaohua W.U. 2016. Mycorrhizal inoculation affects Pb and Cd accumulation and translocation in Pakchoi (*Brassica chinensis* L.). *Pedosphere* 26: 13–2.

Zhou JL. 1999. Zn biosorption by *Rhizopus arrhizus* and other fungi. *Appl Microbiol Biotechnol* 51: 686–693.

Zhuang P, Shu WS, Li Z, Liao B, Li J, and Shao J. 2009. Removal of metals by sorghum plants from contaminated land. *Journal of Environmental Sciences* 21(10): 1432–1437.

Zomlefer WB. 1994. *Guide to Flowering Plant Families*. University of North Carolina Press, Chapel Hill, North Carolina, pp. 65–67.

Zubair M, Shakir M, Ali Q, Rani N, Fatima N, Farooq S, Shafiq S, Kanwal N, Ali F, and Nasir IA. 2016. Rhizobacteria and phytoremediation of heavy metals. *Environ Technol Rev* 5: 112–119.

12

Ionomics vis à vis Heavy Metals Stress and Amelioration

Rashmi Mukherjee,[1] *Mukul Barwant*[2] and *Dwaipayan Sinha*[3,*]

ABSTRACT

The study of the relationship between mineral elements and plants has attracted interest in the scientific field for many years. With the discovery of elements, it was gradually realized that elements play an important role in the overall physiological process of plants. Consequently, scientists categorized elemental nutrients in micro and macroelement depending upon the quantity required by the plants. The deficiency symptoms of each element were also deciphered. However, it was in the last decade, the relationship with elements, their impact, and effects on plants were streamlined more systematically through the concept of ionomics. Ionomics is defined as the sum total of the entire inorganic component of a living system. The concept was proposed largely to evaluate the effect of elements on the physiological process of plants and thus overlapped with the corresponding domains of genomics, proteomics, and metabolomics. Thus ionomics functioned as a bridging concept connecting the domains of genes, proteins, and metabolites. Presently, the concept of ionomics is frequently applied in case of imbalance in nutrients of plants and also for toxicity responses to heavy metals. They are also handy in evaluating the disease conditions. This chapter is an attempt to overview the concept of ionomics. Efforts are made to highlight the elemental nutrition, heavy metal toxicity, and its relationship with ionomics.

[1] Department of Botany, Raja N.L. Khan Women's College (Autonomous) , Gope Palace, Midnapore-721102; West Bengal, INDIA.
[2] Department of Botany, Sanjivani Arts Commerce and Science College Kopargoan, Maharashtra, India.
[3] Department of Botany, Government General Degree College, Mohanpur, Paschim Medinipur, West Bengal-721436, India.
* Corresponding author: dwaipayansinha@hotmail.com
ORCID iD: https://orcid.org/0000-0001-7870-8998

1. Ionomics—an introduction

Ionome refers to the total inorganic component present in a cellular organism. It consists of its mineral nutrition and trace element composition (Baxter et al., 2009). Ionome is an active complex of elements influenced directly by the existing physiological as well as biochemical conditions of the plants which in turn is regulated by environmental and genetic causes (Singh et al., 2013). This concept became was introduced and became of matter of extreme relevance due to the ever increasing interest among scientists to explore the behaviour and effect of mineral elements in a living system. It is studied by quantitative evaluation and instantaneous measurements of the living organism's elemental conformation of, as well as physiological stimuli, developmental condition, and gene induced changes in their composition (Salt et al., 2008). The ionome is characterised as the organism's mineral nutrient content in addition to the trace element framework or the complete ions set (Huang et al., 2016; Lahner et al., 2003). Nutrients are vital for plant development and reproduction (Hererra et al., 2015) which may vary from one season to another (Cruz et al., 2019). This nutrient requirement is provided by the elements in their ionic form. The abundance of nutrients in soil varies greatly depending on the soil composition and is also influenced by nutrient solubility (Baxter, 2009). To comprehend how life functions, it becomes necessary to understand the functions and complexities of elements. The ionome is useful in understanding other cellular systems (e.g., proteome, genome, metabolome, etc.), and plant-nutrient exchange. Ionome, which is the total inorganic constituent, independently exists with three other independent concepts namely genome, transcriptome, and metabolome and are all interrelated to one another having overlapping boundaries with respect to their functions and properties (Salt, 2004). The ionome is also useful for learning electrophysiology, signalling, enzymology, osmoregulation, and transport (Salt et al., 2008) as shown in Fig. 1.

Ionomics is the study of ionomes, which focuses on the quantification of various elements (Pita-Barbosa et al., 2019) and their relationship to physiology, growth, environmental change, and inorganic composition of organs, tissues, and cells. (Singh et al., 2013). Robinson and Pauling in the late 1960s and early 1970s proposed the principle of ionomics (Matsuoka et al., 2020) which states that an organism's metabolite profile provides a rich repository of knowledge that reflects the organism's physiological status (Singh et al., 2013). Ionomics is capable of collecting information about an organism's functional state under various circumstances, which are influenced by genetic and developmental variations, and biotic or abiotic influences. (Salt et al., 2008). Through ionomic study, the effect of gene modification of uptake of minerals can be studied (Watanabe et al., 2015). The world's population is increasing each day and most people are at risk of suffering from mineral deficiency (Gernad et al., 2016; Hwalla et al., 2017; Bird et al., 2017). Micronutrient deficiency has affected nearly 2 billion people throughout the world and forms a chief contributing aspect in the global burden of disease (Gorji and Ghadiri, 2021). Consumption of traditional food plants offers an important scope in the reduction of elemental nutritional deficiency and reduces the scope of non-communicable diseases (Lyons et al., 2020). The leafy vegetables also act as a significant source of mineral nutrients

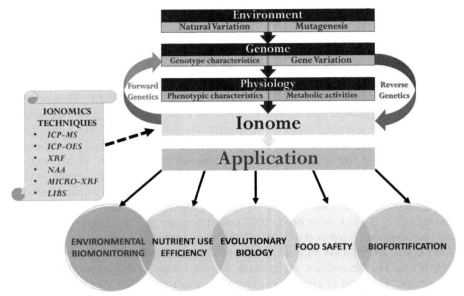

Figure 1: Ionome: Integrating plant genetics, physiology and evolution.

in this regard (Hailu and Addis, 2016). Consequently, research in ionomics becomes very crucial in evaluating the ionome structure of plants especially those which are consumed as foods. As a part of the research on the ionone, identification of various trace elements in food crops and cereals that are essential for their survival, growth and reproduction, and most importantly, the overall yield is very much relevant. Ionomic studies have also been conveniently devised for the study of elemental deficiency and screen of abiotic resistant plants (Fleet et al., 2011). An overview of ionomics highlighting the plants' mineral nutrition is attempted in this chapter. Efforts are made to illustrate various mechanisms involved in the characterization of ionomes in plant system for unravelling their biological effects.

2. History of ionomics

The idea of ionomics in plants was first conceptualized through the unison of the concept of metabolomics with the mineral nutrition of plants. The concept of ionomics was initiated by Robinson and Pauling and was based on the fact that the metabolite profile is a rich source of information which in turn represents an organism's physiological status (Singh et al., 2013). The exploration of the metabolite profile of the organism became feasible with the discovery of sophisticated technologies such as gas chromatography mass-spectrometry (GCMS), liquid chromatography mass spectrometry (LC-MS) (Nam et al., 2019), ultra-performance liquid chromatography (UPLC) (Hu et al., 2018), and proton nuclear magnetic resonance (^{1}H-NMR) (Kim et al., 2021). At present ionomics is further strengthened through association with more sophisticated techniques namely bioinformatics and DNA microarrays (Salt et al., 2008). Lahner and coworkers in 2003 were the first to include all the elements present in an organism within the domain of ionome (Lahner et al., 2003) thereby broadening metallome's domain (Lobinski et al., 2010; Ogra et al., 2015) to include biologically and physiologically relevant nonmetal element. In 2008, Salt and coworkers coined the term 'plant ionome' to encompass all the elements with a motive of discussing and investigating the mechanism which regulates these elements in response to various factors related to environment and genetic level (Salt et al., 2008; Williams and Salt, 2009). It thus opens the door to evaluate the information on the cell's function in both qualitative and quantitative terms (Singh et al., 2013). Plant ionomics was first developed in *Arabidopsis* for the identification of elemental accumulation mutants and characterization of their physiological responses in the environment (Pauli et al., 2018). Gradual involvement of inductively coupled plasma (ICP) spectroscopy coupled to mass spectrometry has enabled scientists to simultaneously, measure many elements in a single sample. Moreover, it requires less sample for analysis due to its high sensitivity thus proving to be a nondestructive approach especially applicable to evaluate small plants (Salt, 2004). ICP coupled to MS can also distinguish between various isotopes of the same element at extremely low concentrations (Zhang et al., 2020). These sophisticated techniques helped to combine Pauling's and Robinson's first thoughts on metabolite profiling based on the general idea of mineral nutrition of plants. Ionomics can thus be an effective tool in the future to examine the physiological complexity of plants and understand the functional intricacies of the genome (Salt et al., 2008).

3. Mineral nutrition in plants

The curiosity about the mineral nutrition of plants cropped up somewhere in the 17th and 18th centuries in the European continent. With the proliferation of chemistry in the 18th century pioneered by Lavoisier (West, 2013), various scientists got a scope to delve into the details of the association of minerals with plants. In the first quarter of the 19th century, the humus theory which stated that soil as the unique source of carbon was refuted by Carl Sprengel (van-der Ploegg et al., 1999). The beginning of the twentieth century and the whole of the nineteenth century experienced remarkable effort towards an investigation of mineral nutrition in plants and the most famous yield response law was proposed by Carl Sprengel by the name 'law of minimum' in 1840 and it was later patronized by Justus von Liebig. According to the law of minimum, a plant can attain optimal growth only if the concentration of required nutrition is at the optimal level (Koch et al., 2019). The law of diminishing yield increase proposed by Eilhard Alfred Mitscherlich is equally important to the mineral nutrition in plants. According to this law, a higher nutrient supply corresponds to a lower increase in yield obtained from fertilization in a plant (Koch et al., 2020). Another law related to plant mineral nutrition is Liebscher's law of optimum. This law states that in a condition where a minimum resource contributes to maximum production, the other factors close to the production are at the optimal level (Parent et al., 2017). These laws formed the platform of fertilization of crops which eventually paved the way for modern agricultural techniques for increased yield. Studies on the mineral nutrition of plants reveal that they require a set of 14 elements for their nutrition besides carbon dioxide, oxygen and water. Out of the 14 mineral elements, six mineral elements namely nitrogen (N), phosphorus (P), calcium (Ca), potassium (K), magnesium (Mg), and sulphur (S) are needed in greater quantities and hence called macronutrients while eight elements namely chlorine (Cl), boron (B), iron (Fe), manganese (Mn), molybdenum (Mo), nickel (Ni), copper (Cu), and zinc (Zn) are needed in smaller amounts and are called as micronutrients (White and Brown, 2010). In addition to it, silicon is also considered to be a beneficial element and is involved in the maintenance of the structural pattern of some plants (Guerriero et al., 2016). The requirement of different macro and micronutrients are shown in Fig. 2.

The deficiency of these elemental nutrients results in alteration of growth in plants (Huang et al., 2020). This deficiency of a given element in a plant may happen due to changes in soil chemistry, plant uptake capacity, chelator accumulation, amount of element that is sequestered in intracellular component (Baxter et al., 2009). Table 1 summarizes all the elemental nutrients required by plants, the form in which their uptake takes place, their functions, the deficiency symptoms caused in plants as well as the toxicity created due to their excess amount.

This total nutrient composition of the plant or rather all the living organisms is called ionome (Huang and Salt, 2016). Ever since the idea was first offered by Lahner in 2003 some two decades ago (Lahner, 2003), significant advancements have been made in the elemental profiling of plants (El-Deftar et al., 2015; Pandotra et al., 2015; Ricachenevsky et al., 2018). Efforts have also been made to identify the genes which are responsible for controlling the ionome (Buescher et al., 2010; Whitt et al., 2020). Plants being sessile have resorted to intricate mechanisms to balance and redistribute the mineral nutrients in their body in response to external stimuli in the form of environmental stress (Campos et al., 2021). These process require sets of proteins that act as transporters or channels (Flowers et al., 2019;

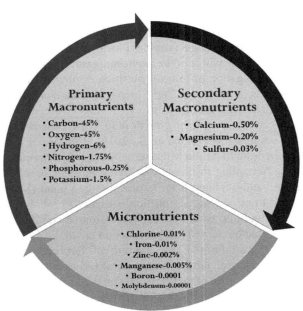

Figure 2: Overview of various plant nutrient requirements.

Table 1: Role macronutrients and micronutrients in plants and their effects.

Nutrient Family	Nutrients	Form for plant uptake	Major function	Effect of Deficiency	Effect of excess	References
MACRONUTRIENTS						
Primary	Carbon	CO_2, HCO_3^-	Plant structures, constituent of carbohydrates, necessary for photosynthesis.	Chlorosis in old leaves, stunted growth, increased seed dormancy.	Thickening of leaves, zebra-like stripes in the leaves.	Turan et al., 2017; de Bang et al., 2021
	Oxygen	H_2O	Constituent of carbohydrates, necessary for respiration, energy production, plant structures.	Depressed growth and yield of dryland species.	Plants produce less biomass, decreased leaf tissue.	Pederson et al., 2021; de Bang et al., 2021
	Hydrogen	H_2O	pH regulation, maintains osmotic balance, constituent of carbohydrates.	Chlorosis, , stunted growth and occasional necrosis..	Inhibit primary root growth.	Luo et al., 2020; de Bang et al., 2021
	Nitrogen	NO_3^-, NH_4^+	constituent of Protein/amino acids/chlorophyll/nucleic acids, cell formation.	Older leaves turns yellow , but the rest of the plant is still light green.	Dark green leaves susceptible to abiotic and biotic stresses.	Comadira et al., 2015; de Bang et al., 2021
	Phosphorous	$H_2PO_4^-$, HPO_4^{2-}, PO_4^{3-}	Cell formation, protein synthesis, metabolism of fat and carbohydrates, important in energy transfer, constituent of coenzymes and metabolic substrates.	The burned appearance of the leaf tips, Aged leaves takes green or reddish purple colour.	Cause micronutrient deficiencies especially iron and zinc.	Carstensen et al., 2018; de Bang et al., 2021
	Potassium	K^+	Water regulation, enzyme activity, involved with photosynthesis and protein synthesis, carbohydrate translocation.	Wilting, chlorosis within th eveins and scorching inward margin of leaf.	Cause deficiencies in magnesium and calcium.	Liu et al., 2013; de Bang et al., 2021

Table 1 contd. ...

...*Table 1 contd.*

Nutrient Family	Nutrients	Form for plant uptake	Major function	Effect of Deficiency	Effect of excess	References
Secondary	Calcium	Ca^{2+}	Root permeability, enzyme activity.	New leaves on the plant's top have a distorted or uneven form.	Reduction in apical growth. Scanty development of fruit.	Chao et al., 2008; de Bang et al., 2021
	Magnesium	Mg^{2+}	Chlorophyll, fat formation and metabolism.	Yellow in between the leaf veins, curling of leaf margins, dead spots.	Causes imbalance with calcium and potassium reduced growth.	Peng et al., 2015; de Bang et al., 2021
	Sulfur	SO_4^{2-}	Protein, amino acid, vitamin and oil formation.	Chlorosis in younger leaves with anthocyanin accumulation, stem and roots become woody, inwardly curving of tip and margins of leaves.	The yellowing of new leaves spreads across the plant at first.	Aarabi et al., 2016; de Bang et al., 2021
MICRONUTRIENTS						
Elements	Chlorine	Cl^-	Chlorophyll formation, enzyme activity, cellular development.	Chlorosis and premature leaf withering.	Death of leaf margins.	Schwenke et al., 2015; Gulzar et al., 2020
	Iron	Fe^{2+}, Fe^{3+}	Enzyme development and activity.	Chlorosis and yellowing between the young leaf's veins, stunted shoots, reduced yield.	Reduced yield, Bronzing of leaves with little brown dots.	Bai et al., 2018; Zaid et al., 2020
	Zinc	Zn^{2+}	Enzyme activity.	Stunted growth, reduced internode length, young leaves are smaller than normal, poor fruit set.	Poor root growth, young leaf chlorosis, excess zinc creates iron deficiency in some plants.	Zhao and Wu, 2017; Gulzar et al., 2020
	Manganese	Mn^{2+}	Enzyme activity and pigmentation, controls several oxidation-reduction systems and photosynthesis.	Chlorosis band on basal leaves and death, decreased cold hardiness, no new lateral roots, inhibited nitrate metabolism.	Tissue injury, older leaves have brown dots surrounded by a chlorotic circular area.	Chen et al., 2019; Gulzar et al., 2020

Elements		Function	Symptoms	Toxicity	References
Boron	H_3BO_3, BO_3^{3-}, B_4O_7	Enzyme activity, important in sugar translocation and carbohydrate metabolism.	Chlorosis, death of apical region, leaves that are distorted and have discoloured spots, stubby roots, inhibited nitrate metabolism.	Toxicity, dark brown speckles or necrosis on edges of older leaves, cupped and wrinkled young leaves which later fall off.	Li et al., 2017; Gulzar et al., 2020
Copper	Cu^{2+}	Enzyme activity, catalyst for respiration.	Chlorosis, death and twisting of leaf tips, short internodes. In young leaves, there is a reduction of turgor pressure.	Reduced vigour, stunted root growth, root damage.	Billard et al., 2014; Gulzar et al., 2020
Molybdenum	$HMoO^{4-}$, MoO_4^{2-}	Enzyme activity and nitrogen fixation in legumes, transforms nitrate to ammonium.	Yellowing of young plants, poor fruit set.	Leaf malformation, chlorosis and low tillering.	Gopal et al., 2016; Gulzar et al., 2020
Nickel	Ni^{2+}	Necessary for proper functioning of urease and seed germination.	Non-viable seeds, leaf chlorosis at the onset of season, foliage reduction of growth, blunting, curling and necrosis of leaf, margins, brittle shoots, diminished root system with dead fibrous roots, rosetting and tree death.	Poor growth, chlorosis, necrosis and wilting.	Wood et al., 2006; Gulzar et al., 2020

Lu et al., 2021). Ionomic studies have resulted in the identification of genes that are involved in elemental sequestration, homoeostasis and other processes in plants (Campos et al., 2021). The classic role of ionome is the reduction in toxicity of heavy metals in plants. The essential and trace elements often act to alleviate the heavy metal toxicity in plants (Singh et al., 2016). It is reported from a study that nitrogen is a key component in reducing cadmium toxicity in plants (Zhang et al., 2014). Potassium and Calcium have also been reported to alleviate cadmium toxicity and help to bring positive responses in stressed plants (Ahmad et al., 2016). Another report says that the application of sulphur results in the alleviation of chromium stress in plants (Kulczycki and Sacała, 2020). Thus, study of ionome is of extreme relevance to assess the plant's nutritional status and also to evaluate the behavior of nutrients concerning the physiological status of the plant.

4. Interaction of heavy metals with plant nutrients

Heavy metals are known to pollute all components of the ecosystem and cause substantial harm to the biome (Jan et al., 2015; Okereafor et al., 2020). These metals are easily assimilated by plants and are considered to be elements of stress in addition to their subsequent channeling down the food chain (Jalmi et al., 2018; Zwolak et al., 2019). Heavy metals destabilize plant nutrition by causing mineral imbalances (Karahan et al., 2019), induce oxidative stress (Seneviratne et al., 2019), cause inactivation and altered functioning of key biomolecules (Emamverdian et al., 2015), and activate signaling cascades in plants (Jalmi et al., 2018; Keyster et al., 2020). It is well-known that nutrient elements are required for the growth of plants (Ågren and Weih, 2020). The imbalance of the nutrient element results in inhibition of growth (White and Brown, 2010) and reduction in yield (Sharma et al., 2013). In this section, the interaction of heavy metals with essential elements would be discussed in brief. Heavy metals interact with the nutrient element in a very interesting way. They may either reduce the contents of essential nutrients in the plant body or in certain cases, cause an increase in their concentration. Table 2 represents the effect of heavy metals on essential mineral nutrients.

On the other way round, the heavy metal stress in a plant can also be ameliorated by supplementation of essential nutrients. There are several reports of alleviation or amelioration of heavy metal toxicity by the addition of essential elements. A report states that cadmium stressed *Nicotiana tabacum* when subjected to treatment with 0.5 mM of K^+, resulted in alleviation of cadmium induced toxicity. This was manifested by a reduction in cadmium uptake and translocation, enhancement of photosynthetic function, stomatal conductance and improvement in contents of K^+, Zn, Cu, Mn, and Fe in both the plant's roots and branches (Wang et al., 2017a). Similar reports have also come up from a recent study on the tomato plant in which potassium was shown to counteract the harmful consequences of cadmium toxicity concerning the photosynthetic efficiency and reduction in cadmium transfer of nutrients from the roots to the aerial portions (Naciri et al., 2021). Another experiment using the algal system *Micrasterias denticulate* states that uptake of chromium was decreased when the cells were pretreated with iron. In addition, Zinc was able to ameliorate chromium toxicity with respect to photosynthetic and respiratory parameters. On

Table 2: List of heavy metal, their concentration, essential element effected and physiological effect.

Heavy Metals	Concentration	Effect on essential element	Plant species	Physiological effect	References
Cadmium	24 mg Cd kg^{-1} and 48 mg Cd kg^{-1} in soils	Nitrogen	*Solanum nigrum*	Reduced nitrogen, nitrate and soluble protein concentration in seedling, flowering and matured stages.	Wang et al., 2008
Cadmium	10–190 mg Cd kg^{-1} in soils	Roots: P, K, Ca, Fe, and Zn. Shoots: Ca, P, K and Cu	*Brassica juncea*	Concentrations of these elements decreased in roots and shoots.	Jiang et al., 2004
Cadmium	1 and 10 μM	K$^+$, Ca^{2+} and Mg^{2+}	*Amaranthus cruentus*	Decrease in K$^+$ concentration in roots, while increase in concentration in the basic stems and young leaves. Increase in content of Ca^{+2} in stem and leaves. Magnesium content in the roots and leaves increases..	Osmolovskaya et al., 2018
Vanadium	0–40 mg L^{-1}	N, P, Zn, K ,Fe, Cd, Pb	*Cucumis Sativus*	Decrease in concentration of the elements along with inhibition of uptake with increase in concentration of vanadium.	Osu et al., 2016
Chromium	0.05, 0.1, 0.2, 0.3 and 0.4 mM	P, Mn, Fe, Cu, Zn,S	*Citrillus* sp.	Increased concentration of Phosphorus and Manganese and decreased concentration of iron, copper, zinc and sulphur in the leaves.	Dube et al., 2003
Chromium	2.5, 5, 10, 25, 50, 75, 100 and 200 mg/l	N, P, K, Mn, Cu, Zn, Fe	*Oryza sativa*	Reduced absorption of minerals.	Sundaramoorthy et al., 2010
Lead	1.5, 3 and 15 mM	Na, K, Ca, P, Mg, Fe, Cu and Zn	*Spinacia oleracea, Triticum aestivum*	Lowering of the nutrients uptake.	Lamhamdi et al., 2013
Copper	25 and 100 mM	P, Fe, Ca	*Rumex japonicus*	Lowering of phosphorus and Iron content. Increase in calcium content in shoots.	Ke et al., 2007
Copper	150 and 300 ppm	Fe, Zn	*Arundo donax*	Reduction in content of Iron and Zinc.	Pietrini et al., 2019
Arsenic	0.2 mM		*Hyper accumulators* *Pteris vittata* *Pteris multifida* *Nonhyperaccumulators* *Pteris ensiformis*	*Hyper accumulators* Increased uptake of potassium and zinc. Decreased uptake of calcium *Non hyper accumulators* Decreased uptake of potassium and zinc.	Liu et al., 2018

the other hand, calcium and gadolinium were able to compensate for the inhibitory effect of lead and cadmium concerning cell division and morphogenetic parameters (Volland et al., 2014). A study on hexavalent chromium stressed *Solanum melongela*, reported that sulphur treatment ameliorated toxic effect on photosystem II (Singh et al., 2017). Another study shows the sulphur intracellular accumulation of hexavalent chromium in tomato, pea and brinjal seedling. Moreover chromium induced growth inhibition was ameliorated by the addition of sulphur. It was further observed that glutathione and hydrogen sulphide are associated with sulphur mediated reduction of toxicity of chromium (Kushwaha and Singh, 2020). It was also reported that phosphate can ameliorate chromium toxicity by reducting the content of chromium in a plant (López-Bucio et al., 2014).

The inhibition of essential element uptake by heavy metals is largely due to their competition across the membranes. It is reported that cadmium can compete with the other divalent cations namely calcium, magnesium, and iron during membrane transport (Llamas et al., 2000). Another study states that paddy plants subjected to nickel stress resulted in depolarization of the plasma membrane with a significant loss of K^+ ions in plants upon long term exposure. Additionally, there is also a loss in water content of the cell possibly due to alteration of membrane potential (Llamas et al., 2008). Another results states that zinc stress induces membrane depolarization in metal susceptible varieties of *Arabidopsis thaliana* (Kenderesová et al., 2012). It was also reported that chromium stress resulted in electrolyte leakage in maize (Wang et al., 2013a). Cobalt in high concentration is also reported to displace iron, manganese, zinc, and copper from physiologically important binding sites which leads to the possibility of reduction in the assimilation of micronutrients (Chatterjee and Chatterjee, 2000). On the other hand, copper under toxic conditions is reported to displace the central magnesium ion of the chlorophyll molecule resulting in altered physiological function (Yruela, 2009). The competition of heavy metals with the essential nutrients to bind with the common binding sites is due to their comparable size. For example, Nickel has a similar character as compared to magnesium, calcium, iron, copper, and zinc. This results in the competition of nickel with the other essential elements for absorption and translocation, ultimately leading to their deficiency (Chen et al., 2009). Nickel also causes iron deficiency by either inhibiting the absorption of the element or immobilizing the same in the roots (Mysliwa-Kurdziel et al., 2004). Moreover, the decline in nutrients may also be due to heavy metal induced impairment of the enzymes. For example, nickel, copper and cadmium inhibit the activity of ferric(III) reductase in cucumber thereby possibly hampering the absorption and translocation of iron (Alcántara et al., 1994). Another report states that cadmium also inhibits H+/K+ antiporters in plants thereby unbalancing the ion equilibrium (Obata et al., 1996). Though there have been mechanisms of inhibition of essential element uptake by the plants suffering from heavy metal stress but there are certain cases where the heavy metal stress can be ameliorated through the action of essential elements or other non-essential elements. For example, silicon can ameliorate heavy metal toxicity either through a reduction in absorption or activity of the metal, or by changing the metal's formation through the addition of a silicon based compound (Emamverdian et al., 2018). Moreover, supplementation of sulphur can boost the sulphur-containing amino acid biosynthesis and eventually glutathione or other low molecular weight thiols which have a strong affinity towards toxic metals (Yadav et al., 2010). Sulphur also participates in the biosynthesis of phytochelatins and metallothionein which also operate as a heavy metal scavenger (Verkleij et al., 2003; Ravilious and Jez, 2012). Potassium also counteracts heavy metal stress by reducing oxidative stress, enhancing the generation of prolines and other phenolics and increasing the antioxidant enzyme activity (Yasin et al., 2018). Thus there is an interesting behaviour among various elements with the heavy metals whose action largely depends on the ambient cellular condition and the concentration of each element present at that point.

5. Inonomics: the connecting link and its necessity

Plants uptake essential and non-essential elements present in the soil through the root system. The accumulated elements (collectively called "the ionome") in the plant system has a long history of research by various scientists and are presently investigated by the emerging and sophisticated field of plant ionomics which was introduced two decades ago. It is now well recognized that numerous causes like species diversity, organs, genes, stress factors, environment have a profound effect on this ionomic profile of a plant (Watanabe et al., 2016). A better understanding of the plant ionome will help in solving many questions related to evolutionary biology, nutrient use efficiency, environmental monitoring, biofortification as well as food safety. Ionomics has the potential to develop into an effective platform for providing a deeper insight into the genetic basis behind the elemental accumulation in plant tissues. Ionomics is very important for understanding plant nutrition through functional genomics. Study reports already exist which suggest that the APR2 gene regulates natural variation in S and Se content in certain plants (Chao et al., 2014); the FRO2 gene is associated with natural variation exhibited by root length under reduced Fe concentration (Satbhai et al., 2017). Ionomics is also capable to throw light on issues related to environmental monitoring, food safety and evolutionary studies as it includes the essential elements (macro and micronutrients), and trace elements that may be toxic when they accumulate in plant tissues. For example, ionomics studies have revealed that the HMA3 (Liu et al., 2017) and HMA4 genes (Chen et al., 2018) control the accumulation of Cd and Zn in shoots whereas the HAC1 gene regulates natural variation in shoot arsenic(As) (Fischer et al., 2021). These highly toxic heave and trace elements are also present in edible plant parts viz. fruits, leaves and seeds (Pita-Barbosa et al., 2019). When ingested by animals and humans, they lead to specific toxicity and diseases (Sharma and Kumar, 2019; Othman et al., 2021). Then comes the scope of disease ionomics, which provides significant evidence about how the minerals/elements and trace elements have a vital role in human and animal health as well as pathogenesis and help in formulating clinical strategies (Zhang et al., 2020). Furthermore, one can have great knowledge about understanding salinity tolerance and evolution through ionomics based approaches. Figure 3 illustrates the linking of ionomics with various domains of life sciences related to plants.

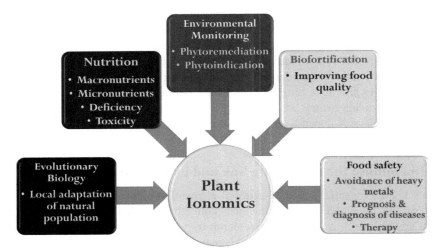

Figure 3: The interlinking network of Plant Ionomics.

6. Ionomics platform: an overview

The study of plant ionomics involves the following steps:

- Plant sample collection
- Sample processing
- Use of analytical instrumentation techniques
- Data analysis
- Bioinformatics
- Performance evaluation of proposed model and experimental validation

Sample collection forms the first step of the study of plant ionomics. In this stage roots, leaves, seeds and shoots of the plants of specific maturity are collected in a specific time or season. The time of harvesting is also very important for sample collection as elements are likely to vary depending upon the maturity of the plant samples. Different instrumentation techniques require various sample pre-processing including digestion, denaturation or centrifugation. It is followed by high throughput quantification of elements (macronutrients, micronutrients or heavy metals) using advanced instrumentation techniques like ICP, Neutron Activation Analysis (NAA) or X-ray fluorescence (XRF). Laser-Induced Breakdown Spectroscopy (LIBS) and Synchrotron based micro X-Ray Fluorescence (micro-SXRF) form the other instrumental platform for analysis. The result generated is then used for statistical analysis whereby data normalization, summarization, visualization and multivariate analysis is applied to identify significantly altered elements. This step is subsequently followed by the application of bioinformatics whereby database search at Purdue Ionomics Information Management System (PiiMS) provides ionomic network analysis as well as develops ionome based models for classification (Huang et al., 2016; Danku et al., 2013). Figure 4 depicts the work plan involved in any plant based ionomics study.

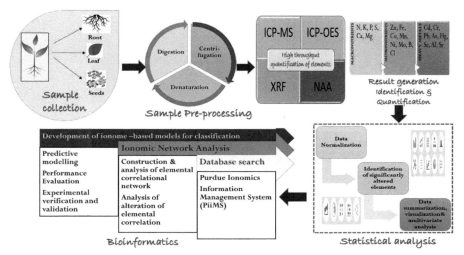

Figure 4: Pictorial representation of work plan for ionomic study in plants.

As earlier discussed, there are several instrumentation techniques that have been used for the study of ionomes in plants to date. However, the following three analytical modalities have been the most frequently involved with plant ionomics studies worldwide:

6.1 Inductively coupled plasma (icp)

ICP ionizes atoms present in the sample/analyte (Calderón-Celis and Encinar, 2019). These atoms are subsequently identified by optical emission spectroscopy (ICP-OES) or known as atomic emission spectroscopy (ICP-AES). ICP tagged with mass spectrometry (ICP-MS) also serves in detecting ionized atoms (Williams and Salt, 2009). A liquid sample is ideal for analysis which should be present as an aerosol. Precautions are required to remove any particulate from the aerosol, as they tend to obstruct the instrument. Samples can be prepared by simple acid digestion or microwave digestion for ICP-MS and ICP-OES. Figure 5 shows the schematic overview of the working principle along with all the functional components of ICP-MS.

6.2 X-ray fluorescence (XRF)

In X-ray fluorescence (XRF), an atom emits secondary or fluorescent X-rays after being excited due to the absorption of high energy X-rays or gamma rays (Chen et al., 2012). The emitted fluorescence X-rays possess energies that are the distinctive signature of the particular atom which emits them. As such, XRF helps in the identification of specific elements from a complex mixture. It is also a novel technique to quantify elements. Extensive and elaborative sample preparation is required for XRF studies which includes either powdering and compressing of samples into a wafer or fusing the samples into a glass (Punshon et al., 2013;

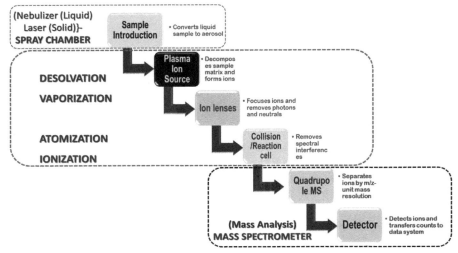

Figure 5: Working principle of ICP-MS.

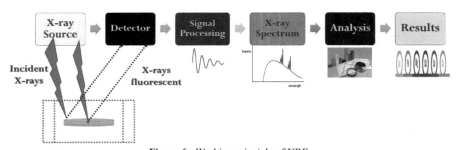

Figure 6: Working principle of XRF.

van der Ent et al., 2019). The basic overview of elemental analyte identification by XRF is depicted in Fig. 6.

6.3 *Neutron activation analysis (naa)*

Atomic nuclei capture free neutrons during the process of neutron activation as a result of which there is the formation of new radioactive nuclei. The newly formed radioactive nuclei degenerate with time. It releases an amount of energy in the form of gamma radiation which if recorded at a specific energy, indicates the occurrence of a precise radionuclide. The obtained gamma-ray spectral signal is processed and subsequently, various elemental concentrations in the sample can be quantified (Singh et al., 2013). NAA thus allows the non-destructive estimation of numerous major trace elements (Bode, 2012). This technique can be performed directly on the sample placed in a polyethylene vial. The working principle of NAA is shown in Fig. 7.

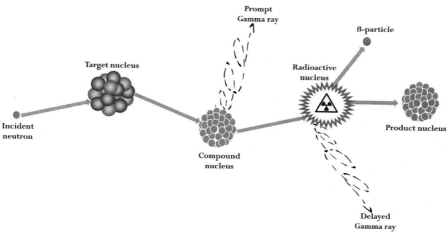

Figure 7: Working principle of NAA.

7. Use of ionomics in the characterization of essential nutrients and heavy metals

Environmental heavy metal contamination has increased due to innumerable anthropogenic activities, incessant industrialization, and current agricultural practices causing toxicity to all the living organisms in some form or the other (Briffa et al., 2020). Indiscriminate use of synthetic fertilizers, pesticides, weedicides, insecticides, etc., has contaminated huge stretches of soil with heavy metals (Tchounwou et al., 2012). As mentioned earlier "Ionome" takes into account the role of mineral macronutrients (Ca, N,S, P, Mg, and K) and micronutrients (Fe, Mn, Cu, Mo, Zn, and Co) for easing the toxicity caused by heavy metals (Singh et al., 2013). Although, the macro and micronutrients are indispensable for plants, however, unfortunately, when the concentration level of these nutrients in the plant body surpasses the requisite level, they become toxic to the plants. On the contrary, nutrients at the optimal concentration limit, play a substantial part in relieving the heavy metals toxicity (Dimkpa and Bindraban, 2016).

Nitrogen [N], the most vital plant nutrient is the primary element of nucleic acids, proteins, vitamins, amino acids and growth regulators. Its major functions include augmenting the photosynthetic ability by enhancing the synthesis of chlorophyll, producing metabolites (e.g., proline, etc.), and by increasing antioxidant activity. All these functions make N a potential element for lessening heavy metal toxicity in plants (Shang et al., 2019). Phosphorous (P) is the main fundamental unit of nucleic acids and cell membranes. It is primarily involved in phosphorylation reactions (Plaxton and Tran, 2011). Reports suggest that it significantly reduces heavy metaltoxic effect not only by diluting its concentration but also increases its immobility (Grobelak and Napora, 2015; Seshadri et al., 2017) through the formation of metal complexes. Furthermore, P also proliferates glutathione (GSH) content and inhibits membrane damage (Jaishankar et al., 2014). Plants require potassium (K) ions to sustain anion–cation equilibrium in cells. K also has a vital monitoring part in protein synthesis (Sustr et al., 2019) as well as in the activation of enzymes (Wang et al., 2013b). Researchers have found oxidative stress (OS) in plants can be modulated by increasing the K nutrient content (Marques et al., 2014). Sulfur (S), a mineral macronutrient, assists as a key building block of numerous vitamins, amino acids, coenzymes, chloroplasts, sulfatides, prosthetic groups, and ferredoxin (Li et al., 2020a). Calcium (Ca) is a ubiquitous intracellular second messenger (Aldon et al., 2018) which regulates various metabolic activities by activating the enzymes and is recognized as the central regulator of growth and development (Hepler, 2005) cytoplasmic streaming, photosynthesis, cell division, and intracellular signalling transduction (Huang et al., 2017). There is a chemical resemblance between Ca and Cadmium (Cd) (Pathak et al., 2020). Camay perhaps arbitrates plants physiological or metabolic alterations induced by Cd and thus can be utilized as an extrinsic element to defend them from Cd induced stress (Farzadfar et al., 2013). This is done through mitigation of growth retardation, translocation, metal uptake regulation, enhancement of photosynthesis, alleviation of oxidative stress, and controlling plant signal transduction (Huang et al., 2017). Magnesium (Mg) is an indispensable macro-nutrient that is known to reduce soil borne heavy metal induced toxicity in plants. Hydrated radii of Mg^{2+} and Al^{3+} are analogous; hence, both of them strive to bind with ion transporters and additional significant biomolecules. It is established that Mg lessens Al toxicity in several plants by modulating physiological processes (Shen et al., 2016; Li et al., 2019) and by reducing Al saturation as well as its activity in the cell wall (Bose et al., 2011) and plasma membrane (Rahman et al., 2018). Interestingly, reduced micromolar Mg concentrations and perhaps augment organic ligands biosynthesis in dicots, thus, strengthening the reduction of Al induced toxicity (Rengel et al., 2015). Mg ameliorates phytotoxicity of other heavy metals as well, e.g., Cd, Zn, etc. This heavy metal induced phytotoxicity alleviation role played by Mg^{2+} in plants may be linked either with physicochemical processes at the cell surface of the root and/or increased H^+-ATPase activity, organic acid anions synthesis and exudation, segregation of metal ions in cellular compartments, in addition to improved defence against free radicals (Janicka-Russak et al., 2012). Table 3 gives an overview of some of the significant recent studies done to understand the role played by macronutrients in alleviating heavy metals induced phytotoxicity.

Heavy metals (Fe, Cu, Zn, Mn, Mo,Co, and Ni) being indispensable micronutrients are vital for the maintenance of structure formation and the colloidal system as well as acid-base equilibrium. They are also integral components of major enzymes, structural proteins, and hormones (Swain and Rautray, 2021). Several trace elements have a direct role in heavy metal availability and its toxic effect which include decreased soil heavy metals solubility, antagonism of heavy metals with trace elements for the identical membrane transporters, and build-up of heavy metal inside vacuoles (Kolarova and Napiórkowski, 2021). On the other hand, dilution of heavy metal concentration along with lessening of heavy metal stress by enhancing the antioxidant defence system of plants are considered as indirect effects of trace elements on heavy metal induced stress (El-Esawi et al., 2020).

Zn, a significant member of the metal transporter family, has been reported to thwart harm produced due to Cd toxicity (Zhou et al., 2019; Kapur et al., 2019). It also augments the antioxidant enzymes activities and binds with the membrane protein thus, preventing Cd to do the same (Genchi et al., 2020). Alkaline

Table 3: Effect of various macronutrients on heavy metal concentration in plants.

Nutrient	Concentration	Effect on heavy metal	Plant species	Mechanism of Action	References
Nitrogen	Supplementing 7.5 mM of N	Reduces Cd induced photosynthesis inhibitory effect	*Helianthus annus*	Increases Rubisco activity and protein content	Panković et al., 2000
	Augmenting N fertilizer by adding 16 mM $(NH4)_2SO_4$	Decreases Cd induced toxic effect	*Sedum alfredii* Hance		Zhu et al., 2010
	Application of N by addingNH+ -N [100 mg kg⁻¹]	Reduces the leaves Cd concentration	*Oryza sativa*	Enhances the plant antioxidant enzyme system	Jalloh et al., 2009
Potassium	K addition at 60 mg kg⁻¹	Decreases the Cd induced toxicity at 25 mg kg⁻¹	*Cucumis sativus* cultivars	Increases the AsA and GSH. Content	Shen et al., 2000
	Application of KCl and K2SO4 (increasing concentration up to 55 mg kg⁻¹)	60–90% increased Cd (at 15 mg/ kg) accumulation in shoots	Different cultivars of *Triticum aestivum* L.	Anions Cl⁻ and SO_4^{2-}escalates Cd uptake by plants	Zhao et al., 2004
	KNO₃ application (in increasing concentration up to 55 mg kg⁻¹)	Marginal increase in the Cd (at 15 mg Kg⁻¹) content	Different cultivars of *Triticum aestivum* L.		Zhao et al., 2004
Phosphorous	33 kg/ha phosporous pentoxide	Removal of Pb, Ni, B, Mn, and Zn in one growing season	Meadow plants	Reduces oxidative end products	Gullap et al., 2014
Sulphur	12 days supplementation with sulphate	Decreased Cd phytotoxicity (100 μM)	*Zea mays* L.	Increasing S-containing defense compounds (GSH and phytochelatins) synthesis	Astolfi et al., 2004
	Supplementation with S at 40 mg/kg	Decreased Cd-induced toxicity	*Brassica campestris* L.	Improves the AsA–GSH cycle	Anjum et al., 2008
	Optimum Sulphate availability (0 and 1 mM)	Lesser Ni accumulation in exposed plants (0.5 mM)	*Brassica napus* L.	Significantly decreases TBARS concentration and oxidative stress	Sujetovienė and Bučytė, 2021

Table 3 contd. ...

...Table 3 contd.

Nutrient	Concentration	Effect on heavy metal	Plant species	Mechanism of Action	References
Calcium	Application of numerous Ca polypeptide concentration	Negative correlation between Cd concentration and Ca application	*Brassia campestris* L.	By competitive inhibition of Ca on Cd	Chen et al., 2019
	Application of $CaCl_2$ (100 mM) for 3 additional days	Cd-induced oxidative stress	*Cicer arietinum* L.	ReducesH_2O_2 and ~ 2.75 fold rise in thioredoxin (Trx) and thioredoxin reductase (TrxR)	Sakouhi et al., 2021
	Ca supplementation	Decreased Cr (VI) uptake in cells and its sequestration into vacuoles	*Solanum lycopersicum* L. and *Solanum melongena* L.	Reverses Cr (VI) mediated inhibition of nitrate reductase (NR) activity	Singh et al., 2020
Magnesium	24 hrs application of 2 mM Mg	Reduced Cd-induced NO production and Cd accumulation in roots	*Panax notoginseng*	Decreases Cd-accumulation by reducing NR-mediated NO production	Li et al., 2020b
	Addition of mM Mg concentrations in the rooting media	Augments Al induced citrate secretion	*Vigna umbellata*	Restores plasma membrane H^+-ATPase activity	Yang et al., 2007
	Supplementation with 10 mM Mg^{2+}	40% lower Cd shoot concentration	*Brassica rapa* L. var. *perviridis*	~ 2-fold increase in shoot growth	Kashem and Kawai, 2007
	Addition of (1–5 µM) Mg concentrations to 2 mM CaCl2 (pH 6.0) medium	Lessens rhizotoxicity by Zn (at 60 µM)	*Triticum aestivum* L., cv. Yecora Rojo and *Raphanus sativus* L., cv. Cherry Belle	Mg2+mediated intracellular physiological protection	Pedler et al., 2004

phosphatase, phospholipase, Cu-Zn superoxide dismutase (CuZnSOD), RNA polymerase, and carboxypeptidase are principal enzymes associated with Zn (Hafeez et al., 2013) and it is involved in the metabolism of carbohydrate, maintaining cell membrane integrity, synthesis of protein, auxin synthesis, formation of pollens and environmental stress tolerance. Fe is essential for chlorophyll synthesis (Hu et al., 2017) as well as redox systems functioning in the chloroplasts (López-Millán et al., 2016). It is also present in several enzymes or enzyme systems like ferredoxin, cytochromes, nitrate reductase, leghaemoglobin, catalase (CAT), FeSOD, ascorbate peroxidase, etc. (Broadley et al., 2007). Copper (Cu) redox-active transition element, actively takes part in photosynthesis (Aguirre and Pilon, 2016), respiration (Yruela, 2009), C and N metabolism (Senovilla et al., 2018), and protection against oxidative stress (Broadley et al., 2007). Plastocyanin and Cu/Zn SOD are two major copper proteins present in plants (Ravet et al., 2011). Manganese (Mn) is a vital constituent of the photosystem II (Schmidt and Husted, 2019). There is the presence of Mn-containing superoxide dismutase (Wang et al., 2017b) and oxalate oxidase (Leplat et al., 2016). Mn acts as the cofactor for 35 distinct enzymes (Schmidt et al., 2016) and is very important for the tricarboxylic acid cycle (Alejandro et al., 2020). Nickel (Ni) is essential for seed germination and functioning of urease (Fabiano et al., 2015) involved in the activity of methyl coenzyme M reductase, glyoxalase, acireductone dioxygenase, superoxide dismutase, NiFe-hydrogenase, acetyl-CoA decarbonylase synthase, carbon monoxide dehydrogenase, and methylene urease (Broadley et al., 2007). Molybdenum (Mo) is present as a cofactor in a few enzymes, e.g., nitrate reductase (Coelho and Romão, 2015), xanthine oxidoreductase (Okamoto et al., 2013), sulfite oxidase (Kirk and Khadanad, 2020), and aldehyde oxidase (Montefiori et al., 2017). Mo has a critical role in the N metabolism of plants (Kovács et al., 2015). Boron (B) in plants helps in the cell wall and reproductive tissue formation (Shireen et al., 2018; Matthes et al., 2020). Chloride (Cl) stabilizes and maintains the membrane potential (Colmenero-Flores et al., 2019) and pH regulation of plant cells (Franco-Navarro et al., 2016).

Apart from the above mentioned heavy metals, other heavy metals like Cd, Pb, Cr, Hg, and As wield lethal consequences on plants and affect their normal metabolic functioning as they are non-degradable and reach plants via soil. As outcompetes phosphate ion when its concentration is higher and enters plants by transporter proteins (Strawn et al., 2018). As not only restricts growth and biomass development (Srivastava and Sharma, 2014) in plant organs but also leads to sterility (Finnegan and Weihua, 2012). Lead (Pb) hampers seed germination, seedling elongation, chlorophyll synthesis, root elongation, and transpiration (Pourrut et al., 2011). It also obstructs Fe and Mg uptake, deficiency of carbon dioxide as a result of stomatal closing (Pourrut et al., 2011), and changes membrane permeability (Dias et al., 2019). Pb uptake results in plant growth retardation with decreased/no yield (Tiwari and Lata, 2018). Cd causes nutrient imbalance (Irfan et al., 2013) in crops and affects photosynthesis (Song et al., 2019), stomatal conductance and transpiration rate (Benavides et al., 2005). Cd interrupts nitrogen (Genchi et al., 2020) as well as competes with Zn, Mn, and Ca transport and lessens their uptake as well as their consequent translocation (Naciri et al., 2021). Cd inhibits electron transport (Xue et al., 2014), hormonal imbalance (Guo et al., 2019), reduced water potential, stomatal conductance, transpiration (Andrade Júnior et al., 2019), and increased oxidative stress (Gill and Tuteja, 2011). In plants, mercury (Hg) is responsible for altering cell membrane permeability (Frick et al., 2013), and causes nutrient deficiency (Manikandan et al., 2015). Hg affects plants' transpiration, photosynthesis, chlorophyll synthesis in addition to water uptake processes (Azevedo and Rodriguez 2012). Chromium (Cr) inhibits the germination of seeds, inhibits plant growth as well as causes leaf necrosis (Wakeel and Xu, 2020). Cr replaces essential nutrients at cation exchange sites (Ongon'g et al., 2020), disrupts the equilibrium of both macro-and micronutrients, increases the production of ROS, and weakens the plant's antioxidant defense mechanism, e.g., SOD, CAT, dehydrogenase, and glucose-6-phosphate (Oliveira, 2012). The various plant ionomics studies performed in the last decade are summarized in Table 4.

8. Future prospects and conclusion

A huge proportion of the population of developing countries across Asia, Latin America, and Africa, is dependent on farming and its allied work (food, fodder, feed, livestock, pisciculture, etc.) as means of their livelihood. As per FAO, 2018, agriculture alone still provides food and employment to the extremely deprived economically backward section of the society (FAO, 2018). Presently, the COVID19 pandemic situation has created a foremost crisis concerning the food and nutrition security of susceptible populations, thereby, aggravating the serious food shortage that already exists. In this context, the conditions of the agricultural land/soil become more significant, as we know that due to various anthropogenic activities such as incessant industrialization or urbanization the soil profile undergoes detrimental alteration. Heavy metals are leached into the soil which bioaccumulates in plants/crops. When these are consumed, they enter animals and humans and harm them. Therefore, innovative plant-recombinant approaches for heavy metal tolerance must be developed. However, the design of such experiments will have to use an interdisciplinary approach along with a combinatorial "omics", approach, i.e., ionomics, transcriptomics, metabolomics, proteomics, etc. It is worth mentioning here ionomics will have to play a leading role in the identification of possible gene(s) which control the uptake mechanism of plants ions, including the quantification of elements, alterations if any, concerning physiological, growth, genetic, and environmental causes. Ionomics also assist in the functional study of genetic networks affecting the ionome in some way or the other. This calls for the improvement and varied ionomics application as a high-efficiency phenotyping platform. It should also intensify its capability to analyze more samples which is presently approximately 10,000 samples/week with a single analytical instrument (Salt et al., 2004). The sampling time also needs to be minimized and focus on measuring multiple elements should be given which will permit researchers to assess the ionome dynamics in totality. Furthermore, most of the ionomics studies to date do not include a large number of species. This should also be taken care of as different species and families are expected to possess distinct elemental uptake mechanisms (Baxter et al., 2009).

To Summarize, one can state that elemental (macronutrients, micronutrients, and heavy metals) amassing is an intricate process that influences nearly all

Table 4: Selected significant ionomics studies performed in the last decade.

Analytical Techniques	Plant species	Elements Quantified	References
ICP-MS/Laser Ablation ICP-MS	*Arabidopsis thaliana*	complete ionomic profile	Buescher et al., 2010
	Vitis vinifera Angelica sinensis, Bupleurum sinensis, Hypericum perforatum, Curcuma longa, Emblica officinalis, Ocimum sanctum, Garcinia cambogia, Panax ginseng, Mucuna pruriens, Pueraria lobata, Schisandra sinensis, Salvia miltiorrhiza, Scutellaria baicalensis, Terminalia arjuna Siraitia grosvenorii, Terminalia chebula, Bacopa monnieri, and Echinacea purpurea, Cola accuminata	Cd, Al, Pb, As, Ni, Ba and Sb	Filipiak-Szok et al., 2015
	Acia Raddiena and Aerva Javanica	Hg, Cd, Pb, As, Ag, Ni, Cr, Au, Zn, Mo, Mn and Cu	Rashed, 2010
	Pisum sativum	Pb	Hanć et al., 2009
	Oryza sativa	As, Co, Cd, Cr, Ni, Cu and Zn	Zhao et al., 2016
	Alpinia oxyphylla and Morinda officinalis	Mg, Na, Al, Zn, Hg, Fe, Tl, Ba, Cr, Ni, Mn, , As, K, Ca, Se, Pb, Cd, Cu, Mo, and V	Patnaik et al., 2019
	Momordica charantia, Hemidesmus indicus, Andropogan zizanioides, Cyathula prostrate, and Withania somnifera	Se, Rb, Sr, Li, V, Co, Ni, Cr, Zn, Ga, As, Mn, Cs, Fe, Cu, Ag, Be, Al and Ba	Singh et al., 2016
	Eichhornia crassipes, Lemna minor, Spirodela polyrhiza	As, Si, Pb, Cr, Cd and Ni	Gaiss et al., 2019
	Zea mays L.	As, Cd, Hg, Sb and Zn	Sasmaz and Sasmaz, 2017
	Carduus nutans, Alyssum saxatile, Onosma sp., Anchusa arvensis; Verbascum thapsus, Centaurea cyanus, Isatis sp., Phlomis sp., Glaucium flavum and Cynoglossum officinale	Sr	

Table 4 contd. ...

Table 4 contd. ...

Analytical Techniques	Plant species	Elements Quantified	References
ICP-OES	*Chiococca alba* (L.) Hitchc., *Cordia salicifolia*, and *Echites peltata*	Zn, Cu, Fe, Co, Cr, Na, and Cd	Tschinkel et al., 2020
	Solanum tuberosum	Zn, Cu, Pb, Cd	Jusufi et al., 2017
	Rubus fruticosus and *Vitis vinifera* Cultivar Rkatsiteli	Cu, Zn, Ni, Pb, As, and Cd	Alagić et al., 2016
	Cymbopogon citratus	Cr, Pb, Ni, and Cd	Pandey et al., 2020
	Harpagophytum procumbens D.C., *Cynara scolymus* L., "devil's claw" - and "espinheira santa" - *Maytenus ilifolia* (Mart) ex Reiss	all the macro and micronutrients	de Aragão et al., 2020
	Avicennia schaueriana Stapf and Leechman ex Moldenke	Al, Mn, Fe, and Zn	Victório et al., 2020
	Cenchrus ciliaris, Zea mays	As, Pb, Fe, Cu, Cr, Mn, Ni, Ti, Cd, and Zn	do Nascimento Júnior et al., 2021
	Ficus carica	Cd, Cu, Mn, Fe, Ni, Cr Zn, and Pb	Ugulu et al., 2016
	Marsilea crenata, Ipomoea aquatica	As, Pb, Cd, and Cr	Ruchuwararak et al., 2019
	Brassica rapa L.	Cd, Cu, Ni, Cr, Pb, and Mn,	Iqbal et al., 2021
XRF	*Brassica napus* and *Festuca arundinacea*	Pb	Mera et al., 2019
	Brassica oleracea	Ni, Cu, Zn, Mn, Cd, Fe, and Pb	Radulescu et al., 2013
	Achillea nobilis (L.), *Tanacetum vulgare* (L.), *Artemisia austriaca* (Pall. ex. Wild.), and *Ambrosia artemisiifolia* (L.)	Cu, Cr, Mn,Ni, Zn, Pb, and Cd,	Minkina et al., 2017
	Phragmites australis Cav.	Pb, Zn, Cd, Cu, Mn, Cr, and Ni	Minkina et al., 2018
	Calotropis procera	Fe, Sr, Mn, and Zn	Almehdi et al., 2019
	Nerium Oleander L.	S, K, Cl, Ca, Fe, Mn, Cu, Br, Zn, Rb, Ba Sr, and Pb	Santos et al., 2019
NAA	*Mentha piperita* L., *Matricaria chamomilla* L., *Plantago lanceolata* L.; *Lavandula angustifolia* Mill.; *Echinacea angustifolia* DC.; *Coriandrum sativum* L.and *Anethum graveolens* L.	Ba, Co, Sb, Cr, Fe, Rb, Mn, Sr, K, Ca, Mg, Na, V, Zn, Al, and As,	Haidu et al., 2017
	Hypnum cupressiforme, Pleurozium schreberi, Hylocomium splendens	Al, Cr, Cd, Cu, Pb, Fe,Ni, Zn, and V	Stihi et al., 2017
	45 medicinal plants belonging to Lamiaceae	Al, Ba, As, Br, Ce, Ca, Cl, Cr, Co, Cs, Hf, Fe, K, Mg, La, Mo, Na, Mn, Rb, Sc, Sb, Sr, Th Zn U, and Sm	Zinicovscaia et al., 2020
	*Artemisia abyssinica, Celematics Longicanda, Thymus schimperi Echinops Kebericho, Millettia ferrugine*and *Lippia Adolensis*	Mg, Ca, Al, Cl, Mn, Na, K, and V	Birhanu et al., 2015
	Fabronia ciliaris (Brid.) and *Leskea angustata* (Tayl.)	As, Cu, Cr, Zn, Se, Pb, Sb, and Cs	Ávila-Pérez et al., 2018

characteristics of plant growth and development. Ionomics provides precise gene identification, permits novel genetic mapping methods to recognize numerous new loci that control this intricate system. A highly efficient and preciseionomics framework provides a feasible system for investigating numerous physiological and biochemical mechanisms which influence ionome simultaneously. Ionomics act as a genetic screen in deciphering various functional and physiological plant activities like biosynthesis of lipids, cytokinins, phloem transport, response to pathogens, bio-accumulation of numerous mineral nutrients and trace elements as well as heavy metals (Cd, Cr, Pb, Sr, Al, Hg). Ionomics, along with supplementary "omics" techniques (e.g., transcriptomics, proteomics, and metabolite profiling) has the prospect to reduce the emergent space between genotype and its associated phenotypes.

References

Aarabi F, Kusajima M, Tohge T, Konishi T, Gigolashvili T, Takamune M, Sasazaki Y, Watanabe M, Nakashita H, Fernie AR, Saito K, Takahashi H, Hubberten HM, Hoefgen R, and Maruyama-Nakashita A. 2016. Sulfur deficiency-induced repressor proteins optimize glucosinolate biosynthesis in plants. *Sci Adv* 2(10): e1601087.

Ågren GI, and Weih M. 2020. Multi-dimensional plant element stoichiometry—looking beyond carbon, nitrogen, and phosphorus. *Front Plant Sci* 11: 23.

Aguirre G, and Pilon M. 2016. Copper delivery to chloroplast proteins and its regulation. *Front Plant Sci* 6: 1250.

Ahmad P, Abdel Latef AA, Abd_Allah EF, Hashem A, Sarwat M, Anjum NA, and Gucel S. 2016. Calcium and potassium supplementation enhanced growth, osmolyte secondary metabolite production, and enzymatic antioxidant machinery in cadmium-exposed chickpea (*Cicer arietinum* L.). *Front Plant Sci* 7: 513.

Alagić SČ, Tošić SB, Dimitrijević MD, Petrović JV, and Medić DV. 2016. The characterization of heavy metals in the grapevine (*Vitis vinifera*) cultivar Rkatsiteli and wild blackberry (*Rubus fruticosus*) from East Serbia by ICP-OES and BAFs. *Commun Soil Sci Plant Anal* 47(17): 2034–45.

Alcántara E, Romera FJ, Cañete M, and De la Guardia MD. 1994. Effects of heavy metals on both induction and function of root Fe (III) reductase in Fe-deficient cucumber (*Cucumis sativus* L.) plants. *J Exp Bot* 45(12): 1893–8.

Aldon D, Mbengue M, Mazars C, and Galaud JP. 2018. Calcium signalling in plant biotic interactions. *Int J Mol Sci* 19(3): 665.

Alejandro S, Höller S, Meier B, and Peiter E. 2020. Manganese in plants: from acquisition to subcellular allocation. *Front Plant Sci* 11: 300.

Almehdi A, El-Keblawy A, Shehadi I, El-Naggar M, Saadoun I, Mosa KA, and Abhilash PC. 2019. Old leaves accumulate more heavy metals than other parts of the desert shrub Calotropis procera at a traffic-polluted site as assessed by two analytical techniques. *Int J Phytoremediation* 21(12): 1254–62.

Andrade Júnior WV, de Oliveira Neto CF, Santos Filho BG, do Amarante CB, Cruz ED, Okumura RS, Barbosa AV, de Sousa DJ, Teixeira JS, and Botelho AD. 2019. Effect of cadmium on young plants of Virola surinamensis. *AoB Plants* 11(3): plz022.

Anjum NA, Umar S, Ahmad A, Iqbal M, and Khan NA. 2008. Sulphur protects Mustard (*Brassica campestris* L.) from cadmium toxicity by improving leaf ascorbate and glutathione. *Plant Growth Regul* 54: 271–279.

Astolfi S, Zuchi S, and Passera C. 2004. Role of sulphur availability on cadmium-induced changes of nitrogen and sulphur metabolism in maize (*Zea mays* L.) leaves. *J Plant Physiol* 161(7): 795–802.

Ávila-Pérez P, Longoria-Gándara LC, García-Rosales G, Zarazua G, and López-Reyes C. 2018. Monitoring of elements in mosses by instrumental neutron activation analysis and total X-ray fluorescence spectrometry. *J Radioanal Nucl Chem* 317(1): 367–80.

Azevedo R, and Rodriguez E. 2012. Phytotoxicity of mercury in plants: a review. *J Bot* 2012: 1.

Bai G, Jenkins S, Yuan W, Graef GL, and Ge Y. 2018. Field-based scoring of soybean iron deficiency chlorosis using RGB imaging and statistical learning. *Front Plant Sci* 9: 1002.

Baxter I. 2009. Ionomics: studying the social network of mineral nutrients. *Curr Opin Plant Biol* 12(3): 381–386.

Billard V, Ourry A, Maillard A, Garnica M, Coquet L, Jouenne T, Cruz F, Garcia-Mina JM, Yvin JC, and Etienne P. 2014. Copper-deficiency in *Brassica napus* induces copper remobilization, molybdenum accumulation and modification of the expression of chloroplastic proteins. *PLoS One* 9(10): e109889.

Bird JK, Murphy RA, Ciappio ED, and McBurney MI. 2017. Risk of deficiency in multiple concurrent micronutrients in children and adults in the United States. *Nutrients* 9(7): 655.

Birhanu WT, Chaueby AK, Teklemariamc TT, Dewud BM, and Funtua II. 2015. Application of instrumental neutron activation analysis (INAA) in the analysis of essential elements in six endemic Ethiopian medicinal plants. *Int J Sci Basic Appl Res* 19: 213–27.

Bode P. 2012. Opportunities for innovation in neutron activation analysis. *J Radioanal Nucl Chem* 291(2): 275–280.

Bose J, Babourina O, and Rengel Z. 2011. Role of magnesium in alleviation of aluminium toxicity in plants. *J Exp Bot* 62(7): 2251–2264.

Briffa J, Sinagra E, and Blundell R. 2020. Heavy metal pollution in the environment and their toxicological effects on humans. *Heliyon* 6(9): p.e04691.

Broadley MR, White PJ, Hammond JP, Zelko I, and Lux A. 2007. Zinc in plants. *New Phytol* 173: 677–702.

Buescher E, Achberger T, Amusan I, Giannini A, Ochsenfeld C, Rus A, Lahner B, Hoekenga O, Yakubova E, Harper JF, and Guerinot ML. 2010. Natural genetic variation in selected populations of Arabidopsis thaliana is associated with ionomic differences. *PloS one* 5(6): e11081.

Calderón-Celis F, and Encinar JR. 2019. A reflection on the role of ICP-MS in proteomics: Update and future perspective. *J Proteomics* 198: 11–17.

Campos AC, van Dijk WF, Ramakrishna P, Giles T, Korte P, Douglas A, Smith P, and Salt DE. 2021. 1,135 ionomes reveals the global pattern of leaf and seed mineral nutrient and trace element diversity in Arabidopsis thaliana. *Plant J*.

Carstensen A, Herdean A, Schmidt SB, Sharma A, Spetea C, Pribil M, and Husted S. 2018. The impacts of phosphorus deficiency on the photosynthetic electron transport chain. *Plant Physiol* 177(1): 271–284.

Chao DY, Baraniecka P, Danku J, Koprivova A, Lahner B, Luo H, Yakubova E, Dilkes B, Kopriva S, and Salt DE. 2014. Variation in sulfur and selenium accumulation is controlled by naturally occurring isoforms of the key sulfur assimilation enzyme adenosine 5′-phosphosulfate reductase2 across the Arabidopsis species range. *Plant Physiol* 166(3): 1593–1608.

Chao L, Bofu P, Weiqian C, Yun L, Hao H, Liang C, Xiaoqing L, Xiao W, and Fashui H. 2008. Influences of calcium deficiency and cerium on growth of spinach plants. *Biol Trace Elem Res* 121(3): 266–275.

Chatterjee J, and Chatterjee C. 2000. Phytotoxicity of cobalt, chromium and copper in cauliflower. *Environ Pollut* 109(1): 69–74.

Chen C, Huang D, and Liu J. 2009. Functions and toxicity of nickel in plants: recent advances and future prospects. *Clean (Weinh)* 37(4-5): 304–13.

Chen H, Rogalski MM, and Anker JN. 2012. Advances in functional X-ray imaging techniques and contrast agents. *Phys Chem Chem Phys* 14(39): 13469–13486.

Chen H, Shu F, Yang S, Li Y, and Wang S. 2019. Competitive inhibitory effect of calcium polypeptides on Cd enrichment of *Brassia campestris* L. *Int J Environ Res Public Health* 16(22): 4472.

Chen ZR, Kuang L, Gao YQ, Wang YL, Salt DE, and Chao DY. 2018. AtHMA4 drives natural variation in leaf Zn concentration of Arabidopsis thaliana. *Front Plant Sci* 9: 270.

Coelho C, and Romão MJ. 2015. Structural and mechanistic insights on nitrate reductases. *Protein Sci* 24(12): 1901–1911.

Colmenero-Flores JM, Franco-Navarro JD, Cubero-Font P, Peinado-Torrubia P, and Rosales MA. 2019. Chloride as a beneficial macronutrient in higher plants: New roles and regulation. *Int J Mol Sci* 20(19): 4686.

Comadira G, Rasool B, Karpinska B, Morris J, Verrall SR, Hedley PE, Foyer CH, and Hancock RD. 2015. Nitrogen deficiency in barley (*Hordeum vulgare*) seedlings induces molecular and metabolic adjustments that trigger aphid resistance. *J Exp Bot* 66(12): 3639–3655.

Cruz AF, Almeida GM, Wadt PG, Pires MD, and Ramos ML. 2019. Seasonal variation of plant mineral nutrition in fruit trees. *Braz Arch Biol Technol* 62.

Danku JM, Lahner B, Yakubova E, and Salt DE. 2013. Large-scale plant ionomics. *Methods Mol Biol* 953: 255–76.

de Aragão Tannus C, de Souza Dias F, Santana FB, Dos Santos DC, Magalhães HI, de Souza Dias F, and Júnior AD. 2020. Multielement determination in medicinal plants and herbal medicines containing *Cynara scolymus* L., Harpagophytum procumbens DC, and Maytenus ilifolia (Mart.) ex Reiss from Brazil using ICP OES. *Biol Trace Elem Res* 13: 1–2.

de Bang TC, Husted S, Laursen KH, Persson DP, and Schjoerring JK. 2021. The molecular-physiological functions of mineral macronutrients and their consequences for deficiency symptoms in plants. *New Phytol* 229(5): 2446–2469.

Dias MC, Mariz-Ponte N, and Santos C. 2019. Lead induces oxidative stress in *Pisum sativum* plants and changes the levels of phytohormones with antioxidant role. *Plant Physiol Biochem* 137: 121–129.

Dimkpa CO and Bindraban PS. 2016. Fortification of micronutrients for efficient agronomic production: a review. *Agron Sustain Dev* 36(1): 7.

do Nascimento Júnior AL, Paiva AD, Souza LD, Souza-Filho LF, Souza LD, Fernandes Filho EI, Schaefer CE, da Silva EF, Fernandes AC, and Xavier FD. 2021. Heavy metals distribution in different parts of cultivated and native plants and their relationship with soil content. *Int J Environ Sci Technol* 18: 1–6.

Dube BK, Tewari K, Chatterjee J, and Chatterjee C. 2003. Excess chromium alters uptake and translocation of certain nutrients in citrullus. *Chemosphere* 53(9): 1147–53.

El-Deftar MM, Robertson J, Foster S, and Lennard C. 2015. Evaluation of elemental profiling methods, including laser-induced breakdown spectroscopy (LIBS), for the differentiation of *Cannabis* plant material grown in different nutrient solutions. *Forensic Sci Int* 251: 95–106.

El-Esawi MA, Sinha RP, Chauhan DK, Tripathi DK, and Pathak J. 2020. Role of ionomics in plant abiotic stress tolerance. pp. 835–860. *In*: Tripathi DK, Singh VP, Chauhan DK, Sharma S, Prasad SM, Dubey NK, and Ramawat N (eds.). *Plant Life Under Changing Environment*. Academic Press.

Emamverdian A, Ding Y, Mokhberdoran F, and Xie Y. 2015. Heavy metal stress and some mechanisms of plant defense response. *Sci World J* 2015: Article ID 756120.

Emamverdian A, Ding Y, Xie Y, and Sangari S. 2018. Silicon mechanisms to ameliorate heavy metal stress in plants. *Bio Med Res Int* 2018: Article ID 8492898..

Fabiano CC, Tezotto T, Favarin JL, Polacco JC, and Mazzafera P. 2015. Essentiality of nickel in plants: a role in plant stresses. *Front Plant Sci* 6: 754.

FAO. 2018. Sustainable Food Systems—Concepts and Framework. Rome: Food and Agriculture Organisation of the United Nations Available from: http://www.fao.org/3/ca2079en/CA2079EN.pdf.Accessed on: 08.05.2021.

Farzadfar S, Zarinkamar F, Modarres-Sanavy SA, and Hojati M. 2013. Exogenously applied calcium alleviates cadmium toxicity in *Matricaria chamomilla* L. plants. *Environ Sci Pollut Res Int* 20(3): 1413–1422.

Filipiak-Szok A, Kurzawa M, and Szłyk E. 2015. Determination of toxic metals by ICP-MS in Asiatic and European medicinal plants and dietary supplements. *J Trace Elem Med Biol* 30: 54–8.

Finnegan P, and Weihua C. 2012. Arsenic toxicity: the effects on plant metabolism. *Front Physiol* 3: 182.

Fischer S, Sánchez-Bermejo E, Xu X, Flis P, Ramakrishna P, Guerinot ML, Zhao FJ, and Salt DE. 2021. Targeted expression of the arsenate reductase HAC1 identifies cell type specificity of arsenic metabolism and transport in plant roots. *J Exp Bot* 72(2): 415–425.

Fleet JC, Replogle R, and Salt DE. 2011. Systems genetics of mineral metabolism. *J Nutr* 141(3): 520–5.

Flowers TJ, Glenn EP, and Volkov V. 2019. Could vesicular transport of Na+ and Cl–be a feature of salt tolerance in halophytes? *Ann Bot* 123(1): 1–8.

Franco-Navarro JD, Brumós J, Rosales MA, Cubero-Font P, Talón M, and Colmenero-Flores JM. 2016. Chloride regulates leaf cell size and water relations in tobacco plants. *J Exp Bot* 67(3): 873–891.

Frick A, Järvå M, Ekvall M, Uzdavinys P, Nyblom M, and Törnroth-Horsefield S. 2013. Mercury increases water permeability of a plant aquaporin through a non-cysteine-related mechanism. *Biochem J* 454(3): 491–499.

Gaiss S, Amarasiriwardena D, Alexander D, and Wu F. 2019. Tissue level distribution of toxic and essential elements during the germination stage of corn seeds (*Zea mays* L.) using LA-ICP-MS. *Environ Pollut* 252: 657–65.

Genchi G, Sinicropi MS, Lauria G, Carocci A, and Catalano A. 2020. The effects of cadmium toxicity. *Int J Environ Res Public Health* 17(11): 3782.

Gernand AD, Schulze KJ, Stewart CP, West KP, and Christian P. 2016. Micronutrient deficiencies in pregnancy worldwide: health effects and prevention. *Nat Rev Endocrinol* 12(5): 274–89.

Gill SS, and Tuteja N. 2011. Cadmium stress tolerance in crop plants: probing the role of sulfur. *Plant Signal Behav* 6(2): 215–222.

Gopal R, Sharma YK, and Shukla AK. 2016. Effect of molybdenum stress on growth, yield and seed quality in black gram. *J Plant Nutr* 39(4): 463–469.

Gorji A, and Ghadiri MK. 2020. The potential roles of micronutrient deficiency and immune system dysfunction in COVID-19 pandemic. *Nutrition* 6: 111047.

Grobelak A, and Napora A. 2015. The chemophytostabilisation process of heavy metal polluted soil. *PLoS One* 10(6): e0129538.

Guerriero G, Hausman JF, and Legay S. 2016. Silicon and the plant extracellular matrix. *Front Plant Sci* 7: 463.

Gullap MK, Dasci M, Erkovan Hİ, Koc A, and Turan M. 2014. Plant growth-promoting rhizobacteria (PGPR) and phosphorus fertilizer-assisted phytoextraction of toxic heavy metals from contaminated soils. *Commun Soil Sci Plant Anal* 45(19): 2593–606.

Gulzar S, Hassan A, and Nawchoo IA. 2020. A review of nutrient stress modifications in plants, alleviation strategies, and monitoring through remote sensing. *In*: Aftab T, and Hakeem KR (eds.). *Plant Micronutrients*. Springer, Cham.

Guo J, Qin S, Rengel Z, Gao W, Nie Z, Liu H, Li C, and Zhao P. 2019. Cadmium stress increases antioxidant enzyme activities and decreases endogenous hormone concentrations more in Cd-tolerant than Cd-sensitive wheat varieties. *Ecotoxicol Environ Saf* 172: 380–387.

Hafeez B, Khanif YM, and Saleem M. 2013. Role of zinc in plant nutrition—a review. *Am J Exp Agric* 3(2): 374–391.

Haidu D, Párkányi D, Moldovan RI, Savii C, Pinzaru I, Dehelean C, and Kurunczi L. 2017. Elemental characterization of Romanian crop medicinal plants by neutron activation analysis. *J Anal Methods Chem* 2017: Article ID 9748413.

Hailu AA, and Addis G. 2016. The content and bioavailability of mineral nutrients of selected wild and traditional edible plants as affected by household preparation methods practiced by local community in Benishangul Gumuz Regional State, Ethiopia. *Int J Food Sci* 2016: 7615853.

Hanć A, Barałkiewicz D, Piechalak A, Tomaszewska B, Wagner B, and Bulska E. 2009. An analysis of long-distance root to leaf transport of lead in *Pisum sativum* plants by laser ablation–ICP–MS. *Int J Environ Anal Chem* 89(8-12): 651–9.

Hepler PK. 2005. Calcium: a central regulator of plant growth and development. *Plant Cell* 17(8): 2142–2155.

Hu W, Pan X, Li F, and Dong W. 2018. UPLC-QTOF-MS metabolomics analysis revealed the contributions of metabolites to the pathogenesis of *Rhizoctonia solani* strain AG-1-IA. *PLoS one* 13(2): e0192486.

Hu X, Page MT, Sumida A, Tanaka A, Terry MJ, and Tanaka R. 2017. The iron–sulfur cluster biosynthesis protein SUFB is required for chlorophyll synthesis, but not phytochrome signaling. *Plant J* 89(6): 1184–1194.

Huang CH, Singh GP, Park SH, Chua NH, Ram RJ, and Park BS. 2020. Early diagnosis and management of nitrogen deficiency in plants utilizing Raman spectroscopy. *Front Plant Sci* 11: 663.

Huang D, Gong X, Liu Y, Zeng G, Lai C, Bashir H, Zhou L, Wang D, Xu P, Cheng M, and Wan J. 2017. Effects of calcium at toxic concentrations of cadmium in plants. *Planta* 245(5): 863–873.

Huang XY, and Salt DE. 2016. Plant ionomics: from elemental profiling to environmental adaptation. *Mol Plant* 9(6): 787–797.

Hwalla N, Al Dhaheri AS, Radwan H, Alfawaz HA, Fouda MA, Al-Daghri NM, Zaghloul S, and Blumberg JB. 2017. The prevalence of micronutrient deficiencies and inadequacies in the Middle East and approaches to interventions. *Nutrients* 9(3): 229.

Iqbal Z, Abbas F, Ibrahim M, Qureshi TI, Gul M, and Mahmood A. 2021. Assessment of heavy metal pollution in *Brassica* plants and their impact on animal health in Punjab, Pakistan. *Environ Sci Pollut Res Int* 10: 1–1.

Irfan M, Hayat S, Ahmad A, and Alyemeni MN. 2013. Soil cadmium enrichment: Allocation and plant physiological manifestations. *Saudi J Biol Sci* 20(1): 1–10.

Jaishankar M, Tseten T, Anbalagan N, Mathew BB, and Beeregowda KN. 2014. Toxicity, mechanism and health effects of some heavy metals. *Interdiscip Toxicol* 7(2): 60–72.

Jalloh MA, Chen J, Zhen F, and Zhang G. 2009. Effect of different N fertilizer forms on anti-oxidant capacity and grain yield of rice growing under Cd stress. *J Hazard Mater* 162: 1081–1085.

Jalmi SK, Bhagat PK, Verma D, Noryang S, Tayyeba S, Singh K, Sharma D, and Sinha AK. 2018. Traversing the links between heavy metal stress and plant signaling. *Front Plant Sci* 9: 12.

Jan AT, Azam M, Siddiqui K, Ali A, Choi I, and Haq QM. 2015. Heavy metals and human health: mechanistic insight into toxicity and counter defense system of antioxidants. *Int J Mol Sci* 16(12): 29592–630.

Janicka-Russak M, Kabala K, and Burzynski M. 2012. Different effect of cadmium and copper on H+-ATPase activity in plasma membrane vesicles from *Cucumis sativus* roots. *J Exp Bot* 63: 4133–4142.

Jiang XJ, Luo YM, Liu Q, Liu SL, and Zhao QG. 2004. Effects of cadmium on nutrient uptake and translocation by Indian Mustard. *Environ Geochem Health* 26(2): 319–24.

Jusufi K, Stafilov T, Vasjari M, Korca B, Halili J, and Berisha A. 2017. Measuring the presence of heavy metals and their bioavailability in potato crops around Kosovo's power plants. *Fresen Environ Bull* 26(2): 1682–6.

Kapur D, and Singh KJ. 2019. Zinc alleviates cadmium induced heavy metal stress by stimulating antioxidative defense in soybean [*Glycine max* (L.) Merr.] crop. *J Appl Nat Sci* 11(2): 338–345.

Karahan F, Ozyigit II, Saracoglu IA, Yalcin IE, Ozyigit AH, and Ilcim A. 2019. Heavy metal levels and mineral nutrient status in different parts of various medicinal plants collected from eastern Mediterranean region of Turkey. *Biol Trace Elem Res* 9: 1–4.

Kashem MA, and Kawai S. 2007. Alleviation of cadmium phytotoxicity by magnesium in Japanese mustard spinach. *Soil Sci Plant Nutr* 53: 246–251.

Ke W, Xiong ZT, Chen S, and Chen J. 2007. Effects of copper and mineral nutrition on growth, copper accumulation and mineral element uptake in two Rumex japonicus populations from a copper mine and an uncontaminated field sites. *Environ Exp Bot* 59(1): 59–67.

Kenderešová L, Staňová A, Pavlovkin J, Ďurišová E, Nadubinská M, Čiamporová M, and Ovečka M. 2012. Early Zn2+-induced effects on membrane potential account for primary heavy metal susceptibility in tolerant and sensitive *Arabidopsis* species. *Ann Bot* 110(2): 445–59.

Keyster M, Niekerk LA, Basson G, Carelse M, Bakare O, Ludidi N, Klein A, Mekuto L, and Gokul A. 2020. Decoding heavy metal stress signalling in plants: towards improved food security and safety. *Plants* 9(12): 1781.

Kim HS, Kim ET, Eom JS, Choi YY, Lee SJ, Lee SS, Chung CD, and Lee SS. 2021. Exploration of metabolite profiles in the biofluids of dairy cows by proton nuclear magnetic resonance analysis. *PloS one* 16(1): e0246290.

Kirk ML and Kc K. 2020. Molybdenum and tungsten cofactors and the reactions they catalyze. pp. 313–342. *In*: Transition Metals and Sulfur–A Strong Relationship for Life 9. De Gruyter.

Koch M, Naumann M, Pawelzik E, Gransee A, and Thiel H. 2020. The importance of nutrient management for potato production part I: Plant Nutrition and Yield. *Potato Res* 63(1): 97–119.

Kolarova N, and Napiórkowski P. 2021. Trace elements in aquatic environment. Origin, distribution, assessment and toxicity effect for the aquatic biota. *Ecohydrol Hydrobiol*. https://doi.org/10.1016/j.ecohyd.2021.02.002.

Kovács B, Puskás-Preszner A, Huzsvai L, Lévai L, and Bódi É. 2015. Effect of molybdenum treatment on molybdenum concentration and nitrate reduction in maize seedlings. *Plant Physiol Biochem* 96: 38–44.

Kulczycki G, and Sacała E. 2020. Sulfur application alleviates chromium stress in maize and wheat. *Open Chem* 18(1): 1093–104.

Kushwaha BK, and Singh VP. 2020. Glutathione and hydrogen sulfide are required for sulfur-mediated mitigation of Cr (VI) toxicity in tomato, pea and brinjal seedlings. *Physiol Plant* 168(2): 406–21.

Lahner B, Gong J, Mahmoudian M, Smith EL, Abid KB, Rogers EE, Guerinot ML, Harper JF, Ward JM, McIntyre L, and Schroeder JI. 2003. Genomic scale profiling of nutrient and trace elements in Arabidopsis thaliana. *Nat Biotechnol* 21(10): 1215–21.

Lamhamdi M, El Galiou O, Bakrim A, Nóvoa-Muñoz JC, Arias-Estévez M, Aarab A, and Lafont R. 2013. Effect of lead stress on mineral content and growth of wheat (*Triticum aestivum*) and spinach (*Spinacia oleracea*) seedlings. *Saudi J Biol Sci* 20(1): 29–36.

Leplat F, Pedas PR, Rasmussen SK, and Husted S. 2016. Identification of manganese efficiency candidate genes in winter barley (*Hordeum vulgare*) using genome wide association mapping. *BMC genomics* 17(1): 805.

Li D, Ma W, Wei J, Mao Y, Peng Z, Zhang J, Kong X, Han Q, Fan W, Yang Y, and Chen J. 2019. Magnesium promotes root growth and increases aluminum tolerance via modulation of nitric oxide production in Arabidopsis. *Plant Soil*, 1–13.

Li Q, Gao Y, and Yang A. 2020a. Sulfur homeostasis in plants. *Int J Mol Sci* 21(23): 8926.

Li D, Xiao S, Ma WN, Peng Z, Khan D, Yang Q, Wang X, Kong X, Zhang B, Yang E, Rengel Z, Wang J, Cui X, and Chen Q. 2020b. Magnesium reduces cadmium accumulation by decreasing the nitrate reductase-mediated nitric oxide production in Panax notoginseng roots. *J Plant Physiol* 248: 153131.

Li M, Zhao Z, Zhang Z, Zhang W, Zhou J, Xu F, and Liu X. 2017. Effect of boron deficiency on anatomical structure and chemical composition of petioles and photosynthesis of leaves in cotton (*Gossypium hirsutum* L.). *Sci Rep* 7(1): 4420.

Liu CH, Chao YY, and Kao CH. 2013. Effect of potassium deficiency on antioxidant status and cadmium toxicity in rice seedlings. *Bot Stud* 54(1): 2.

Liu H, Zhao H, Wu L, Liu A, Zhao FJ, and Xu W. 2017. Heavy metal ATPase 3 (HMA3) confers cadmium hypertolerance on the cadmium/zinc hyperaccumulator *Sedum plumbizincicola*. *New Phytol* 215(2): 687–698.

Liu X, Feng HY, Fu JW, Chen Y, Liu Y, and Ma LQ. 2018. Arsenic-induced nutrient uptake in As-hyperaccumulator *Pteris vittata* and their potential role to enhance plant growth. *Chemosphere* 198: 425–31.

Llamas A, Ullrich CI, and Sanz A. 2000. Cd2+ effects on transmembrane electrical potential difference, respiration and membrane permeability of rice (*Oryza sativa* L.) roots. *Plant soil* 219(1): 21–8.

Llamas A, Ullrich CI, and Sanz A. 2008. Ni^{2+} toxicity in rice: effect on membrane functionality and plant water content. *Plant Physiol Biochem* 46(10): 905–10.

Lobinski R, Becker JS, Haraguchi H, and Sarkar B. 2010. Metallomics: Guidelines for terminology and critical evaluation of analytical chemistry approaches (IUPAC Technical Report). *Pure Appl Chem* 82(2): 493–504.

López-Bucio J, Hernández-Madrigal F, Cervantes C, Ortiz-Castro R, Carreón-Abud Y, and Martínez-Trujillo M. 2014. Phosphate relieves chromium toxicity in *Arabidopsis thaliana* plants by interfering with chromate uptake. *Biometals* 27(2): 363–70.

López-Millán AF, Duy D, and Philippar K. 2016. Chloroplast iron transport proteins–function and impact on plant physiology. *Front Plant Sci* 7: 178.

Lu C, Yuan F, Guo J, Han G, Wang C, Chen M, and Wang B. 2021. Current understanding of role of vesicular transport in salt secretion by salt glands in recretohalophytes. *Int J Mol Sci* 22(4): 2203.

Luo S, Calderon-Urrea A, Jihua YU, Liao W, Xie J, Lv J, Feng Z, and Tang Z. 2020. The role of hydrogen sulfide in plant alleviates heavy metal stress. *Plant Soil* 449(1): 1–0.

Lyons G, Dean G, Tongaiaba R, Halavatau S, Nakabuta K, Lonalona M, and Susumu G. 2020. Macro-and micronutrients from traditional food plants could improve nutrition and reduce non-communicable diseases of islanders on atolls in the South Pacific. *Plants* 9(8): 942.

Maillard A, Etienne P, Diquélou S, Trouverie J, Billard V, Yvin JC, and Ourry A. 2016. Nutrient deficiencies modify the ionomic composition of plant tissues: a focus on cross-talk between molybdenum and other nutrients in *Brassica napus*. *J Exp Bot* 67(19): 5631–41.

Manikandan R, Sahi SV, and Venkatachalam P. 2015. Impact assessment of mercury accumulation and biochemical and molecular response of Mentha arvensis: a potential hyperaccumulator plant. *Sci World J* 2015: 715217.

Marques DJ, Broetto F, Ferreira MM, Lobato AK, Ávila FW, and Pereira FJ. 2014. Effect of potassium sources on the antioxidant activity of eggplant[1]. *Bras Ciênc Solo* 38(6): 1836–42.

Matsuoka K. 2020. Methods for nutrient diagnosis of fruit trees early in the growing season by using simultaneous multi-element analysis. *Hort J* 2020: UTD-R006.

Matthes MS, Robil JM, and McSteen P. 2020. From element to development: the power of the essential micronutrient boron to shape morphological processes in plants. *J Exp Bot* 71(5): 1681–1693.

Mera MF, Rubio M, Perez CA, Cazon S, Merlo M, and Munoz SE. 2019. SR induced micro-XRF for studying the spatial distribution of Pb in plants used for soil phytoremediation. *Radiat Phys Chem* 154: 69–73.

Minkina TM, Mandzhieva SS, Chaplygin VA, Bauer TV, Burachevskaya MV, Nevidomskaya DG, Sushkova SN, Sherstnev AK, and Zamulina IV. 2017. Content and distribution of heavy metals in herbaceous plants under the effect of industrial aerosol emissions. *J Geochem Explor* 174: 113–20.

Minkina T, Fedorenko G, Nevidomskaya D, Fedorenko A, Chaplygin V, and Mandzhieva S. 2018. Morphological and anatomical changes of *Phragmites australis* Cav. due to the uptake and accumulation of heavy metals from polluted soils. *Sci Total Environ* 636: 392–401.

Montefiori M, Jørgensen FS, and Olsen L. 2017. Aldehyde oxidase: reaction mechanism and prediction of site of metabolism. *Acs Omega* 2(8): 4237–4244.

Myśliwa-Kurdziel B, Prasad MNV, and Strzałtka K.(2004. Photosynthesis in heavy metal stressed plants. pp. 146–181. *In*: Prasad MNV (eds.). *Heavy Metal Stress in Plants*. Springer, Berlin, Heidelberg.

Naciri R, Lahrir M, Benadis C, Chtouki M, and Oukarroum A. 2021. Interactive effect of potassium and cadmium on growth, root morphology and chlorophyll a fluorescence in tomato plant. *Sci Rep* 11(1): 1–10.

Nam KH, Kim HJ, Pack IS, Kim HJ, Chung YS, Kim SY, and Kim CG. 2019. Global metabolite profiling based on GC–MS and LC–MS/MS analyses in ABF3-overexpressing soybean with enhanced drought tolerance. *Applied Biological Chemistry* 2019 Dec; 62(1): 1–9.

Obata H, Inoue N, and Umebayashi M. 1996. Effect of Cd on plasma membrane ATPase from plant roots differing in tolerance to Cd. *Soil Sci Plant Nutr* 42(2): 361–6.

Ogra Y. 2015. Development of metallomics research on environmental toxicology. *J Pharm Soc Jpn* 135(2): 307–14.

Okamoto K, Kusano T, and Nishino T. 2013. Chemical nature and reaction mechanisms of the molybdenum cofactor of xanthine oxidoreductase. *Curr Pharm Des* 19(14): 2606–2614.

Okereafor U, Makhatha M, Mekuto L, Uche-Okereafor N, Sebola T, and Mavumengwana V. 2020. Toxic metal implications on agricultural soils, plants, animals, aquatic life and human health. *Int J Environ Res Public Health* 17(7): 2204.

Oliveira H. 2012. Chromium as an environmental pollutant: insights on induced plant toxicity. *J Bot* 2012: 1.

Ongon'g RO, Edokpayi JN, Msagati TAM, Tavengwa NT, Ijoma GN, and Odiyo JO. 2020. The potential health risk associated with edible vegetables grown on Cr (VI) Polluted soils. *Int J Environ Res Public Health* 17(2): 470.

Osmolovskaya NG, Dung VV, Kudryashova ZK, Kuchaeva LN, and Popova NF. 2018. Effect of cadmium on distribution of potassium, calcium, magnesium, and oxalate accumulation in *Amaranthus cruentus* L. Plants. *Russ J Plant Physiol* 65(4): 553–62.

Osu Charles I, and Onyema MO. 2016. Vanadium inhibition capacity on nutrients and heavy metal uptake by *Cucumis Sativus*. *J Am Sci*, 12.

Othman L, Nafadi A, Alkhalid SH, and Mazraani N. 2021. Arsenic poisoning due to high consumption of canned sardines in Jeddah, Saudi Arabia. *Cureus* 13(1): e12780.

Pandey J, Sarkar S, Verma RK, and Singh S. 2020. Sub-cellular localization and quantitative estimation of heavy metals in lemongrass plants grown in multi-metal contaminated tannery sludge. *S Afr J Bot* 131: 74–83.

Pandotra P, Viz B, Ram G, Gupta AP, and Gupta S. 2015. Multi-elemental profiling and chemo-metric validation revealed nutritional qualities of *Zingiber officinale*. *Ecotoxicol Environ Saf* 114: 222–31.

Panković D, Plesničar M, Arsenijević-Maksimović I, Petrović N, Sakač Z, and Kastori R. 2000. Effects of nitrogen nutrition on photosynthesis in Cd-treated sunflower plants. *Ann Bot* 86(4): 841–7.

Parent SÉ, Leblanc MA, Parent AC, Coulibali Z, and Parent LE. 2017. Site-specific multilevel modeling of potato response to nitrogen fertilization. *Front Environ Sci* 5: 81.

Pathak J, Ahmed H, Kumari N, Pandey A, Rajneesh, and Sinha RP. 2020. Role of calcium and potassium in amelioration of environmental stress in plants. *In:* Roychoudhury A, and Tripathi DK (eds.). *Protective Chemical Agents in the Amelioration of Plant Abiotic Stress: Biochemical and Molecular Perspectives.* John Wiley and Sons.

Patnaik PS, Ramanaiah M, and Ramaraju B. 2019. Quantitative determination of essential and trace element content of some medicinal plants by ICP-MS technique. *Res J Pharm Technol* 12(4): 1595–600.

Pauli D, Ziegler G, Ren M, Jenks MA, Hunsaker DJ, Zhang M, Baxter I, and Gore MA. 2018. Multivariate analysis of the cotton seed ionome reveals a shared genetic architecture. *G3 (Bethesda)* 8(4): 1147–60.

Pedersen O, Sauter M, Colmer TD, and Nakazono M. 2021. Regulation of root adaptive anatomical and morphological traits during low soil oxygen. *New Phytol* 229(1): 42–49.

Pedler JF, Kinraide TB, and Parker DR. 2004. Zinc rhizotoxicity in wheat and radish is alleviated by micromolar levels of magnesium and potassium in solution culture. *Plant Soil* 259: 191–199.

Peng HY, Qi YP, Lee J, Yang LT, Guo P, Jiang HX, and Chen LS. 2015. Proteomic analysis of *Citrus sinensis* roots and leaves in response to long-term magnesium-deficiency. *BMC Genomics* 16: 253.

Pietrini F, Carnevale M, Beni C, Zacchini M, Gallucci F, and Santangelo E. 2019. Effect of different copper levels on growth and morpho-physiological parameters in giant reed (*Arundo donax* L.) in Semi-Hydroponic Mesocosm Experiment. *Water* 11(9): 1837.

Pita-Barbosa A, Ricachenevsky FK, and Flis PM. 2019. One "OMICS" to integrate them all: ionomics as a result of plant genetics, physiology and evolution. *Theor Exp Plant Physiol* 31(1): 71–89.

Plaxton WC, and Tran HT. 2011. Metabolic adaptations of phosphate-starved plants. *Plant Physiol* 156(3): 1006–1015.

Pourrut B, Shahid M, Dumat C, Winterton P, and Pinelli E. 2011. Lead uptake, toxicity, and detoxification in plants. *Rev Environ Contam Toxicol* 213: 113–136.

Punshon T, Ricachenevsky FK, Hindt M, Socha AL, and Zuber H. 2013. Methodological approaches for using synchrotron X-ray fluorescence (SXRF) imaging as a tool in ionomics: examples from Arabidopsis thaliana. *Metallomics* 5(9): 1133–1145.

Radulescu C, Stihi C, Popescu IV, Dulama ID, Chelarescu ED, and Chilian A. 2013. Heavy metal accumulation and translocation in different parts of *Brassica oleracea* L. *Rom J Phys* 58(9-10): 1337–54.

Rahman MA, Lee SH, Ji HC, Kabir AH, Jones CS, and Lee KW. 2018. Importance of mineral nutrition for mitigating aluminum toxicity in plants on acidic soils: current status and opportunities. *Int J Mol Sci* 19(10): 3073.

Rashed MN. 2010. Monitoring of contaminated toxic and heavy metals, from mine tailings through age accumulation, in soil and some wild plants at Southeast Egypt. *J Hazard Mater* 178(1-3): 739–46.

Ravet K, Danford FL, Dihle A, Pittarello M, and Pilon M. 2011. Spatiotemporal analysis of copper homeostasis in *Populus trichocarpa* reveals an integrated molecular remodeling for a preferential allocation of copper to plastocyanin in the chloroplasts of developing leaves. *Plant Physiol* 157(3): 1300–12.

Ravilious GE, and Jez JM. 2012. Structural biology of plant sulfur metabolism: from assimilation to biosynthesis. *Nat Prod Rep* 29(10): 1138–52.

Rengel Z, Bose J, Chen Q, and Tripathi BN. 2016. Magnesium alleviates plant toxicity of aluminium and heavy metals. *Crop Pasture Sci* 66(12): 1298–307.

Ricachenevsky FK, Punshon T, Lee S, Oliveira BH, Trenz TS, Maraschin FD, Hindt MN, Danku J, Salt DE, Fett JP, and Guerinot ML. 2018. Elemental profiling of rice FOX lines leads to characterization of a new Zn plasma membrane transporter, OsZIP7. *Front Plant Sci* 9: 865.

Ruchuwararak P, Intamat S, Tengjaroenkul B, and Neeratanaphan L. 2019. Bioaccumulation of heavy metals in local edible plants near a municipal landfill and the related human health risk assessment. *Hum Ecol Risk Assess* 25(7): 1760–72.

Sakouhi L, Kharbech O, Massoud MB, Gharsallah C, Hassine SB, Munemasa S, Murata Y, and Chaoui A. 2021. Calcium and ethylene glycol tetraacetic acid mitigate toxicity and alteration of gene expression associated with cadmium stress in chickpea (*Cicer arietinum* L.) shoots. *Protoplasma.* doi: 10.1007/s00709-020-01605-x.

Salt DE. 2004. Update on plant ionomics. *Plant Physiol* 136(1): 2451–6.

Salt DE, Baxter I, and Lahner B. 2008. Ionomics and the study of the plant ionome. *Annu Rev Plant Biol* 59: 709–33.

Santos RS, Sanches FA, Leitão RG, Leitão CC, Oliveira DF, Anjos MJ, and Assis JT. 2019. Multielemental analysis in *Nerium Oleander* L. leaves as a way of assessing the levels of urban air pollution by heavy metals. *Appl Radiat Isotopes* 152: 18–24.

Sasmaz M, and Sasmaz A. 2017. The accumulation of strontium by native plants grown on Gumuskoy mining soils. *J Geochem Explor* 181: 236–42.

Satbhai SB, Setzer C, Freynschlag F, Slovak R, Kerdaffrec E, and Busch W. 2017. Natural allelic variation of FRO2 modulates Arabidopsis root growth under iron deficiency. *Nat Commun* 8(1): 15603.

Schmidt SB, Jensen PE, and Husted S. 2016. Manganese deficiency in plants: the impact on photosystem II. *Trends Plant Sci* 21(7): 622.

Schmidt SB, and Husted S. 2019. The biochemical properties of manganese in plants. *Plants* 8(10): 381.

Schwenke GD, Simpfendorfer SR, and Collard BCY. 2015. Confirmation of chloride deficiency as the cause of leaf spotting in durum wheat grown in the Australian northern grains region. *Crop Pasture Sci* 66: 122–134.

Seneviratne M, Rajakaruna N, Rizwan M, Madawala HM, Ok YS, and Vithanage M. 2019. Heavy metal-induced oxidative stress on seed germination and seedling development: a critical review. *Environ Geochem Health* 41(4): 1813–31.

Senovilla M, Castro-Rodríguez R, Abreu I, Escudero V, Kryvoruchko I, Udvardi MK, Imperial J, and González-Guerrero M. 2018. Medicago truncatula copper transporter 1 (Mt COPT 1) delivers copper for symbiotic nitrogen fixation. *New Phytol* 218(2): 696–709.

Seshadri B, Bolan NS, Choppala G, Kunhikrishnan A, Sanderson P, Wang H, Currie LD, Tsang DCW, Ok YS, and Kim G. 2017. Potential value of phosphate compounds in enhancing immobilization and reducing bioavailability of mixed heavy metal contaminants in shooting range soil. *Chemosphere* 184: 197–206.

Shang Y, Hasan M, Ahammed GJ, Li M, Yin H, and Zhou J. 2019. Applications of nanotechnology in plant growth and crop protection: a review. *Molecules (Basel, Switzerland)* 24(14): 2558.

Sharma A, Patni B, Shankhdhar D, and Shankhdhar SC. 2013. Zinc–an indispensable micronutrient. *Physiol Mol Biol Plants* 19(1): 11–20.

Sharma A, and Kumar S. 2019. Arsenic exposure with reference to neurological impairment: an overview. *Rev Environ Health* 34(4): 403–414.

Shen J, Song L, Müller K, Hu Y, Song Y, Yu W, Wang H, and Wu J. 2016. Magnesium alleviates adverse effects of lead on growth, photosynthesis, and ultrastructural alterations of *Torreya grandis* seedlings. *Front Plant Sci* 7: 1819.

Shen W, Nada K, and Tachibana S. 2000. Involvement of polyamines in the chilling tolerance of cucumber cultivars. *Plant Physiol* 124(1): 431–40.

Shireen F, Nawaz MA, Chen C, Zhang Q, Zheng Z, Sohail H, Sun J, Cao H, Huang Y, and Bie Z. 2018. Boron: functions and approaches to enhance its availability in plants for sustainable agriculture. *Int J Mol Sci* 19(7): 1856.

Singh M, Kushwaha BK, Singh S, Kumar V, Singh VP, and Prasad SM. 2017. Sulphur alters chromium (VI) toxicity in *Solanum melongena* seedlings: role of sulphur assimilation and sulphur-containing antioxidants. *Plant Physiol Biochem* 112: 183–92.

Singh NK, Raghubanshi AS, Upadhyay AK, and Rai UN. 2016. Arsenic and other heavy metal accumulation in plants and algae growing naturally in contaminated area of West Bengal, India. *Ecotoxicol Environ Saf* 130: 224–33.

Singh S, Parihar P, Singh R, Singh VP, and Prasad SM. 2016. Heavy metal tolerance in plants: role of transcriptomics, proteomics, metabolomics, and ionomics. *Front Plant Sci* 6: 1143.

Singh S, Mohan Prasad S, and Pratap Singh V. 2020. Additional calcium and sulfur manages hexavalent chromium toxicity in *Solanum lycopersicum* L. and *Solanum melongena* L. seedlings by involving nitric oxide. *J Hazard Mater* 2020 398: 122607.

Singh UM, Sareen P, Sengar RS, and Kumar A. 2013. Plant ionomics: a newer approach to study mineral transport and its regulation. *Acta Physiol Plant* 35(9): 2641–2653.

Song X, Yue X, Chen W, Jiang H, Han Y, and Li X. 2019. Detection of cadmium risk to the photosynthetic performance of *Hybrid Pennisetum*. *Front Plant Sci* 10: 798.

Srivastava S, and Sharma YK. 2014. Arsenic induced changes in growth and metabolism of black gram seedlings (*Vigna mungo* L.) and the role of phosphate as an ameliorating agent. *Environ Process* 1(4): 431–445.

Stihi C, Popescu IV, Frontasyeva M, Radulescu C, Ene A, Culicov O, Zinicovscaia I, Dulama ID, Cucu-Man S, Todoran R, and Gheboianu AI. 2017. Characterization of heavy metal air pollution in Romania using moss biomonitoring, neutron activation analysis, and atomic absorption spectrometry. *Anal Lett* 50(17): 2851–8.

Strawn DG. 2018. Review of interactions between phosphorus and arsenic in soils from four case studies. *Geochem Trans* 19(1): 1–13.

Sujetovienė G, and Bučytė J. 2021. Effects of nickel on morpho-physiological parameters and oxidative status in *Brassica napus* cultivars under different sulphur levels. *Biometals* 34: 415–421.

Sundaramoorthy P, Chidambaram A, Ganesh KS, Unnikannan P, and Baskaran L. 2010. Chromium stress in paddy: (i) nutrient status of paddy under chromium stress; (ii) phytoremediation of chromium by aquatic and terrestrial weeds. *C R Biol* 333(8): 597–607.

Sustr M, Soukup A, and Tylova E. 2019. Potassium in root growth and development. *Plants* 8(10): 435.

Swain S, and Rautray TR. 2021. Estimation of trace elements, antioxidants, and antibacterial agents of regularly consumed indian medicinal plants. *Biol Trace Elem Res* 199: 1185–1193.

Tchounwou PB, Yedjou CG, Patlolla AK, and Sutton DJ. 2012. Heavy metal toxicity and the environment. *Experientia Supplementum* 101: 133–164.

Tiwari S, and Lata C. 2018. Heavy metal stress, signaling, and tolerance due to plant-associated microbes: an overview. *Front Plant Sci* 9: 452.

Tschinkel PF, Melo ES, Pereira HS, Silva K, Arakaki DG, Lima NV, Fernandes MR, Leite L, Melo ES, Melnikov P, and Espindola PR. 2020. The hazardous level of heavy metals in different medicinal plants and their decoctions in water: a public health problem in Brazil. *Biomed Res Int* 2020: Article ID 1465051.

Turan M, Kitir N, Elkoca E, Uras D, Ünek C, Nikerel E, Özdemir BS, Tarhan L, Eşitken A, Yildirim E, and Mokhtari NE. 2017. Nonsymbiotic and symbiotic bacteria efficiency for legume growth under different stress conditions. pp. 387–404. *In*: Zaidi A, Khan M, and Musarrat J (eds.). *Microbes for Legume Improvement*. Springer, Cham.

Ugulu I, Unver MC, and Dogan Y. 2016. Determination and comparison of heavy metal accumulation level of *Ficus carica* bark and leaf samples in Artvin, Turkey. *Oxid Commun* 39(1): 765–75.

van der Ent A, Echevarria G, Pollard J, and Erskine D. 2019. X-ray fluorescence ionomics of herbarium collections. *Sci Rep* 9(1): 4746.

van der Ploeg RR, Böhm W, and Kirkham MB. 1999. On the origin of the theory of mineral nutrition of plants and the law of the minimum. *Soil Sci Soc Am J* 63(5): 1055–62.

Verkleij JAC, Sneller FEC, and Schat H. 2003. Metallothioneins and phytochelatins: ecophysiological aspects. pp. 163–176. *In*: Abrol YP, and Ahmad A (eds.). *Sulphur in Plants*. Springer, Dordrecht.

Victório CP, Dos Santos MS, de Mello MC, Bento JP, da Costa Souza M, Simas NK, and de Oliveira Arruda RD. 2020. The presence of heavy metals in *Avicennia schaueriana* Stapf & Leechman ex Moldenke leaf and epicuticular wax from different mangroves around Sepetiba Bay, Rio de Janeiro, Brazil. *Environ Sci Pollut Res* 27(19): 23714–29.

Volland S, Bayer E, Baumgartner V, Andosch A, Lütz C, Sima E, and Lütz-Meindl U. 2014. Rescue of heavy metal effects on cell physiology of the algal model system Micrasterias by divalent ions. *J Plant Physiol* 171(2): 154–63.

Wakeel A, and Xu M. 2020. Chromium morpho-phytotoxicity. *Plants* 9(5): 564.

Wang L, Zhou Q, Ding L, and Sun Y. 2008. Effect of cadmium toxicity on nitrogen metabolism in leaves of *Solanum nigrum* L. as a newly found cadmium hyperaccumulator. *J Hazard Mater* 154(1-3): 818–25.

Wang R, Gao F, Guo BQ, Huang JC, Wang L, and Zhou YJ. 2013a. Short-term chromium-stress-induced alterations in the maize leaf proteome. *Int J Mol Sci* 14(6): 11125–44.

Wang M, Zheng Q, Shen Q, and Guo S. 2013b. The critical role of potassium in plant stress response. *Int J Mol Sci* 14(4): 7370–7390.

Wang X, Shi M, Hao P, Zheng W, and Cao F. 2017a. Alleviation of cadmium toxicity by potassium supplementation involves various physiological and biochemical features in *Nicotiana tabacum* L. *Acta Physiol Plant* 39(6): 132.

Wang W, Zhang X, Deng F, Yuan R, and Shen F. 2017b. Genome-wide characterization and expression analyses of superoxide dismutase (SOD) genes in *Gossypium hirsutum*. *BMC Genomics* 18(1): 376.

Watanabe T, Urayama M, Shinano T, Okada R, and Osaki M. 2015. Application of ionomics to plant and soil in fields under long-term fertilizer trials. *Springer Plus* 4(1): 1–3.

Watanabe T, Maejima E, Yoshimura T, Urayama M, Yamauchi A, Owadano M, Okada R, Osaki M, Kanayama Y, and Shinano T. 2016. The ionomic study of vegetable crops. *PLoS One* 11(8): e0160273.

West JB. 2013. The collaboration of Antoine and Marie-Anne Lavoisier and the first measurements of human oxygen consumption. *Am J Physiol Lung Cell Mol Physiol* 305(11): L775–85.

White PJ, and Brown PH. 2010. Plant nutrition for sustainable development and global health. *Ann Bot* 105(7): 1073–80.

Whitt L, Ricachenevsky FK, Ziegler GZ, Clemens S, Walker E, Maathuis FJ, Kear P, and Baxter I. 2020. A curated list of genes that affect the plant ionome. *Plant Direct* 4(10): e00272.

Williams L, and Salt DE. 2009. The plant ionome coming into focus. *Curr Opin Plant Biol* 12(3): 247.

Wood BW, Reilly CC, and Nyczepir AP. 2006. Field deficiency of nickel in trees: Symptoms and causes. *Acta Hort* 721.

Xue Z, Gao H, and Zhao S. 2014. Effects of cadmium on the photosynthetic activity in mature and young leaves of soybean plants. *Environ Sci Pollut Res Int* 21(6): 4656–4664.

Yadav SK. 2010. Heavy metals toxicity in plants: an overview on the role of glutathione and phytochelatins in heavy metal stress tolerance of plants. *S Afr J Bot* 76(2): 167–79.

Yang JL, You JF, Li YY, Wu P, and Zheng SJ. 2007. Magnesium enhances aluminum-induced citrate secretion in rice bean roots (*Vigna umbellata*) by restoring plasma membrane H+-ATPase activity. *Plant Cell Physiol* 48: 66–73.

Yasin NA, Zaheer MM, Khan WU, Ahmad SR, Ahmad A, Ali A, and Akram W. 2018. The beneficial role of potassium in Cd-induced stress alleviation and growth improvement in *Gladiolus grandiflora* L. *Int J Phytoremediation* 20(3): 274–83.

Yruela I. 2009. Copper in plants: acquisition, transport and interactions. *Funct Plant Biol* 36(5): 409–430.

Zaid A, Ahmad B, Jaleel H, Wani SH, and Hasanuzzaman M. 2020. A critical review on iron toxicity and tolerance in plants: role of exogenous phytoprotectants. *In*: Aftab T, and Hakeem KR (eds.). Plant *Micronutrients*. Springer, Cham.

Zhang C, Hiradate S, Kusumoto Y, Morita S, Koyanagi TF, Chu Q, and Watanabe T. 2021. Ionomic responses of local plant species to natural edaphic mineral variations. *Front Plant Sci*, 12.

Zhang F, Wan X, and Zhong Y. 2014. Nitrogen as an important detoxification factor to cadmium stress in poplar plants. *J Plant Interact* 9(1): 249–58.

Zhang Y, Xu Y, and Zheng L. 2020. Disease ionomics: understanding the role of ions in complex disease. *Int J Mol Sci* 21(22): 8646.

Zhao K, and Wu Y. 2017. Effects of Zn deficiency and bicarbonate on the growth and photosynthetic characteristics of four plant species. *PLoS One* 12(1): e0169812.

Zhao X, Wei J, Shu X, Kong W, and Yang M. 2016. Multi-elements determination in medical and edible Alpinia oxyphylla and Morinda officinalis and their decoctions by ICP-MS. *Chemosphere* 164: 430–435.

Zhao ZQ, Zhu YG, Li HY, Smith SE, and Smith FA. 2004. Effects of forms and rates of potassium fertilizers on cadmium uptake by two cultivars of spring wheat (*Triticum aestivum* L.). *Environ Int* 29(7): 973–8.

Zhou Z, Zhang B, Liu H, Liang X, Ma W, Shi Z, and Yang S. 2019. Zinc effects on cadmium toxicity in two wheat varieties (*Triticum aestivum* L.) differing in grain cadmium accumulation. *Ecotoxicol Environ Saf* 183: 109562.

Zhu E, Liu D, Li JG, Li TQ, Yang XE, He ZL, and Stoffella PJ. 2010. Effect of nitrogen fertilizer on growth and cadmium accumulation in *Sedum alfredii* Hance. *J Plant Nutr* 34(1): 115–26.

Zinicovscaia I, Gundorina S, Vergel K, Grozdov D, Ciocarlan A, Aricu A, Dragalin I, and Ciocarlan N. 2020. Elemental analysis of Lamiaceae medicinal and aromatic plants growing in the Republic of Moldova using neutron activation analysis. *Phytochem Lett* 35: 119–27.

Zwolak A, Sarzyńska M, Szpyrka E, and Stawarczyk K. 2019. Sources of soil pollution by heavy metals and their accumulation in vegetables: A review. *Water Air Soil Pollut* 230(7): 19.

13

Understanding Heavy Metal Stress in Plants Through Mineral Nutrients

Marya Khan,[1] *Ummey Aymen,*[1] *Ashiq Hussain Mir,*[2]
Anupam Tiwari,[1] *Sheo Mohan Prasad,*[3] *Joginder Singh,*[4]
Praveen C Ramamurthy,[5] *Rachana Singh,*[3,*] *Simranjeet Singh*[5]
and *Parul Parihar*[1,3,*]

ABSTRACT

Heavy metals such as cadmium, lead, iron, manganese, copper, cobalt, zinc, arsenic are major sources of environmental pollution especially in areas that witness high anthropogenic activities. Although these are essential nutrients required for the growth and development of plants, however,their excess concentrations have adverse effects on plants and attribute towards agricultural loss worldwide. Heavy metals increase oxidative stress and the production of ROS resulting in the impairment of important physiological and biochemical processes in plants that influence growth, metabolism, and senescence. Hence, the impact of increase, as well as deficiency in any mineral nutrient in plants, needs to be evaluated to understand the stress mechanisms and their better management. This can be achieved through ionomics, which involves the study of all mineral nutrition and trace elements. The chapter below lists the heavy metal uptake mechanisms and their various interactions with plant nutrients and the possible role of macro and micronutrients in alleviating heavy metal toxicity.

---

[1] Department of Botany, School of Bioengineering and Biosciences, Lovely Professional University, Phagwara-144411, Punjab, India.

[2] Department of Zoology, School of Bioengineering and Biosciences, Lovely Professional University, Phagwara-144411, Punjab, India.

[3] Department of Botany, University of Allahabad, Prayagraj, 211003, UP, India.

[4] Department of Biotechnology, Lovely Professional University, Phagwara (Punjab) - 144411, India.

[5] Interdisciplinary Centre for Water Research (ICWaR) Indian Institute of Science, Bangalore - 560012, India.

* Corresponding authors: parulprhr336@gmail.com; rachanaranjansingh@gmail.com

1. Introduction

Elements, proteins, metabolites, and nucleic acids form the essential components of the living cell since they take part in every process of an organism. Understanding the basic and functional dynamics of elements is hence regarded crucial for understanding the basics of life. Ionome may be defined as the study of all the mineral nutrients as well as trace elements present in an organism (Salt et al., 2008). It is a vast elemental network controlled by plant physiology and biochemistry which in turn is governed by environmental and genetic factors (Baxter, 2010). Mineral nutrition and deficiencies in plants can be calculated through ionome. The study of ionome is termed ionomics. Ever since the ionomics concept has been developed (Lahner et al., 2003), its application to various plant species has been taking place examples of which are *Arabidopsis thaliana* (Chao et al., 2014; Huang and Salt, 2016), *Zea mays* (Baxter et al., 2014), *Oryza sativa* (Pinson et al., 2015), *Brassica napus* (Thomas et al., 2016), *Glycine max* (Ziegler et al., 2013). This concept has also been applied to *Saccharomyces cerevisiae* which is yeast (Yu et al., 2012), and also to human cells (Malinouski et al., 2014). High throughput ionomics in combination with genomic approaches have been employed to understand the various genetic mechanisms that are involved in the regulation of plant ionome. The most common approaches include mutant screening which identifies genes that play a key role in the accumulation of elements (Kamiya et al., 2015; Hindt et al., 2017), and natural variation screening which helps to detect alleles that account for ionome differences in varying genotypes (Huang and Salt, 2016; Yan et al., 2016; Chen et al., 2018). Out of these two natural variation screening has proved very helpful in the identification of genes that are associated with high or low element concentrations (Huang and Salt, 2016).

Plant growth and productivity are highly compromised by any sort of micro, macro deficiency, or trace element deficiency. It must be noted that all trace elements and mineral nutrients are necessary for processes involved in the growth and development of plants, however, their concentrations exceeding the normal levels can prove toxic to plants. Ionome includes the role of mineral nutrients (Ca, K, P, N, Cu, Zn, Mo, S, Mg, Fe, Co, Mn, No) in reducing toxicity caused by heavy metals. The measurement of the composition of elements and any change in them due to some sort of stimuli is included in the study of ionomics. Changes in ionome can be both direct which includes alteration in soil nutrient level and damage in ion transporters and indirect which includes alteration in the structure of cell wall (Salt et al., 2008). Since heavy metals interact with nutrient elements, they impact their uptake and distribution resulting in mineral deficiency or their excess which hinders the normal growth and development of plants.

2.1 *Metal permeability, transport, and efflux in plants*

Heavy metals being non-biodegradable cannot be removed from the environment by natural means. Heavy metals accumulate in the soil via anthropogenic activities like smelting, fertilizer usage, sewage, and industrial waste disposal (Aydinalp and Marinova, 2009). These activities result in groundwater metal leaching and metal accumulation in soil surface (Gupta and Ali, 2002; Ali and Aboul-Enein, 2006;

Ali et al., 2009; Gupta and Ali, 2012; Dağhan et al., 2015; Basheer, 2018). Some of the heavy metals are immobile while certain metals are mobile and hence plant roots can take them up through endocytosis or diffusion or metal transporters (Ozturk, 1989; Ali and Jain, 2004; Ashraf et al., 2010; Sabir et al., 2015; Dehghani et al., 2016; Alharbi et al., 2018; Burakova et al., 2018). Trace elements such as nickel, zinc and copper which act as cofactors several enzymes are required by plants. However certain metals from pesticides like cadmium and zinc hardly benefit plants and their increased concentrations can be toxic for plants (Gough and HT, 1979; Gücel et al., 2009; Ashraf et al., 2010; Ali et al., 2017b). Plants act as natural bio accumulators by extracting and taking up heavy metals from soil irrespective of their requirement for proper growth (Ali and Aboul-Enein, 2006; Yilmaz et al., 2009; Gücel et al., 2009; Celik et al., 2010; Haribabu and Sudha, 2011; Ali et al., 2016a,b,c,d,e). The heavy metal accumulation and plant tolerance rate vary between different species. Heavy metal toxicity leads to a wide range of physiological and metabolic changes in plants (Villiers et al., 2011). Heavy metal stress is visually expressed in plants through stunted growth, leaf necrosis, low rate of seed germination, turgor loss, and impairment in photosynthetic apparatus (Sharma and Dubey, 2007). Transpiration, uptake of water, transport, nutrient metabolism are other factors of plants that are affected by heavy metals stress including essential metal uptake (Poschenrieder and Barcel, 2004; Benavides et al., 2005). Usually, root cells accumulate heavy metals via Casparian strips blockage or trapping through cell walls. Heavy metal tolerance can be attained either by excluding the mechanisms concerned with uptake via roots or by metal compartmentalization, efflux, and detoxification following their uptake. Since the scientific community is taking a keen interest in phytoremediation nowadays, metal uptake has become a very important issue.

Transport proteins are regarded as critical for the transport of heavy metals. These include (CPx-type) ATPases which not only perform a very important part in the transport of heavy metals in higher plants but also aid in metal ion homeostasis and device strategies to combat heavy metal stress. Similarly, CDF (Cation diffusion facilitator) and Nramp (Natural resistance linked macrophage protein family) also play a critical role in heavy metal transport (Williams et al., 2000; Memon, 2016; Merlot et al., 2021). Several cations such as manganese, cadmium, zinc, and iron are transported via ZIP gene family transporters. ZIP transporters are found in different cell organelles and they take part in the homeostasis of Zn and regulate adaptation of plant to soils with low as well as high Zn concentrations (Tiong et al., 2014). Apart from plants, the ZIP family is found in other species as well. These play a major role in heavy metal homeostasis and heavy metal stress. Other than transporting divalent cations it was also found that with the loading of zinc and iron in the plant roots certain ZIP proteins also get activated. It has been reported that in plants ZIP transporters were involved in cellular uptake of zinc as tested in yeast complementation test (Fu et al., 2017). The first ever characterised ZIP protein was seen in Arabidopsis. ZIP1 and ZIP3 are induced at the times of zinc stress and are principally located in roots. In roots and shoots of Arabidopsis AtZIP4 is present which is accounted for zinc nutrition since its expression is initiated with zinc restriction (Manara, 2012).

NRAMP metal transporters are present in animals, bacteria, and fungi apart from plants and are responsible for the translocation of a series of metal ions (Nevo and

Nelson, 2006). These transporters are expressed in roots and shoots of both plasma as well as tonoplast membranes in plants. NRAMP transporters are mainly involved in the transport of Fe and Mn in plant cells (Curie et al., 2000; Lanquar et al., 2010; Castaings et al., 2016; Li et al., 2019). Thomine et al. (2000) reported that these metal transporters in plants are coded by Nramp genes and AtNramp in *A. thaliana* is accountable for relocation of nutrient metal like Fe as well as toxic metals like Cd. In *A. thaliana*, identification of six Nramp genes has been done which are further classified into 2 subfamilies. First group consist of AtNRAMP1 and AtNRAMP6 and second group is comprised by AtNRAMP2 to AtNRAMP5 (Mäser et al., 2001). AtNRAMP1 is present in the Golgi apparatus as well as both the plasma and intracellular membranes (Cailliatte et al., 2010; Agorio et al., 2017) and mediates the transport of iron, Cadmium, and manganese. AtNRAMP2 is present in the trans-Golgi network and under low Mn concentration, its activation is witnessed in the roots of plants (Gao et al., 2018). AtNRAMP3 and AtNRAMP4 are localised in the tonoplast membrane and are involved in the transport of Mn from vacuoles to chloroplasts. Identification of 13 NRAMP genes in the soybean genome has been done so far (Qin et al., 2017). OsNramp1, OsNramp2, and OsNramp3 are three homologs of NRAMP protein expressed in various tissues of *Oryza sativa* and are revealed to be involved in the transport of metal ions (Belouchi et al., 1997). In the transgenic *Arabidopsis thaliana* plant when AtNramp1 was overexpressed, it resulted in increased tolerance to iron toxicity (Curie et al., 2000). NRAMP also mediate the transport of Cd in plants (Pottier et al., 2015). Takahashi et al. (2011) through an experiment with yeast revealed that AtNramp1, AtNramp3, and AtNramp4 in the *A. thaliana* plant take a major part in the influx of Cd. A similar Cd influx mechanism was seen in *Oryza sativa* by OsNramp1, which is an iron transporter protein (Takahashi et al., 2011).

The transport of copper in plants is mediated by the CTR (The copper transporters) family. High affinity is exhibited by these copper transporters and these are found in plasma membranes of several organisms, they possess conserved N and C terminus and precisely bind to Cu(I) (Puig and Thiele, 2002; Blaby-Haas and Merchant, 2012). CTR transporters have been studied well in Arabidopsis and are involved in plants growth as well as the development of pollen and also take part in cu ion transport.

Metals are translocated from roots to shoots by using xylem and phloem in the form of chelated complexes along with malate and citrate organic compounds. Metals are transported across the plant by different kinds of metal transporters. The Ptype ATPases utilise the energy released through exergonic ATP hydrolysis reaction and these mainly deal with heavy metal cations. These ATPases are present in all living organisms and besides protons such as K^+, Na^+, Mg^{2+}, Ca^{2+}, Cu^{2+}, Cd^{2+} they are known to a carry number of ions which is very important in plants. The CPx-type ATPases or heavy metals ATPases aid in soft metal cation transport. These function as efflux pumps by promoting metal ion removal from cells. During the period of plant growth, HMAs are not only responsible for uptake and transport of essential metal ions such as Cu^{2+} and Zn^{2+} but also aid in non-essential ions uptakes such as plumbum (Pb^{2+}) and cadmium (Cd^{2+}) (Takahashi et al., 2012; Migocka et al., 2015). They load Zn and Cd ions from surrounding tissues into the xylem.

HMA family member of *Arabidopsis*, AtHMA4 when overexpressed results in increased zinc and cadmium tolerance as well as increased rate of translocation from roots to shoots in plants (Verret et al., 2004; Mills et al., 2003). An important part in translocation of zinc and cadmium into shoots is played by P1B-ATPase members, AtHMA4 and AtHMA2 in *A. thaliana* as well as AhHMA4 (AtHMA4 homolog) in *Arabidopsis halleri* (Baekgaard et al., 2010; Siemianowski et al., 2014). OsHMA3 is a P1B type ATPase 3 in *Oryza sativa* which sequesters Cd in the root cell vacuoles and is predicted to be a regulator of Cd transport in the xylem (Ueno et al., 2010; Miyadate et al., 2011). Similarly, another member OsHMA9 (*Oryza sativa* heavy metal ATPases 9) belonging to P1B-type ATPase family is involved in the efflux of Pb, Zn, and Cu (Lee et al., 2007).

MATE (multidrug and toxin extrusion) transporter family consists of membrane-localised efflux protein efflux. These are responsible for poisonous compounds and multidrug transport from the cell (Schaaf et al., 2005). A protein namely FDR3 belonging to this family is expressed in the roots of *Arabidopsis halleri* and *Thlaspi caerulescens, the gene that encodes it aids in heavy metal translocation* (Talke et al., 2006; van de Mortel et al., 2006; Krämer et al., 2007).

To sequester or detoxify metal ions, metal compartmentalisation or transportation of metals to vacuoles is done by plants. This transport is carried out by certain specified transporters located in the tonoplast. ABC transporter family is divided into two subfamilies the MRP (multidrug resistance-associated protein) and PDR (pleiotropic drug resistance transporter) which sequesters and transports metal ions to the vacuole. Phytochelatin and cadmium complex is the most common type of vacuole transport that is efficiently carried out by ABC transporters. MRP'S are responsible for the transport of the PC–Cd (phytochelatins–cadmium) complex across the tonoplast in plants (Gigolash-vili and Kopriva, 2014). The first vacuolar ABC transporter that has been reported is HMT1 and is involved in vacuolar transportation of phytochelatin–cadmium (PC–Cd) complexes which is Mg-ATP dependent in nature (Ortiz et al., 1995; Kuriakose and Prasad, 2008). Two MRP transporters viz AtMRP1 and AtMRP2 are responsible for vacuolar transport of PC–Cd complexes in *A. thaliana* (Lu et al., 1997, 1998), and these transporters also grant metal tolerance.

CDF family (cation diffusion facilitator) also termed as MTP (metal tolerance protein) takes part in the transportation of metals from the cytosol to vacuoles (Krämer et al., 2007; Montanini et al., 2007). Members of this family consist of 6 transmembrane domains, a C terminal that aids cation efflux and a histidine-rich region which is believed to be a metal sensor and exists between transmembrane IV and V (Mäser et al., 2001; Kawachi et al., 2008). These transporters have been divided into four groups out of which the most important are groups I and III (Blaudez et al., 2003). MTP8 (group I), MTP1 (group III), and MTP11 (group I) showed a higher expression in *A. halleri* and *T. caerulescens* compared to those of non-hyperaccumulators (Becher et al., 2004; Talke et al., 2006; van de Mortel et al., 2006). In leaves of *A. halleri* AhMTP1 protein was highly expressed when zinc was supplied exogenously (Dräger et al., 2004). Similarly, the AtMTP1 protein in *A. thaliana* showed zinc transport activity (Lan et al., 2013). Overexpression of AtMTP1 protein in *A. thaliana* promoted tolerance to zinc (Ovecka and Takác, 2014). MTP11 and MTP8 in *A. thaliana* provided tolerance to Mn (Delhaize et al.,

1993) which proves that these proteins do have a role in metal tolerance as well. Additionally, other transporters that are involved in metal ion transportation from the cytosol to the vacuole are NRAMP and HMA transporters. However, HMA's take part in detoxification mechanisms as well owing to their overexpression, as seen in *Arabidopsis thaliana* (Morel et al., 2009).

2.2 Efflux of heavy metals from cell

An eminent approach by plants to alleviate heavy metal toxicity is achieved by the efflux of heavy metals. Efflux transporters are involved in the excretion of heavy metal ions to the cell exterior hence playing a role in metal detoxification (Singh et al., 2015). Aquaporins Lsi1 takes part both in uptake as well efflux of As(III) (Zhao et al., 2010). Apart from this, intrinsic protein of *Oryza sativa*, OsNIP2:1, OsNIP3;2 (nodulin 26-like intrinsic proteins 2;1, 3;2) along with aquaporins from other plants like *A. thaliana* AtNIP3;1, AtNIP5;1, AtNIP6;1, and AtNIP7;1 (nodulin 26-like intrinsic protein 3;1, 5;1, 6;1, 7;1) and 4;1 aquaporin gene from *Pteris vittata* (PvTIP4;1) have been reported to be involved in As(III) bidirectional transport (Bienert et al., 2008; Xu et al., 2015; He et al., 2016). As(III) efflux permease (ACR3) which is an efflux transporter is responsible for external extrusion of As(III) in yeast (Wysocki et al., 1997). The ACR3 protein is also found in *Pteris vitata* which is a hyperaccumulator of Arsenic however this protein is not found in flowering plants and is also not found in rice plants. The P1B-type HMA's (Heavy metal ATPases), being a metal-transporting trans membranal protein, play a vital role in the homeostasis of metals. OsHMA9 is an *Oryza sativa* heavy metal ATPase 9 and is known to take part in the efflux of Cu, Zn, and Pd respectively (Lee et al., 2007). Similarly, a member of the MTP (metal tolerance protein) family namely CsMTP9 (Cucumber gene) which is an antiporter bounded by plasma membrane is involved in the efflux of Cd^{2+} and Mn^{2+} from endodermis into vascular tissues by H^+ ion influx (Migocka et al., 2015).

2.3 Heavy metal interaction with plant nutrients

Heavy metals are required by plants in the form of micronutrients to carry out various biological and physiological functions which include maintenance of the membranal and cellular structure and function, stress tolerance, biosynthesis of nucleic acids, growth hormones, chlorophyll, proteins, and metabolism of certain secondary metabolites such as lipids and carbohydrates (Päivöke and Simola, 2001; Tu and Ma, 2005). However,certain heavy metals such as cadmium and chromium hinder the normal functioning of micronutrients leading to impairment of growth and development in plants.

To date, no biological role that contributes towards plant physiological development has been assigned to chromium (Reale et al., 2016). Excessive levels of Cr in plant tissues lead to a number of biochemical, physiological, and morphological processes in plants (UdDin et al., 2015; Kamran et al., 2016). Hindrances are reported in the growth and important metabolic processes of plants that are under chromium toxicity (Shanker et al., 2009). Cr toxicity results in stunted plant growth due to structural modifications in chloroplasts and cell membrane resulting in leaf chlorosis, damaged root cells, reduction in pigment concentration, a disorder in mineral nutrition and water relations, negative effects on transpiration, nitrogen assimilation, and alteration in the activities of different enzymes (Cervantes et al., 2007; Reale et al., 2016; Ali et al., 2015; Farooq et al., 2016; Anjum et al., 2017). The toxic effects of chromium are probably due to excessive ROS (reactive oxygen species) production in plants thereby causing redox imbalance (Anjum et al., 2017). The uptake of nutrients in plants is affected by heavy metal stress via interaction with essential minerals. Cr forms insoluble compounds thereby restricting nutrient uptake in soil (Chigonum et al., 2019). Heavy metals when exceeding normal levels of concentrations result in inhibition of nutrient uptake (Osu et al., 2016). Higher the level of Cr, lower is the uptake of mineral nutrients like Fe, Mg, Ca, and P by the formation of insoluble complexes and blockage of sorption sites (Kabata-pendias and Szteke et al., 2015; Osu et al., 2016). It has been reported by Dube et al. (2003) that in *Citrullus* plant with an increase in transport of Cr to different parts, there is an increase in concentration levels of Mn and P however, there is alleviation in S, Zn, Fe, and Cu concentrations in leaves indicating that nutrient imbalance is caused by Cr. Similarly, *Oryza sativa* when exposed to severe Cr concentrations showed a great decline in micro (Cu, Fe, Zn, Mn) as well as macronutrient (K, N, P) uptake (Sundaramoorthy et al., 2010). Similarly, Biddappa and Bopaiah (2007) reported that in *Cocos nucifera* plants uptake of essential nutrients (P, Mn, Cu, Zn, Mg, Fe) is greatly affected due to Cr (VI) and Cr (III) toxicity. Similar results have been seen in other plants such as *Oryza sativa* where Cr impacts the uptake of N, P, K. (Khan, 2001), *Salsola Kali* in which interference in uptake of mineral nutrients such as P, K, Mn, Fe, Ca, Mg has been reported (Gardea-Torresdey et al., 2005), *Brassica oleracea* where Cr toxicity leads to translocation of mineral nutrients (S, Mn, Cu, P, Zn) (Chatterjee and Chatterjee, 2000) and in *Amaranthus viridism* where chromium has negative effects on Fe, Cu, Mn, and Zn uptake (Liu et al., 2008). Hence Cr does decrease or inhibit the uptake of many mineral nutrients in plants. The reduction in uptake of plant nutrients under Cr toxicity is probably because of poor growth of roots and incapability of penetration or due to impairment in translocation of essential elements because binding sites that are physiologically important for translocation witness nutrient displacement (Mengel et al., 1987; Shahzad et al., 2018). Cr decreases the plasma membrane H^+ ATPase activity which can be accounted as another reason that attributes towards a decrease in nutrient uptake (Shanker et al., 2003).

Cadmium is regarded as a highly poisonous element having high mobility in plant soil systems. In agriculture, Cd has been widely studied owing to its toxic potentials in both humans and plants (Liao et al., 2015). The sources which lead to Cd increase in agricultural soils are sewage irrigation, chemical fertiliser usage, atmospheric deposition, and sludge application (Inglezakis et al., 2014; Liang et al., 2017; Dharma-Wardana, 2018). Cd at low concentrations can be toxic to plants and its effects in various species of plants have been studied extensively (Adriano et al., 2013; Hasanuzzaman, 2013; Martins et al., 2013; Rizwan et al., 2017a). These toxic effects in plants are attributed directly to Cd or the displacement of essential nutrients by Cd or either by a reduction in essential element uptake leading to their deficiency (Alloway, 2008). It has been reported that Cd impairs the uptake of Fe as well as its root to shoot translocation (Hodoshima et al., 2007; Meda et al., 2007; Gao et al., 2011). In another study, it was reported that in *Arabidopsis* Cd inhibits the Fe uptake while in peanut higher accumulation of Cd is seen under Fe deficiency (Su et al., 2014; He et al., 2017). Several plant species including lettuce (Monteiro et al., 2009), *Brassica napus* (Mendoza-Cózatl et al., 2008), *Thlaspi caerulescens* (Küpper and Kochian, 2010), flax plants (Douchiche et al., 2012), and tomato plants (Bertoli et al., 2012) have been a subject of studies related to Cd toxicity which reveal that there was an increased concentration of Fe in roots as compared to shoots suggesting that Fe uptake by roots was unaffected however root to shoot translocation of Cd was greatly affected. *Oryza sativa* showed an antagonistic relation between Cd and Mn in which Mn application reduced Cd toxicity (Rahman et al., 2016; Wang et al., 2018). Majority of the studies conducted have shown increased concentrations of Cd lead to alleviation in Cu uptake (Carpena et al., 2003; Redondo-Gómez et al., 2010). In the aquatic plant, *Ceratophyllum demers* copper concentration was decreased with an increase in Cd content (Andresen et al., 2016). Yujing et al. (2008) reported that copper application reduced Cd concentrations in roots and shoots of *Oryza sativa*. Cu and Cd use the same transporters therefore decreased Cu uptake is probably due to competition between Cu and Cd for transporters (Burzyński et al., 2005). Zn competes with Cd uptake in plants and being similar chemically they use the same transporters as well (Wong and Cobbett, 2009; Wong et al., 2009). Fahad et al. (2015) reported a 50% decline in Cd concentrations when Zn fertilisers were applied to soil comparatively with no Zn application. Cd toxicity also leads to alleviation in Ca content possibly due to competition as reported in different plants (Bertoli et al., 2012). Decrease in K content was observed in different plants such as white lupin (Carpena et al., 2003), tumbleweed (*Salsola kali*) (de la Rosa et al., 2004), flax (Douchiche et al., 2012), soybean (Drazic and Mihailovic, 2005), *Matricaria chamomilla* (Kovacik et al., 2009) under Cd stress, this decline is because of decrease in the uptake of K or increased levels of K in shoots (Drazic and Mihailovic, 2005).

Lead is another toxic heavy metal and is one of the major causes of environmental pollution (Sidhu et al., 2016). It has no role in plant nutrition as well as enzyme activities. Lead enters via the surrounding environment in plants (Islam et al., 2008; Uzu et al., 2010). Lead is known to cause damage to mineral nutrients concentrations in various plant tissues (Kabata-Pendias and Pendias, 1992; Eun et al., 2000) since cations such as (K^+, Mg^{2+}, Cu^{2+}, Ca^{2+}) compete with it for entry into the root system (Sharma et al., 2005). Lead toxicity negatively impacts plant growth and restricts the essential mineral uptake in plants (Lamhamdi et al., 2013). It was reported by Yilmaz et al. (2009) that in eggplant high Pb levels hindered the uptake of all mineral nutrients (Ca, K, Cu, Mn, Mg, Fe, K, P, Zn). Sinha et al. (2006) reported that when Cabbage (*Brassica oleracea*) that was grown in refined sand with all the required nutrients and was subjected to increased lead supply it led to an increase in Zn concentrations however there was a decline in other nutrients such as P, S, Cu, and Mn. In yet another study when two genotypes (EV-1098 and EV-77) of maize were studied to check various parameters caused by lead PbSO4 (0.01, 0.1, and 1.0 mg L^{-1}) which included nutrient accumulation (K^+ and Cu^{2+}) as well, it was observed that in both genotypes there was a subsequent reduction in K^+ and Cu^{2+} concentrations both in roots and shoots (Ahmad et al., 2011). Pb can get accumulated in the leaves of plants over time in a concentration dependent manner (kabata-Pendias and pendias, 1992; Nwosu et al., 1995). This can be held accountable for the decline in nutrient uptake since a high build-up of Pb occurs in leaves that reduce concentrations of other nutrients.

Arsenic (As) is considered a widely distributed and concerning toxic metalloid in the environment (Tu and Ma, 2002; American Agency for Toxic substances and Disease Registry (ATSDR) 2007). It is ubiquitous and harmful to almost every form of life. As is considered a non-essential as well as toxic element to plants (Zhao et al., 2009). It exists primarily as As (V) and As (III) which are its inorganic forms (Tripathi et al., 2007). Plants usually take up Arsenic in the form of AS (V). As (V) proves toxic to plants in several ways such as disturbances in physiological balance in plants (Stoeva et al., 2005), growth inhibition (Stoeva and Bineva, 2003), and ultimately plant ceases to exist. As influences the nutrient uptake (Both micro and macro) by competing for transporters with other nutrients for instance it competes with phosphorus for uptake (Mokgalaka-Matlala et al., 2008) since both of them have the same chemical properties and electronic configurations and hence compete for similar root uptake transporters (Meharg and Hartley-Whitaker, 2002; Smith et al., 2010). Low levels of As in *Pteris vittata* increased the concentration of both K and P (Tu and Ma, 2005) while As concentrations in *Pisum sativum* decreased P content in shoots but increased Mn, Mg, and Zn in roots (Paivoke and Simola, 2001). In a research by Kumar et al. (2015) on *Wrightia aborea* seedlings to check As impact on nutrient elements it was revealed that macronutrients showed different responses to As such as Ca content increased both in roots and shoots, although higher in shoots which were similar to results by Li et al. (2006) in *Pteris vittata* and Mallick et al. (2011) in *Zea mays* where they witnessed Ca increased under As stress, K concentrations increased in roots which might be a mechanism to counter balance anions produced as a result of As toxicity as indicated by Tu and Ma (2005) since K is a dominant cation in plants known to balance anions (Marschner, 1995) and high levels of As decreased P and S content in *Wrightia aborea*. In another study on sorghum (Shaibur et al., 2008), mesquite (Mokgalaka-Matlata et al., 2008), and white lupine (Vazquez et al., 2008) a decrease in micronutrient supply was reported by the application of As(V) and As(III), their application inhibited P uptake due to impairment of cell metabolism. However, white lupine was found to As resistant. Hence mineral nutrient responses to As stress in plants keeps varying and differ from species to species.

2.4 Role of nitrogen, phosphorus, potassium, calcium, sulphur, and magnesium in alleviating heavy metal toxicity

Nitrogen is one of the most important macronutrient elements. It plays a significant role in the growth and development of plants by taking part in important metabolic processes which include photosynthesis, distribution of nutrients, and plant biomass production (Makino, 2011). N deficiency in plants leads to a series of negative effects in plants such as reduction in chlorophyll content and enzyme activity, decreased rate of photosynthesis, and respiration and decreased crop yield (Lin et al., 2011a;

Luo et al., 2012). To increase crop yield practice of application of N fertiliser has been carried out for decades all over the world (Peng et al., 2002). Nitrogen being the major component of nucleic acids, hormones, proteins, and vitamins is regarded as an essential nutrient. It is capable of reducing heavy metal toxicity owing to its various abilities that include increased production of chlorophyll which in turn promotes the photosynthetic capacity, intensification of antioxidant enzyme activity, and synthesis of GSH, proline that are N-containing metabolites (Sharma and Dietz, 2006; Lin et al., 2011b). Deficiency of N leads to increased production of ROS thereby resulting in toxicity in *Oryza sativa* (Lin et al., 2011a). N when supplemented in optimal level (7.5 mM) to sunflower plants, promoted Rubisco activity and led to an increase in the protein content thereby alleviating Cd inhibitory effect on photosynthesis (Pankovic et al., 2000). In a study by Zaid et al. (2018) on the combined action of Nitrogen and Methyl jasmonate on *Mentha arvensis* under Cd stress, it was found that they alleviated Cd stress by increasing mineral nutrient content, decreasing production of ROS, and reducing Cd translocation from root to shoots. Oxidative damage caused by Cd was inhibited by exogenous N by alleviation in the accumulation of malondialdehyde (MDA) and H_2O_2 in poplar plants (Zhang et al., 2014a). Similarly, in *Sedum* toxicity induced by cadmium was decreased by supplementation of N fertiliser (16 mM) (NH4) 2SO4 (Zhu et al., 2011). The reducing potential not only depends on concentrations but also on the N source, indicating its positive role in the alleviation of heavy metal stress, for instance Cd concentrations were reduced to below 100 mg kg^{-1} in leaves of *Oryza sativa* when supplemented with NH_4^+-N (Jalloh et al., 2009) but when it was supplemented in the form of NO_3^-N it increased Cd concentrations.

Potassium (K) is an important nutrient element in plants owing to the part it plays in important physiological and biochemical processes required by plants to deal with stresses both biotic and abiotic (Wang et al., 2013). Potassium plays a key role in numerous processes such as turgor regulation during stomatal movement in guard cells (Marschner, 2011). Several other biochemical processes are also regulated by K which include protein synthesis, activation of enzymes, stomatal movements, phloem transport, anion-cation balance, metabolism of carbohydrates and stress resistance (Marschner, 2011; Tripathi et al., 2014). Oxidative stress can be minimised by improving the nutritional content of potassium in plants (Shen et al., 2000). A study was conducted by Yasin et al. (2018) on *Gladiolus grandiflora* under Cd stress in which it was observed that K application alleviated the uptake of Cd thereby reducing Cd stress by an improved accumulation of nutrients, secondary metabolites and antioxidants and reduction in oxidative stress, improvement was due to alleviation in contents of H_2O_2 and MDA. Zhao et al. (2014) reported that application of K$^+$ increased the yield of summer maize grain by 9.9–14.9% comparatively with those crops sans any K$^+$ application. K also showed differential effects when applied on soil in the form of KNO3, K2SO4, and KCl at 55, 110 and 166 mg.K.kg^{-1} respectively. KCl and K_2SO_4 increased the accumulation of Cadmium by 60–90% in shoots when applied in increasing order of application (0–55 mg). KNO3 when applied in similar concentrations increased cadmium content slightly emphasizing tolerance to Cd stress (Zhao et al., 2004).

Sulphur (S) is an essential macronutrient that not only aids in growth and development in plants but also takes part in stress tolerance mechanisms (Khan et al., 2008; Gill et al., 2011). Sulphur has many important roles in plants which include the formation of cysteine and methionine that are S-containing amino acids, chlorophyll, protein, and vitamin synthesis, and also synthesise GSH which is a part of the stress-tolerant mechanism (Rausch and Wachter, 2005; Bouranis et al., 2008; Spadaro et al., 2010). S supply in adequate amounts to crop plants intensifies the growth and photosynthetic potential (Scherer et al., 2008). It is regarded as an essential nutrient for almost all organisms since it is a part of many major functions in plants (Kopriva et al., 2015). Generally, N, P, and K are regarded as essential nutrients for crop plants but S too is very important for better yield of crops and is now gaining the 4th position of an important macronutrient after N, P, K (Jamal et al., 2010; TSI, 2020). Uptake of some essential nutrients like N, P, K and Boron may also increase by the application of sulphur in the soil (Singh et al., 2018). Sulphur upregulates the S-assimilation pathway thereby reducing Cd toxicity hence suggesting its role in heavy metal toxicity (Wangeline et al., 2004). Ethylene signalling is also regulated by sulphur which further indicated its role in heavy metal stress (Masood et al., 2012). In many plants such as *Triticum aestivum* (Khan et al., 2007), Arabidopsis (Howarth et al., 2003), and on *B. juncea* (Wangeline et al., 2004) S alleviated Cd stress by enhancing activities of ATPS (ATP-sulfurylase) and SAT (serineacetyl transferase). Another study was carried out by Singh et al. (2020b) to check the efficacy of Sulphur in combination with Calcium on *Brassica juncea* under Cd stress. Results revealed that both these elements have a protective role in alleviating Cd stress since their application intensified enzyme activities by improving the AsA/DHA and GSH/GSSG ratios and can be more effective for stress reduction when applied in combination (S+Ca). In yet another study by Singh et al. (2020a) that S and Ca were able to manage Cr toxicity in both *Solanum lycopersicum* L. and *Solanum melongena* L. by stimulation of Cr sequestration in vacuoles through intensification of glutathione-S-transferase activity thereby reducing its uptake however nitric oxide was an important part of this process.

Phosphorus has an important role in enzymatic regulation and energy transformation and therefore is considered an essential nutrient that is involved in the growth and development of plants (Schulze et al., 2006). P is a part of many important biomolecules that are required for the normal functioning of plant cells, these include lipids, aldehydes, nucleic acids, and proteins (Zhang et al., 2014b). It plays an important role in numerous metabolic processes such as photosynthesis, respiration, nucleic acid synthesis, and the maintenance of membrane integrity (Vance et al., 2003; Zhang et al., 2014a,b). Apart from playing a critical role in a wide number of cellular processes such as biomolecule synthesis, maintenance of membrane structures, and high energy molecule formation, it is known for its role in the metabolism of carbohydrates, activation or inactivation of enzymes and also aids cell division (Razaq et al., 2017). P is involved in increasing the rate of photosynthesis by maintaining turgidity in cells and a high-water potential in leaves (Waraich et al., 2011). P is required in almost all biochemical pathways since it forms an important part of the compounds that carry energy (both ATP and ADP) (Mosali et al., 2006). Phosphorus is involved in the alleviation of heavy metal toxicity either by the formation of metal-phosphate complexes hence decreasing mobility or by metal dilution (Sarwar et al., 2010). It has also been reported to increase the content of GSH in plants which protects against membrane damage, thereby confirming its part in stress tolerance (Wang et al., 2009). In a pot experiment by Arshad et al. (2016) to check the efficacy of P in the reduction of Cd stress (100 μM Cd) in wheat plants it was found that P application (0, 10, and 20 kg ha^{-1}) improved gaseous exchange and antioxidant enzyme activity in wheat and decreased the Cd concentration in the shoot of the plant and hence contribute toward the alleviation of Cd stress. In yet another research a hydroponic experiment was carried out by Dai et al. (2017) to check the impact of P on two mangrove species *Kandelia obovata* (salt exclusion species) and *Avicennia marina* (salt excretion species) under Cd stress. Results revealed that P application immobilised Cd accumulation and regulated content of proline, photosynthetic pigment, and synthesized GSH and phytochelatins in the leaves leading to a conclusion that P can mediate Cd detoxification.

Ca is involved in regulating a wide range of plant physiological processes hence forming a major plant macronutrient (Yang et al., 2015). It is involved in processes such as cell differentiation, elongation and division, cytoplasmic streaming, cell polarization, and gravitropism. It maintains the structures of the cell walls and cell membranes and regulates cell metabolism (Taiz and Zeiger, 2010). Ca (calcium) is involved in the regulation of metabolic activities as well as enzyme activation. Ca has been found to lower heavy metal toxicity (Suzuki, 2005; Farzadfar et al., 2013). Suzuki in 2005 reported the reduction of Cd concentration from 46.7 to 17.4 μg in seedlings of *Arabidopsois* through 30 mM of Ca. In *Vicia faba* plants Cd toxicity adversely affected the growth rate of the plant which was however reduced through Ca application by decreasing the chromosomal aberration frequency and enlacement in the mitotic index (El-Ashry et al., 2012). Ca (80 and 160 mg kg^{-1}) has also been reported to alleviate Ni toxicity (20 and 40 mg kg^{-1}) in *Oryza sativa*, 160 mg kg^{-1} of Ca reduced the adverse effects caused by Ni on all physiological parameters measured (plant growth, rate of photosynthesis and transpiration, content of chlorophyll and stomatal conductance) and also reduced Ni translocation to roots upto 42% when compared to untreated plants which showed 62% translocation (Aziz et al., 2015). A greenhouse experiment was carried out by Lwalaba et al. (2017) to check the efficacy of Ca in two different barley genotypes Ya66 (tolerant) and Ea52 (sensitive) under Co stress, results showed that Ca application enhanced antioxidant potential thereby alleviated Cobalt toxicity however, Co tolerance was greater in Yan66 than Ea52.

Magnesium (Mg) is an important mineral nutrient in plants, 75% of Mg present in leaves is involved in the synthesis of proteins and about 15–20% of Mg is associated with photosynthetic pigments (White et al., 2009). It acts as an enzyme cofactor for several enzymes involved in photosynthetic carbon fixation (Cakmak et al., 2008; Maathuis et al., 2009; Hermans et al., 2013). Mg (magnesium) forms an important constituent for the biosynthesis of chlorophyll and plays an active role in plants under metal stress. Magnesium at a concentration of 10 mM reduced the toxicity caused by cadmium (0.25 μM) by 40% in Japanese mustard spinach (Kashem and Kawai, 2007). Reduction in metal stress in magnesium is not by metal uptake inhibition but primarily due to amplification of antioxidant enzymes (Chou et al., 2011). Al toxicity in plants was alleviated in a wide variety of plant species by high concentration (millimolar) of Mg (Bose et al., 2011; Chen and Ma, 2013). In both rice and wheat Al toxicity was reduced by Mg which was probably due to the decrease in the activity of aluminium at the surface of the root-cell plasma membrane and reduction in the saturation of Al at apoplasmic binding sites (Kinraide et al., 2004; Watanabe and Okada, 2005). In a study conducted by Shen et al. (2016) to check the effect of magnesium (Mg^{2+}) (1040 mg kg^{-1} Mg^{2+}) on lead toxicity (Pb^{2+}) (0, 700 and 1400 mg Pb^{2+} per kg^{-1}) in *Torreya grandis* seedlings, results revealed that 1040 mg kg^{-1} Mg^{2+} improved the seedling growth by enhancing chlorophyll content and chloroplast development hence increasing rate of photosynthesis. Mg application also increased oxidative activity thereby giving results in alleviation of Pb toxicity.

2.5 Role of iron, manganese, molybdenum, and zinc in alleviating heavy metal toxicity

Iron (Fe) is an essential micronutrient required by all forms of life including plants. It plays a major role in various physiological and biochemical processes taking place in plants (Römheld and Nikolic, 2006). It is required in the normal growth and development of plants by being a part of many processes such as respiration and photosynthesis (Welch and Graham, 2004). Iron is also known to alleviate heavy metal stress. It reduces Cd induced oxidative stress in Indian mustard and maintained chlorophyll contents and stabilises chloroplast and thylakoid complexes (Qureshi et al., 2010). Iron (Fe) being an essential micronutrient, any deficiency in its content in crops can lead to serious issues such as stunted growth, and a decrease in yield and nutritional content (Ghasemi et al., 2012; Ghasemi et al., 2014; Bashir et al., 2018). Many studies have shown that Fe can alleviate Cr toxicity as well as toxicity by other metals in various species of plants (Hussain et al., 2018; Kobayashi et al., 2019; Zaheer et al., 2019). In a study conducted by Feng et al. (2013) on the effects of Fe in *T. latifolia* roots under Pb stress, results revealed that iron was able to reduce Pb toxicity. Increased application of Fe in the *Typha latifolia* plant led to decreased uptake and decreased transport of both lead and cadmium (Rodriguez-Hernandez et al., 2015). A research was conducted by Zaheer et al. (2020a) to study the impact of Fe chelated with lysine on *Brassica napus* under Cr stress, results showed that negative effects of Cr toxicity were reversed and increased in growth parameters, gaseous exchange, chlorophyll content, and activity of antioxidant enzymes was seen. Another study by Basheer et al. (2018) on rice under Cd toxicity revealed that Fe-lys foliar application increased the growth and biomass potential and also physiological and biochemical attributes in rice plants.

Manganese is both a heavy metal as well as an essential micronutrient. In plants, it is a part of photosynthetic enzymes and proteins. Its deficiency is harmful tochloroplasts since it impacts the water-splitting system of photosystem II (PSII), which provides electrons that are important in photosynthesis (Buchanan, 2000; White and Greenwood, 2013). However, its exceeding amounts cause great damage to the photosynthetic apparatus (Mukhopadhyay and Sharma, 1991). Therefore, Mn has dual roles in plants; at normal low levels it aids certain metabolic processes and acts as an essential micronutrient and has toxic effects if exists in higher levels (Kochian et al., 2004; Ducic and Polle, 2005). Mn plays a role in the alleviation of heavy metal stress. Mn can reduce the toxicity of Cd in plants by interacting with Cd (Zornoza et al., 2010). It was shown by Cd toxicity was reduced to a certain level by Mn by decreasing the uptake of cadmium in maize (Pal'ove-Balang et al., 2006). In *Phytolacca americana* reduction of toxicity caused by Cd through the application of Mn was followed by a notable reduction of Cd content in all plant organs (Peng et al., 2008). Mn was able to alleviate Cd toxicity in Cacao plants (CCN 51 genotype) (Oliveira et al., 2020). In yet another study by Rahman et al. (2016) on *Oryza sativa* seedlings under Cd stress, results showed that by application of Mn exogenously (0.3 mM $MnSO_4$), there was partial recovery of chlorosis, waterloss, imbalance in nutrients, and inhibition of growth induced by cadmium and also reduction in oxidative damage and lipid peroxidation. This was achieved through reduced uptake and translocation of cadmium.

Molybdenum (Mo) is a trace element in higher plants and is known to play important role in a number of physiological and biochemical processes in plants such as synthesis of chlorophyll and endogenous hormones, root growth, photosynthesis, maintenance of the integrity of ultra-structure, and chloroplast configuration, and N assimilation (Sun et al., 2009; Imran et al., 2019; Rana et al., 2020a; Rana et al., 2020b). Mo has shown resistance against many abiotic stresses such as low temperature (Sun et al., 2009), salinity (Zhang et al., 2012), ammonium stress (Imran et al., 2020a) via the promotion of oxidative stress tolerance mechanism against these. Mo application has also been shown to alleviate Cd stress in some plants such as *Brassica napus* (Ismael et al., 2018) and *Ricinus communis* (Ali et al., 2018). Ina study by Imran et al. (2020b) to check the efficacy of Mo against Cd stress in two fragrant rice cultivars viz GXZ (Guixiangzhan) and MXZ-2 (Meixiangzhan-2), results revealed that dry biomass was increased by 73.24% in GXZ and MXZ-2 by 58.09% through Mo application suggesting a reduction in Cd induced toxicity. The ability to alleviate Cd stress is because it can reduce Cd uptake and oxidative stress by lowering H_2O_2 and malondialdehyde levels and preventing electrolyte leakage. Yet another experiment by Imran et al. (2021) on two fragrant rice cultivars Guixiangzhan and Meixiangzhan-2 was carried to check grain yield and 2-acetyl-1-pyrroline (2AP) contents under Cd stress (0 and 100 mgkg^{-1}) and efficacy of Mo (0 and 0.15 mgkg^{-1}) application, results revealed that Mo application enhanced the content of 2AP and increased grain yield by 87.71% and 83.51% in Meixiangzhan-2 and in Guixiangzhan by 75.05% and 67.94% hence Mo alleviated Cd stress and improved grain quality as well as yield.

Zinc (Zn) is the 24th most commonly found transition heavy metal present in the earth (Alloway, 2013). In plants, it forms an essential micronutrient (Alloway, 2008a). Zn is known to promote growth and development in plants (Hansch and Mendel, 2009). It also takes part in major metabolic processes such as activation of enzymes that are important for the synthesis of proteins and the metabolism of nucleic acids and lipids (Bonnet et al., 2000). It alleviates heavy metal toxicity by restricting its availability since it plays a vital role in heavy metal homeostasis (Appenroth, 2010). Zn alleviates Cd toxicity in plants and also improves uptake of Zn (Hafeez et al., 2013). Cd induced ROS production in plants has been shown to decrease by exogenous application of Zn (Wu and Zhang, 2002; Zhao et al., 2005; Rizwan et al., 2017b). Cd toxicity was improved by Zn in wheat (Zhao et al., 2011) and *Oryza sativa* (Hassan et al., 2005). Zn was also able to reduce Cu stress in rice seedlings by the intensification of Antioxidant mechanisms against Cu induced oxidative stress (Thounaojam et al., 2014).

It was reported by Poblaciónes et al. (2017) that the content of protein was intensified by foliar Zn application in plants under Cd stress. This enhancement of protein content will further lead to the enhancement of mineral accumulation such as Zn and Mg (Marschner, 1995). This can be accounted as a mechanism for dealing with metal stress by increased accumulation of protective substances. Foliar treatments of Zn-lys promoted the growth in *Oryza sativa* and alleviated oxidative stress induced by Cr by an intensification of photosynthesis, anti-oxidant defence mechanism and uptake of Zn (Hussain et al., 2018). Zn-lys was also found to alleviate Cr toxicity in *Brassica napus* by enhancing the activities of antioxidant enzymes and reducing the uptake and accumulation of chromium from waste water (Zaheer et al., 2020b).

3. Conclusion

Plants are a major source of nutrition worldwide. Heavy metal toxicity takes a huge toll on the environment. Heavy metals accumulate in the earth's surface and are taken up by plants hence making their way into the food chain thereby deteriorating the environment and influencing human health. Now when the world population is increasing, there is an urgent need to address this phenomenon by devising strategies that will help in the alleviation of metal toxicity. Ionomics can be helpful in heavy metal stress management. Although abundant research has been done to understand various stress alleviation mechanisms in plants more is still to be done. Also, the role of various nutrients needs to be further evaluated which can help in the future development of stress-tolerant transgenic plants.

References

Adriano DC. 2013. Trace Elements in the Terrestrial Environment. Springer Science and Business Media.

Agorio A, Giraudat J, Bianchi MW, Marion J, Espagne C, Castaings L, Lelièvre F, Curie C, Thomine S, and Merlot S. 2017. Phosphatidylinositol 3-phosphate–binding protein AtPH1 controls the localization of the metal transporter NRAMP1 in Arabidopsis. *Proceedings of the National Academy of Sciences* 114(16): E3354–E3363.

Ahmad MSA, Ashraf M, Tabassam Q, Hussain M, and Firdous H. 2011. Lead (Pb)-induced regulation of growth, photosynthesis, and mineral nutrition in maize (*Zea mays* L.) plants at early growth stages. *Biological Trace Element Research* 144(1): 1229–1239.

Alharbi OM, Khattab RA, and Ali I. 2018. Health and environmental effects of persistent organic pollutants. *Journal of Molecular Liquids* 263: 442–453. Studies. *Journal of Molecular Liquids* 215: 671–679.

Ali I, and Jain CK. 2004. Advances in arsenic speciation techniques. *International Journal of Environmental Analytical Chemistry* 84(12): 947–964.

Ali I, and Aboul-Enein HY. 2006. *Instrumental Methods in Metal ion Speciation*. CRC Press.

Ali I, Aboul-Enein HY, Gupta VK, and Chromatography, N. 2009. *Pharmaceutical and Environmental Analyses*. Wiley, Hoboken.

Ali I, Alothman ZA, and Al-Warthan A. 2016a. Sorption, kinetics and thermodynamics studies of atrazine herbicide removal from water using iron nano-composite material. *International Journal of Environmental Science and Technology* 13(2): 733–742.

Ali I, Al-Othman ZA, and Al-Warthan A. 2016b. Removal of secbumeton herbicide from water on composite nanoadsorbent. *Desalination and Water Treatment* 57(22): 10409–10421.

Ali I, Al-Othman ZA, and Alharbi, O. 2016c. Uptake of pantoprazole drug residue from water using novel synthesized composite iron nano adsorbent. *Journal of Molecular Liquids* 218: 465–472.

Ali I, AL-Othman ZA, and Alwarthan A. 2016d. Green synthesis of functionalized iron nano particles and molecular liquid phase adsorption of ametryn from water. *Journal of Molecular Liquids* 221: 1168–1174.

Ali I, AL-Othman ZA, and Alwarthan A. 2016e. Synthesis of composite iron nano adsorbent and removal of ibuprofen drug residue from water. *Journal of Molecular Liquids* 219: 858–864.

Ali N, and Hadi F. 2018. CBF/DREB transcription factor genes play role in cadmium tolerance and phytoaccumulation in *Ricinus communis* under molybdenum treatments. *Chemosphere* 208: 425–432. [CrossRef] [PubMed].

Ali S, Bharwana SA, Rizwan M, Farid M, Kanwal S, Ali Q, Ibrahim M, Gill RA, and Khan MD. 2015. Fulvic acid mediates chromium (Cr) tolerance in wheat (*Triticum aestivum* L.) through lowering of Cr uptake and improved antioxidant defense system. *Environmental Science and Pollution Research*, 22(14): 10601–10609.

Alloway BJ. 2008, February. Copper and Zinc in soils: Too little or too much. In *NZ Trace Elements Group Conference, Waikato University, Hamilton.*

Alloway BJ. 2008. Zinc in Soils and Crop Nutrition. 2nd Edition, IZA and IFA, Brussels, Belgium and Paris, France.

Alloway BJ. 2013. Heavy metals and metalloids as micronutrients for plants and animals. *In*: Alloway B (eds.). Heavy Metals in Soils. Environmental Pollution, vol 22. Springer, Dordrecht. https://doi.org/10.1007/978-94-007-4470-7_7.

Anjum SA, Ashraf U, Imran KHAN, Tanveer M, Shahid M, Shakoor A, and Longchang, WANG. 2017. Phyto-toxicity of chromium in maize: oxidative damage, osmolyte accumulation, anti-oxidative defense and chromium uptake. *Pedosphere* 27(2): 262–273.

Appenroth KJ. 2010. Definition of "Heavy Metals" and their role in biological systems. *In*: Sherameti I, and Varma A (eds.). Soil Heavy Metals. Soil Biology, vol 19. Springer, Berlin, Heidelberg. https://doi.org/10.1007/978-3-642-02436-8_2.

Arshad M, Ali S, Noman A, Ali Q, Rizwan M, Farid M, and Irshad, MK. 2016. Phosphorus amendment decreased cadmium (Cd) uptake and ameliorates chlorophyll contents, gas exchange attributes, antioxidants, and mineral nutrients in wheat (*Triticum aestivum* L.) under Cd stress. *Archives of Agronomy and Soil Science* 62(4): 533–546.

Ashraf M, Ozturk M, and Ahmad, MSA (eds.). 2010. *Plant Adaptation and Phytoremediation* (p. 481). Dordrecht, The Netherlands: Springer.

ATSDR. 2007. Toxicological profile for arsenic. US Department of Health and Human Services, Agency for Toxic Substances and Disease Registry, Atlanta.

Aydinalp C, and Marinova S. 2009. The effects of heavy metals on seed germination and plant growth on alfalfa plant (*Medicago sativa*). *Bulgarian Journal of Agricultural Science* 15(4): 347–350.

Aziz H, Sabir M, Ahmad HR, Aziz T, Zia-ur-Rehman M, Hakeem KR, and Ozturk M. 2015. Alleviating effect of calcium on nickel toxicity in rice. *CLEAN–Soil, Air, Water* 43(6): 901–909.

Bækgaard L, Mikkelsen MD, Sørensen DM, Hegelund JN, Persson DP, Mills RF, Yang Z, Husted S, Andersen JP, Buch-Pedersen MJ, and Schjoerring JK. 2010. A combined zinc/cadmium sensor and zinc/cadmium export regulator in a heavy metal pump. *Journal of Biological Chemistry* 285(41): 31243–31252.

Basheer AA. 2018. Chemical chiral pollution: impact on the society and science and need of the regulations in the 21st century. *Chirality* 30(4): 402–406.

Bashir A, Rizwan M, Ali S, ur Rehman MZ, Ishaque W, Riaz MA, and Maqbool A. 2018. Effect of foliar-applied iron complexed with lysine on growth and cadmium (Cd) uptake in rice under Cd stress. *Environmental Science and Pollution Research* 25(21): 20691–20699.

Baszyński T, Wajda L, Krol M, Wolińska D, Krupa Z, and Tukendorf A. 1980. Photosynthetic activities of cadmium-treated tomato plants. *Physiologia Plantarum* 48(3): 365–370.

Baxter I. 2010. Ionomics: The functional genomics of elements. *Briefings in Functional Genomics* 9(2): 149–156.

Baxter IR, Ziegler G, Lahner B, Mickelbart MV, Foley R, Danku J, Armstrong P, Salt DE, and Hoekenga OA. 2014. Single-kernel ionomic profiles are highly heritable indicators of genetic and environmental influences on elemental accumulation in maize grain (*Zea mays*). *PLOS one* 9(1): e87628.

Becher M, Talke IN, Krall L, and Krämer U. 2004. Cross-species microarray transcript profiling reveals high constitutive expression of metal homeostasis genes in shoots of the zinc hyperaccumulator *Arabidopsis halleri*. *The Plant Journal* 37(2): 251–268.

Belouchi A, Kwan T, and Gros P. 1997. Cloning and characterization of the OsNramp family from *Oryza sativa*, a new family of membrane proteins possibly implicated in the transport of metal ions. *Plant Molecular Biology* 33(6): 1085–1092.

Benavides MP, Gallego SM, and Tomaro ML. 2005. Cadmium toxicity in plants. *Brazilian Journal of Plant Physiology* 17(1): 21–34.

Bertoli AC, Cannata MG, Carvalho R, Bastos ARR, Freitas MP, and dos Santos Augusto A. 2012. *Lycopersicon esculentum* submitted to Cd-stressful conditions in nutrition solution: nutrient contents and translocation. *Ecotoxicology and Environmental Safety* 86: 176–181.

Biddappa CC, and Bopaiah MG. 2007. Effect of heavy metals on the distribution of P, K, Ca, Mg and micronutrients in the cellular constituents of coconut leaf. *Journal of Plantation Crops* 17: 1–9.

Bienert GP, Thorsen M, Schüssler MD, Nilsson HR, Wagner A, Tamás MJ, and Jahn TP. 2008. A subgroup of plant aquaporins facilitate the bi-directional diffusion of As (OH) 3 and Sb (OH) 3 across membranes. *BMC Biology* 6(1): 1–15.

Blaby-Haas CE, and Merchant SS. 2012. The ins and outs of algal metal transport. *Biochimica et Biophysica Acta (BBA)-Molecular Cell Research* 1823(9): 1531–1552.

Blaude D, Kohler A, Martin F, Sanders D, and Chalot M. 2003. Poplar metal tolerance protein 1 confers zinc tolerance and is an oligomeric vacuolar zinc transporter with an essential leucine zipper motif. *The Plant Cell* 15(12): 29112928.

Bonnet M, Camares O, and Veisseire P. 2000. Effects of zinc and influence of *Acremonium lolii* on growth parameters, chlorophyll a fluorescence and antioxidant enzyme activities of ryegrass (*Lolium perenne* L. cv Apollo). *Journal of Experimental Botany* 51(346): 945–953.

Bose J, Babourina O, and Rengel Z. 2011. Role of magnesium in alleviation of aluminium toxicity in plants. *Journal of Experimental Botany* 62(7): 2251–2264.

Bouranis DL, Buchner P, Chorianopoulou SN, Hopkins L, Protonotarios VE, Siyiannis VF, and Hawkesford MJ. 2008. Responses to sulfur limitation in maize. *In*: Khan NA, Singh S, and Umar S (eds.). Sulfur Assimilation and Abiotic Stress in Plants. Springer, Berlin, Heidelberg. https://doi.org/10.1007/978-3-540-76326-0_1.

Buchanan B, Gruissem W, and Jones, R. 2000. Biochemistry and Molecular Biology of Plants, American Society of Plant Physiologists. Courier Companies. *Inc. Waldorf, MD*.

Burakova EA, Dyachkova TP, Rukhov AV, Tugolukov EN, Galunin EV, Tkachev AG, and Ali I. 2018. Novel and economic method of carbon nanotubes synthesis on a nickel magnesium oxide catalyst using microwave radiation. *Journal of Molecular Liquids* 253: 340–346.

Burritt DJ. 2008. Glutathione metabolism in bryophytes under abiotic stress. pp. 303–316. *In*: Khan NA, Singh S, and Umar S (eds.). Sulfur Assimilation and Abiotic Stress in Plants. Springer, Berlin, Heidelberg.

Burzyński M, Migocka M, and Kłobus G. 2005. Cu and Cd transport in cucumber (*Cucumis sativus* L.) root plasma membranes. *Plant Science* 168(6): 1609–1614.

Cailliatte R, Schikora A, Bria JF, Mari S, and Curie C. 2010. High-affinity manganese uptake by the metal transporter NRAMP1 is essential for Arabidopsis growth in low manganese conditions. *The Plant Cell* 22(3): 904–917.

Cakmak I, and Kirkby EA. 2008. Role of magnesium in carbon partitioning and alleviating photooxidative damage. *Physiologia Plantarum* 133(4): 692–704.

Carpena RO, Vázquez S, Esteban E, Fernández-Pascual M, de Felipe MR, and Zornoza P. 2003. Cadmium-stress in white lupin: effects on nodule structure and functioning. *Plant Physiology and Biochemistry* 41(10): 911–919.

Castaings L, Caquot A, Loubet S, and Curie C. 2016. The high-affinity metal transporters NRAMP1 and IRT1 team up to take up iron under sufficient metal provision. *Scientific Reports* 6(1): 1–11.

Celik S, Yucel E, Celik S, Gucel, and Ozturk M. 2010. Carolina poplar (Populus x canadensis Moench) as a biomonitor of trace elements in Black Sea region of Turkey. *Journal of Environmental Biology* 31(1): 225.

Cervantes C, and Campos-García J. 2007. Reduction and efflux of chromate by bacteria. *In*: Nies DH, and Silver S (eds.). Molecular Microbiology of Heavy Metals. *Microbiology Monographs*, vol 6. Springer, Berlin, Heidelberg. https://doi.org/10.1007/7171_2006_087.

Chao DY, Chen Y, Chen J, Shi S, Chen Z, Wang C, Danku JM, Zhao FJ, and Salt DE. 2014. Genome-wide association mapping identifies a new arsenate reductase enzyme critical for limiting arsenic accumulation in plants. *PLoS Biol* 12(12): e1002009.

Chatterjee J, and Chatterjee C. 2000. Phytotoxicity of cobalt, chromium and copper in cauliflower. *Environmental Pollution* 109(1): 69–74.

Chen ZC, and Ma JF. 2013. Magnesium transporters and their role in Al tolerance in plants. *Plant and Soil* 368(1): 51–56.

Chen ZR, Kuang L, Gao YQ, Wang YL, Salt DE, and Chao DY. 2018. AtHMA4 drives natural variation in leaf Zn concentration of *Arabidopsis thaliana*. *Frontiers in Plant Science* 9: 270.

Chigonum WJ, Ikenna OC, and John CU. 2019. Dynamic impact of chromium on nutrient uptake from soil by fluted pumpkin (*Telfairia occidentalis*). *Am J Biosci Bioeng* 7: 1–9.

Chou TS, Chao YY, Huang WD, Hong CY, and Kao CH. 2011. Effect of magnesium deficiency on antioxidant status and cadmium toxicity in rice seedlings. *Journal of Plant Physiology* 168(10): 1021–1030.

Curie C, Alonso JM, JEAN ML, Ecker JR, and Briat JF. 2000. Involvement of NRAMP1 from *Arabidopsis thaliana* in iron transport. *Biochemical Journal* 347(3): 749–755.

Dağhan H, Öztürk M, Hakeem KR, Sabir M, and Mermut AR. 2015. Soil pollution in Turkey and remediation methods. *Soil Remediation and Plants: Prospects And Challenges*, pp. 287–312.

Dai M, Lu H, Liu W, Jia H, Hong H, Liu J, and Yan C. 2017. Phosphorus mediation of cadmium stress in two mangrove seedlings *Avicennia marina* and *Kandelia obovata* differing in cadmium accumulation. *Ecotoxicology and Environmental Safety* 139: 272–279.

de la Rosa G, Peralta-Videa JR, Montes M, Parsons JG, Cano-Aguilera I, and Gardea-Torresdey JL. 2004. Cadmium uptake and translocation in tumbleweed (Salsola kali), a potential Cd-hyperaccumulator desert plant species: ICP/OES and XAS studies. *Chemosphere* 55(9): 1159–1168.

Dehghani MH, Sanaei D, Ali I, and Bhatnagar A. 2016. Removal of chromium (VI) from aqueous solution using treated waste newspaper as a low-cost adsorbent: kinetic modeling and isotherm. *Journal of Molecular Liquids* 215: 671–679.

Delhaize E, Ryan PR, and Randall PJ. 1993. Aluminum tolerance in wheat (*Triticum aestivum* L.) (II. Aluminum-stimulated excretion of malic acid from root apices). *Plant Physiology* 103(3): 695–702.

Dharma-Wardana MWC. 2018. Fertilizer usage and cadmium in soils, crops and food. *Environmental Geochemistry and Health* 40(6): 2739–2759.

Douchiche O, Chaïbi W, and Morvan C. 2012. Cadmium tolerance and accumulation characteristics of mature flax, cv. Hermes: contribution of the basal stem compared to the root. *Journal of Hazardous Materials* 235: 101–107.

Dräger DB, Desbrosses-Fonrouge AG, Krach C, Chardonnens AN, Meyer RC, Saumitou-Laprade P, and Krämer U. 2004. Two genes encoding *Arabidopsis halleri* MTP1 metal transport proteins co-segregate with zinc tolerance and account for high MTP1 transcript levels. *The Plant Journal* 39(3): 425–439.

Drazic G, and Mihailovic N. 2005. Modification of cadmium toxicity in soybean seedlings by salicylic acid. *Plant Science* 168(2): 511–517.

Dube BK, Tewari K, Chatterjee J, and Chatterjee C. 2003. Excess chromium alters uptake and translocation of certain nutrients in citrullus. *Chemosphere* 53(9): 1147–1153.

Ducic T, and Polle A. 2005. Transport and detoxification of manganese and copper in plants. *Braz J Plant Physiol* 17: 103–112.

El-Ashry ZM, and Mohamed FI. 2012. Protective effects of some antioxidant metals against chromosomal damage induced by cadmium in *Vicia faba* plants. *International Journal of Agricultural Research* 7(8): 376–387.

Eun SO, Shik Youn H, and Lee Y. 2000. Lead disturbs microtubule organization in the root meristem of Zea mays. *Physiologia Plantarum* 110(3): 357–365.

Fahad S, Hussain S, Khan F, Wu C, Saud S, Hassan S, Ahmad N, Gang D, Ullah A, and Huang J. 2015. Effects of tire rubber ash and zinc sulfate on crop productivity and cadmium accumulation in five rice cultivars under field conditions. *Environmental Science and Pollution Research* 22(16): 12424–12434.

Farooq MA, Ali S, Hameed A, Bharwana SA, Rizwan M, Ishaque W, Farid M, Mahmood K, and Iqbal Z. 2016. Cadmium stress in cotton seedlings: physiological, photosynthesis and oxidative damages alleviated by glycinebetaine. *South African Journal of Botany* 104: 61–68.

Farzadfar S, Zarinkamar F, Modarres-Sanavy SAM, and Hojati M. 2013. Exogenously applied calcium alleviates cadmium toxicity in *Matricaria chamomilla* L. plants. *Environmental Science and Pollution Research* 20(3): 1413–1422.

Feng H, Qian Y, Gallagher FJ, Wu M, Zhang W, Yu L, Zhu Q, Zhang K, Liu CJ, and Tappero R. 2013. Lead accumulation and association with Fe on *Typha latifolia* root from an urban brownfield site. *Environmental Science and Pollution Research* 20(6): 3743–3750.

Fu XZ, Zhou X, Xing F, Ling LL, Chun CP, Cao L, Aarts MG, and Peng LZ. 2017. Genome-wide identification, cloning and functional analysis of the zinc/iron-regulated transporter-like protein (ZIP) gene family in trifoliate orange (*Poncirus trifoliata* L. Raf.). *Frontiers in Plant Science* 8: 588.

Gao C, Wang Y, Xiao DS Qiu CP, Han DG, Zhang XZ, Wu T, and Han ZH. 2011. Comparison of cadmium-induced iron-deficiency responses and genuine iron-deficiency responses in Malus xiaojinensis. *Plant Science* 181(3): 269–274.

Gao H, Xie W, Yang C, Xu J, Li J, Wang H, Chen X, and Huang CF. 2018. NRAMP2, a trans-Golgi network-localized manganese transporter, is required for Arabidopsis root growth under manganese deficiency. *New Phytologist* 217(1): 179–193.

Gardea-Torresdey JL, De la Rosa G, Peralta-Videa JR, Montes M, Cruz-Jimenez G, and Cano-Aguilera I. 2005. Differential uptake and transport of trivalent and hexavalent chromium by tumbleweed (Salsola kali). *Archives of Environmental Contamination and Toxicology* 48(2): 225–232.

Ghasemi S, Khoshgoftarmanesh AH, Hadadzadeh H, and Jafari M. 2012. Synthesis of iron-amino acid chelates and evaluation of their efficacy as iron source and growth stimulator for tomato in nutrient solution culture. *Journal of Plant Growth Regulation* 31(4): 498–508.

Ghasemi S, Khoshgoftarmanesh AH, Afyuni M, and Hadadzadeh H. 2014. Iron (II)–amino acid chelates alleviate salt-stress induced oxidative damages on tomato grown in nutrient solution culture. *Scientia Horticulturae* 165: 91–98.

Gigolashvili T, and Kopriva S. 2014. Transporters in plant sulfur metabolism. *Frontiers in Plant Science* 5: 442.

Gill SS, Khan NA, Anjum NA, and Tuteja N. 2011. Amelioration of cadmium stress in crop plants by nutrients management: morphological, physiological and biochemical aspects. *Plant Stress* 5(1): 1–23.

Gough LP, Shacklette HT, and Case AA. 1979. B. Element concentrations toxic to plants, animals, and man. US Govt Print Off, 1466.

Guecel S, Oeztuerk M, Yuecel E, Kadis C, and Güvensen A. 2009. Studies on trace metals in soils and plants growing in the vicinity of copper mining area-Lefke, Cyprus. *Fresenius Environ Bull* 18: 360–368.

Gupta VK, and Ali I. 2002. Encyclopedia of surface and colloid science. *Marcel Dekker, New York*, pp. 136–166.

Gupta VK, and Ali I. 2012. Environmental water: advances in treatment, remediation and recycling. Newnes.

Hafeez B, Khanif YM, and Saleem M. 2013. Role of zinc in plant nutrition-a review. *Journal of Experimental Agriculture International*, pp. 374–391.

Hänsch R, and Mendel RR. 2009. Physiological functions of mineral micronutrients (cu, Zn, Mn, Fe, Ni, Mo, B, cl). *Current Opinion in Plant Biology* 12(3): 259–266.

Haribabu TE, and Sudha PN. 2011. Effect of heavy metals copper and cadmium exposure on the antioxidant properties of the plant *Cleome gynandra*. *International Journal of Plant, Animal and Environmental Sciences* 1(2): 80–87.

Hasanuzzaman M. 2013. *Cadmium: characteristics, sources of exposure, health and environmental effects.* Nova Publishers.

Hassan MJ, Zhang G, Wu F, Wei K, and Chen Z. 2005. Zinc alleviates growth inhibition and oxidative stress caused by cadmium in rice. *Journal of Plant Nutrition and Soil Science* 168(2): 255–261.

He BY, Yu DP, Chen Y, Shi JL, Xia Y Li, QS Wang, LL Ling L, and Zeng EY. 2017. Use of low-calcium cultivars to reduce cadmium uptake and accumulation in edible amaranth (*Amaranthus mangostanus* L.). *Chemosphere* 171: 588–594.

He Z, Yan H, Chen Y, Shen H, Xu W, Zhang H, Shi L, Zhu YG, and Ma M. 2016. An aquaporin Pv TIP 4; 1 from *Pteris vittata* may mediate arsenite uptake. *New Phytologist* 209(2): 746–761.

Hermans C, Conn SJ, Chen J, Xiao Q, and Verbruggen N. 2013. An update on magnesium homeostasis mechanisms in plants. *Metallomics* 5(9): 1170–1183.

Hindt MN, Akmakjian GZ, Pivarski KL, Punshon T, Baxter I, Salt DE, and Guerinot ML. 2017. BRUTUS and its paralogs, BTS LIKE1 and BTS LIKE2, encode important negative regulators of the iron deficiency response in *Arabidopsis thaliana*. *Metallomics* 9(7): 876–890.

Hodoshima H, Enomoto Y, Shoji K, Shimada H, Goto F, and Yoshihara T. 2007. Differential regulation of cadmium-inducible expression of iron-deficiency-responsive genes in tobacco and barley. *Physiologia Plantarum* 129(3): 622–634.

Howarth JR, Domínguez-Solís JR, Gutiérrez-Alcalá G, Wray JL, Romero LC, and Gotor C. 2003. The serine acetyltransferase gene family in *Arabidopsis thaliana* and the regulation of its expression by cadmium. *Plant Molecular Biology* 51(4): 589–598.

Huang XY, and Salt DE. 2016. Plant ionomics: from elemental profiling to environmental adaptation. *Molecular Plant* 9(6): 787–797.

Hussain A, Ali S, Rizwan M, ur Rehman MZ, Hameed A, Hafeez F, Alamri SA, Alyemeni MN, and Wijaya L. 2018. Role of zinc–lysine on growth and chromium uptake in rice plants under Cr stress. *Journal of Plant Growth Regulation* 37(4): 1413–1422.

Imran M, Hu C, Hussain S, Rana MS, Riaz M, Afzal J, Aziz O, Elyamine AM, Ismael MAF, and Sun X. 2019. Molybdenum-induced effects on photosynthetic efficacy of winter wheat (*Triticum aestivum* L.) under different nitrogen sources are associated with nitrogen assimilation. *Plant Physiology and Biochemistry* 141: 154–163.

Imran M, Sun X, Hussain S, Ali U, Rana MS, Rasul F, Shaukat S, and Hu C. 2020a. Molybdenum application regulates oxidative stress tolerance in winter wheat under different nitrogen sources. *Journal of Soil Science and Plant Nutrition* 20(4): 1827–1837.

Imran M, Hussain S, El-Esawi MA, Rana MS, Saleem MH, Riaz M, Ashraf U, Potcho MP, Duan M, Rajput IA, and Tang X. 2020b. Molybdenum supply alleviates the cadmium toxicity in fragrant rice by modulating oxidative stress and antioxidant gene expression. *Biomolecules* 10(11): 1582.

Imran M, Hussain S, Rana MS, Saleem MH, Rasul F, Ali KH, Potcho MP, Pan S, Duan M, and Tang X. 2021. Molybdenum improves 2-acetyl-1-pyrroline, grain quality traits and yield attributes in fragrant rice through efficient nitrogen assimilation under cadmium toxicity. *Ecotoxicology and Environmental Safety* 211: 111911.

Inglezakis VJ, Zorpas AA, Karagiannidis A, Samaras P, Voukkali I, and Sklari S. 2014. European Union legislation on sewage sludge management. *Fresenius Environmental Bulletin* 23(2A): 635–639.

Islam E, Liu D, Li T, Yang X, Jin X, Mahmood Q, Tian S, and Li J. 2008. Effect of Pb toxicity on leaf growth, physiology and ultrastructure in the two ecotypes of *Elsholtzia argyi*. *Journal of Hazardous Materials* 154(1-3): 914–926.

Ismael MA, Elyamine AM, Zhao YY, Moussa MG, Rana MS, Afzal J, Imran M, Zhao XH, and Hu CX. 2018. Can selenium and molybdenum restrain cadmium toxicity to pollen grains in *Brassica napus*? *International Journal of Molecular Science* 19: 2163.

Jalloh MA, Chen J, Zhen F, and Zhang G. 2009. Effect of different N fertilizer forms on antioxidant capacity and grain yield of rice growing under Cd stress. *Journal of Hazardous Materials* 162(2-3): 1081–1085.

Jamal A, Moon YS, and Zainul Abdin M. 2010. Sulphur-a general overview and interaction with nitrogen. *Australian Journal of Crop Science* 4(7): 523.

Kabata-Pendias A, and Pendias H. 1992. Trace elements in soils and plants–CRC Press. *Boca Raton, FL*, p. 356.

Kabata-Pendias A, and Szteke B. 2015. *Trace elements in abiotic and biotic environments* (p. 468). Taylor & Francis.

Kamiya T, Borghi M, Wang P, Danku JM, Kalmbach L, Hosmani PS, Naseer S, Fujiwara T, Geldner N, and Salt DE. 2015. The MYB36 transcription factor orchestrates *Casparian* strip formation. *Proceedings of the National Academy of Sciences* 112(33): 10533–10538.

Kamran M, Eqani SAMAS, Katsoyiannis A, Xu R, Bibi S, Benizri E, and Chaudhary H. 2016. Phytoextraction of chromium (Cr) and influence of *Pseudomonas putida* on *Eruca sativa* growth. *J Geochem Explor. https://doi. org/10.1016/jgexplo201609005.*

Kashem MA, and Kawai S. 2007. Alleviation of cadmium phytotoxicity by magnesium in Japanese mustard spinach. *Soil Science and Plant Nutrition* 53(3): 246–251.

Kawachi M, Kobae Y, Mimura T, and Maeshima M. 2008. Deletion of a histidine-rich loop of AtMTP1, a vacuolar Zn2+/H+ antiporter of *Arabidopsis thaliana*, stimulates the transport activity. *Journal of Biological Chemistry* 283(13): 8374–8383.

Khan AG. 2001. Relationships between chromium biomagnification ratio, accumulation factor, and mycorrhizae in plants growing on tannery effluent-polluted soil. *Environment International* 26(5-6): 417–423.

Khan NA, Samiullah Singh S, and Nazar R. 2007. Activities of antioxidative enzymes, sulphur assimilation, photosynthetic activity, and growth of wheat (*Triticum aestivum*) cultivars differing in yield potential under cadmium stress. *Journal of Agronomy and Crop Science* 193(6): 435–444.

Khan NA, Singh S, and Umar S (eds.). 2008. *Sulfur assimilation and abiotic stress in plants*. Berlin: springer.

Kinraide TB, Pedler JF, and Parker DR. 2004. Relative effectiveness of calcium and magnesium in the alleviation of rhizotoxicity in wheat induced by copper, zinc, aluminum, sodium, and low pH. *Plant and Soil* 259(1): 201–208.

Kobayashi T, Nozoye T, and Nishizawa NK. 2019. Iron transport and its regulation in plants. *Free Radical Biology and Medicine* 133: 11–20.

Kochian LV, Hoekenga OA, and Pineros MA. 2004. How do crop plants tolerate acid soils? Mechanisms of aluminum tolerance and phosphorous efficiency. *Annu Rev Plant Biol* 55: 459–493.

Kopriva S, Calderwood A, Weckopp SC, and Koprivova A. 2015. Plant sulfur and big data. *Plant Science* 241: 1–10.

Kováčik J, Gruz J, Hedbavny J, Klejdus B, and Strnad M. 2009. Cadmium and nickel uptake are differentially modulated by salicylic acid in *Matricaria chamomilla* plants. *Journal of Agricultural and Food Chemistry* 57(20): 9848–9855.

Krämer U, Talke IN, and Hanikenne M. 2007. Transition metal transport. *FEBS Letters* 581(12): 2263–2272.

Kumar D, Singh VP, Tripathi DK, Prasad SM, and Chauhan DK. 2015. Effect of arsenic on growth, arsenic uptake, distribution of nutrient elements and thiols in seedlings of wrightia arborea (Dennst.) Mabb. *International Journal of Phytoremediation* 17(2): 128–134.

Küpper H, and Kochian LV. 2010. Transcriptional regulation of metal transport genes and mineral nutrition during acclimatization to cadmium and zinc in the Cd/Zn hyperaccumulator, *Thlaspi caerulescens* (Ganges population). *New Phytologist* 185(1): 114–129.

Kuriakose SV, and Prasad MNV. 2008. Cadmium stress affects seed germination and seedling growth in *Sorghum bicolor* (L.) Moench by changing the activities of hydrolyzing enzymes. *Plant Growth Regulation* 54(2): 143–156.

Lahner B, Gong J, Mahmoudian M, Smith EL, Abid KB, Rogers EE, Guerinot ML, Harper JF, Ward JM, McIntyre L, and Schroeder JI. 2003. Genomic scale profiling of nutrient and trace elements in *Arabidopsis thaliana*. *Nature Biotechnology* 21(10): 1215–1221.

Lamhamdi M, El Galiou O, Bakrim A, Nóvoa-Muñoz JC, Arias-Estévez M, Aarab A, and Lafont R. 2013. Effect of lead stress on mineral content and growth of wheat (*Triticum aestivum*) and spinach (*Spinacia oleracea*) seedlings. *Saudi Journal of Biological Sciences* 20(1): 29–36.

Lan HX, Wang ZF, Wang QH, Wang MM, Bao YM, Huang J, and Zhang HS. 2013. Characterization of a vacuolar zinc transporter OZT1 in rice (*Oryza sativa* L.). *Molecular Biology Reports* 40(2): 1201–1210.

Lanquar V, Ramos MS, Lelièvre F, Barbier-Brygoo H, Krieger-Liszkay A, Krämer U, and Thomine S. 2010. Export of vacuolar manganese by AtNRAMP3 and AtNRAMP4 is required for optimal photosynthesis and growth under manganese deficiency. *Plant Physiology* 152(4): 1986–1999.

Lee S, Kim YY, Lee Y, and An G. 2007. Rice P1B-type heavy-metal ATPase, OsHMA9, is a metal efflux protein. *Plant Physiology* 145(3): 831–842.

Li J, Wang Y, Zheng L, Li Y, Zhou X, Li J, Gu D, Xu E, Lu Y, Chen X, and Zhang W. 2019. The intracellular transporter AtNRAMP6 is involved in Fe homeostasis in Arabidopsis. *Frontiers in Plant Science* 10: 1124.

Li WX, Chen TB, Huang ZC, Lei M, and Liao XY. 2006. Effect of arsenic on chloroplast ultrastructure and calcium distribution in arsenic hyperaccumulator *Pteris vittata* L. *Chemosphere* 62(5): 803–809.

Liang J, Feng C, Zeng G, Zhong M, Gao X, Li X, He X, Li X, Fang Y, and Mo D. 2017. Atmospheric deposition of mercury and cadmium impacts on topsoil in a typical coal mine city, Lianyuan, China. *Chemosphere* 189: 198–205.

Liao QL, Liu C, Wu HY, Jin Y, Hua M, Zhu BW, Chen K, and Huang L. 2015. Association of soil cadmium contamination with ceramic industry: A case study in a Chinese town. *Science of the Total Environment* 514: 26–32.

Lin T, Zhu X, Zhang F, and Wan X. 2011b. The detoxification effect of nitrogen on cadmium stress in *Populus yunnanensis*. *Botany Research Journal* 4(1): 13–19.

Lin YL, Chao YY, Huang WD, and Kao CH. 2011a. Effect of nitrogen deficiency on antioxidant status and Cd toxicity in rice seedlings. *Plant Growth Regulation* 64(3): 263–273.

Liu D, Zou J, Wang M, and Jiang W. 2008. Hexavalent chromium uptake and its effects on mineral uptake, antioxidant defence system and photosynthesis in *Amaranthus viridis* L. *Bioresource Technology* 99(7): 2628–2636.

Lu YP, Li ZS, and Rea PA. 1997. AtMRP1 gene of Arabidopsis encodes a glutathione S-conjugate pump: isolation and functional definition of a plant ATP-binding cassette transporter gene. *Proceedings of the National Academy of Sciences* 94(15): 8243–8248.

Lu YP, Li ZS, Drozdowicz YM, Hörtensteiner S, Martinoia E, and Rea PA. 1998. AtMRP2, an Arabidopsis ATP binding cassette transporter able to transport glutathione S-conjugates and chlorophyll catabolites: functional comparisons with AtMRP1. *The Plant Cell* 10(2): 267–282.

Luo BF, Du ST, Lu KX, Liu WJ, Lin XY, and Jin CW. 2012. Iron uptake system mediates nitrate-facilitated cadmium accumulation in tomato (*Solanum lycopersicum*) plants. *Journal of Experimental Botany* 63(8): 3127–3136.

Lwalaba JLW, Zvobgo G, Fu L, Zhang X, Mwamba TM, Muhammad N, Mundende RPM, and Zhang G. 2017. Alleviating effects of calcium on cobalt toxicity in two barley genotypes differing in cobalt tolerance. *Ecotoxicology and Environmental Safety* 139: 488–495.

Maathuis FJ. 2009. Physiological functions of mineral macronutrients. *Current Opinion in Plant Biology* 12(3): 250–258.

Makino A. 2011. Photosynthesis, grain yield, and nitrogen utilization in rice and wheat. *Plant Physiology* 155(1): 125–129.

Malinouski M, Hasan NM, Zhang Y, Seravalli J, Lin J, Avanesov A, Lutsenko S, and Gladyshev VN. 2014. Genome-wide RNAi ionomics screen reveals new genes and regulation of human trace element metabolism. *Nature Communications* 5(1): 1–11.

Mallick S, Sinam G, and Sinha S. 2011. Study on arsenate tolerant and sensitive cultivars of *Zea mays* L.: differential detoxification mechanism and effect on nutrients status. *Ecotoxicology and Environmental Safety* 74(5): 1316–1324.

Manara A. 2012. Plant responses to heavy metal toxicity. pp. 27–53. *In: Plants and Heavy Metals*. Springer, Dordrecht.

Marschner H. 1995. Mineral nutrition of higher plants. 2nd (eds.) Academic Press. *New York*, pp. 15–22.

Marschner H. 2011. Marschner's mineral nutrition of higher plants. Academic press.

Martins LL, Reis R, Moreira I, Pinto F, Sales J, and Mourato M. 2013. Antioxidative response of plants to oxidative stress induced by cadmium. *Cadmium: Characteristics, Sources of Exposure, Health and Environmental Effects*, pp. 87–112.

Mäser P, Thomine S, Schroeder JI, Ward JM, Hirschi K, Sze H, Talke I, Amtmann A, Maathuis FJ, Sanders D, and Harper JF. 2001. Phylogenetic relationships within cation transporter families of Arabidopsis. *Plant Physiology* 126(4): 1646–1667.

Masood A, Iqbal N, and Khan NA. 2012. Role of ethylene in alleviation of cadmium-induced photosynthetic capacity inhibition by sulphur in mustard. *Plant, Cell & Environment* 35(3): 524–533.

Meda AR, Scheuermann EB, Prechsl UE, Erenoglu B, Schaaf G, Hayen H, Weber G, and von Wirén N. 2007. Iron acquisition by phytosiderophores contributes to cadmium tolerance. *Plant Physiology* 143(4): 1761–1773.

Meharg AA, and Hartley-Whitaker J. 2002. Arsenic uptake and metabolism in arsenic resistant and nonresistant plant species. *New Phytologist* 154(1): 29–43.

Memon AR. 2016. Metal hyperaccumulators: mechanisms of hyperaccumulation and metal tolerance. pp. 239–268. *In: Phytoremediation*. Springer, Cham.

Mendoza-Cózatl DG, Butko E, Springer F, Torpey JW, Komives EA, Kehr J, and Schroeder JI. 2008. Identification of high levels of phytochelatins, glutathione and cadmium in the phloem sap of *Brassica napus*. A role for thiol-peptides in the long-distance transport of cadmium and the effect of cadmium on iron translocation. *The Plant Journal* 54(2): 249–259.

Mengel K, and Kirkby EA. 1987. Principles of plant nutrition. Bern. *International Potash Institute*, pp. 687–695.

Merlot S., Garcia de la Torre VS, and Hanikenne M. 2021. Physiology and molecular biology of trace element hyperaccumulation. *In:* van der Ent A, Baker AJ, Echevarria G, Simonnot MO, and Morel JL (eds.). Agromining: Farming for Metals. *Mineral Resource Reviews*. Springer, Cham. https://doi.org/10.1007/978-3-030-58904-2_8.

Migocka M, Papierniak A, Kosieradzka A, Posyniak E, Maciaszczyk-Dziubinska E, Biskup R, Garbiec, A and Marchewka, T. 2015. Retracted: Cucumber metal tolerance protein Cs MTP 9 is a plasma membrane H+-coupled antiporter involved in the Mn2+ and Cd2+ efflux from root cells. *The Plant Journal* 84(6): 1045–1058.

Mills RF, Krijger GC, Baccarini PJ, Hall JL, and Williams LE. 2003. Functional expression of AtHMA4, a P1B-type ATPase of the Zn/Co/Cd/Pb subclass. *The Plant Journal* 35(2): 164–176.

Miyadate H, Adachi S, Hiraizumi A, Tezuka K, Nakazawa N, Kawamoto T, Katou K, Kodama I, Sakurai K, Takahashi H, and Satoh-Nagasawa N. 2011. OsHMA3, a P1B-type of ATPase affects root-to-shoot cadmium translocation in rice by mediating efflux into vacuoles. *New Phytologist* 189(1): 190–199.

Mokgalaka-Matlala NS, Flores-Tavizon E, Castillo-Michel H, Peralta-Videa JR, and Gardea-Torresdey JL. 2008. Toxicity of arsenic (III) and (V) on plant growth, element uptake, and total amylolytic activity of mesquite (*Prosopis juliflora* x P. velutina). *International Journal of Phytoremediation* 10(1): 47–60.

Montanini B, Blaudez D, Jeandroz S, Sanders D, and Chalot M. 2007. Phylogenetic and functional analysis of the Cation Diffusion Facilitator (CDF) family: improved signature and prediction of substrate specificity. *BMC Genomics* 8(1): 1–16.

Monteiro MS, Santos C, Soares AMVM, and Mann RM. 2009. Assessment of biomarkers of cadmium stress in lettuce. *Ecotoxicology and Environmental Safety* 72(3): 811–818.

Morel M, Crouzet J, Gravot A, Auroy P, Leonhardt N, Vavasseur A, and Richaud P. 2009. AtHMA3, a P1B-ATPase allowing Cd/Zn/co/Pb vacuolar storage in Arabidopsis. *Plant Physiology* 149(2): 894–904.

Mosali J, Desta K, Teal RK, Freeman KW, Martin KL, Lawles JW, and Raun WR. 2006. Effect of foliar application of phosphorus on winter wheat grain yield, phosphorus uptake, and use efficiency. *Journal of Plant Nutrition* 29(12): 2147–2163.

Mukhopadhyay MJ, and Sharma A. 1991. Manganese in cell metabolism of higher plants. *The Botanical Review* 57(2): 117–149.

Nevo Y, and Nelson N. 2006. The NRAMP family of metal-ion transporters. *Biochimica et Biophysica Acta (BBA)-Molecular Cell Research* 1763(7): 609–620.

Nwosu JU, Harding AK, and Linder G. 1995. Cadmium and lead uptake by edible crops grown in a silt loam soil. *Bulletin of Environmental Contamination and Toxicology* 54(4): 570–578.

Oliveira BRM, de Almeida AAF, Pirovani CP, Barroso JP, Neto CHDC, Santos NA, Ahnert D, Baligar VC, and Mangabeira PAO. 2020. Mitigation of Cd toxicity by Mn in young plants of cacao, evaluated by the proteomic profiles of leaves and roots. *Ecotoxicology* 29(3): 340–358.

Ortiz DF, Ruscitti T, McCue KF, and Ow DW. 1995. Transport of metal-binding peptides by HMT1, a fission yeast ABC-type vacuolar membrane protein. *Journal of Biological Chemistry* 270(9): 4721–4728.

Osu Charles I, and Onyema MO. 2016. Vanadium inhibition capacity on nutrients and heavy metal uptake by *Cucumis Sativus*. *J Am Sci*, 12.

Ovečka M, and Takáč T. 2014. Managing heavy metal toxicity stress in plants: biological and biotechnological tools. *Biotechnology Advances* 32(1): 73–86.

Öztürk MA (ed.). 1989. *Plants and Pollutants in Developed and Developing Countries*. Ege University.

Päivöke AE, and Simola LK. 2001. Arsenate toxicity to *Pisum sativum*: mineral nutrients, chlorophyll content, and phytase activity. *Ecotoxicology and Environmental Safety* 49(2): 111–121.

Pal'ove-Balang P, Kisová A, Pavlovkin J, and Mistrík I. 2006. Effect of manganese on cadmium toxicity in maize seedlings. *Plant Soil Environ* 52: 143–149.

Panković D, Plesničar M, Arsenijević-Maksimović I, Petrović N, Sakač Z, and Kastori R. 2000. Effects of nitrogen nutrition on photosynthesis in Cd-treated sunflower plants. *Annals of Botany* 86(4): 841–847.

Peng K, Luo C, You W, Lian C, Li X, and Shen Z. 2008. Manganese uptake and interactions with cadmium in the hyperaccumulator—Phytolacca *Americana* L. *Journal of Hazardous Materials* 154(1-3): 674–681.

Peng SB, Huang JL, Zhong XH, Yang JC, Wang GH, Zou YB, Zhang FS, Zhu QS, Buresh R, and Witt C. 2002. Challenge and opportunity in improving fertilizer-nitrogen use efficiency of irrigated rice in China. *Agricultural Sciences in China* 1(7): 776–785.

Pinson SRM, Tarpley L, Yan W, Yeater K, Lahner B, Yakubova E, Huang XY, Zhang M, Guerinot ML, and Salt DE. 2015. Worldwide genetic diversity for mineral element concentrations in rice grain. *Crop Science* 55(1): 294–311.

Poblaciones MJ, Damon P, and Rengel Z. 2017. Foliar zinc biofortification effects in *Lolium rigidum* and *Trifolium subterraneum* grown in cadmium-contaminated soil. *Plos one* 12(9): e0185395.

Poschenrieder C, and Barceló J. 2004. Water relations in heavy metal stressed plants. *In*: Prasad MNV (eds.). Heavy Metal Stress in Plants. Springer, Berlin, Heidelberg. https://doi.org/10.1007/978-3-662-07743-6_10.

Pottier M, Oomen R, Picco C, Giraudat J, Scholz-Starke J, Richaud P, Carpaneto A, and Thomine S. 2015. Identification of mutations allowing Natural Resistance Associated Macrophage Proteins (NRAMP) to discriminate against cadmium. *The Plant Journal* 83(4): 625–637.

Puig S, and Thiele DJ. 2002. Molecular mechanisms of copper uptake and distribution. *Current Opinion in Chemical Biology* 6(2): 171–180.

Qin L, Han P, Chen L, Walk TC, Li Y, Hu X, Xie L, Liao H, and Liao X. 2017. Genome-wide identification and expression analysis of NRAMP family genes in soybean (*Glycine max* L.). *Frontiers in Plant Science* 8: 1436.

Qureshi MI, D'Amici GM, Fagioni M, Rinalducci S, and Zolla L. 2010. Iron stabilizes thylakoid protein–pigment complexes in Indian mustard during Cd-phytoremediation as revealed by BN-SDS-PAGE and ESI-MS/MS. *Journal of Plant Physiology* 167(10): 761–770.

Rahman A, Nahar K, Hasanuzzaman M, and Fujita M. 2016. Manganese-induced cadmium stress tolerance in rice seedlings: Coordinated action of antioxidant defense, glyoxalase system, and nutrient homeostasis. *Comptes Rendus Biologies* 339(1112): 462–474.

Rana MS, Hu CX, Shaaban M, Imran M, Afzal J, Moussa MG, Elyamine AM, Bhantana P, Saleem MH, Syaifudin M, and Kamran M. 2020a. Soil phosphorus transformation characteristics in response to molybdenum supply in leguminous crops. *Journal of Environmental Management* 268: 110610.

Rana MS, Sun X, Imran M, Ali S, Shaaban M, Moussa MG, Khan Z, Afzal J, Binyamin R, Bhantana P, and Alam M. 2020b. Molybdenum-induced effects on leaf ultra-structure and rhizosphere phosphorus transformation in *Triticum aestivum* L. *Plant Physiology and Biochemistry* 153: 20–29.

Rausch T, and Wachter A. 2005. Sulfur metabolism: a versatile platform for launching defence operations. *Trends in Plant Science* 10(10): 503–509.

Razaq M, Zhang P, and Shen HL. 2017. Influence of nitrogen and phosphorous on the growth and root morphology of Acer mono. *PloS one* 12(2): e0171321.

Reale L, Ferranti F, Mantilacci S, Corboli M, Aversa S, Landucci F, Baldisserotto C, Ferroni L, Pancaldi S, and Venanzoni R. 2016. Cyto-histological and morpho-physiological responses of common duckweed (*Lemna minor* L.) to chromium. *Chemosphere* 145: 98–105.

Redondo-Gómez S, Mateos-Naranjo E, and Andrades-Moreno L. 2010. Accumulation and tolerance characteristics of cadmium in a halophytic Cd-hyperaccumulator, Arthrocnemum macrostachyum. *Journal of Hazardous Materials* 184(1-3): 299–307.

Rizwan M, Ali S, Adrees M, Ibrahim M, Tsang DC, Zia-ur-Rehman M, Zahir ZA, Rinklebe J, Tack FM, and Ok YS. 2017a. A critical review on effects, tolerance mechanisms, and management of cadmium in vegetables. *Chemosphere* 182: 90–105.

Rizwan M, Ali S, Hussain A, Ali Q, Shakoor MB, Zia-ur-Rehman M, Farid M, and Asma M. 2017b. Effect of zinc-lysine on growth, yield and cadmium uptake in wheat (*Triticum aestivum* L.) and health risk assessment. *Chemosphere* 187: 35–42.

Rodriguez-Hernandez MC, Bonifas I, Alfaro-De la Torre MC, Flores-Flores JL, Bañuelos-Hernández B, and Patiño-Rodríguez O. 2015. Increased accumulation of cadmium and lead under Ca and Fe deficiency in Typha latifolia: A study of two pore channel (TPC1) gene responses. *Environmental and Experimental Botany* 115: 38–48.

Römheld V, and Nikolic M. 2006. Iron. pp. 329–350. *In*: Barker AV, and Pilbeam DJ (eds.). Handbook of Plant Nutrition. Boca Raton: CRC Press.

Sabir M, Waraich EA, Hakeem KR, Öztürk M, Ahmad HR, and Shahid M. 2014. Phytoremediation: mechanisms and adaptations. *Soil Remediation and Plants: Prospects and Challenges* 85: 85–105.

Salt DE, Baxter I, and Lahner B. 2008. Ionomics and the study of the plant ionome. *Annu Rev Plant Biol* 59: 709–733.

Sarwar N, Malhi SS, Zia MH, Naeem A, Bibi S, and Farid G. 2010. Role of mineral nutrition in minimizing cadmium accumulation by plants. *Journal of the Science of Food and Agriculture* 90(6): 925–937.

Schaaf G, Schikora A, Häberle J, Vert G, Ludewig U, Briat JF, Curie C, and von Wirén N. 2005. A putative function for the Arabidopsis Fe–phytosiderophore transporter homolog AtYSL2 in Fe and Zn homeostasis. *Plant and Cell Physiology* 46(5): 762–774.

Scherer HW. 2008. Impact of Sulfur on N2 Fixation of Legumes. *In*: Khan NA, Singh S, and Umar S (eds.). Sulfur Assimilation and Abiotic Stress in Plants. Springer, Berlin, Heidelberg. https://doi.org/10.1007/978-3-540-76326-0_3.

Schulze J, Temple G, Temple SJ, Beschow H, and Vance CP. 2006. Nitrogen fixation by white lupin under phosphorus deficiency. *Annals of Botany* 98(4): 731–740.

Shahzad B, Tanveer M, Rehman A, Cheema SA, Fahad S, Rehman S, and Sharma A. 2018. Nickel; whether toxic or essential for plants and environment-A review. *Plant Physiology and Biochemistry* 132: 641–651.

Shaibur MR, Kitajima N, Sugawara R, Kondo T, Alam S, Huq SI, and Kawai S. 2008. Critical toxicity level of arsenic and elemental composition of arsenic-induced chlorosis in hydroponic sorghum. *Water, Air, and Soil Pollution* 191(1): 279–292.

Shanker AK, Sudhagar R, and Pathmanabhan G. 2003. Growth, phytochelatin SH and antioxidative response of sunflower as affected by chromium speciation. In *2nd International Congress of Plant Physiology on Sustainable Plant Productivity Under Changing Environment, New Delhi*.

Shanker AK, Djanaguiraman M, and Venkateswarlu B. 2009. Chromium interactions in plants: current status and future strategies. *Metallomics* 1(5): 375–383.

Sharma P, and Dubey RS. 2005. Lead toxicity in plants. *Brazilian Journal of Plant Physiology* 17(1): 35–52.

Sharma P, and Dubey RS. 2007. Involvement of oxidative stress and role of antioxidative defense system in growing rice seedlings exposed to toxic concentrations of aluminum. *Plant Cell Reports* 26(11): 2027–2038.

Sharma SS, and Dietz KJ. 2006. The significance of amino acids and amino acid-derived molecules in plant responses and adaptation to heavy metal stress. *Journal of Experimental Botany* 57(4): 711–726.

Shen J, Song L, Müller K, Hu Y, Song Y, Yu W, Wang H, and Wu J. 2016. Magnesium alleviates adverse effects of lead on growth, photosynthesis, and ultrastructural alterations of *Torreya grandis* seedlings. *Frontiers in Plant Science* 7: 1819.

Shen W, Nada K, and Tachibana S. 2000. Involvement of polyamines in the chilling tolerance of cucumber cultivars. *Plant Physiology* 124(1): 431–440.

Sidhu GPS, Singh HP, Batish DR, and Kohli RK. 2016. Effect of lead on oxidative status, antioxidative response and metal accumulation in *Coronopus didymus*. *Plant Physiology and Biochemistry* 105: 290–296.

Siemianowski O, Barabasz A, Kendziorek M, Ruszczyńska A, Bulska E, Williams LE, and Antosiewicz DM. 2014. HMA4 expression in tobacco reduces Cd accumulation due to the induction of the apoplastic barrier. *Journal of Experimental Botany* 65(4): 1125–1139.

Singh AP, Dixit G, Mishra S, Dwivedi S, Tiwari M, Mallick S, Pandey V, Trivedi PK, Chakrabarty D, and Tripathi RD. 2015. Salicylic acid modulates arsenic toxicity by reducing its root to shoot translocation in rice (*Oryza sativa* L.). *Frontiers in Plant Science* 6: 340.

Singh R, Parihar P, and Prasad SM. 2020b. Sulphur and calcium attenuate arsenic toxicity in *Brassica* by adjusting ascorbate–glutathione cycle and sulphur metabolism. *Plant Growth Regulation* 91(2): 221–235.

Singh S, Parihar P, Singh R, Singh VP, and Prasad SM. 2016. Heavy metal tolerance in plants: role of transcriptomics, proteomics, metabolomics, and ionomics. *Frontiers in Plant Science* 6: 1143.

Singh S, Sharma PK, and Rai A. 2018. Response of Sesame (*Sesamum indicum* L.) to sulphur and boron in upland red soil of Vindhyan zone. *Journal of the Indian Society of Soil Science* 66(4): 432–435.

Singh S, Prasad SM, and Singh VP. 2020a. Additional calcium and sulfur manages hexavalent chromium toxicity in *Solanum lycopersicum* L. and *Solanum melongena* L. seedlings by involving nitric oxide. *Journal of Hazardous Materials* 398: 122607.

Sinha P, Dube BK, Srivastava P, and Chatterjee C. 2006. Alteration in uptake and translocation of essential nutrients in cabbage by excess lead. *Chemosphere* 65(4): 651–656.

Spadaro D, Yun BW, Spoel SH, Chu C, Wang YQ, and Loake GJ. 2010. The redox switch: dynamic regulation of protein function by cysteine modifications. *Physiologia Plantarum* 138(4): 360–371.

Stoeva N, and Bineva, T. 2003. Oxidative changes and photosynthesis in oat plants grown in As-contaminated soil. *Bulg J Plant Physiol* 29(1-2): 87–95.

Stoeva N, Berova M, and Zlatev Z. 2005. Effect of arsenic on some physiological parameters in bean plants. *Biologia Plantarum* 49(2): 293–296.

Su Y, Liu J, Lu Z, Wang X, Zhang Z, and Shi G. 2014. Effects of iron deficiency on subcellular distribution and chemical forms of cadmium in peanut roots in relation to its translocation. *Environmental and Experimental Botany* 97: 40–48.

Sun X, Hu C, Tan Q, Liu J, and Liu H. 2009. Effects of molybdenum on expression of cold-responsive genes in abscisic acid (ABA)-dependent and ABA-independent pathways in winter wheat under low-temperature stress. *Annals of Botany* 104(2): 345–356.

Sundaramoorthy P, Chidambaram A, Ganesh KS, Unnikannan P, and Baskaran L. 2010. Chromium stress in paddy: (i) nutrient status of paddy under chromium stress; (ii) phytoremediation of chromium by aquatic and terrestrial weeds. *Comptes Rendus Biologies* 333(8): 597–607.

Suzuki N. 2005. Alleviation by calcium of cadmium-induced root growth inhibition in *Arabidopsis* seedlings. *Plant Biotechnology* 22(1): 19–25.

Taiz L, and Zeiger E. 2010. Plant Physiology, 5e. Sunderland, MA, USA: Sinauer.

Takahashi R, Ishimaru Y, Senoura T, Shimo H, Ishikawa S, Arao T, Nakanishi H, and Nishizawa NK. 2011. The OsNRAMP1 iron transporter is involved in Cd accumulation in rice. *Journal of Experimental Botany* 62(14): 4843–4850.

Takahashi R, Bashir K, Ishimaru Y, Nishizawa NK, and Nakanishi H. 2012. The role of heavy-metal ATPases, HMAs, in zinc and cadmium transport in rice. *Plant Signaling & Behavior* 7(12): 1605–1607.

Talke IN, Hanikenne M, and Krämer U. 2006. Zinc-dependent global transcriptional control, transcriptional deregulation, and higher gene copy number for genes in metal homeostasis of the hyperaccumulator *Arabidopsis halleri*. *Plant Physiology* 142(1): 148–167.

The Sulphur Institute. About Sulphur. The Sulphur Institute, Washington, DC 2010.

Thomas CL, Alcock TD, Graham NS, Hayden R, Matterson S, Wilson L, Young SD, Dupuy LX, White PJ, Hammond JP, and Danku JMC. 2016. Root morphology and seed and leaf ionomic traits in a *Brassica napus* L. diversity panel show wide phenotypic variation and are characteristic of crop habit. *BMC Plant Biology* 16(1): 1–18.

Thomine S, Wang R, Ward JM, Crawford NM, and Schroeder JI. 2000. Cadmium and iron transport by members of a plant metal transporter family in Arabidopsis with homology to Nramp genes. *Proceedings of the National Academy of Sciences* 97(9): 4991–4996.

Thounaojam TC, Panda P, Choudhury S, Patra HK, and Panda SK. 2014. Zinc ameliorates copper-induced oxidative stress in developing rice (*Oryza sativa* L.) seedlings. *Protoplasma* 251(1): 61–69.

Tiong J, McDonald GK, Genc Y, Pedas P, Hayes JE, Toubia J, Langridge P, and Huang CY, 2014. H v ZIP 7 mediates zinc accumulation in barley (*Hordeum vulgare*) at moderately high zinc supply. *New Phytologist* 201(1): 131–143.

Tripathi DK, Singh VP, Chauhan DK, Prasad SM, and Dubey NK. 2014. Role of macronutrients in plant growth and acclimation: recent advances and future prospective. *Improvement of Crops in the Era of Climatic Changes*, pp. 197–216.

Tripathi RD, Srivastava S, Mishra S, Singh N, Tuli R, Gupta DK, and Maathuis FJ. 2007. Arsenic hazards: strategies for tolerance and remediation by plants. *Trends in Biotechnology* 25(4): 158–165.

Tu C, and Ma LQ. 2002. Effects of arsenic concentrations and forms on arsenic uptake by the hyperaccumulator ladder brake. *Journal of Environmental Quality* 31(2): 641–647.

Tu C, and Ma LQ. 2005. Effects of arsenic on concentration and distribution of nutrients in the fronds of the arsenic hyperaccumulator *Pteris vittata* L. *Environmental Pollution* 135(2): 333–340.

UdDin I, Bano A, and Masood S. 2015. Chromium toxicity tolerance of *Solanum nigrum* L. and *Parthenium hysterophorus* L. plants with reference to ion pattern, antioxidation activity, and root exudation. *Ecotoxicology and Environmental Safety* 113: 271–278.

Ueno D, Yamaji N, Kono I, Huang CF, Ando T, Yano M, and Ma JF. 2010. Gene limiting cadmium accumulation in rice. *Proceedings of the National Academy of Sciences* 107(38): 16500–16505.

Uzu G, Sobansk S, Sarret G, Munoz M, and Dumat C. 2010. Foliar lead uptake by lettuce exposed to atmospheric fallouts. *Environmental Science & Technology* 44(3): 1036–1042.

van de Mortel JE, Villanueva LA, Schat H, Kwekkeboom J, Coughlan S, Moerland PD, van Themaat EVL, Koornneef M, and Aarts MG. 2006. Large expression differences in genes for iron and zinc homeostasis, stress response, and lignin biosynthesis distinguish roots of *Arabidopsis thaliana* and the related metal hyperaccumulator *Thlaspi caerulescens*. *Plant Physiology* 142(3): 1127–1147.

Vance CP, Uhde-Stone C, and Allan DL. 2003. Phosphorus acquisition and use: critical adaptations by plants for securing a nonrenewable resource. *New Phytologist* 157(3): 423–447.

Vázquez S, Esteban E, and Carpena RO. 2008. Evolution of arsenate toxicity in nodulated white lupine in a long-term culture. *Journal of Agricultural and Food Chemistry* 56(18): 8580–8587.

Verret F, Gravot A, Auroy P, Leonhardt N, David P, Nussaume L, Vavasseur A, and Richaud P. 2004. Overexpression of AtHMA4 enhances root-to-shoot translocation of zinc and cadmium and plant metal tolerance. *FEBS Letters* 576(3): 306–312.

Villiers F, Ducruix C, Hugouvieux V, Jarno N, Ezan E, Garin J, Junot C, and Bourguignon J. 2011. Investigating the plant response to cadmium exposure by proteomic and metabolomic approaches. *Proteomics* 11(9): 1650–1663.

Wang H, Zhao SC, Liu RL, Zhou W, and Jin JY. 2009. Changes of photosynthetic activities of maize (*Zea mays* L.) seedlings in response to cadmium stress. *Photosynthetica* 47(2): 277–283.

Wang M, Yang Y, and Chen W. 2018. Manganese, zinc, and pH affect cadmium accumulation in rice grain under field conditions in southern China. *Journal of Environmental Quality* 47(2): 306–311.

Wang M, Zheng Q, Shen Q, and Guo S. 2013. The critical role of potassium in plant stress response. *International Journal of Molecular Sciences* 14(4): 7370–7390.

Wangeline AL, Burkhead JL, Hale KL, Lindblom SD, Terry N, Pilon M, and Pilon-Smits EA. 2004. Overexpression of ATP sulfurylase in Indian mustard: effects on tolerance and accumulation of twelve metals. *Journal of Environmental Quality* 33(1): 54–60.

Waraich EA, Ahmad R, and Ashraf MY. 2011. Role of mineral nutrition in alleviation of drought stress in plants. *Australian Journal of Crop Science* 5(6): 764–777.

Watanabe T, Hanan JS, Room PM, Hasegawa T, Nakagawa H, and Takahashi W. 2005. Rice morphogenesis and plant architecture: measurement, specification and the reconstruction of structural development by 3D architectural modelling. *Annals of Botany* 95(7): 1131–1143.

Welch RM, and Graham RD. 2004. Breeding for micronutrients in staple food crops from a human nutrition perspective. *Journal of Experimental Botany* 55(396): 353–364.

White PJ, and Broadley MR. 2009. Biofortification of crops with seven mineral elements often lacking in human diets–iron, zinc, copper, calcium, magnesium, selenium and iodine. *New Phytologist* 182(1): 49–84.

White PJ, and Greenwood DJ. 2013. Properties and management of cationic elements for crop growth. *Soil Conditions and Plant Growth* 12: 160–94.

Williams LE, Pittman JK, and Hall JL. 2000. Emerging mechanisms for heavy metal transport in plants. *Biochimica et Biophysica Acta (BBA)-Biomembranes* 1465(1-2): 104–126.

Wong CKE, and Cobbett CS. 2009. HMA P-type ATPases are the major mechanism for root-to-shoot Cd translocation in *Arabidopsis thaliana*. *New Phytologist* 181(1): 71–78.

Wong CKE, Jarvis RS, Sherson SM, and Cobbett CS. 2009. Functional analysis of the heavy metal binding domains of the Zn/Cd-transporting ATPase, HMA2, in *Arabidopsis thaliana*. *New Phytologist* 181(1): 79–88.

Wu F, and Zhang G. 2002. Alleviation of cadmium-toxicity by application of zinc and ascorbic acid in barley. *Journal of Plant Nutrition* 25(12): 2745–2761.

Wysocki R, Bobrowicz P, and Ułaszewski S. 1997. The *Saccharomyces cerevisiae* ACR3 gene encodes a putative membrane protein involved in arsenite transport. *Journal of Biological Chemistry* 272(48): 30061–30066.

Xu W, Dai W, Yan H, Li S, Shen H, Chen Y, Xu H, Sun Y, He Z, and Ma M. 2015. Arabidopsis NIP3; 1 plays an important role in arsenic uptake and root-to-shoot translocation under arsenite stress conditions. *Molecular Plant* 8(5): 722–733.

Yan J, Wang P, Wang P, Yang M, Lian X, Tang Z, Huang CF, Salt DE, and Zhao FJ. 2016. A loss-of-function allele of OsHMA3 associated with high cadmium accumulation in shoots and grain of Japonica rice cultivars. *Plant, Cell & Environment* 39(9): 1941–1954.

Yang S, Wang F, Guo F, Meng JJ, Li XG, and Wan SB. 2015. Calcium contributes to photoprotection and repair of photosystem II in peanut leaves during heat and high irradiance. *Journal of Integrative Plant Biology* 57(5): 486–495.

Yasin NA, Zaheer MM, Khan WU, Ahmad SR, Ahmad A, Ali A, and Akram W. 2018. The beneficial role of potassium in Cd-induced stress alleviation and growth improvement in *Gladiolus grandiflora* L. *International Journal of Phytoremediation* 20(3): 274–283.

Yilmaz K, Akinci İE, and Akinci S. 2009. Effect of lead accumulation on growth and mineral composition of eggplant seedlings (*Solanum melongena*). *New Zealand Journal of Crop and Horticultural Science* 37(3): 189–199.

Yu D, Danku JM, Baxter I, Kim S, Vatamaniuk OK, Vitek O, Ouzzani M, and Salt DE. 2012. High-resolution genome-wide scan of genes, gene-networks and cellular systems impacting the yeast ionome. *BMC Genomics* 13(1): 1–25.

Yujing CUI, Zhang X, and Yongguan ZHU. 2008. Does copper reduce cadmium uptake by different rice genotypes? *Journal of Environmental Sciences* 20(3): 332–338.

Zaccheo P, Crippa L, and Pasta VDM. 2006. Ammonium nutrition as a strategy for cadmium mobilisation in the rhizosphere of sunflower. *Plant and Soil* 283(1): 43–56.

Zaheer IE, Ali S, Rizwan M, Abbas Z, Bukhari SAH, Wijaya L, Alyemeni MN, and Ahmad P. 2019. Zinc-lysine prevents chromium-induced morphological, photosynthetic, and oxidative alterations in spinach irrigated with tannery wastewater. *Environmental Science and Pollution Research* 26(28): 28951–28961.

Zaheer IE, Ali S, Saleem MH, Imran M, Alnusairi GS, Alharbi BM, Riaz M, Abbas Z, Rizwan M, and Soliman MH. 2020a. Role of iron–lysine on morpho-physiological traits and combating chromium toxicity in rapeseed (*Brassica napus* L.) plants irrigated with different levels of tannery wastewater. *Plant Physiology and Biochemistry* 155: 70–84.

Zaheer IH, Ali S, Saleem MH, Ashraf MAA, Ali Q, Abbas Z, Rizwan M, El-Sheikh MA, Alyemeni MN, and Wijaya L. 2020b. Zinc-lysine supplementation mitigates oxidative stress in rapeseed (*Brassica napus* L.) by preventing phytotoxicity of chromium, when irrigated with tannery wastewater. *Plants* 9(9): 1145.

Zaid A, and Mohammad F. 2018. Methyl jasmonate and nitrogen interact to alleviate cadmium stress in *Mentha arvensis* by regulating physio-biochemical damages and ROS detoxification. *Journal of Plant Growth Regulation* 37(4): 1331–1348.

Zhang F, Wan X, and Zhong Y. 2014a. Nitrogen as an important detoxification factor to cadmium stress in poplar plants. *Journal of Plant Interactions* 9(1): 249–258.

Zhang M, Hu C, Zhao X, Tan Q, Sun X, Cao A, Cui M, and Zhang, Y. 2012. Molybdenum improves antioxidant and osmotic-adjustment ability against salt stress in Chinese cabbage (*Brassica campestris* L. ssp. Pekinensis). *Plant Soil* 355: 375–383.

Zhang Z, Liao H, and Lucas WJ. 2014b. Molecular mechanisms underlying phosphate sensing, signaling, and adaptation in plants. *Journal of Integrative Plant Biology* 56(3): 192–220.

Zhao AQ, Tian XH, Lu WH, Gale WJ, Lu XC, and Cao YX. 2011. Effect of zinc on cadmium toxicity in winter wheat. *Journal of Plant Nutrition* 34(9): 1372–1385.

Zhao FJ, Ma JF, Meharg AA, and McGrath SP. 2009. Arsenic uptake and metabolism in plants. *New Phytologist* 181(4): 777–794.

Zhao FJ, McGrath SP, and Meharg AA. 2010. Arsenic as a food chain contaminant: mechanisms of plant uptake and metabolism and mitigation strategies. *Annual Review of Plant Biology* 61: 535–559.

Zhao S, He P, Qiu S, Jia L, Liu M, Jin J, and Johnston AM. 2014. Long-term effects of potassium fertilization and straw return on soil potassium levels and crop yields in north-central China. *Field Crops Research* 169: 116–122.

Zhao ZQ, Zhu YG, Li HY, Smith SE, and Smith FA. 2004. Effects of forms and rates of potassium fertilizers on cadmium uptake by two cultivars of spring wheat (*Triticum aestivum*, L.). *Environment International* 29(7): 973–978.

Zhao ZQ, Zhu YG, Kneer R, and Smith SE. 2005. Effect of zinc on cadmium toxicity-induced oxidative stress in winter wheat seedlings. *Journal of Plant Nutrition* 28(11): 1947–1959.

Zhu E, Liu D, Li JG, Li TQ, Yang XE, He ZL, and Stoffella PJ. 2010. Effect of nitrogen fertilizer on growth and cadmium accumulation in *Sedum alfredii* Hance. *Journal of Plant Nutrition* 34(1): 115–126.

Ziegler G, Terauchi A, Becker A, Armstrong P, Hudson K, and Baxter I. 2013. Ionomic screening of field-grown soybean identifies mutants with altered seed elemental composition. *The Plant Genome* 6(2): 1–9.

Zornoza P, Sánchez-Pardo B, and Carpena RO. 2010. Interaction and accumulation of manganese and cadmium in the manganese accumulator *Lupinus albus*. *Journal of Plant Physiology* 167(13): 1027–1032.

14

Genomics (Characterization of Genes) and Molecular Aspects of Metal Tolerance and Hyperaccumulation

Aparna Pandey,[1] *Sheo Mohan Prasad*[1,]* and *Jitendra Kumar*[1,2,]*

ABSTRACT

The ever increasing concentration of heavy metals in the environment causes serious disorder in ecological balance that is manifesting in decreasing the growth and productivity of plants. The world's scientists are developing strategies to cope up with this heavy metal stress episode. In this concern OMICS approach has proven to be a panacea that includes an elaborative approach to study metal stress tolerance in plants. It elucidates the plant genes for metal tolerance by studying the constitutive gene over-expression in response to heavy metals and thus in the identification of metal tolerant genes. However, antioxidant enzymes play an important role in metal tolerance response thus, identification of genes involved in metal stress tolerance is carried *via* genetic screens, also real-time analysis of antioxidant enzymes is essential for further researches. At the same time recent studies report, the important role of microRNAs in heavy metal stress tolerance as well as epigenetic changes such as DNA methylation that are carried forward in next generations conferring metal stress tolerance in plants. In addition to above mentioned aspects, heavy metal stress-mediated MAPK signaling response against heavy metal stress in plants also becomes a matter of concern during tolerance strategies of plants. Therefore, keeping the above facts into consideration this chapter is an attempt to elucidate the current omics approaches towards plant tolerance against heavy metal stress.

[1] Ranjan Plant Physiology and Biochemistry Laboratory, Department of Botany, University of Allahabad, Prayagraj, U.P. 211002 (INDIA).
[2] Institute of Engineering and Technology, Dr. Shakuntala Misra National Rehabilitation University, MohanRoad, Lucknow, U.P. 226017 (INDIA).
Email: apsaa24@gmail.com
* Corresponding authors: profsmprasad@gmail.com; jitendradhuria@gmail.com

1. Introduction

To overcome the heavy metal stress, the plants execute responses at genetic, morphological, biochemical, and physiological levels. The fields which deal with the study of gene expressions, metabolites, proteins, trace elements as stress tolerance mechanisms are called genomics, metabolomics, proteomics, and ionomics, respectively. The increasing industrialisation, development of factories based on heavy metals such as for the development of pesticides, batteries, fuels, etc., have led to the increment of heavy metals in the agricultural fields *via* their industrial effluents. However, these heavy metal stress cause a huge negative impact on the plants' growth. This chapter discusses the activation of genes, the heavy metal transporters (present at vacuoles, plasmamembrane), and various epigenetic changes which take place in the plants under different kinds of heavy metal stresses. It also throws light on the techniques such as microarray, GC-MS, NMR, etc., which help researchers in the quantification of mRNA expressions, metabolites, and proteins produced under the stressed conditions. This chapter aims to provide the important aspects of this knowledge which can be used in genetic engineering to lead to the development of plant varieties that will be able to combat different heavy metal stresses.

1.1 'OMICS' approaches for metal stress tolerance in plants

The OMICS approach signifies the combination of genomics, transcriptomics, proteomics, metabolomics, ionomics, and phenomics to understand the physiological and biochemical responses at genetic levels in plants against the stressed conditions (Chaudhary et al., 2019). Metabolomics deals with characterisation and quantification of low molecular weight molecules or metabolites under stress (Singh et al., 2016), proteomics involves the study of the expression of proteins at a large scale in stressed conditions. Plants adapt themselves to the stressed conditions by regulating gene expressions, proteins, and metabolites (Chaudhary et al., 2019).

Researches in the field of genomics are contributing to the development of new varieties which are more tolerant to abiotic stress (Chaudhary et al., 2019). Molecular markers are used in the identification of genes that help in genetics and breeding in plants (Kordrostami and Rahimi, 2015). Microarray helps in the identification of variable expression of transcripts of a gene under abiotic stressed conditions of salinity and drought induced ABA, etc. (Baillo et al., 2019), this also elucidates the function of important genes in stress tolerance.

To understand the proteins expressed under stress conditions, techniques such as mass spectrometry (MS) along with computational methods, matrix-assisted laser desorption ionization-time of flight (MALDI TOF), two-dimensional (2-D) gel electrophoresis, western blot, and ELISA assisted with bioinformatics tools are widely used. Label free quantitation procedures reveal protein expressions lines (Kim et al., 2019). However, understanding about the metabolites generated (especially low molecular weight molecules) under stress conditions are done in metabolomics. Metabolomics are carried through NMR (nuclear magnetic resonance), GC-MS (Gas Chromatography-Mass Spectrometry) and CE-MS (capillary electrophoresis–Mass spectrometry (Salem et al., 2020) (Fig. 1). Inductively Coupled Plasma-Mass

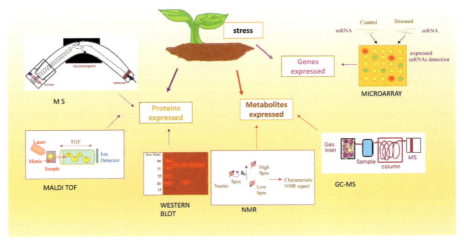

Figure 1: Illustrating the application of different OMICS approaches through different techniques in understanding plants' response to stress.

Spectrometry (ICP-MS) and ICP-Atomic Emission Spectrometry (ICP-AES) detects the presence of metals, non-metals and metalloids in the stress exposed plants, such studies are included in the ionomics. Furthermore, phenomics gives a detailed analysis of phenotypic variation in the stressed plants.

The genome and phenome (a set of all phenotypes) studies performed with individuals or with large populations are complementary to each other (Diouf et al., 2020). The tolerant phenotypes are good genomic resources to identify the tolerant alleles through high throughput sequencing. Genotyping efficiency is improved with advanced sequencing technologies in comparison to (phenotyping methods) quantitative traits characterisation specific to stress tolerance. However, advanced imaging systems, sensors, automation, and computational resources have enabled the broad level plant phenotype study (Yang et al., 2020).

1.2 Plant genes for metal tolerance

Several studies have been conducted on the plant genes involved in heavy metal transports, homeostasis, and tolerance which may be applied in the genetic modifications of plants in phytoremediation. Based on the similarities in the structures, temporal and semantic expressions, Metal tolerance proteins or Plant CDF family members are divided into seven groups (i.e., groups 1,5,6,7,8,9 and 12). Out of the twelve metal tolerance proteins genes which are found in *Arabidopsis*, *AtMTP1* and *AtMTP3* (of group 1) participate in the Zn homeostasis. They are involved in its transport and are present on the vacuolar membrane (Gao et al., 2020). Similarly, proteins AtMTP5 and AtMTP12 are involved in Zn transport at Golgi (Fig. 2). AtMTP8 is an Mn transporter and determines Mn transport. AtMTP11 also participates in Mn transport and tolerance at prevacuolar compartments or trans-Golgi.

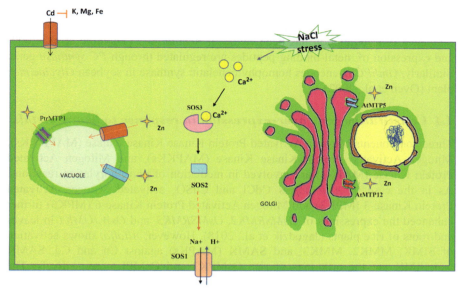

Figure 2: The figure shows different transporters which are involved in the storage of heavy metals in vacuoles and pathways executed through SOS3 protein in the activation of Na^+/H^+ antiporter under NaCl stress.

However, 72 metal transporters and 22 PtMTPs from *Populus trichocarpa* v 3.0 genome have been sequenced and analyzed. Under the influence of different heavy metal stresses the expression profiles of the whole *PtrMTP* family has been explored in various parts of the plant *via* yeast assay thus giving an understanding of heavy metal transports in the plants (Gao et al., 2020) and most *PtrMTP* genes are induced by more than two heavy metal ions in various parts of the plants. In Poplar, PtrMTP1 is a homo-oligomer present at vacuolar membrane involved in Zn tolerance and transport (Fig. 2). PtrMTP6 has been shown to transport three different heavy metal ions, i.e., Co, Fe, and Mn in yeast cells whereas PtrMTP8.1, PtrMTP9, and PtrMTP10.4 behave as Mn transporters.

1.3 Plant genotypic differences under metal enriched conditions

Metal enriched conditions

Toxic metal ions show different abilities to bind the various ligands such as metal ions which form stable complexes with oxygen (O) belong to Class A whereas those heavy metals forming complexes with sulphur (S) and nitrogen (N) come under Class B. Due to increasing industrialisation thus release of industrial wastes, application of fertilizers, pesticides, etc., have contributed to the irreversible persistence of heavy metals in agricultural fields (Verma et al., 2021). The heavy metals hamper the growth of photoautotrophs present in the environment. Plants express several heavy metal genes and proteins to tolerate such excess stress conditions. The heavy metals such as cadmium interferes with uptake of K, Mg, Fe and irreversibly replacing micronutrients Cu, Zn, essentially required by enzymes involved in DNA replication, transcription and translation in the plants. However, expression of Cd resistance gene *AtNramp* which encodes metal transporters to regulate Cd accumulation has been found in *Arabidopsis* (Thomine et al., 2000) and expression of metal-binding peptides are regulated through *PC synthase* gene. Similarly, *GmhPCS1* encodes homophytochelatin synthase in soybean *Glycine max* plant (Oven et al., 2002).

1.4 Constitutive gene over-expression in response to heavy metals

Three components Mitogen Activated Protein Kinase Kinase Kinase (MAPKKKs), Mitogen Activated Protein Kinase Kinase (MAPKKs), and Mitogen Activated Protein Kinase (MAPKs) are involved in mediation of phosphorylation reactions. As in the case of *Arabidopsis* $CdCl_2$ and $CuSO_4$ activated Mitogen Activated Protein Kinase-3 (MPK3) and Mitogen Activated Protein Kinase-6 (MPK6) further enhanced the expressions of *OsMSRMK2*, *OsMSRMK3*, and *OsWJUMK1* in leaves and roots of rice plants (Jagodzik et al., 2018). However, *Alfalfa*, shows activation of SIMK, MMK2, MMK3, and SAMK (MAPKs) against Cu and Cd. SAMK and SIMK are the orthologs of rice OsMPK3 and OsMPK6, respectively. The specificity in the signaling is seen in the cascade on induction of SIMK *via* SIMKK by Cu and not of MMK2 and MMK3. MAP kinase genes are involved against the Al^{3+} tolerance. In wheat, under Al^{3+} exposure transmission of signals is supposed because of the activation of MAPK, MAPK activates myelin basic protein (MBP) in rice. Upregulation of MAPKs (*MAPKKK7, MAPK6, MAPK18*, and *MAPK20*) is reported in Pb stress (Jalmi et al., 2018). Arsenite stress induces *OsMKK4* and *OsMPK3* transcripts in leaves and roots of rice. The activation of MAPK *via* participation of nitric oxide (NO) signaling is seen under As and Cr stress (Rao et al., 2011).

1.5 Identification of metal tolerant genes in plants

Real time analysis for antioxidant enzymes

Stressed conditions cause excessive ROS generation which damage *via* oxidation of cell organelles and significant biomolecules such as nucleic acids, lipids, proteins inside the cell. To be tolerant against the damage caused by oxidative stress, plants possess high antioxidant enzymes (Verma et al., 2021). Quantitative Real Time Analysis has revealed that expression of antioxidant enzymes genes of CAT, POD, and APX (*IbAPX, IbSOD, IbPOD*, and *IbCAT3*) were high in tolerant sweet potato and enhanced in fenugreek seedlings under heavy metal stress (Cd, Cr, Pb) (Alaraidh et al., 2018; Tang et al., 2019). However, Pb caused the highest expression of these enzymes around 1.33 to 1.63 fold than in untreated seedlings. In this study, QTL mapping and GWAS together identify a chromosomal locus of a particular trait (Chaudhary et al., 2019).

1.6 Genetic screens to identify metal stress genes

On screening of salt overly sensitive (*sos*) mutants of *Arabidopsis* which showed root curling and elongation, the regulatory cascade involved in ion balance has been revealed. The SOS pathway, detects and promotes Ca^{2+} signals, SOS3, a calcineurin B-type calcium binding protein responds to intracellular Ca^{2+} concentrations, further SOS3 activates protein kinase (SnRK3-type SOS2 protein kinase), the kinase activates membrane ion transporters such as H^+/Ca^{2+} antiporter CAX1, vacuolar H^+-ATPase and plasmamembrane Na^+/H^+ antiporter SOS1 (Gong et al., 2000) which help plant to maintain ionic balance.

Genetic screens enable us to identify the mutations in the regulatory genes which regulate the expression of a reporter gene. To identify the particular gene regulated *via* some pathway different reporter gene constructs enable to isolate and identify the specific pathways which control these reporter genes (Nakashima et al., 2009). The variable expression level of osmotic response genes has been identified through non-destructive bioluminescence imaging which has given an insight into *cos*, *los* and *hos* mutants which show their constitutive, low, or high expressions, respectively (Papdi et al., 2009). Luciferase screens also help to identify significant regulatory components of salt, cold, and ABA signaling pathways (Min et al., 2020). The RD29A-LUC reporter system has been used in the identification of several alleles of transcriptional activator FIERY2 which control the expression of genes with DRE/CRT-type *cis* elements in their promoters (Xiong et al., 2002). However, several abiotic stresses utilise the ABA-dependent pathway to induce the *Arabidopsis RCI2A* gene. Further, changes in *RCI2A-LUC* activity enabled the identification of mutants with impaired tolerance to freezing, dehydration, and salt stress. *Arabidopsis* redox imbalanced (*rimb*) mutants have been unable to induce 2-cysteine peroxiredoxin A (*2CPA*) gene transcription which has been detected *via* transgenic *Arabidopsis* containing *2CPA-LUC* reporter gene (Papdi et al., 2009).

1.7 Approaches to identify genes for metal tolerance

It is interesting to know whether locus specific changes take place against heavy metal stress responses. In rice seedlings, gene expressions under heavy metal stress have been studied through Reverse Transcription–Polymerase Chain Reaction (RT)-PCR. The nine genes *OsHMA(1-9)* are heavy metal transporters, two genes *Tos17* and *Osr42* are upregulated, they show epigenetic changes, and others are involved in DNA replication, translation, and other processes (heat shock proteins, HSP70) (Cong et al., 2019). The epigenetic changes under Cu, Cd, Cr, and Hg stress have been studied through semi-quantitative RT-PCR analysis of above mentioned genes.

1.8 Gene regulation during metal stress tolerance in plants

In order to investigate the mutations caused by heavy metal stress response, genome wide mapping is performed. High density linkage maps available for various important crop plants (such as *Triticum aestivum* L., *Oryza sativa* L., *Hordeum vulgare* L., *Solanum tuberosum* L., *Lycopersicom esculentum* L., *Brassica* sp., *Glycine max* L., *Carica papaya* L., etc.) easily helps to detect any mutations through bands detections (Yu et al., 2020). Microarray helps in the identification of genetic changes in the plants with big genomes whose genomic sequence information, marker densities are not known. Microarray had led to identifying a 523 bp deletion in the AtHKT1 gene in a salt hypersensitive mutant (Papdi et al., 2009). The deletion and inactivation of AtHKT1 confirmed its role in the salt sensitivity of mutant through segregation analysis and mutant complementation (Gong et al., 2004). With increasing advancement in technologies for sequencing of genomes, it is easy to do mapping of mutations and polymorphism in plant species.

1.9 Regulation of microRNAs during heavy metal stress

MicroRNAs are 20–24 nucleotide long, non-coding RNAs which perform the targeted mRNA degradation thus regulating gene expression at a post-transcriptional level under heavy metal stressed conditions (Waititu et al., 2020). Therefore, different miRNA families are regulated temporally and spatially based on the target and physiology of miRNA. Transcriptome analysis gives the details of the regulation of conserved microRNAs in heavy metal stress responses. Heavy metals such as Hg, Pb, and Cr have been shown to affect the upregulation and downregulation of miRNA expression (Jalmi et al., 2018). This throws light towards possible significant roles of miRNAs in heavy metal stress tolerance (Noman et al., 2019). The miRNAs also regulate signaling pathways. Under exposure to Cd stress, down-regulation of 12 miRNAs have been found such as miR192 which targets ABC transporter transcripts that sequester Cd under stress thus mutant plants overexpressing miR192 showed reduced seedling growth under Cd stress (Jalmi et al., 2018). Similarly, miR159 and miR166 which target heavy metal detoxification genes such as Cu–Zn superoxide dismutase (*CSD*) (which detoxifies superoxide radicals) are downregulated in the stressed plants. The downregulation of *miR156, miR395, miR398, miR159* has been seen under Al stress in rice whereas similar miRNAs showed upregulation in maize plants (Liu et al., 2019). However, studies have also revealed about As responsive downregulation of *miR172* in rice and mustard. The upregulation of miRNAs under stress includes *miR441, miR393, miR397, miR408 and miR399, miR166, miR168* under Al stress in rice (Djami-Tchatchou et al., 2017). Interestingly, ROS also leads to downregulation of *miR397* that targets the laccase enzyme to promote positive regulation of lignin biosynthesis to overcome lipid peroxidation. Genes that participate in the protection of plants under stress are promoted via the downregulation of miRNAs which degrade them such as synthesis of glutathione and phytochelatins are enhanced by promoting genes *ATP sulfurylase* and *SULTR2:1* (Jalmi et al., 2018).

1.10 Epigenome and metal stress tolerance in plants

Epigenetic changes regulate responses against heavy metal stress, mainly DNA methylation. However, in rice after Al exposure more expression of the methylated genes (methylated cytosines) have been reported which are present 1000 bps upstream and downstream of transcription site, associated with Al tolerance (He et al., 2020). To understand the specific effects of methylated regions in gene expression, research groups removed the duplicated genes and SNP (single nucleotide polymorphism), they found more methylation in the more tolerant variety (Nipponbare) of rice, therefore, revealing the important role of methylation under gene expression in Al stress tolerance.

1.11 DNA Methylation

DNA Methylation, at promoters, repress gene expression whereas in the gene body activates gene expression (Zhang et al., 2018) thus it regulates gene expressions without changing DNA sequence, thus being an epigenetic way. Compared to the salinity susceptible wheat cultivar JN177 the salinity tolerant wheat cultivar SR3 showed enhanced expression of the *TaFLS1* gene (a flavonol synthase gene) due to low methylation at its promoter (Wang et al., 2014). DNA methylation enables plants to adapt against salinity and heavy metal stress.

Multidrug resistance-associated protein (MRP) type ATP-binding cassette (ABC) transporters play a major role in heavy metal clearance (Granitzer et al., 2020). DNA methylation of promoter associated with TaABCCs and Heavy Metal ATPase 2 (TaHMA2) being low lead to high expressions in heavy metal resistant varieties. Similarly, Al-tolerant barley plants show higher expression of the *HvAACT1* gene which plays a major role in citrate efflux from roots *via* Al-activated citrate transporter1 (HvAACT1) (Zhou et al., 2013). DNA demethylation is mediated by: (1) DNA glycosylases (active demethylation) and (2) by replacing unmodified cytosines with methylated cytosines (passive demethylation). The DNA demethylation promotes the tolerance against drought in barley seedlings which is seen through enhanced expression of *HvDME* (DME-family DNA glycosylase) (Kong et al., 2020). Other DNA glycosylases such as REPRESSOR OF SILENCING 1 (ROS1) and DMEMER (DME) are well studied in *Arabidopsis* (Kong et al., 2020).

DNA demethylation in multiretrotransposon-like (MRL) sequence upstream of *HvAACT1* caused enhanced expressions however low level DNA methylation resulted high expressions of *HvAACT1* thus providing high tolerance in barley. The reasons behind regulation of gene because of methylation are that binding of transcription factors are affected by DNA methylation (Zhang et al., 2018).

The calmodulin (Calcium binding protein) pathway regulates Al stress (Munir et al., 2016). The calmodulin gene insertion in the *Saccharomyces cerevisiae* strains improved their tolerance against Al stress. Studies have also reported the combined role of *STAR1* and *STAR2* proteins in Al detoxification, as *STAR1* activates *STAR2* which expresses a transmembrane domain that forms a bacterial-type ABC transporter (Gallo-Franco et al., 2020). *ART1* is a transcription factor (zinc finger protein), regulating several genes involved in Al stress responses (Yamaji et al., 2009). Coupled Restriction Enzyme Digestion and Random Amplification (CRED-RA) in maize plants have shown that Al stress leads to the transposition of long terminal repeat retrotransposons (LTR) and DNA hypermethylation (Taspinar et al., 2018). Using the same technique, in wheat cultivars, Pour et al. (2019) reported higher methylation at high Al stressed conditions and lower methylation under low Al exposures. Liquid chromatography (RP-HPLC) results showed higher DNA methylation in tolerant ones than compared to less tolerant plants (Johnston et al., 2005). However, MSAP showed high demethylation in roots of tolerant and non-tolerant plants. Results of metAFLP (methylation—sensitive Amplified Length Polymorphism) showed very less methylation variation tolerant and sensitive varieties concluding towards the involvement of a small part of the genome in DNA methylation (Niedziela, 2018). Multi-retrotransposon-like (MRL) insertion has significantly enhanced the expression of *HvAACT1* gene coding for citrate efflux transporter under Al stress.

1.12 Genetic approaches to improve metal tolerance in plants: Candidate markers/genes for Cd tolerance

As a stress response, up-regulation of genes associated with enzymatic anti-oxidants, involves the accumulation of heavy metals in vacuoles, regulating biochemical and physiological processes to tolerate metal stress conditions. Next generation sequencing such as high-throughput transcriptome or RNA Sequencing (RNA-Seq) help in the identification of gene expressions (Sahu et al., 2020). Under Cd stress, cDNA-AFLP analysis has enabled us to understand that 52 genes regulate processes such as cellular metabolism, photosynthetic process, transcription factors under Cd stress. Similarly, through transcriptome analysis, it has been possible to detect 1172 regulatory genes under Cd stress (Sun et al., 2019). Microarray enables us to understand the up and down regulation of genes in tolerant and sensitive Cd stress barley plants (Gul et al., 2016).

Furthermore, in *Populus tremula* L. proteins involved in nitrogen metabolism, post-translational modifications of proteins are expressed under Cd stress (Kieffer et al., 2009). MALDI-TOF/TOF MS have been employed for proteomic analysis in the roots, anti-oxidant defense proteins are produced in higher amounts under Cd stress in *Solanum torvum*. Quantitative proteomic assay (iTRAQ) and Fluorescence 2-D difference gel electrophoresis (2D-DIGE) identify enhancement in the amounts of proteins linked with plant's response under stress such as those involved in oxidative stress regulation in *Brassica junceae* (Arruda et al., 2011). Digital expression analysis (DGA) is helping to understand molecular processes occurring in a plant's cell under stress.

1.13 Genetic variability and adaptation to stress

Signalling networks executed by various stresses ultimately lead to the expression of the specific genes which generate tolerance against the stress. However, first, it is important to discover the receptors that perceive stresses. AtHK1 as previously discussed behaves as a receptor, it is a two-component hybrid histidine kinase that senses osmotic stress (Urao et al., 1999). G-protein complexes also sense stress signals which are further transmitted through second messengers such as Ca, H_2O_2, a phospholipid which regulates protein kinase phosphorylations through CDPK (Calcium dependent protein kinase) and MAPK (Mitogen-activated Protein Kinases) (Jagodzik et al., 2018).

1.14 Genetic variation in metal stress perception: MAPK Signaling in Heavy Metal Stress

Mitogen Activated Protein Kinases (MAPKs) perform signaling under different kinds of stresses out of which heavy metal stress highly utilises this cascade. MAPKs are activated on the binding of specific metal ligands and also *via* ROS molecules generated during stress as a protective mechanism (Jalmi et al., 2018). As discussed earlier that ROS also activate MAPKs, the reduction in activation of MAPK on treatment with glutathione proves that it is ROS which participate in iron triggered pathway. Zn has also activated MBP *via* MAPK through ROS in rice (Jagodzik et al., 2018). Both abiotic and biotic stress generated ROS execute two independent MAPK cascades, i.e., MEKK1-MKK4/5-MPK3/6 and MEKK1-MKK2-MPK4/6 in *Arabidiopsis* (Jalmi and Sinha, 2015). Under biotic stress, MEKK1-MKK4/5-MPK3/6 pathway activates NADPH oxidase enhancing ROS production, H_2O_2 produced, in turn, activates MPK3 and MPK6. Being a reactive oxygen species, still, H_2O_2 is known to regulate oxidative stress through nitric oxide in two cyanobacteria (Verma and Prasad, 2021). Thus, MAPK pathways regulate the promotion of ROS production. OXI1-MPK6 is activated through ROS thus it promotes ROS production in turn (Jalmi and Sinha, 2015). Biotic and Cd stress stimulated FLS2 and receptor like kinase (RLKs) activate MAPK cascade, i.e., MEKK1-MKK4/MKK5-MPK3/MPK6 (Jagodzik et al., 2018).

1.15 Importance of genetic diversity to manage metal stress

Osr42 has been significantly up-regulated under heavy metal treatments of Cu, Cr, and Hg however exhibited no expression under Cd treatments. Furthermore, it is noticed that two *TE* (transposable elements) genes and the *Tos17* gene have shown contrasting and the most conspicuous activation of expression patterns, respectively under Cd exposure (Cong et al., 2019). Genes such as *homeoboxgene, DNA-binding protein, elongation factor, HSP70,* and *SNF-FZ14* are upregulated under heavy metal stress whereas *YF25* has shown no expression (Cd) to down regulation (Cu, Cr, and Hg) under heavy metal stress exposed conditions (Cong et al., 2019). Heavy metal ATPases show differential up regulation according to specific heavy metal stress such as *OsHMA1* in Cd and Hg stress, *OsHMA2* under Cu stress, *OsHMA5* in Cu, Cd, and Hg stress, *OsHMA6* and *OsHMA7* under Cu, Cr, Cd and Hg stress, *OsHMA9* in Cd and Hg stress (Li et al., 2015). This shows that various genes show altered expression under different heavy metal stress. The up-regulated expressions of *OsHMAs* indicate its role in heavy metal stress tolerance in plants. The high upregulation of *OsHMA1* under Hg stress in rice plants suggests its significant role in Hg transport in them. However, high expression of *OsHMA2* under Cu stress indicates its involvement in the Cu transport, similarly *OsHMA4* is present on vacuolar membrane sequesters under Cu stress (Martinoia, 2018). The high expression of *OsHMA5* occurs under Zn, Fe, Mn, Cd, and Hg stress (Mani and Sankaranarayanan, 2018). An important alleviating role of *OsHMA3* in Cd stress is known in the rice (Rizwan et al., 2016) whereas; *OsHMA9* is involved in Hg efflux.

An important concept of epigenetics indicates variations in the gene expression without changes in the DNA sequence. There is also the transgenerational inheritance of the differential gene expression behaviour of plants. Some genes show the inheritance of expressed states (*OsHMA2, OsHMA5, OsHMA8, OsHMA9, Tos17A,* Homeobox gene, and *HSP70*) but other genes (*OsHMA1, OsHMA2, OsHMA6, OsHMA7 Tos17B, Osr42,* and *SNF-FZ14*) show a more up-regulated expression in more than 50% progeny plants which have suggested towards genetic memory response on re-exposed heavy metal stress (Cong et al., 2019). Thus, the heavy metals exposure induced DNA methylation changes are also present in the next successive generations even in the absence of heavy metals exposure and siRNA (small interfering RNA) might have important roles in the transgenerational memory response of heavy metals exposure. Still, studies need to be done to find out the mechanism behind the transgenerational memory as a heavy metal stress response in the plants.

2. Conclusion

The application of recombinant technology in the insertion of genes through transcriptomics studies is a great application in the coming times, to enhance the heavy metal stress tolerance of plants. Similarly, studies that have provided evidence related to DNA methylation, other epigenetic changes, miRNA degradation, activation of heavy metal transporters under heavy metal stress will provide insight into the development of heavy metal stress tolerant varieties necessarily required to promote crop productivity to meet the food demand.

References

Alaraidh IA, Alsahli AA, and Abdel Razik ES. 2018. Alteration of antioxidant gene expression in response to heavy metal stress in *Trigonella foenum-graecum* L. *S Afr J Bot* 115: 90–93.

Arruda SCC, Barbosa HS, Azevedo RA, and Arruda MAZ. 2011. Two-dimensional difference gel electrophoresis applied for analytical proteomics: Fundamentals and applications to the study of plant proteomics. *The Analyst* 136(20): 4119–26.

Baillo EH, Kimotho RN, Zhang Z, and Xu P. 2019. Transcription factors associated with abiotic and biotic stress tolerance and their potential for crops improvement. *Genes* 10(10): 771.

Chaudhary J, Khatri P, Singla P, Kumawat S, Kumari A, Vinaykumar R, Vikram, Jindal SK, Kardile H, Kumar R, Sonah H, and Deshmukh R. 2019. Advances in Omics Approaches for Abiotic Stress Tolerance in Tomato. *Biology* 8(90): 1–19.

Cong W, Miao Y, Xu L, Zhang Y, Yuan C, Wang J, Zhuang T, Lin X, Jiang L, Wang N, Ma J, Sanguinet K, Liu B, Rustgi S, and Ou X. 2019. Transgenerational memory of gene expression changes induced by heavy metal stress in rice (*Oryza sativa* L.). *BMC Plant Biol* 19: 282.

Diouf I, Derivot L, Koussevitzky S, Carretero Y, Bitton F, Moreau L, and Causse M. 2020. Genetic basis of phenotypic plasticity and genotype × environment interactions in a multi-parental tomato population. *J Exp Bot* 71(18): 5365–5376.

Djami-Tchatchou AT, Sanan-Mishra N, Ntushelo K, and Dubery IA. 2017. Functional roles of microRNAs in agronomically important plants-potential as targets for crop improvement and protection. *Front Plant Sci* 8: 378.

Gallo-Franco JJ, Sosa CC, Ghneim-Herrera T, and Quimbaya M. 2020. Epigenetic control of plant response to heavy metal stress: a new view on aluminum tolerance. *Front Plant Sci* 11: 602625.

Gao Y, Yang F, Liu J, Xie W, Zhang L, Chen Z, Peng Z, Ou Y, and Yao Y. 2020. Genome-wide identification of metal tolerance protein genes in *Populus trichocarpa* and their roles in response to various heavy metal stresses. *Int J Mol Sci* 21: 1680.

Gong D, Guo Y, Schumaker KS, and Zhu J. 2000. The SOS3 family of calcium sensors and SOS2 family of protein kinases in *Arabidopsis*. *Plant Physiol* 134(3): 919–26.

Gong J, Waner D, Horie T, Li S, Horie R, Abid K, and Schroeder JI. 2004. Microarray-based rapid cloning of an ion accumulation deletion mutant in *Arabidopsis thaliana*. *Proc Natl Acad Sci* 101(43): 15404–9.

Granitzer S, Ellinger I, Khan R, Gelles K, Widhalm R, Hengstschlager M, Zeisler H, Tupova L, Ceckova M, Salzer H, and Gundacker C. 2020. *In vitro* function and *in situ* localization of Multidrug Resistance-associated Protein (MRP)1 (*ABCC1*) suggest a protective role against methyl mercury-induced oxidative stress in the human placenta. *Arch Toxicol* 94: 3799–3817.

Gul A, Ahad A, Akhtar S, Ahmad Z, Rashid B, and Hussain T. 2016. Microarray: gateway to unravel the mystery of abiotic stresses in plants. *Biotechnol Lett* 38: 527–543.

He C, Zhang HY, Zhang YX, Fu P, You LL, Xiao WB, Wang ZH, Song HY, Huang YJ, and Liao JL. 2020. Cytosine methylations in the promoter regions of genes involved in the cellular oxidation equilibrium pathways affect rice heat tolerance. *BMC Genomics* 21(1): 560.

Jagodzik P, Tajdel-Zielinska M, Ciesla A, Marczak M, and Ludwikow A. 2018. Mitogen-activated protein kinase cascades in plant hormone signaling. *Front Plant Sci* 9: 1387.

Jalmi S, and Sinha AK. 2015. ROS mediated MAPK signaling in abiotic and biotic stress-striking similarities and differences. *Front Plant Sci* 6(769).

Jalmi SK, Bhagat PK, Verma D, Noryang S, Tayyeba S, Singh K, Sharma D, and Sinha AK. 2018. Traversing the links between heavy metal stress and plant signaling. *Front Plant Sci* 9(12).

Johnston JW, Hardling K, Bremner DH, Souch GR, Green JE, Lynch, Grout BWW, and Benson EE. 2005. HPLC analysis of plant DNA methylation: A study of critical methodological factors. *Plant Physiol Biochem* 43(9): 844–53.

Kieffer P, Schroder P, Dommes J, Hoffmann L, Renault J, and Hausman JF. 2009. Proteomic and enzymatic response of poplar to cadmium stress. *J Proteom* 72(3): 379–396.

Kim SW, Gupta R, Min CW, Lee SH, Cheon YE, Meng QF, Jang JW, Hong CE, Lee JY, Jo IH, and Kim ST. 2019. Label-free quantitative proteomic analysis of *Panax ginseng* leaves upon exposure to heat stress. *J Ginseng R* 43(1): 143–153.

Kong L, Liu Y, Wang X, and Chang C. 2020. Insight into the role of epigenetic processes in abiotic and biotic stress response in wheat and barley. *Int J Mol Sci* 21: 1480.

Kordrostami M, and Rahimi M. 2015. Molecular markers in plants: Concepts and applications. *Zhinitik Hizarahi Sivvum* 13: 4024–4031.

Li D, Xu X, Hu X, Liu Q, Wang Z, Zhang H, Wang H, Wei M, Wang H, Liu H, and Li C. 2015. Genome-wide analysis and heavy metal-induced expression profiling of the HMA gene family in *Populus trichocarpa*. *Front Plant Sci* 6: 1149.

Liu X, Zhang X, Sun B, Hao L, Liu C, and Zhang D. 2019. Genome-wide identification and comparative analysis of drought-related microRNAs in two maize inbred lines with contrasting drought tolerance by deep sequencing. *PLOS ONE* 14(7): e0219176.

Mani A, and Sankaranarayanan K. 2018. Heavy metal and mineral element-induced abiotic stress in rice plant. *In*: Shah F, Khan ZH, and Iqbal A (eds.). *Rice Crop - Current Developments*. IntechOpen Limited, United Kingdom.

Martinoia E. 2018. Vacuolar Transporters - Companions on a Longtime Journey. *Plant Physiol* 176(2): 1384–1407.

Min MK, Kim R, Moon S, Lee Y, Han S, Lee S, and Kim B. 2020. Selection and functional identification of a synthetic partial ABA agonist, S7. *Sci Rep* 10(4): 1–10.

Munir S, Liu H, Xing Y, Hussain S, Ouyang B, Zhang Y, Li H, and Ye Z. 2016. Overexpression of calmodulin-like (ShCML44) stress-responsive gene from *Solanum habrochaites* enhances tolerance to multiple abiotic stresses. *Sci Rep* 6: 31772.

Nakashima K, Ito Y, and Yamaguchi-Shinozaki K. 2009. Transcriptional regulatory networks in response to abiotic stresses in *Arabidopsis* and grasses. *Plant Physiol* 149(1): 88–95.

Niedziela A. 2018. The influence of Al^{3+} on DNA methylation and sequence changes in the triticale ($\times$ *Triticosecale* Wittmack) genome. *J Appl Genetics* 59: 405–417.

Noman A, Sanaullah T, Khalid N, Islam W, Baloch SK, Irshad HMK, and Aqeel M. 2019. Crosstalk between plant miRNA and heavy metal toxicity. pp. 145–168. *In*: Sablok G (ed.). *Plant Metallomics and Functional Omics*. Vol. 1. Springer Nature Switzerland AG.

Oven M, Page JE, Zenk MH, and Kutchan TM. 2002. Molecular characterisation of the homo-phytochelatin synthase of soybean *Glycine max*. *J Biol Chem* 277(7): 4747–4754.

Papdi C, Joseph MP, Pérez I, Vidal S, and Szabados L. 2009. Genetic technologies for the identification of plant genes controlling environmental stress responses. *Funct Plant Biol* 36(8).

Pour AH, Özkan G, Nalci ÖB, and Haliloğlu K. 2019. Estimation of genomic instability and DNA methylation due to aluminum (Al) stress in wheat (*Triticum aestivum* L.) using iPBS and CRED-iPBS analyses. *Turk J Bot* 43: 27–37.

Rao KP, Vani G, Kumar K, Wankhede DP, Misra M, Gupta M, and Sinha AK. 2011. Arsenic stress activates MAP kinase in rice roots and leaves. *Arch Biochem Biophys* 506(1): 73–82.

Rizwan M, Ali S, Adrees M, Rizvi H, Ziar-ur-Rehman M, Hannan F, Qayyum MF, Hafeez F, and Ok YS. 2016. Cadmium stress in rice: toxic effects, tolerance mechanisms, and management: a critical review. *Environ Sci Pollut Res* 23: 17859–17879.

Sahu PK, Sao R, Mondal S, Vishwakarma G, Gupta SK, Kumar V, Singh S, Sharma D, and Das BK. 2020. Next generation sequencing based forward genetic approaches for identification and mapping of causal mutations in crop plants: a comprehensive review. *Plants* 9(10): 1355.

Salem MA, Perez de Souza L, Serag A, Fernie AR, Farag MA, Ezzat SM, and Alseekh S. 2020. Metabolomics in the context of plant natural products research: from sample preparation to metabolite analysis. *Metabolites* 10(1): 37.

Singh S, Parihar P, Singh R, Singh VP, and Prasad SM. 2016. Heavy metal tolerance in plants: Role of transcriptomics, proteomics, metabolomics, and ionomics. *Front Plant Sci* 6: 1–36.

Sun L, Wang J, Song K, Sun Y, Qin Q, and Xue Y. 2019. Transcriptome analysis of rice (*Oryza sativa* L.) shoots responsive to cadmium stress. *Sci Rep* 9: 10177.

Tang J, Wang SQ, Hu KD, Huang Z, Li Y, Han Z, Chen X, Hu L, Yao G, and Zhang H. 2019. Antioxidative capacity is highly associated with the storage property of tuberous roots in different sweet potato cultivars. *Sci Rep* 9: 11141.

Taspinar MS, Aydin M, Sigmaz B, Yagci S, Arslan E, and Agar G. 2018. Aluminum-induced changes on DNA damage, DNA methylation and LTR retrotransposon polymorphism in maize. *Arab J Sci Eng* 43: 123–131.

Thomine S, Wang R, Ward JM, Crawford NM, and Schroeder JI. 2000. Cadmium and iron transport by members of a plant metal transporter family in Arabidopsis with homology to Nramp genes. *Procd Natl Acad Science USA* 97(9) 4991–4996.

Urao T, Yakubov B, Satoh R, Yamaguchi-Shinozaki K, Seki M, and Hirayama T. 1999. A transmembrane hybrid-type histidine kinase in arabidopsis functions as an osmosensor. *The Plant Cell* 11(9): 1743–54.

Verma N, Pandey A, Tiwari S, and Prasad SM. 2021. Calcium mediated nitric oxide responses: Acquisition of nickel stress tolerance in cyanobacterium *Nostoc muscorum* ATCC 27893. *Biochem Biophys Rep* 26: 100953.

Verma N, and Prasad SM. 2021. Regulation of redox homeostasis in cadmium stressed rice field cyanobacteria by exogenous hydrogen peroxide and nitric oxide. *Sci Rep* 11(1): 1–14.

Waititu JK, Zhang C, Liu J, and Wang H. 2020. Plant non-coding RNAs: origin, biogenesis, mode of action and their roles in abiotic stress. *Int J Mol Sci* 21(21): 8401.

Wang M, Qin L, Xie C, Li W, Yuan J, Kong L, Xia G, and Liu S. 2014. Induced and constitutive DNA methylation in a salinity-tolerant wheat introgression line. *Plant Cell Physiol* 55(7): 1354–1365.

Xiong L, Lee H, Ishitani M, Tanaka Y, Stevenson B, Koiwa H, Bressan RA, Hasegawa PM, and Zhu JK. 2002. Repression of stress-responsive genes by FIERY2, a novel transcriptional regulator in Arabidopsis. *Procd Natl Acad Science USA* 99(16): 10899–10904.

Yamaji N, Huang CF, Nagao S, Yano M, Sato Y, Nagamura Y, and Ma JF. 2009. A zinc finger transcription factor ART1 regulates multiple genes implicated in aluminum tolerance in rice. *Plant Cell* 21(10): 3339–3349.

Yang W, Feng H, Zhang X, Zhang J, Doonan JH, Batchelor WD, Xiong L, and Yan J. 2020. Crop phenomics and high-throughput phenotyping: past decades, current challenges, and future perspectives. *Mol Plant* 13(2): 187–214.

Yu L, Zhang F, Culma C, Lin S, Niu Y, Zhang T, Yang Q, Smith M, and Hu J. 2020. Construction of high density linkage maps and identification of quantitative trait loci associated with Verticillium wilt resistance in autotetraploid Alfalfa (*Medicago sativa* L.). *Plant Dis* 104(5).

Zhang H, Lang Z, and Zhu J. 2018. Dynamics and function of DNA methylation in plants. *Nat Rev Mol Cell Biol* 19: 489–506.

Zhou G, Delhaize E, Zhou M, and Ryan PR. 2013. The barley MATE gene, HvAACT1, increases citrate efflux and Al($3+$) tolerance when expressed in wheat and barley. *Ann Bot* 112(3): 603–612.

15

Stress-inducible Proteins and their Roles under Heavy Metal Stress

Ummey Aymen,[1] *Marya Khan,*[1] *Anuradha Patel,*[2] *Sanjesh Tiwari,*[2]
Aman Deep Raju,[2] *Sheo Mohan Prasad,*[2] *Rachana Singh*[2,*]
and *Parul Parihar*[1,2,*]

ABSTRACT

Industrialization is introducing various harmful chemicals into our environment and the plants being sessile have adopted different strategies to prevent the accumulation of these harmful chemicals within them or to detoxify them. This whole defense mechanism is being regulated at the genomic level where overexpression or silencing of certain genes leads to such responses that make the plant resistant to various stress conditions like heavy metal stress here. Genes, themselves are regulated by transcription factors and upon expression again give proteins which are either transcription factors to activate other genes or enzymes required in a metabolic or defense pathway. Thus, tracing the proteomic activity can help us understand how the pathways operate and enable us to make much better changes or inculcate the same proteomic activity in the plants that are metal stress sensitive. Omics tools have advanced to a great level because of the evolving technology from simple gel electrophoresis to much advanced gel-free based methods and novel spectroscopic techniques and all this is making our research more accurate and a number of cross-links are comprehended from the data of proteomics and are easily sorted by the biostatistical tools, hence making our study easier and quick. Heavy metal detection and their effects on the proteomes are being studied at molecular levels and the second generation proteomic tools are promising towards many new discoveries in proteomics.

[1] Department of Botany, School of Bioengineering and Biosciences, Lovely Professional University, Phagwara-144411, Punjab, India.
[2] Ranjan Plant Physiology and Biochemistry Laboratory, Department of Botany, University of Allahabad, Prayagraj, 211003, UP, India.
* Corresponding authors: rachanaranjansingh@gmail.com; parulprhr336@gmail.com

1. Proteins in metal stress tolerance

Heavy metals can be defined in terms of chemical definitions as the elements that possess specific gravity of more than 5 (Hossain and Komatsu, 2013; Paul et al., 2017). But, in general terms, heavy metals are usually referred to as the elements which are toxic like cadmium (Cd), Chromium (Cr), Zinc (Zn), Copper (Cu), etc., and even metalloids that are hazardous like Boron (B), Arsenic (As), etc. (Hossain et al., 2012; Kumar et al., 2018). These elements are taken up by the plant in ionic form in which they are present in the soil and enter the plant cells through ion channels that are not selective and hence allow all the ions to pass through (Rostami and Azhdarpoor, 2019). As soon as these toxic elements are taken up by the cells, they start disturbing the cell homeostasis by causing oxidative stress and ROS increase (Sharma and Dietz, 2009; Rizvi and Khan, 2018) affecting vital plant functions. This can be observed physically in the plant by the morphological changes that occur like necrotic spots, lesions, stunted growth, etc., and the extra ion intake leads to blockage of ions that are essential for the plant photosynthesis and respiration (Xiong, 1997; Munzuroglu and Geckil, 2002; Gomes, 2011; Sorrentino et al., 2018; Zhang et al., 2020). As a result of this, plant defense adapted strategies like reducing the ion intake through roots which is the main supply of ion increase inside the plant, and also by increasing ion concentration in various plant parts relatively so that plants get adapted to high ionic concentrations (Clemens et al., 2002; Gupta and Diwan, 2017). Scientists have always tried to use the genomic (Bohnert et al., 2006; Dutta et al., 2018) and proteomic (Ghosh et al., 2017; Pirzadah et al., 2019; Parihar et al., 2019) data to solve heavy metal stress faced by the plant but we have still not been able to connect the dots as changes in genes do not perfectly change the protein levels that we targeted and thus more research and experimental work are required to solve the heavy metal stress and to understand the pathway in more detail to analyse if other pathways are interconnected at the same time (Gygi et al., 1999a; Novaković et al., 2018). Heavy metal and plant response inter relationship has been studied extensively and in the past year too, various studies were carried out and focus being placed over stress alleviation strategies like Hydrogen sulphide (H_2S) by acting as a signalling molecule (Zanganeh et al., 2019; Luo et al., 2020), Nitric oxide (NO) by reducing oxidative stress (Sharma et al., 2020; Wei et al., 2020a; Wei et al., 2020b), mi-RNAs (Ding et al., 2020; Anjali and Sabu, 2020), tomato metallocarboxypeptidase inhibitor (TCMP)-1 (Manara et al., 2020), etc., and it is not only the latter one that comes under the domain of proteomics but above mentioned all researches are directly or indirectly associated with proteomics because H_2S and NO being signalling molecules will operate their pathways through some mechanism which will require the biochemical catalysts (enzymes) at each stage (Li et al., 2012) and hence provide us with the opportunity to study and alter the proteomics to give the pathway a proper direction to tolerate the heavy metal stress condition. Similarly, micro-RNAs work post-transcriptionally and hence again associated with proteomics. To study and analyse the proteomic pathways, high-level technologies are required, and since now, sophisticated tools being invented, the proteomic studies are improving but recalling the most common technique is the gel electrophoresis which lays out to us the total protein content which is further analysed by spectroscopic studies. From these studies, we have come to know the proteomic content change in a cell or even in an organelle that occurs (Hossain and Komatsu, 2013) and some estimations are as follows:

Cytoplasmic proteins that increased: APX, ascorbate peroxidase; ACCO, 1-aminocyclopropane-1-carboxylicacid oxidase; CAT, catalase; Cysteine; CS, cysteine synthase; GST, Glutathione S-transferase; GR, glutathione reductase; Gly-I, glyoxalaseI; GS, glutamine synthetase; Glutamine; Glutamate; GSH, reduced glutathione; MTs, metallothioneins; MDAR, monodehydroascorbate reductase; PK, Pyruvate kinase; Prx, peroxidoxin; PR, pathogenesis-related; Trx, thioredoxin; SAMS, S-adenosyl-L-methionine synthetase; TPI, triosephosphate isomerase; vBPO, vanadium-dependent bromoperoxidases; NADP(H)-oxido-reductase; SHMT, Serine hydroxymethyltransferase; PC, phytochelatin; PGM, phosphoglucomutase; G6PI, Glucose-6-phosphate isomerase; BiP, Binding immunoglobulin protein; HSP70, heat shock protein 70; STI-1, Stress-induced-phosphoprotein 1.

Cytoplasmic proteins that decreased: Organic Hydroperoxides.

Cytoplasmic proteins whose levels both increased and decreased: G3PDH, glyceraldehyde-3-phosphate dehydrogenase; ENO, enolase.

Mitochondrial proteins that increased: ATPase β, ATP synthase subunit beta; ACO, aconitase; CSy, citrate synthase; FDH, formate dehydrogenase; MDH, malate dehydrogenase; PDH, pyruvate dehydrogenase; SD, succinate dehydrogenase.

Mitochondrial proteins whose levels both increased and decreased: AH, aconitate hydratase.

Proteins that increased in vacuole include CAX, cation/protonex changer; ABC transporter (ATP-binding cassette transporter).

Chloroplast proteins that increased: OEE-1, oxygen-evolving enhancer protein-1; PsaD, photosystem I protein D; PSI-IVA (Photosystem); RuBisCO LSU-binding α, β, Ribulose-1,5-bisphosphate carboxylase oxygenase large subunit-binding α, β; FCP, Fucoxanthin-chlorophyll a/c binding protein); Cu/Zn-SOD, Copper/zinc superoxide dismutase.

Chloroplast proteins that decreased: LHC, light harvesting complex; LhcII-4, light-harvesting chlorophyll-a/b binding protein; PSII-OEC2, photosystem II oxygen-evolving complex protein 2; RuBisCO LSU and SSU, Ribulose-1,5-bisphosphate carboxylase oxygenase large subunit, and a small subunit.

Chloroplast proteins whose levels both increased and decreased: OEE-2, oxygen-evolving enhancer protein-2; RuBisCO activase; PSII-(photosystem II).

Upon analysing these heavy metal associated proteins (HMP), scientists realized that these metalloproteins contain certain heavy metal associated domains (HMA) which actually detoxify the heavy metal stress by binding to it through two of its cysteine residues within the conserved HMA domain of ~ 30 amino acids and further transports or directly detoxifies the heavy metal ion like copper, zinc, cadmium, etc.

(Bulland Cox, 1994; Gitschier et al., 1998; Tehseen et al., 2010; Zhanget al., 2018; Nikolić and Tomašević, 2020; Lei et al., 2021).

Separate studies started to be conducted in each species and upon each metal associated protein, as was performed by Li et al., 2020 and they observed these stress tolerance proteins in the case of *Arabidopsis* and Rice accounting to about 55 and 46 HMPs respectively and naming them as AtHMP 1–55 and OsHMP 1–46 serially as we go from chromosome 1–5 in case of *Arabidopsis* and 1–6 in case of rice. These HMPs are not similar to one other in terms of structure and vary widely in their metal associated domains and motifs and were hence categorized into six clades: HPPs (heavy metal-associated plant proteins), ATX1-like (Puig et al., 2007), HIPPs (heavy metal-associated isoprenylated plant proteins) (de Abreu-Neto et al., 2013), and P1B-ATPase (Pedersen et al., 2012).

These proteins regulate metal stress inflicted upon the plants but are themselves regulated by some proteins, that is, transcription factors (Joshi et al., 2019). The HMPs are hence differently expressing, some of them expressing constitutively (8 OsHMPs) [even among those, their expression levels varied (OsHMP37 showing the highest expression, while OsHMP28 being expressed only in the roots)] while others upon induction by stresses like heavy metal stress (OsHMP09, OsHMP18, OsHMP22) increased to sufficient levels equally in all the tissues (Li et al., 2020). The ongoing research is also specifically putting a focus upon the evolution of these HMPs from monocots to dicots and may further help to increase their functionality in the stress-sensitive crops.

2. Protein oxidative modifications

Heavy metal contamination has found its way inside living organisms, be it marine life or plant life, which ultimately reaches us through the complex food webs that operate in the biosphere. Anthropogenic activities are considered a major reason for this degradation of the environment but we also possess the capability to reverse the effects and scientists have been carefully examining how tolerant species emerge even in such harsh conditions and how the genomes and proteomes of such species actually respond. Plants being the first step of the ladder of a generalized food chain is the first that needs to be saved and analysed. Many scientists have reported cases where unknowingly people consumed food that was heavy metal contaminated and people have been adversely affected, pregnant women and children being at most risk (Xu and Thornton, 1985; Boon and Soltanpour, 1992; Kumar and Soni, 2007; Abadin et al., 2007; Uwah et al., 2011). Thus, quick solutions are required in addition to mass education about these phenomena.

Among all the changes that occur in a plant against various stresses, the first and foremost change is oxidative stress even though ROS are essential for maintaining the ROS-antioxidant balance of a plant. But when the balance is disturbed by the toxic elements making their way into the plant tissues either by hyperaccumulation through the normal ion passage or by getting bound to the functional groups of entering plant acids (Morel et al., 1986; Sharma and Dubey, 2005; Singh et al., 2019). However, studies have shown that metal accumulation depends on the type of plant species and soil factors involved and in addition to this, cultivars of the same species have also been found to respond differently to the heavy metal toxicity because their trace element accumulation occurs in different concentrations which point to the fact that molecular biochemistry and omic tools are critical for understanding and solving the problems here (Alexander et al., 2006; Nabulo et al., 2011; Baig et al., 2020).

Although heavy metal is the stress here, it is the oxidative stress that arises due to the metal toxicity which affects the biomolecular machinery leading to negative impacts on essential physiological processes (Shinozaki and Yamaguchi-Shinozaki, 2000; Zhu, 2001; Dubey et al., 2018; Zhang et al., 2020). ROS species are produced inside plant cells almost during every biological event as the sunlight absorbed excites and produces high energy electrons which automatically give rise to elements with unpaired electrons and have the potential to create a chain reaction which could oxidize whole biomembranes but in a balanced cell environment, free radicals just do the signalling part and increase the antioxidant response which keeps a cell healthy and properly functional and these signalling free radicals include superoxide radical anion ($O2^{\cdot-}$), the hydroxyl radical ($HO^{\cdot}$), and peroxyl radicals ($ROO^{\cdot}$), hydrogen peroxide (H_2O_2), singlet oxygen O_2, hypochlorous acid (HOCl) and peroxynitrite ($ONOO^{-}$) (Fridovich, 1999; Betteridge, 2000; Ježek and Hlavatá, 2005; Saed-Moucheshi et al., 2014; Schuch et al., 2017; Hasanuzzaman et al., 2020). Free radicals when out of balance cause DNA damage through mutations, breakage, degradation, or even their winding with amino acid residues of proteins (Silva et al., 2019). However, their impact on proteins causing damage also occurs separately by causing breakage of the peptide bonds, charge impairment, loss of enzymatic function, or becoming prone to undergo proteolysis (Scandalios, 2005; Banerjee and Roychoudhury, 2018). Free radicals themselves do not just operate on their own but are cross-linked with various other signalling pathways as is seen in the case of calcium and hydrogen peroxide interaction wherein it is the stress that elevates the calcium levels which then further elevates H_2O_2 levels (Yang and Poovaiah, 2002; Kolupaev et al., 2017; Khan et al., 2018; Siddiqui et al., 2020). ROS maintenance is essential for plant homeostasis and for this the anti-oxidant defense mechanism is the ultimate tool and this defense mechanism is largely composed of various enzymes which sequester these free radicals or make them ready to be detoxified and these enzymes being chemically proteins and hence the proteome content of a plant is vital for analysing the stress conditions under which the plant cell operates and why some plants are better able to cope up with the heavy metal stress. Although vitamins also constitute the anti-oxidant defense system but proteins like superoxide dismutase (SOD), catalase (CAT), ascorbate peroxidase (APX), glutathione (GSH), heat-shock proteins, HSPs, several NADPH-oxidases, etc., and the ones mentioned above are among the main antioxidants that have proved in making a plant heavy metal tolerant (You and Chan, 2015; Ranjan et al., 2021). This has been proved by making heavy metal stress-sensitive plants transgenic and overexpressing the genes that produced these proteins leading to enhanced stress tolerance and making the plants tolerant species (Table 1). Proteomic tools have analysed these proteins and such proteins have been overexpressed in metal sensitive plants and the results came out to be promising by reducing the ROS concentrations while increased antioxidant levels decreased the overall stress that the plant faced. In addition, proteins like metallothionein's (e.g., Met1, Met2, Met3, Met4, and Met5), transport proteins (copper transport protein

Table 1: Effect of heavy metal stress on the expression of protein and the genes involved.

Type of heavy metal stress	Plant used	Protein overexpressed	Gene involved	Effects against stress	References
1. Cadmium	*Avena strigosa*	Superoxide dismutase, Glutathione reductase and Catalase	SOD, GR, CAT	Due to this protein accumulation, the plant becomes cadmium tolerant	Uraguchi et al., 2006
	Crotalaria juncea	Superoxide dismutase, Glutathione reductase and Catalase	SOD, GR, CAT	Due to this protein accumulation, the plant becomes cadmium tolerant	Uraguchi et al., 2006
	Brassica juncea and Nicotiana tabacum	Catalase	BjCET3	Transgenic tobacco showed improved growth and development under cadmium stress when the gene was overexpressed that gave increased catalase levels	Wei et al., 2009
	Thlaspi caerulescens and Nicotiana tabacum	Catalase, Superoxide dismutase and Glutathione reductase	CAT, SOD, GR,	The hyperaccumulator, *Thlaspi caerulescens* accumulated these proteins and hence maintained normal physiological functioning during metal stress as opposed to non-hyperaccumulator, *Nicotiana tabacum*	Boominathan and Doran, 2003
	Arabidopsis	MT1 proteins	MT1a, MT1b, and MT1c	The plants were cadmium tolerant but when the genes were knocked down, it led to cadmium hypersensitivity	Zimeri et al., 2005
2. Nickel	*Alyssum bertolonii and Nicotiana tabacum*	Catalase and Superoxide dismutase	SOD and CAT	The hyperaccumulator, *Alyssum bertolonii* upon accumulating these stress proteins could become resistant to negative effects of ROS production upon Ni stress while this coping mechanism was absent in the non-hyperaccumulator, *Nicotiana tabacum*	Boominathan and Doran, 2002

	Organism	Protein	Gene	Description	Reference
	Brassica juncea	Superoxide dismutase, Ascorbate peroxidase, catalase, Glutathione reductase, Monodehydroascorbate reductase, Dehydroascorbate reductase, and Glutathione-S-transferase	SOD, APX, CAT, GR, MDHAR, DHAR, GST	The overexpression of these enzymes was caused by Salicylic acid induction that regulated the Ni intake by enhancing AsA–GSH cycle	Zaid et al., 2019
	Salicornia brachiata and Nicotiana tabacum	SbMYB15	MYB (Myelobalastoma) genes	This transcription factor from the succulent, *Salicornia brachiate* made tobacco stress resistant towards Ni and Cd by increasing antioxidants and decreasing ROS	Sapara et al., 2019
	Arabidopsis Thaliana	zinc finger transcription factor	ZAT11	The overexpression caused both negative as well as positive regulation by positively regulating root growth while negative accumulation of Nickel occurred.	Liu et al., 2014
	Hordeum vulgare, Arabidopsis and Nicotiana	Nicotianamine synthase	NAS1	The protein overexpression in tobacco and Arabidopsis due to barley NAS1 gene increased Nicotianamine concentration and made them Nickel stress tolerant	Kim et al., 2005
3. Zinc	*Arabidopsis helleri and Arabidopsis thaliana*	Metallothionein 2b and 3, Ascorbate peroxidase, Monodehydroascorbate reductase 4, P1B-ATPase, Nramp, Zrt and Irt-like protein, cation diffusion facilitator	MTs, APX, MDAR4, HMA, NRAMP, ZIP, CDF	These proteins provide high metal tolerance to the hyperaccumulator, *Arabidopsis helleri* due to which it is unaffected by ROS while non-hyperaccumulator, *Arabidopsis thaliana* is negatively affected by heavy metal stress	Chiang et al., 2006

Table 1 contd. ....

...Table 1 cont.

Type of heavy metal stress	Plant used	Protein overexpressed	Gene involved	Effects against stress	References
	Hybrid aspen (*Populus tremula × tremuloides*)	Metallothionein 2b	MT2b	Upon heavy metal stress, the plant accumulated cadmium and zinc most in its foliar regions to cope up with the stress suggesting mechanism for these two metals among many others and a proportional concentration of Metallothionein 2b got accumulated	Hassinen et al., 2009
	Oryza sativa	Catalase, Peroxidase, Ascorbate peroxidase, Metallothioneins, Zinc finger transcription factors	CAT, POD, APX, MTs, ZNF genes, OsMT1a	Transgenic plants made from the Metallothionein, MT1a expression led to zinc tolerance and accumulation along with an increase in other antioxidant activity and zinc finger transcription factors	Yang et al., 2009
	Arabidopsis thaliana	AtSAP10 (SAP family)	SAPs	In the transgenic variant, the overexpressed protein provided stress tolerance to the plant which included not only heavy metal stress but tolerance to high temperature, cold and ABA	Dixit and Dhankher, 2011
	Lobularia maritima and Nicotiana tabacum	Stress associated proteins (LmSAP), Snakin/GASA proteins, and metallothioneins	LmSAP, GASA gene family, MTs	The SAP protein over-expression detoxified the ROS production and also activated various other transcription factors that regulated the stress environment	Saad et al., 2018b
4. Arsenic	*Medicago sativa and Nicotiana tabacum*	Mitochondrial heat shock protein	MsHSP23	Transgenic tobacco overexpressing the mitochondrial heat shock proteins has shown arsenic and salt tolerance	Lee et al., 2012

Arabidopsis thaliana	Class III peroxidases	OsPRX38	Overexpression of the proteins lead to increased antioxidant activity and arsenic stress was tolerated and a lignin biosynthesis pathway was also found to support the same	Kidwai et al., 2019
Arabidopsis thaliana	γ-glutamylcysteine synthetase	γ-ECS	The enzyme catalyses the formation of phytochelatins and after overexpressing the protein by inserting a bacterial γ-ECS, levels of glutathione (GSH) and PCs increased and made the plant arsenic resistant	Li et al., 2005
Arabidopsis thaliana	Phytochelatins (PCs) and glutathione (GSH)	*AsPCS1* and *GSH1*	The gene from garlic yeast, *AsPCS1*, and baker's yeast, *GSH1* when inserted in the model plant caused Phytochelatins (PCs) and glutathione (GSH) overexpression providing tolerance against arsenic and cadmium	Guo et al., 2008b
Arabidopsis thaliana	Glutaredoxins (Grxs)	OsGrx_C7 and OsGrx_C2.1	The protein overexpression led to decreased ROS concentrations and maintained the redox balance that was disturbed by arsenic stress and hence providing tolerance	Verma et al., 2016
Medicago sativa	Superoxide dismutase, Glutathione reductase and Catalase	SOD, GR, CAT	When the plant was subjected to lead stress, it accumulated a number of antioxidants, transporters, and transcription factors, thus making the plant tolerant against lead stress	Xu et al., 2017
5. Lead				

Table 1 cont. ...

...*Table 1 cont.*

Type of heavy metal stress	Plant used	Protein overexpressed	Gene involved	Effects against stress	References
	Hordeum spontaneum and *Oryza sativa*	CBL-interacting protein kinase	HsCIPK	The ectopic overexpression of CIPK genes from the barley in rice leads to increased tolerance towards leads and other metal and abiotic stresses	Pan et al., 2018
	Triticum aestivum	Heavy Metal ATPase 2 and ATP-Binding Cassette	TaHMA2 and TaABCC2/3/4	The lead resistant varieties showed increased expression of these detoxification transporters and thus lead tolerance is associated with this protein overexpression	Shafiq et al., 2019
	Trigonella foenum-graecum	Superoxide dismutase, Glutathione reductase and Catalase	SOD, GR, CAT	Overexpression of these antioxidant genes lead to heavy metal stress tolerance in the plant including resistance against lead toxicity	Alaraidh et al., 2018
	Triticum aestivum	Superoxide dismutase, Glutathione reductase, Catalase, Ascorbate peroxidase and Glutathione peroxidase	SOD, GR, CAT, APX, GPX	When exposed to lead stress, the cultivars that accumulated high levels of these antioxidants proved to be more tolerant	Navabpour et al., 2020

CCH), Rar1, PUB1 (Poly uridylate-binding protein), MTP3 (Metal transporter 3), ZIF1 (Zinc-Induced Facilitator 1), VIT2 (Vacuolar Iron Transporter), CAX4 (Cation Exchanger 4), cation/H^+ antiporter, ZIP (ZRT-IRT-like proteins), Nramp (Natural resistance-associated macrophage proteins) and CDF (Cation Diffusion Facilitators) families also increased their levels and further decreased the stress levels by either eating up the ROS species or the heavy metal ions by means of chelation and further compartmentalization in the vacuoles (Clemens, 2001; Cobbett and Goldsbrough, 2002; Arrivault et al., 2006; Haydon and Cobbett, 2007; Mei et al., 2009; Yang and Chu, 2011; Zhang et al., 2012; Saad et al., 2018a; Sarma et al., 2018; Santos et al., 2019). Against this, it was observed that when the antioxidant protein levels decreased, it led to stress hypersensitivity as in the case of *Arabidopsis*, MT1 family genes were suppressed and this led to decreased MT1a and MT1c levels which overall had a negative impact on the plant's stress tolerance levels to cadmium (Zimeri et al., 2005; Zhi et al., 2020).

Another stress-tolerant protein family is the SAP (stress associated protein) family and it is not only found in plants but fungi and animals as well and has proved effective against alleviating not only metal stress but other abiotic stresses as well like salinity, submergence, injury, drought, cold, etc. (Saad et al., 2010; Giri et al., 2013; Dixit et al., 2018; He et al., 2019). These proteins possess A20/AN1 zinc finger domain and have been well studied in the monocot and dicot models of rice and Arabidopsis respectively (Mukhopadhyay et al., 2004; Dixit and Dhankher, 2011). These proteins have been taken into much consideration because plants produce them during multiple stresses and are able to cope up with stress conditions which point to the fact that these proteins might be signalling agents for various pathways (Dixit and Dhankher, 2011; Ben Saad et al., 2012; Kim et al., 2015; Ghneim-Herrera et al., 2017; Ben Saad et al., 2018). Their exact mechanism is not yet illustrated and their analysis could get us into various stress coping mechanisms and thus further help in fighting against the abiotic and environmental stress conditions.

3. Proteomics of Hyperaccumulator plants

Hyperaccumulators are the plants that are able to tolerate the heavy metal toxic environments by accumulating the excess metal ions within their tissues without any negative impact on the physiological functions. The molecular strategies that bring up the result are not fully understood but omics are making things clearer and while genomics is one of the best approaches, yet we rely on transcriptomic and proteomic data to analyse our studies because proteins are the master elements that coordinate most interactions of a cell, be it with DNA, among themselves or with other metabolites (Zhang and Kuster, 2019). Microarrays, gel electrophoresis, spectroscopy, and their various techniques show us real-time how the proteome content varies within a cell before and after the application of heavy metal stress or what changes occur when protein contents are changed within cells. In addition, we get an idea about the differences between a heavy metal sensitive plant and a heavy metal tolerant plant and how a hyperaccumulator plant is able to accumulate these heavy metals without getting negatively affected (Vasupalli et al., 2020; Terzi and Yıldız, 2021).

Through our studies, we have seen that hyperaccumulating heavy metals are quite common among the members of the family Brassicaceae (about 500 species) (Krämer, 2010; Dar et al., 2018) and the metal that usually accumulated was Nickel (Baker et al., 2000). Some common features of these plants were observed to be: quick uptake of the roots and further transport, efficient translocation, and ultimately the way it deals with heavy metal ions, that is, systemic handling of the metal ions by chelation, sequestration, or detoxification (Clemens et al., 2002; Kumar et al., 2017a; Mathur and Chauhan, 2020). Metal accumulation was found to be often associated with ROS cope up, activation of genes that led to the formation of such proteins that prevented disease, and expression of various metal transporters. Before proteomic analysis, microarrays provide a comparison of the proteomes that we desire as was seen in the case of one non-hyperaccumulator species, *Arabidopsis thaliana*, and two Cd/Zn hyperaccumulators, *Noccaea caerulescens* and *Arabidopsis helleri* (Becher et al., 2004; Weber et al., 2004; van de Mortel et al., 2006, 2008). These studies suggest that all hyperaccumulators expressed stress proteins constitutively while the non-accumulating species showed different levels of gene expression (Plessl et al., 2010). Even though here the levels are analysed on the basis of mRNA transcripts but MHX transporter protein is proteomic evidence since MHX transcripts analysed in two *Arabidopsis* species, *Arabidopsis helleri* and *A. thaliana* were similar but the protein concentrations were more in the case of the latter, that is the hyperaccumulator species concluding that it is a post-transcriptional modification that makes only one of the species hyperaccumulator (Elbaz et al., 2006). This suggests that changes at the proteome level are equally important and essential to study and analyse as are RNA (transcriptomics) or DNA (genomics) level studies. This point can be further strengthened by mentioning a study performed by Zhao et al. (2011) where they observed the cadmium hyperaccumulator *Phytolacca americana* before and after metal stress induction and found that the RNA polymerase of the plastids performed differently in the two cases. This, in turn, was due to the reason that Sig1, which is the sigma factor for this RNA polymerase got altered by metal stress in respect of its isoelectric point. In fact, proteome studies are the ultimate guide since many processes like protein folding, protein/protein interactions and their stability after translation are essential to determine whether the genes and RNAs associated, did ultimately prove functional.

Since a focus has been laid and evidence have been provided that post-transcriptional and post-translational changes are necessary to understand what really gets expressed within a cell and how a plant becomes a hyperaccumulator relative to the other species or varieties, the whole proteome content began to be taken into account. This includes even the proteins whose efficiency is not altered by metal stress but their concentrations before and after the induction of stress vary from lower to higher as in the case of photosynthesis associated proteins like photosystems, ATPase complex, RuBisCO (ribulose bisphosphate carboxylase/oxygenase), OEC (oxygen evolving complex) proteins, cytochrome complex, etc. (Farinati et al., 2009; Duquesnoy et al., 2009; Bona et al., 2010). It has also been observed that hyperaccumulators possess a tendency to accumulate large amounts of metal within their tissues which results in the plants feeling nutrient deficiency even though the metal concentrations are normal like for zinc (van de Mortel et al., 2006).

Moreover, such response to different metal stress ions is not the same since cadmium toxicity caused increased photosynthetic protein content in hyperaccumulator *Lonicera japonica* but decreased their content in *P. americana* and *N. caerulescens* (Hossain and Komatsu, 2013; Jia et al., 2013).

One of the key aspects where hyperaccumulators differ from non-hyperaccumulators is the property of chelation. A lot of plant proteins are able to bind with these toxic metal ions and transport them to places where they get detoxified and GSH is the most important one among others like phytochelatins and nicotianamine synthase, etc. (Becher et al., 2004; Weber et al., 2004; Ingle et al., 2005; van deMortel et al., 2006, 2008; Alvarez et al., 2009; Farinati et al., 2009; Tuomainen et al., 2010; Zeng et al., 2011; Zhao et al., 2011; Deinlein et al., 2012; Schneider et al., 2013; Nazir et al., 2020). Therefore, a general analysis has been that hyperaccumulators accumulate relatively much higher levels of heavy metal stress proteins which either detoxify the heavy metals or act as transcription factors to give rise to various other stress-related proteins.

4. Proteomic markers for metal stress

As a plant is subjected to heavy metal stress, various changes begin to occur right from morphology to physiology. All biochemical reactions within a cell operate due to the activity of specific biocatalysts, that is enzymes and which are chemically proteins, and moreover, they form the basic structural units of all cellular and subcellular entities. Therefore, as the homeostasis of a cell is disturbed, the biochemistry of a cell is the first to get affected which governs both the physiology and morphology of a plant. Now, since proteins or the proteomic part of a cell provides a broad image of the changes occurring within a plant cell before and after the metal stress, thus they act as markers (Alharbi, 2020) or proteomic biomarkers for determining the effects of metal stress and enable us to compare the differences between a normal cell and stress affected cell. In addition to this, a functional analysis of proteomes of hyperaccumulators and non-hyperaccumulators can be made which will pave the way for us to be able to make a stress-sensitive plant hyper-accumulative and hence heavy metal tolerant. This omic tool, however, is not as easy to operate as it sounds because the techniques employed need high-tech technologies which are very costly and each step needs proper handling, since any impurity could alter the whole outcome. But, considering the present advancements in technology, the IT sector have put little space for such problems and the general walk through into the process would involve simply selecting the plant tissue that needs to be analysed for heavy metal stress impact, obtaining the proteome content through various processes like fractionation, lysis, enrichment, dialysis, depletion, etc., separating the various types of proteins (Ali et al., 2020) within the proteome based upon their charge and weight by means of gel electrophoresis like 2D-DIGE (Two-dimensional difference gel electrophoresis), identifying the proteins that need to be addressed according to the type of metal stress induced and extracting them by means of a process of digestion and ultimately the proteins of interest or the proteomic biomarkers, which will enable us to understand the heavy metal effects, will undergo spectroscopic analysis like MALDI/TOF (Matrix-assisted laser desorption/ionization-time of flight) followed by verification through ELISA testing. SELDI-TOF (Surface-enhanced laser desorption/ionization time-of-flight) has been used at times as in the case of arsenic (He et al., 2003; Zhai et al., 2005) and lead (Zhai et al., 2005) stress. The data is fed to a computerized analysis and various proteomic databases like ProMEX or Mascot identify them and the biostatistical tools like Statistical Analysis System (SAS) software or Statistical package for social sciences (SPSS) help in the determination of the heavy metal stress impact (Alharbi, 2020). This whole process is based on the principle that the proteomic content of a cell in normal conditions undergoes prominent change upon heavy metal stress exposure and analysing these concentration changes will lead scientists in their research towards making sensitive plants tolerant to metal stress (Parihar et al., 2019). This integrative approach of biotechnology and computational biostatistics has already empowered proteomic research exponentially while at the same time decreasing the use of manpower (Piasecka et al., 2019). The most crucial step in the whole process occurs when the protein content has to be extracted because various metabolites (starches, wall polysaccharides, phenols, etc.) interfere during this process and efficient extraction is faltered which worsens when the protein extract is homogenised and some more protein content is lost in the process (Rose et al., 2004; Isaacson et al., 2006). Proteomic analyses include both qualitative as well as quantitative study and for this reason, two-dimensional gel electrophoresis (2-DE) is the most preferred and usually it is the gel-based techniques that are employed and rarely the gel-free ones (Beranova-Giorgianni, 2003). But even in gel-based techniques, limitations arise like the proteins being diverse in abundance and characteristics and such problems have been solved by efforts like protein enrichment techniques (Bandow, 2010) although not been applied in this field yet.

Non-Reproducibility is another issue and DIGE (difference gel electrophoresis) has been employed to address the issue (Minden et al., 2009). This step is often followed by MALDI-TOF and has been studied in the case of cadmium metal-induced Poplar plants (Kieffer et al., 2008). Some Manganese and cadmium stress-induced plants had their proteomes analysed by blue native (BN)-PAGE followed by SDS-PAGE providing direct quantification of protein content (Führs et al., 2008; Fagioni et al., 2009). Besides these, one of the well-known gel-free methods is Multidimensional Protein Identification Technology (MudPIT) but is not fit for quantitative analyses, and for this, various quantitative techniques have been developed like isotope-coded affinity tag (ICAT) (Kennedy, 2002; Ningbo et al., 2020), isobaric tags for relative and absolute quantitation (iTRAQ) (Zieske, 2006; Farooq et al., 2018; Liu et al., 2020a) which has been successfully employed for detection of boron stress (Patterson et al., 2007), stable-isotope labelling by amino acids in cell culture (SILAC) (Mann, 2006; Fryzova et al., 2017), N-terminal labelling (Wildes and Wells, 2010), isotope-coded protein label (ICPL) (Kellermann, 2008) and even label-free methods (Grossmann et al., 2010). These second-generation tools are also proving to be efficient for making proteomic biomarkers one of the best markers for metal stress analysis. Thus, understanding metal stress and its effects on plants in terms of proteomics is well established and promising and yet needs to inculcate better error free steps for precise quantitative and qualitative results and in addition to this, focussing on the various second-generation technologies can prove to be a booster in this field.

5. Molecular responses to metal stress

Metals are not a threat to the plants themselves but it is the unbalanced concentrations that create toxic environments for the plant and the anthropogenic activities like industrialization, energy production, military operations, mining, etc., have worsened the conditions (Nedelkoska and Doran, 2000; Hall 2002; Dong et al., 2010). These excess metal ions create toxicity through various ways like creating oxidized environments in the form of ROS species or reducing sugars or may alter the protein conformations by attaching to their sulfhydryl groups or taking the place of an essential metal ion that is required for normal and basic cell functioning (van Assche and Clíjsters, 1990; Capuana 2011; Roychoudhury et al., 2021). Studies have shown that heavy metals inhibit chlorophyll formation by inhibiting the enzymes involved in their biosynthesis and hence again points to the fact that heavy metal impact on proteins at a molecular level is responsible for the macro level impact on both physiology and morphology of a plant (van Assche and Clíjsters, 1990; Lanaras et al., 1993; Meers et al., 2010). These biomolecular studies display the impact of heavy metal stress on the genetic materials, proteins, lipids, carbohydrates, etc., but the focus is being laid upon the proteomics because it is the proteome that has a direct interaction with the heavy metals and not the genetic material and unlike carbohydrates and lipids, proteins form the structural as well as the functional basis of cell in the form of building blocks and as biocatalysts respectively. Different metal ions and their concentrations have different biochemical responses in different plant species and even in different tissue parts. For instance, the Mustard plant (*Brassica juncea*) can tolerate quite good levels of heavy metals like chromium, zinc, copper, boron, nickel, cadmium, selenium, etc., while a common sensitive plant cannot (Palmer et al., 2001; Smirnova et al., 2006; Akbaş et al., 2009; Chaudhry et al., 2020; Kolbert et al., 2020) and many cases of copper have been recorded to cause serious cellular injuries in tissues like leaves, stem, etc. (Gupta and Kalra, 2006; Saleem et al., 2020; Peco et al., 2020). Copper concentration in the soil has drastically increased in recent years from 5 to 50 mg/ml and yet lead is considered to be the most dangerous heavy metal due to its presence everywhere, being toxic even in extremely little quantities, high toxicity, and entering the plant quite easily (Salt et al., 1998; Wu et al., 2003; Amari et al., 2017; Nas et al., 2018). Generally, the proteome content of a cell increases as metal stress is induced and upon biomolecular analysis, scientists observed that ROS species increased which in turn led to antioxidant response increase and the letter response was much more than the former in the plants that could better tolerate the stress conditions and various enzymes involved in Krebs cycle, phytochelatin and glutathione synthesis increased along with various heat shock proteins and this metal-protein interaction could be traced and manipulated only by precise proteomic studies (Mishra et al., 2006).

Molecular studies also revealed that hyperaccumulators possess certain mechanistic strategies which enable them to make the toxic metals non-toxic as the metals cannot be degraded but can be changed into some forms that are non-toxic (Kumar et al., 2017b; Wang et al., 2020) and for this many methods have been employed by such plants like sequestering the toxic metals, transporting them to vacuoles or altogether removing them (Choppala et al., 2014; Peng et al., 2020) and this they do by means of various antioxidants and proteomic contents. Due to this property of certain plants, they are being employed for the process of phytoremediation wherein a hyperaccumulator plant is selected for a certain toxic heavy metal that it detoxifies the most and are grown in the agricultural fields or the land that needs to be freed of the toxic heavy metal (Cunningham and Ow, 1996; Singh et al., 2019) and these hyperaccumulators have been observed to be allelopathic against various fungi which makes proteomics all the more important here in order to analyse which proteins were involved in the detoxification process and did the same proteins prove toxic to the pathogens or did they act as transcription factors for some other interlinked pathway. It was observed that amino acids and organic acids present in the xylem sap are responsible for the transport of these metal ions through the plant and this happens through chelation as was observed in one of plant genus *Alyssum* that could accumulate more than the normal concentration of Nickel within the xylem sap due to the presence of amino acid, Histidine (Krämer et al., 1996; AkkusOzen and Yaman, 2017; Chaudhary et al., 2018). One more amino acid that has been found to be a copper transporter is Nicotianamine which is a methionine derivative (Pich and Scholz, 1996; Jain et al., 2018; Jin et al., 2019). The latter has also been found to act as a metal transporter through the phloem (Curie et al., 2009; Palmer and Stangoulis, 2018). This mechanistic approach is employed while the metals are being taken up the by plant but as soon as the metal ions reach the cytoplasm, they need to be either compartmentalized or chelated once again if they evaded the process before. The proteins that are majorly handling this situation are phytochelatins (glutathione oligomers) (Cobbett, 2000; Talebi et al., 2019; Filiz et al., 2019) and metallothionein's (cysteine rich proteins) (Chaudhary et al., 2018; Liu et al., 2020b). Glutathione and glutathione transferases (GST) of tau and phi subfamilies form another mechanistic detoxification strategy through which are formed glutathione compounds being covalently attached which are then transported to vacuoles by ABC (ATP-dependent tonoplast) transporters (Kang et al., 2004; Gao et al., 2020; Estévez and Hernández, 2020). Since these mechanisms are important to guide us about which proteins are vital and mainstream for heavy metal tolerance, much focus has been laid on the molecular interactions and the core proteins that make such resistance possible and the next chapter deals with the same in detail.

6. Molecular mechanisms involved in metal uptake, toxicity, and detoxification in higher plants

Metal uptake is a must for the survival of a plant because all physiological processes require enzymes and almost all enzymes need metals in the form of their co-factors or at times for catalysing a particular reaction. The process, however, is gradient-regulated and plants grow the best in a soil where all the elements are present in optimum concentrations (Zuo et al., 2018). This technologically growing world is making our lives easier but the hazards that come along with it have created havoc in the environment. These negative consequences are not only confined to land but aquatic life as well and for bringing back the balance in our environments, scientists are performing molecular studies to determine what mechanisms make plants absorb these heavy metals due to which their physiology gets badly affected and why some plants are able to cope up with the conditions even in severe stress conditions. Many transgenic crops have been designed where some protein contents overexpressed and led to metal detoxification and ROS scavenging (Li et al., 2017; Ai et al., 2018; Sun et al., 2019). Thus, proteomic markers enable us to modify stress-sensitive plants and make them tolerant and also present and further studies promise to enable us to design proteomic biomarkers that are specialized towards heavy metals and according to our mechanistic approach.

Chelation is one of the main strategies employed by heavy metal tolerant plants to detoxify metals and for this, two major proteins that play a role are PCs (phytochelatins) (Clemens et al., 1999; Cobbett, 2000) and MTs (Metallothioneins) (Robinson et al., 1997; Chaudhary et al., 2018; Liu et al., 2020b). PCs are not only found in plants but microbes as well and have been mostly associated with cadmium stress regulation by bringing the plant into homeostasis state back after stress-induced (Cobbett, 2000; Guo et al., 2008a; Emamverdian et al., 2015; Hasan et al., 2017).

Phytochelatins are made up of three amino acids: glutamine (Glu), cysteine (Cys), and glycine (Gly) and are synthesized from glutathione (GSH) in larger amounts as soon as the heavy metal stress is induced (Yadav, 2010) and act as strong chelators to which the heavy metals bind. On the basis of these proteins, synthetic PCs have been biosynthesized in an attempt to bind to heavy metals and the research is ongoing (Shukla et al., 2013).

Metallothioneins is another class of proteins that are rich in the amino acid, cysteine and are found in abundance in both animals and plants (Singh et al., 2015). Their function is also similar to PCs but their method is sequestration instead of chelation and they try to segregate the metal ions in a manner that heavy metals do not damage the cellular machinery or lead to the creation of ROS. Their function however is not only limited to heavy metal stress but has interconnectivity with various other stress-induced pathways like phytohormonal, UV, starvation, viral infection, etc., and hence regulation of MTs would eventually lead to the regulation of other stress pathways as well (Kushwaha et al., 2015). Arabidopsis plant when made to overexpress some MTs showed enhanced tolerance to metals like copper and cadmium (overexpressing pigeonpea CcMT1) and cadmium alone (due to simultaneous overexpression of garlic MTs, AsMT2b, and the *Colocasia esculenta* MT, CeMT2b) (Zhang et al., 2006).

Heat shock proteins (HSPs) is another group of proteins whose expression, both in terms of transcripts and protein level increases at the time of heavy metal stress and is an essential family of proteins that are required even under normal conditions (Ghosh et al., 2018; Shende, 2019). Their main function is to prevent the protein aggregation that happens rarely under normal conditions but quite frequently under stress environments and hence their selective protein degradation has a great role in maintaining the homeostasis of a cell. They have been grouped into many groups on the basis of their molecular weight (HSP100, HSP90, HSP70, HSP40, small heat shock proteins, sHSP) (Lal, 2010; Al-Whaibi, 2011). Their role in metal stress has been observed in a number of cases as in the case of *Chenopodium album*, HSP26.13p proves to be tolerant against heat stress as well as metal stress (Nickel, Copper, and Cadmium) (Hasan et al., 2017).

Metals, either chelated or sequestered need to be out of the plant system, and this step is accomplished by compartmentalization or efflux systems (Sharma et al., 2016a,b). Although active efflux systems are more prominent in the case of microbes the process of compartmentalization follows the basic pathway in a more evolved sense and the transporters been studied in plants are heavy metal (or CPx-type) ATPases, CDF (cation diffusion facilitator), NRAMPs (natural resistance-associated macrophage proteins), ABCs (ATP-binding cassettes), CAXes (cation exchangers), and ZIPs (Zrt-Irt like proteins) (Sharma et al., 2016a,b; Srivastava et al., 2017).

Thus, the whole process of heavy metal intake through absorption followed by transport along xylem and then accumulation within tissues and ultimately detoxification needs proteins at each level in the form of enzymes, antioxidants, and transporters and thus proteomic studies are essential to make heavy metal sensitive plants tolerant and making plants detoxify metals in such a manner that they prove harmless to the whole environment. Studies of a much broader perspective in terms of proteomics are going on and advancing further due to emerging novel technologies.

7. Gel electrophoresis as a tool to discover metal stress-regulated proteins

Gel electrophoresis is one of the ancient protein separation techniques that formed the basis of proteomics even at the time when the word proteomics had not emerged (Mosa et al., 2017; Jorrín-Novo et al., 2018; Ivarie et al., 2019). The technique is modified in a number of ways to make it more advanced and errorless but even now the principles are relevant and are an essential step in the process of proteomic analysis. Charge and mass are the main two properties that distinguish proteins and exploiting these two properties, the process of 2D-gel electrophoresis operates, that is, by firstly separating the thousands of proteins from a protein mixture on the basis of their isoelectric points (First dimension—isoelectric focusing (IEF) and next with the help of SDS-polyacrylamide gel, which acts as a coating over those separated proteins so that they become uniformly charged and are now separated and distinguished only on the basis of molecular weights (Second dimension—SDS-PAGE) (Mohanty et al., 2017). This qualitative and quantitative separation of proteins has enabled scientists to analyse what effects do different metal ions have on the wide variety of proteins that are present in a cell and how their abundance increases and decreases with the change in the type of metal stress. Since proteomics is relatively a newer field as compared to genomics and transcriptomics, yet it has evolved into a wide field of its own and its need came to arise from the facts when stress was being studied and its effects being analysed by means of genomic expression or the microarray analysis, it could not justify the post-translational results or we can say, the protein content was quite different from what the transcriptional or the genetic data predicted (Miklos and Maleszka, 2001). This led to the need for data at translational and post-translational levels and gave rise to proteomics although gel electrophoresis was being used to separate proteins for proteomics, gel electrophoresis became the basis and even now when heavy metal stress research is going on, gel electrophoresis is a key technique that is to be employed (Pandey and Mann, 2000; Lilley et al., 2002; Rappsilber and Mann, 2002). Thus, 2D gel electrophoresis in combination with mass spectroscopy gives the proteome information of a tissue which was otherwise impossible to find but inspite of such a simple technique or a single gel providing all the information, it does have limitations (Gygi et al., 2000; Meleady, 2018). The technique is error prone and slight changes give a wide range of variations and thus the exact influence of heavy metals cannot be totally relied on and usually, an average of a number of times is taken into account which makes it difficult to simultaneously detect changes in a number of proteins by a number of metal ions. The other problem is the staining techniques which could have an effect on the proteins and hence our data gets altered (Patton, 2002; Vesterberg and Hansen, 2019). But efforts have been made to remove these errors and many new supplementations have been done giving rise to isotope-coded affinity tagging (ICAT) (Gygi et al., 1999b; Fryzova et al., 2017), differential gel exposure (DifExpo) (Monribot-Espagne and Boucherie, 2002), differential gel electrophoresis (DIGE) (Hmmier and Dowling, 2018), etc. These advancements over the conventional gel electrophoresis have proved extremely useful in proteome profiling without affecting the protein sensitivity and are also reproducible, DIGE proving the most to fit the latter properties and is even today being used in a wide range of proteomic analysis techniques. It is a protein labelling technique and fluorescent dyes, Cy3 and Cy5 are being used (Ünlü et al., 1997; Fernández-Cisnal et al., 2017). It has been used to study effects of toxicity in microbes, fungi, and plants in addition to the effects of different chemicals on the biology and chemistry within the cells of a model organism *in vivo* as well as *in vitro* and Cancer study has particularly benefited a lot from DIGE protein profiling (Yan et al., 2002; Hu et al., 2003; Ruepp et al., 2002; Zhou et al., 2002; Friedman et al., 2004; Chou et al., 2018; Sharma and Kumar, 2021). Thus, gel electrophoresis laid the foundation for proteomics and is not only used for detecting and studying the effects of heavy metals and the response of metal stress proteins but various ongoing scientific studies are being benefited by it.

8. Conclusion

After exploring a little aspect of proteomics in terms of heavy metal detection and their impact upon the cellular proteome and hence affecting all the cellular activities, we can surely rely on the conclusion that understanding the proteomics of a cell will ultimately help us in making a plant tolerant towards stress. Analysing the differences between hyperaccumulators and non-hyperaccumulators will enable us to cause expression of those stress regulating proteins that are absent in the latter. Moreover, the biotechnological advancements has opened a gateway of synthetic biomolecular technology and upon analysing the detoxification mechanisms of stress tolerant plants, scientists can devise such synthetic biomolecules or cause their expression within the stress sensitive plants by some post translational or post transcriptional changes so that gene silencing or transcript silencing risk is avoided. Moreover, many gel-based and gel-free techniques used have already proved significant in the researches done and are still being used with such modifications that time and resources, are being saved. A lot of researches are still going on under Proteomics and this relatively new omic tool is progressing towards making our plants tolerant to the increasing environmental stress conditions.

References

Abadin H, Ashizawa A, Llados F, and Stevens YW. 2007. Toxicological profile for lead.

Ai TN, Naing AH, Yun BW, and Kim C. 2018. Overexpression of RsMYB1 enhances heavy metal stress tolerance in transgenic petunia by elevating the transcript levels of stress tolerant and antioxidant genes. *BioRxiv*, p. 286849.

Akbaş H, Dane F, and Meriç Ç. 2009. Effect of nickel on root growth and the kinetics of metal ions transport in onion (*Allium cepa*) root.

Akkus Ozen S, and Yaman M. 2017. Examination of correlation between histidine, sulfur, cadmium, and cobalt absorption by *Morus* L., *Robinia pseudoacacia* L., and *Populus nigra* L. *Communications in Soil Science and Plant Analysis* 48(10): 1212–1220.

Alaraidh IA, Alsahli AA, and Razik EA. 2018. Alteration of antioxidant gene expression in response to heavy metal stress in *Trigonella foenum-graecum* L. *South African Journal of Botany* 115: 90–93.

Alexander PD, Alloway BJ, and Dourado AM. 2006. Genotypic variations in the accumulation of Cd, Cu, Pb and Zn exhibited by six commonly grown vegetables. *Environmental Pollution* 144(3): 736–745.

Alharbi RA. 2020. Proteomics approach and techniques in identification of reliable biomarkers for diseases. *Saudi Journal of Biological Sciences* 27(3): 968–974.

Ali A, Bhat BA, Rathe GA, Malla BA, and Ganie SA. 2020. Proteomic studies of micronutrient deficiency and toxicity. pp. 257–284. *In*: Aftab T, and Hakeem KR (eds.). *Plant Micronutrients*. Springer, Cham.

Alvarez S, Berla BM, Sheffield J, Cahoon RE, Jez JM, and Hicks LM. 2009. Comprehensive analysis of the *Brassica juncea* root proteome in response to cadmium exposure by complementary proteomic approaches. *Proteomics* 9(9): 2419–2431.

Al-Whaibi MH. 2011. Plant heat-shock proteins: a mini review. *Journal of King Saud University-Science* 23(2): 139–150.

Amari T, Ghnaya T, and Abdelly C. 2017. Nickel, cadmium and lead phytotoxicity and potential of halophytic plants in heavy metal extraction. *South African Journal of Botany* 111: 99–110.

Anjali NN, and Sabu KK. 2020. Role of miRNAs in abiotic and biotic stress management in crop plants. pp. 513–532. *In*: *Sustainable Agriculture in the Era of Climate Change*. Springer, Cham.

Arrivault S, Senger T, and Krämer U. 2006. The *Arabidopsis* metal tolerance protein AtMTP3 maintains metal homeostasis by mediating Zn exclusion from the shoot under Fe deficiency and Zn oversupply. *The Plant Journal* 46(5): 861–879.

Baig MA, Qamar S, Ali AA, Ahmad J, and Qureshi MI. 2020. Heavy metal toxicity and tolerance in crop plants. pp. 201–216. *In*: *Contaminants in Agriculture*. Springer, Cham.

Baker AJM. 2000. Metal hyperaccumulator plants: a review of the ecology and physiology of a biological resource for phytoremediation of metal-polluted soils. *Phytoremediation of Contaminated Soil and Water* 85–107.

Bandow JE. 2010. Comparison of protein enrichment strategies for proteome analysis of plasma. *Proteomics* 10(7): 1416–1425.

Banerjee A, and Roychoudhury A 2018. Abiotic stress, generation of reactive oxygen species, and their consequences: an overview. *Revisiting the Role of Reactive Oxygen Species (ROS) in Plants: ROS Boon or Bane for Plants*, pp. 23–50.

Becher M, Talke IN, Krall L, and Krämer U. 2004. Cross-species microarray transcript profiling reveals high constitutive expression of metal homeostasis genes in shoots of the zinc hyperaccumulator *Arabidopsis halleri*. *The Plant Journal* 37(2): 251–268.

Ben Saad R, Fabre D, Mieulet D, Meynard D, Dingkuhn M, Al-Doss ABDULLAH, Guiderdoni E, and Hassairi A. 2012. Expression of the *Aeluropus littoralis* AlSAP gene in rice confers broad tolerance to abiotic stresses through maintenance of photosynthesis. *Plant, Cell & Environment* 35(3): 626–643.

Beranova-Giorgianni S. 2003. Proteome analysis by two-dimensional gel electrophoresis and mass spectrometry: strengths and limitations. *TrAC Trends in Analytical Chemistry* 22(5): 273–281.

Betteridge DJ. 2000. What is oxidative stress? *Metabolism* 49(2): 3–8.

Bohnert HJ, Gong Q, Li P, and Ma S. 2006. Unraveling abiotic stress tolerance mechanisms–getting genomics going. *Current Opinion in Plant Biology* 9(2): 180–188.

Bona E, Cattaneo C, Cesaro P, Marsano F, Lingua G, Cavaletto M, and Berta G. 2010. Proteomic analysis of *Pteris vittata* fronds: two arbuscular mycorrhizal fungi differentially modulate protein expression under arsenic contamination. *Proteomics* 10(21): 3811–3834.

Boominathan R, and Doran PM. 2002. Ni-induced oxidative stress in roots of the Ni hyperaccumulator, *Alyssum bertolonii*. *New Phytologist* 156(2): 205–215.

Boominathan R, and Doran PM. 2003. Cadmium tolerance and antioxidative defenses in hairy roots of the cadmium hyperaccumulator, *Thlaspi Caerulescens*. *Biotechnology and Bioengineering* 83(2): 158–167.

Boon DY, and Soltanpour PN. 1992. Lead, cadmium, and zinc contamination of aspen garden soils and vegetation (Vol. 21, No. 1, pp. 82–86). American Society of Agronomy, Crop Science Society of America, and Soil Science Society of America.

Bull PC, and Cox DW. 1994. Wilson disease and Menkes disease: new handles on heavy-metal transport. *Trends in Genetics* 10(7): 246–252.

Capuana M. 2011. Heavy metals and woody plants-biotechnologies for phytoremediation. *iForest-Biogeosciences and Forestry* 4(1): 7.

Chaudhary K, Agarwal S, and Khan S. 2018. Role of phytochelatins (PCs), metallothioneins (MTs), and heavy metal ATPase (HMA) genes in heavy metal tolerance. pp. 39–60. *In: Mycoremediation and Environmental Sustainability*. Springer, Cham.

Chaudhry H, Nisar N, Mehmood S, Iqbal M, Nazir A, and Yasir M. 2020. Indian mustard *Brassica juncea* efficiency for the accumulation, tolerance, and translocation of zinc from metal contaminated soil. *Biocatalysis and Agricultural Biotechnology* 23: 101489.

Chiang HC, Lo JC, and Yeh KC. 2006. Genes associated with heavy metal tolerance and accumulation in Zn/Cd hyperaccumulator *Arabidopsis halleri*: a genomic survey with cDNA microarray. *Environmental Science & Technology* 40(21): 6792–6798.

Choppala G, Saifullah Bolan N, Bibi S, Iqbal M, Rengel Z, Kunhikrishnan A, Ashwath N, and Ok YS. 2014. Cellular mechanisms in higher plants governing tolerance to cadmium toxicity. *Critical Reviews in Plant Sciences* 33(5): 374–391.

Chou HC, Lu CH, Su YC, Lin LH, Yu HI, Chuang HH, Tsai YT, Liao EC, We, YS, Yang YT, and Chien YA. 2018. Proteomic analysis of honokiol-induced cytotoxicity in thyroid cancer cells. *Life Sciences* 207: 184–204.

Clemens S, Kim EJ, Neumann D, and Schroeder JI. 1999. Tolerance to toxic metals by a gene family of phytochelatin synthases from plants and yeast. *The EMBO Journal* 18(12): 3325–3333.

Clemens S. 2001. Molecular mechanisms of plant metal tolerance and homeostasis. *Planta* 212(4): 475–486.

Clemens S, Palmgren MG, and Krämer U. 2002. A long way ahead: understanding and engineering plant metal accumulation. *Trends in Plant Science* 7(7): 309–315.

Cobbett C, and Goldsbrough P. 2002. Phytochelatins and metallothioneins: roles in heavy metal detoxification and homeostasis. *Annual Review of Plant Biology* 53(1): 159–182.

Cobbett CS. 2000. Phytochelatin biosynthesis and function in heavy-metal detoxification. *Current Opinion in Plant Biology* 3(3): 211–216.

Cunningham SD, and Ow DW. 1996. Promises and prospects of phytoremediation. *Plant Physiology* 110(3): 715.

Curie C, Cassin G, Couch D, Divol F, Higuchi K, Le Jean M, Misson J, Schikora A, Czernic P, and Mari S. 2009. Metal movement within the plant: contribution of nicotianamine and yellow stripe 1-like transporters. *Annals of Botany* 103(1): 1–11

Dar MI, Naikoo MI, Green ID, Sayeed N, Ali B, and Khan FA. 2018. Heavy metal hyperaccumulation and hypertolerance in *Brassicaceae*. pp. 263–276. *In*: Hasanuzzaman M, Nahar K, and Fujita M (eds.). *Plants Under Metal and Metalloid Stress*. Springer, Singapore.

de Abreu-Neto JB, Turchetto-Zolet AC, de Oliveira LFV, Bodanese Zanettini MH, and Margis-Pinheiro M. 2013. Heavy metal-associated isoprenylated plant protein (HIPP): characterization of a family of proteins exclusive to plants. *The FEBS Journal* 280(7): 1604–1616.

Deinlein U, Weber M, Schmidt H, Rensch S, Trampczynska A, Hansen TH, Husted S, Schjoerring JK, Talke IN, Krämer U, and Clemens, S. 2012. Elevated nicotianamine levels in *Arabidopsis halleri* roots play a key role in zinc hyperaccumulation. *The Plant Cell* 24(2): 708–723.

Ding Y, Ding L, Xia Y, Wang F, and Zhu C. 2020. Emerging roles of microRNAs in plant heavy metal tolerance and homeostasis. *Journal of Agricultural and Food Chemistry* 68(7): 1958–1965.

Dixit A, Tomar P, Vaine E, Abdullah H, Hazen S, and Dhankher OP. 2018. A stress-associated protein, AtSAP13, from *Arabidopsis thaliana* provides tolerance to multiple abiotic stresses. *Plant, Cell & Environment* 41(5): 1171–1185.

Dixit AR, and Dhankher OP. 2011. A novel stress-associated protein 'AtSAP10' from *Arabidopsis thaliana* confers tolerance to nickel, manganese, zinc, and high temperature stress. *PLoS one* 6(6): e20921.

Dong X, Li C, Li J, Wang J, Liu S, and Ye B. 2010. A novel approach for soil contamination assessment from heavy metal pollution: A linkage between discharge and adsorption. *Journal of Hazardous Materials* 175(1-3): 1022–1030.

Dubey S, Shri M, Gupta A, Rani V, and Chakrabarty D. 2018. Toxicity and detoxification of heavy metals during plant growth and metabolism. *Environmental Chemistry Letters* 16(4): 1169–1192.

Duquesnoy I, Goupil P, Nadaud I, Branlard G, Piquet-Pissaloux A, and Ledoigt G. 2009. Identification of *Agrostis tenuis* leaf proteins in response to As (V) and As (III) induced stress using a proteomics approach. *Plant Science* 176(2): 206–213.

Dutta S, Mitra M, Agarwal P, Mahapatra K, De S, Sett U, and Roy S. 2018. Oxidative and genotoxic damages in plants in response to heavy metal stress and maintenance of genome stability. *Plant Signaling & Behavior* 13(8): e1460048.

Elbaz B, Shoshani-Knaani NOA, David-Assael ORA, Mizrachy-Dagri TALYA, Mizrahi K, Saul H, Brook E, Berezin I, and Shaul O. 2006. High expression in leaves of the zinc hyperaccumulator *Arabidopsis halleri* of AhMHX, a homolog of an *Arabidopsis thaliana* vacuolar metal/proton exchanger. *Plant, Cell & Environment* 29(6): 1179–1190.

Emamverdian A, Ding Y, Mokhberdoran F, and Xie Y. 2015. Heavy metal stress and some mechanisms of plant defense response. *The Scientific World Journal*.

Estévez IH, and Hernández MR. 2020. Plant glutathione S-transferases: An overview. *Plant Gene*, p. 100233.

Fagioni M, D'Amici GM, Timperio AM, and Zolla L. 2009. Proteomic analysis of multiprotein complexes in the thylakoid membrane upon cadmium treatment. *Journal of Proteome Research* 8(1): 310–326.

Farinati S, DalCorso G, Bona E, Corbella M, Lampis S, Cecconi D, Polati R, Berta G, Vallini G, and Furini A. 2009. Proteomic analysis of *Arabidopsis halleri* shoots in response to the heavy metals cadmium and zinc and rhizosphere microorganisms. *Proteomics* 9(21): 4837–850.

Farooq MA, Zhang K, Islam F, Wang J, Athar HU, Nawaz A, Ullah Zafar Z, Xu J, and Zhou W. 2018. Physiological and itraq-based quantitative proteomics analysis of methyl jasmonate–induced tolerance in *Brassica napus* Under Arsenic Stress. *Proteomics* 18(10): 1700290.

Fernández-Cisnal R, García-Sevillano MA, Gómez-Ariza JL, Pueyo C, López-Barea J, and Abril N. 2017. 2D-DIGE as a proteomic biomarker discovery tool in environmental studies with *Procambarus clarkii*. *Science of The Total Environment* 584: 813–827.

Filiz E, Saracoglu IA, Ozyigit II, and Yalcin B. 2019. Comparative analyses of phytochelatin synthase (PCS) genes in higher plants. *Biotechnology & Biotechnological Equipment* 33(1): 178–194.

Fridovich I. 1999. Fundamental aspects of reactive oxygen species, or what's the matter with oxygen? *Annals of the New York Academy of Sciences* 893(1): 13–18.

Friedman DB, Hill S, Keller JW, Merchant, NB, Levy SE, Coffey RJ, and Caprioli RM. 2004. Proteome analysis of human colon cancer by two-dimensional difference gel electrophoresis and mass spectrometry. *Proteomics* 4(3): 793–811.

Fryzova R, Pohanka M, Martinkova P, Cihlarova H, Brtnicky M, Hladky J, and Kynicky J. 2017. Oxidative stress and heavy metals in plants. *Reviews of Environmental Contamination And Toxicology Volume* 245: 129–156.

Führs H, Hartwig M, Molina LEB, Heintz D, Van Dorsselaer A, Braun HP, and Horst WJ. 2008. Early manganese-toxicity response in *Vigna unguiculata* L.—a proteomic and transcriptomic study. *Proteomics* 8(1): 149–159.

Gao J, Chen B, Lin H, Liu Y, Wei Y, Chen F, and Li W. 2020. Identification and characterization of the glutathione S-Transferase (GST) family in radish reveals a likely role in anthocyanin biosynthesis and heavy metal stress tolerance. *Gene* 743: 144484.

Ghneim-Herrera T, Selvaraj MG, Meynard D, Fabre D, Peña A, Ben Romdhane W, Ben Saad R, Ogawa S, Rebolledo MC, Ishitani M, and Tohme J. 2017. Expression of the *Aeluropus littoralis* AlSAP gene enhances rice yield under field drought at the reproductive stage. *Frontiers in Plant Science* 8: 994.

Ghosh D, Lin Q, Xu J, and Hellmann HA. 2017. How plants deal with stress: exploration through proteome investigation. *Frontiers in Plant Science* 8: 1176.

Ghosh S, Sarkar P, Basak P, Mahalanobish S, and Sil PC. 2018. Role of heat shock proteins in oxidative stress and stress tolerance. *Heat Shock Proteins And Stress*, pp. 109–126.

Giri J, Dansana PK, Kothari KS, Sharma G, Vij S, and Tyagi AK. 2013. SAPs as novel regulators of abiotic stress response in plants. *Bioessays* 35(7): 639–648.

Gitschier J, Moffat B, Reilly D, Wood WI, and Fairbrother WJ. 1998. Solution structure of the fourth metal-binding domain from the Menkes copper-transporting ATPase. *Nature Structural Biology* 5(1): 47–54.

Gomes EJR. 2011. *Genotoxicity and cytotoxicity of Cr (VI) and Pb2+ in Pisum sativum* (Doctoral dissertation, Universidade de Aveiro (Portugal)).

Grossmann J, Roschitzki B, Panse C, Fortes C, Barkow-Oesterreicher S, Rutishauser D and Schlapbach R. 2010. Implementation and evaluation of relative and absolute quantification in shotgun proteomics with label-free methods. *Journal of Proteomics* 73(9): 1740–1746.

Guo J, Dai X, Xu W, and Ma M. 2008b. Overexpressing GSH1 and AsPCS1 simultaneously increases the tolerance and accumulation of cadmium and arsenic in *Arabidopsis thaliana*. *Chemosphere* 72(7): 1020–1026.

Guo WJ, Meetam M, and Goldsbrough PB. 2008a. Examining the specific contributions of individual *Arabidopsis* metallothioneins to copper distribution and metal tolerance. *Plant Physiology* 146(4): 1697–1706.

Gupta P, and Diwan B. 2017. Bacterial exopolysaccharide mediated heavy metal removal: a review on biosynthesis, mechanism and remediation strategies. *Biotechnology Reports* 13: 58–71.

Gupta UC, and Kalra YP. 2006. Residual effect of copper and zinc from fertilizers on plant concentration, phytotoxicity, and crop yield response. *Communications in Soil Science and Plant Analysis* 37(15-20): 2505–2511.

Gygi SP, Rochon Y, Franza BR, and Aebersold R. 1999a. Correlation between protein and mRNA abundance in yeast. *Molecular and Cellular Biology* 19(3): 1720–1730.

Gygi SP, Rist B, Gerber SA, Turecek F, Gelb MH, and Aebersold R. 1999b. Quantitative analysis of complex protein mixtures using isotope-coded affinity tags. *Nature Biotechnology* 17(10): 994–999.

Gygi SP, Corthals GL, Zhang Y, Rochon Y, and Aebersold R. 2000. Evaluation of two-dimensional gel electrophoresis-based proteome analysis technology. *Proceedings of the National Academy of Sciences* 97(17): 9390–9395.

Hall JÁ. 2002. Cellular mechanisms for heavy metal detoxification and tolerance. *Journal of Experimental Botany* 53(366): 1–11.

Hasan M, Cheng Y, Kanwar MK, Chu XY, Ahammed GJ, and Qi ZY. 2017. Responses of plant proteins to heavy metal stress—a review. *Frontiers in Plant Science* 8: 1492.

Hasanuzzaman M, Bhuyan MHM, Parvin K, Bhuiyan TF, Anee TI, Nahar K, Hossen M, Zulfiqar F, Alam M, and Fujita M. 2020. Regulation of ROS metabolism in plants under environmental stress: A review of recent experimental evidence. *International Journal of Molecular Sciences* 21(22): 8695.

Hassinen V, Vallinkoski VM, Issakainen S, Tervahauta A, Kärenlampi S, and Servomaa K. 2009. Correlation of foliar MT2b expression with Cd and Zn concentrations in hybrid aspen (Populus tremula× tremuloides) grown in contaminated soil. *Environmental Pollution* 157(3): 922–930.

Haydon MJ, and Cobbett CS. 2007. Transporters of ligands for essential metal ions in plants. *New Phytologist* 174(3): 499–506.

He QY, Yip TT, Li M, and Chiu JF. 2003. Proteomic analyses of arsenic-induced cell transformation with SELDI-TOF ProteinChip® technology. *Journal of Cellular Biochemistry* 88(1): 1–8.

He X, Xie S, Xie P, Yao M, Liu W, Qin L, Liu Z, Zheng M, Liu H, Guan M, and Hua W. 2019. Genome-wide identification of stress-associated proteins (SAP) with A20/AN1 zinc finger domains associated with abiotic stresses responses in *Brassica napus*. *Environmental and Experimental Botany* 165: 108–119.

Hmmier A, and Dowling P. 2018. DIGE analysis software and protein identification approaches. pp. 41–50. *In*: Kay Ohlendieck (ed.). *Difference Gel Electrophoresis*. Humana Press, New York, NY.

Hossain MA, Piyatida P, da Silva JAT, and Fujita M. 2012. Molecular mechanism of heavy metal toxicity and tolerance in plants: central role of glutathione in detoxification of reactive oxygen species and methylglyoxal and in heavy metal chelation. *Journal of Botany* Volume 2012, Article ID 872875, 37 pages doi: 10.1155/2012/872875.

Hossain Z, and Komatsu S. 2013. Contribution of proteomic studies towards understanding plant heavy metal stress response. *Frontiers in Plant Science* 3: 310.

Hu YI, Wang G, Chen GY, Fu XIN, and Yao SQ. 2003. Proteome analysis of *Saccharomyces cerevisiae* under metal stress by two-dimensional differential gel electrophoresis. *Electrophoresis* 24(9): 1458–1470.

Ingle RA, Smith JAC, and Sweetlove LJ. 2005. Responses to nickel in the proteome of the hyperaccumulator plant *Alyssum lesbiacum*. *Biometals* 18(6): 627–641.

Isaacson T, Damasceno CM, Saravanan RS, He Y, Catalá C, Saladié M, and Rose JK. 2006. Sample extraction techniques for enhanced proteomic analysis of plant tissues. *Nature Protocols* 1(2): 769.

Ivarie RD, Gelfand DH, Jones PP, O'Farrell PZ, Polisky BA, Steinberg RA, and O'Farrell PH. 2019. Biological applications of two-dimensional gel electrophoresis. pp. 369–384. *In: Electrofocusing and Isotachophoresis*. De Gruyter.

Jain S, Muneer S, Guerriero G, Liu S, Vishwakarma K, Chauhan DK, Dubey NK, Tripathi DK, and Sharma S. 2018. Tracing the role of plant proteins in the response to metal toxicity: a comprehensive review. *Plant Signaling & Behavior* 13(9): e1507401.

Ježek P, and Hlavatá L. 2005. Mitochondria in homeostasis of reactive oxygen species in cell, tissues, and organism. *The International Journal of Biochemistry & Cell Biology* 37(12): 2478–2503.

Jia L, He X, Chen W, Liu Z, Huang Y, and Yu S. 2013. Hormesis phenomena under Cd stress in a hyperaccumulator—*Lonicera japonica* Thunb. *Ecotoxicology* 22(3): 476–485.

Jin C, Sun Y, Shi Y, Zhang Y, Chen K, Li Y, Liu G, Yao F Cheng D, Li J, and Zhou J. 2019. Branched-chain amino acids regulate plant growth by affecting the homeostasis of mineral elements in rice. *Sci China Life Sci* 62: 1107–1110.

Jorrín-Novo JV, Valledor-González L, Castillejo-Sánchez MA, Sánchez-Lucas R, Gómez-Gálvez IM, López-Hidalgo C, Guerrero-Sánchez VM, Gómez MCM, Martin ICM, Carvalho K, and González APM. 2018. Proteomics analysis of plant tissues based on two-dimensional gel electrophoresis. pp. 309–322. *In*: Sanchez-Moreiras AM, and Reigosa MR (eds.). *Advances in Plant Ecophysiology Techniques*. Springer, Cham.

Joshi R, Dkhar J, Singla-Pareek SL, and Pareek A. 2019. Molecular mechanism and signaling response of heavy metal stress tolerance in plants. pp. 29–47. *In*: Srivastava S, Srivastava AK, and Penna S (ed.). *Plant-metal interactions*. Springer, Cham.

Kang SY, Lee JU, Moon SH, and Kim KW. 2004. Competitive adsorption characteristics of Co2+, Ni2+, and Cr3+ by IRN-77 cation exchange resin in synthesized wastewater. *Chemosphere* 56(2): 141–147.

Kellermann J. 2008. ICPL—isotope-coded protein label. pp. 113–123. *In*: Anton Posch (ed.). *2D PAGE: Sample Preparation and Fractionation*. Humana Press.

Kennedy S. 2002. The role of proteomics in toxicology: identification of biomarkers of toxicity by protein expression analysis. *Biomarkers* 7(4): 269–290.

Khan TA, Yusuf M, and Fariduddin Q. 2018. Hydrogen peroxide in regulation of plant metabolism: Signalling and its effect under abiotic stress. *Photosynthetica* 56(4): 1237–1248.

Kidwai M, Dha YV, Gautam N, Tiwari M, Ahmad IZ, Asif MH, and Chakrabarty D. 2019. *Oryza sativa* class III peroxidase (OsPRX38) overexpression in *Arabidopsis thaliana* reduces arsenic accumulation due to apoplastic lignification. *Journal of Hazardous Materials* 362: 383–393.

Kieffer P, Dommes J, Hoffmann L, Hausman JF, and Renaut J. 2008. Quantitative changes in protein expression of cadmium-exposed poplar plants. *Proteomics* 8(12): 2514–2530.

Kim GD, Cho YH, and Yoo SD. 2015. Regulatory functions of evolutionarily conserved AN1/A20-like Zinc finger family proteins in *Arabidopsis* stress responses under high temperature. *Biochemical and Biophysical Research Communications* 457(2): 213–220.

Kim S, Takahash M, Higuchi K, Tsunoda K, Nakanishi H, Yoshimura E, Mori S, and Nishizawa NK. 2005. Increased nicotianamine biosynthesis confers enhanced tolerance of high levels of metals, in particular nickel, to plants. *Plant and Cell Physiology* 46(11): 1809–1818.

Kolbert Z, Oláh D, Molnár Á, Szőllősi R, Erdei L, and Ördög A. 2020. Distinct redox signalling and nickel tolerance in *Brassica juncea* and *Arabidopsis thaliana*. *Ecotoxicology and Environmental Safety* 189: 109989.

Kolupaev YE, Firsova EN, Yastreb TO, and Lugovaya AA. 2017. The participation of calcium ions and reactive oxygen species in the induction of antioxidant enzymes and heat resistance in plant cells by hydrogen sulfide donor. *Applied Biochemistry and Microbiology* 53(5): 573579.

Krämer U, Cotter-Howells JD, Charnock JM, Baker AJ, and Smith JAC. 1996. Free histidine as a metal chelator in plants that accumulate nickel. *Nature* 379(6566): 635–638.

Krämer U. 2010. Metal hyperaccumulation in plants. *Annual Review of Plant Biology* 61: 517–534.

Kumar A, Singh N, Pandey R, Gupta VK, and Sharma B. 2018. Biochemical and molecular targets of heavy metals and their actions. pp. 297–319. *In*: Mahendra RaiAvinash P, IngleSerenella Medici (eds.). *Biomedical applications of metals*. Springer, Cham.

Kumar B, Smita K, and Flores LC. 2017a. Plant mediated detoxification of mercury and lead. *Arabian Journal of Chemistry* 10: S2335–S2342.

Kumar N, and Soni H. 2007. Characterization of heavy metals in vegetables using inductive coupled plasma analyzer (ICPA). *Journal of Applied Sciences and Environmental Management* 11(3).

Kumar SS, Kadier A, Malyan SK, Ahmad A, and Bishnoi NR. 2017b. Phytoremediation and rhizoremediation: uptake, mobilization, and sequestration of heavy metals by plants. *Plant-microbe Interactions in Agro-ecological Perspectives*, pp. 367–394.

Kushwaha A, Rani R, Kumar S, and Gautam A. 2015. Heavy metal detoxification and tolerance mechanisms in plants: Implications for phytoremediation. *Environmental Reviews* 24(1): 39–51.

Lal N. 2010. Molecular mechanisms and genetic basis of heavy metal toxicity and tolerance in plants. pp. 35–58. *In*: Ashraf M, Ozturk M, and Ahmad SA (eds.). Ahmad *Plant adaptation and phytoremediation*. Springer, Dordrecht.

Lanaras T, Moustakas M, Symeonidis L, Diamantoglou S, and Karataglis S. 1993. Plant metal content, growth responses and some photosynthetic measurements on field-cultivated wheat growing on ore bodies enriched in Cu. *Physiologia Plantarum* 88(2): 307–314.

Lee KW, Cha JY, Kim KH, Kim YG, Lee BH, and Lee SH. 2012. Overexpression of alfalfa mitochondrial HSP23 in prokaryotic and eukaryotic model systems confers enhanced tolerance to salinity and arsenic stress. *Biotechnology Letters* 34(1): 167–174.

Lei GJ, Yamaji N, and Ma JF. 2021. Two metallothionein genes highly expressed in rice nodes are involved in distribution of Zn to the grain. *New Phytologist* 229(2): 1007–1020.

Li J, Zhang M, Sun J, Mao X, Wang J, Liu H, Zheng H, Li X, Zhao H, and Zou D. 2020. Heavy metal stress-associated proteins in rice and *Arabidopsis*: Genome-wide identification, phylogenetics, duplication, and expression profiles analysis. *Frontiers in Genetics* 11: 477.

Li S, Yang X, Yang S, Zhu M, and Wang X. 2012. Technology prospecting on enzymes: application, marketing, and engineering. *Computational and Structural Biotechnology Journal* 2(3): e201209017.

Li Y, Dhankher OP, Carreira L, Balish RS, and Meagher RB. 2005. Arsenic and mercury tolerance and cadmium sensitivity in *Arabidopsis* plants expressing bacterial γ-glutamylcysteine synthetase. *Environmental Toxicology and Chemistry: An International Journal* 24(6): 1376–1386.

Li Z, Han X, Song X, Zhang Y, Jiang J, Han Q, Liu M, Qiao G, and Zhuo R. 2017. Overexpressing the *Sedum alfredii* Cu/Zn superoxide dismutase increased resistance to oxidative stress in transgenic *Arabidopsis*. *Frontiers in Plant Science* 8: 1010.

Lilley KS, Razzaq A, and Dupree P. 2002. Two-dimensional gel electrophoresis: recent advances in sample preparation, detection, and quantitation. *Current Opinion in Chemical Biology* 6(1): 46–50.

Liu S, Li Y, Liu L, Min J, Liu W, Li X, Pan X, Lu X, and Deng Q. 2020a. Comparative proteomics in rice seedlings to characterize the resistance to cadmium stress by high-performance liquid chromatography–tandem mass spectrometry (HPLC-MS/MS) with isobaric tag for relative and absolute quantitation (iTRAQ). *Analytical Letters* 53(5): 807–820.

Liu XM, An J, Han HJ, Kim S, Lim CO, Yun DJ, and Chung WS. 2014. ZAT11, a zinc finger transcription factor, is a negative regulator of nickel ion tolerance in *Arabidopsis*. *Plant Cell Reports* 33(12): 2015–2021.

Liu Y, Kang T, Cheng JS, Yi, YJ, Han JJ, Cheng HL, Li Q, Tang N, and Liang MX. 2020b. Heterologous expression of the metallothionein PpMT2 gene from *Physcomitrella patens* confers enhanced tolerance to heavy metal stress on transgenic *Arabidopsis* plants. *Plant Growth Regulation* 90(1): 63–72.

Luo S, Calderon-Urrea A, Jihua YU, Liao W, Xie J, Lv J, Feng Z, and Tang Z. 2020. The role of hydrogen sulfide in plants alleviates heavy metal stress. *Plant and Soil* 449(1): 1–10.

Manara A, Fasani E, Molesini B, DalCorso G, Pennisi F, Pandolfini T, and Furini A. 2020. The tomato metallocarboxypeptidase inhibitor I, which interacts with a heavy metal-associated isoprenylated protein, is implicated in plant response to cadmium. *Molecules* 25(3): 700.

Mann M. 2006. Functional and quantitative proteomics using SILAC. *Nature reviews Molecular cell Biology* 7(12): 952–958.

Mathur J, and Chauhan P. 2020. Mechanism of toxic metal uptake and transport in plants. pp. 335–349. *In*: Mishra K, Tandon PK, and Srivastava S (eds.). *Sustainable Solutions for Elemental Deficiency and Excess in Crop Plants*. Springer, Singapore.

Meers E, Van Slycken S, Adriaensen K, Ruttens A, Vangronsveld J, Du Laing G, Witters N, Thewys T, and Tack FMG. 2010. The use of bio-energy crops (*Zea mays*) for 'phytoattenuation' of heavy metals on moderately contaminated soils: a field experiment. *Chemosphere* 78(1): 35–41.

Mei H, Cheng NH, Zhao J, Park S, Escareno RA, Pittman JK, and Hirschi KD. 2009. Root development under metal stress in *Arabidopsis thaliana* requires the H+/cation antiporter CAX4. *New Phytologist* 183(1): 95–105.

Meleady P. 2018. Two-dimensional gel electrophoresis and 2D-DIGE. *Difference Gel Electrophoresis*, pp. 3–14.

Miklos GLG, and Maleszka R. 2001. Protein functions and biological contexts. *PROTEOMICS: International Edition* 1(2): 169–178.

Minden JS, Dowd SR, Meyer HE, and Stühler K. 2009. Difference gel electrophoresis. *Electrophoresis* 30(S1): S156–S161.

Mishra S, Srivastava S, Tripathi RD, Kumar R, Seth CS, and Gupta DK. 2006. Lead detoxification by coontail (*Ceratophyllum demersum* L.) involves induction of phytochelatins and antioxidant system in response to its accumulation. *Chemosphere* 65(6): 1027–1039.

Mohanty AK, Yadav ML, and Choudhary S. 2017. Gel electrophoresis of proteins and nucleic acids. pp. 233–246. *In*: Srivastava N, and Megha Pande (ed.). *Protocols in Semen Biology (comparing assays)*. Springer, Singapore.

Monribot-Espagne C, and Boucherie H. 2002. Differential gel exposure, a new methodology for the two-dimensional comparison of protein samples. *Proteomics* 2(3): 229–240.

Morel JL, Mench M, and Guckert A. 1986. Measurement of Pb 2+, Cu 2+ and Cd 2+ binding with mucilage exudates from maize (*Zea mays* L.) roots. *Biology and Fertility of Soils* 2(1): 29–34.

Mosa KA, Ismail A, and Helmy M. 2017. Omics and system biology approaches in plant stress research. pp. 21–34. *In*: Ramanjulu Sunkar (ed.). *Plant Stress Tolerance*. Springer, Cham.

Mukhopadhyay A, Vij S, and Tyagi AK. 2004. Overexpression of a zinc-finger protein gene from rice confers tolerance to cold, dehydration, and salt stress in transgenic tobacco. *Proceedings of the National Academy of Sciences* 101(16): 6309–6314.

Munzuroglu O, and Geckil HİKMET. 2002. Effects of metals on seed germination, root elongation, and coleoptile and hypocotyl growth in *Triticum aestivum* and *Cucumis sativus*. *Archives of Environmental Contamination and Toxicology* 43(2): 203–213.

Nabulo G, Black CR, and Young SD. 2011. Trace metal uptake by tropical vegetables grown on soil amended with urban sewage sludge. *Environmental Pollution* 159(2): 368–376.

Nas FS, and Ali M. 2018. The effect of lead on plants in terms of growing and biochemical parameters: a review. *MOJ Eco Environ Sci* 3(4): 265–268.

Navabpour S, Yamchi A, Bagherikia S, and Kafi H. 2020. Lead-induced oxidative stress and role of antioxidant defense in wheat (*Triticum aestivum* L.). *Physiology and Molecular Biology of Plants* 26(4): 793–802.

Nazir MM, Ulhassan Z, Zeeshan M, Ali S, and Gill MB. 2020. Toxic metals/metalloids accumulation, tolerance, and homeostasis in *Brassica* oilseed species. pp. 379–408. *In*: Hasanuzzaman M (ed.). *The Plant Family Brassicaceae*. Springer, Singapore.

Nedelkoska TV, and Doran PM. 2000. Characteristics of heavy metal uptake by plant species with potential for phytoremediation and phytomining. *Minerals Engineering* 13(5): 549–561.

Nikolić M, and Tomašević V. 2020. Implication of the plant species belonging to the *Brassicaceae* family in the metabolization of heavy metal pollutants in urban settings. *Polish Journal of Environmental Studies* 30(1): 523–534.

Ningbo G, Xiaoqian R, Haijun Z, Rong C, Xiaoyao S, Yun L, Baoqin Z, and Jiping C. 2020. A review on the application of proteomic approaches in environmental toxicology. *Asian Journal of Ecotoxicology* (4): 88–98.

Novaković L, Guo T, Bacic A, Sampathkumar A, and Johnson KL. 2018. Hitting the wall—sensing and signaling pathways involved in plant cell wall remodeling in response to abiotic stress. *Plants* 7(4): 89.

Palmer CE, Warwick S, and Keller W. 2001. Brassicaceae (Cruciferae) family, plant biotechnology, and phytoremediation. *International Journal of Phytoremediation* 3(3): 245–287.

Palmer LJ, and Stangoulis JC. 2018. Changes in the elemental and metabolite profile of wheat phloem sap during grain filling indicate a dynamic between plant maturity and time of day. *Metabolites* 8(3): 53.

Pan W, Shen J, Zheng Z, Yan X, Shou J, Wang W, Jiang L, and Pan J. 2018. Overexpression of the tibetan plateau annual wild barley (*Hordeum spontaneum*) HsCIPKs enhances rice tolerance to heavy metal toxicities and other abiotic stresses. *Rice* 11(1): 1–13.

Pandey A, and Mann M. 2000. Proteomics to study genes and genomes. *Nature* 405(6788): 837–846.

Parihar P, Singh S, Singh R, Rajasheker G, Rathnagiri P, Srivastava RK, Singh VP, Suprasanna P, Prasad, SM and Kishor PK. 2019. An integrated transcriptomic, proteomic, and metabolomic approach to unravel the molecular mechanisms of metal stress tolerance in plants. pp. 1–28. *In*: Srivastava S, Srivastava AK, and Suprasanna P (eds.). *Plant-Metal Interactions*. Springer, Cham.

Patterson J, Ford K, Cassin A, Natera S, and Bacic A. 2007. Increased abundance of proteins involved in phytosiderophore production in boron-tolerant barley. *Plant Physiology* 144(3): 1612–1631.

Patton WF. 2002. Detection technologies in proteome analysis. *Journal of Chromatography B* 771(1-2): 3–31.

Paul JM, Jimmy J, Therattil JM, Regi L, and Shahana S. 2017. Removal of heavy metals using low-cost adsorbents. *IOSR J Mech Civ Eng* 14(03): 48–50.

Peco JD, Sandalio LM, Higueras P, Olmedilla A, and Campos JA. 2020. Characterization of the biochemical basis for copper homeostasis and tolerance in *Biscutella auriculata* L. *Physiologia Plantarum* 173(1): 167–179.

Pedersen CN, Axelsen KB, Harper JF, and Palmgren MG. 2012. Evolution of plant P-type ATPases. *Frontiers in Plant Science* 3: 31.

Peng JS, Guan YH, Lin XJ, Xu XJ, Xiao L, Wang HH, and Meng S. 2020. Comparative understanding of metal hyperaccumulation in plants: a mini-review. *Environmental Geochemistry and Health*, pp. 1–9.

Piasecka A, Kachlicki P, and Stobiecki M. 2019. Analytical methods for detection of plant metabolomes changes in response to biotic and abiotic stresses. *International Journal of Molecular Sciences* 20(2): 379.

Pich A, and Scholz G. 1996. Translocation of copper and other micronutrients in tomato plants (*Lycopersicon esculentum* Mill.): nicotianamine-stimulated copper transport in the xylem. *Journal of Experimental Botany* 47(1): 41–47.

Pirzadah TB, Malik B, and Hakeem KR. 2019. Integration of "Omic" approaches to unravel the heavy metal tolerance in plants. *Essentials of Bioinformatics, Volume III*, pp. 79–92.

Plessl M, Rigola D, Hassinen VH, Tervahauta A, Kärenlampi S, Schat H, Aarts MG, and Ernst D. 2010. Comparison of two ecotypes of the metal hyperaccumulator *Thlaspi caerulescens* (J. & C. PRESL) at the transcriptional level. *Protoplasma* 239(1): 81–93.

Puig S, Mira H, Dorcey E, Sancenón V, Andrés-Colás N, Garcia-Molina A, Burkhead JL, Gogolin KA, Abdel-Ghany SE, Thiele DJ, and Ecker JR. 2007. Higher plants possess two different types of ATX1-like copper chaperones. *Biochemical and Biophysical Research Communications* 354(2): 385–390.

Ranjan A, Sinha R, Sharma TR, Pattanayak A, and Singh AK. 2021. Alleviating aluminum toxicity in plants: implications of reactive oxygen species signalling and crosstalk with other signaling pathways. *Physiologia Plantarum*.

Rappsilber J, and Mann M. 2002. Is mass spectrometry ready for proteome-wide protein expression analysis? *Genome Biology* 3(8): 1–5.

Rizvi A, and Khan MS. 2018. Heavy metal induced oxidative damage and root morphology alterations of maize (*Zea mays* L.) plants and stress mitigation by metal tolerant nitrogen fixing *Azotobacter chroococcum*. *Ecotoxicology and Environmental Safety* 157: 9–20.

Robinson BH, Brooks RR, Howes AW, Kirkman JH, and Gregg PEH. 1997. The potential of the high-biomass nickel hyperaccumulator *Berkheya coddii* for phytoremediation and phytomining. *Journal of Geochemical Exploration* 60(2): 115126.

Rose JK, Bashir S, Giovannoni JJ, Jahn MM, and Saravanan RS. 2004. Tackling the plant proteome: practical approaches, hurdles and experimental tools. *The Plant Journal* 39(5): 715–733.

Rostami S, and Azhdarpoor A. 2019. The application of plant growth regulators to improve phytoremediation of contaminated soils: A review. *Chemosphere* 220: 818–827.

Roychoudhury A, Singh A, Aftab T, Ghosal P, and Banik N. 2021. Seedling priming with sodium nitroprusside rescues *Vigna radiata* from salinity stress-induced oxidative damages. *Journal of Plant Growth Regulation*, pp. 1–11.

Ruepp SU, Tonge RP, Shaw J, Wallis N, and Pognan F. 2002. Genomics and proteomics analysis of acetaminophen toxicity in mouse liver. *Toxicological Sciences* 65(1): 135–150.

Saad RB, Zouari N, Ramdhan WB, Azaza J, Meynard D, Guiderdoni E, and Hassairi A. 2010. Improved drought and salt stress tolerance in transgenic tobacco overexpressing a novel A20/AN1 zinc-finger "AlSAP" gene isolated from the halophyte grass *Aeluropus littoralis*. *Plant Molecular Biology* 72(1–2): 171.

Saad RB, Hsouna AB, Saibi W, Hamed KB, Brini F, and Ghneim-Herrera T. 2018a. A stress-associated protein, LmSAP, from the halophyte *Lobularia maritima* provides tolerance to heavy metals in tobacco through increased ROS scavenging and metal detoxification processes. *Journal of Plant Physiology* 231: 234–243.

Saad RB, Farhat-Khemakhem A, Halima NB, Hamed KB, Brini F, and Saibi W. 2018b. The LmSAP gene isolated from the halotolerant *Lobularia maritima* improves salt and ionic tolerance in transgenic tobacco lines. *Functional Plant Biology* 45(3): 378–391.

Saed-Moucheshi A, Pakniyat H, Pirasteh-Anosheh H, and Azooz MM. 2014. Role of ROS as signaling molecules in plants. pp. 585–620. *In*: Parvaiz Ahmad (ed.). *Oxidative Damage to Plants*. Academic Press.

Saleem MH, Ali S, Rehman M, Rana MS, Rizwan M, Kamran M, Imran M, Riaz M, Soliman MH, Elkelish A, and Liu L. 2020. Influence of phosphorus on copper phytoextraction via modulating cellular organelles in two jute (*Corchorus capsularis* L.) varieties grown in a copper mining soil of Hubei Province, China. *Chemosphere* 248: 126032.

Salt DE, Smith RD, and Raskin I. 1998. Phytoremediation. *Annual Review of Plant Biology* 49(1): 643–668.

Santos CS, Deuchande T, and Vasconcelos MW. 2019. Molecular aspects of iron nutrition in plants. pp. 125–156. *In*: Ulrich Lüttge, Francisco M Cánovas, María-Carmen Risueño, and Christoph Leuschner (eds.). *Progress in Botany Vol. 81*. Springer, Cham.

Sapara KK, Khedia J, Agarwal P, Gangapur DR, and Agarwal PK. 2019. SbMYB15 transcription factor mitigates cadmium and nickel stress in transgenic tobacco by limiting uptake and modulating antioxidative defence system. *Functional Plant Biology* 46(8): 702–714.

Sarma RK, Gowtham I, Bharadwaj RKB, Hema J, and Sathishkumar R. 2018. Recent advances in metal induced stress tolerance in plants: possibilities and challenges. *Plants Under Metal and Metalloid Stress*, pp. 1–28.

Scandalios JG. 2005. Oxidative stress: molecular perception and transduction of signals triggering antioxidant gene defenses. *Brazilian Journal of Medical and Biological Research* 38(7): 995–1014.

Schneider T, Persson DP, Husted S, Schellenberg M, Gehrig P, Lee Y, Martinoia E, Schjoerring JK, and Meyer S. 2013. A proteomics approach to investigate the process of Z n hyperaccumulation in *Noccaea caerulescens* (J & C. P resl) FK M eyer. *The Plant Journal* 73(1): 131–142.

Schuch AP, Moreno NC, Schuch NJ, Menck CFM, and Garcia CCM. 2017. Sunlight damage to cellular DNA: Focus on oxidatively generated lesions. *Free Radical Biology and Medicine* 107: 110–124.

Shafiq S, Zeb Q, Ali A, Sajjad Y, Nazir R, Widemann E, and Liu L. 2019. Lead, cadmium, and zinc phytotoxicity alter DNA methylation levels to confer heavy metal tolerance in wheat. *International Journal of Molecular Sciences* 20(19): 4676.

Sharma A, Soares C, Sousa B, Martins M, Kumar V, Shahzad B, Sidhu GP, Bali AS, Asgher M, Bhardwaj R, and Thukral AK. 2020. Nitric oxide-mediated regulation of oxidative stress in plants under metal stress: a review on molecular and biochemical aspects. *Physiologia Plantarum* 168(2): 318–344.

Sharma P, and Dubey RS. 2005. Lead toxicity in plants. *Brazilian Journal of Plant Physiology* 17(1): 35–52.

Sharma P, and Kumar K. 2021. Phytoproteomics: A New Approach to Decipher Phytomicrobiome Relationships. *Phytomicrobiome Interactions and Sustainable Agriculture*, pp. 15–31.

Sharma S, Rana S, Thakkar A, Baldi A, Murthy RSR, and Sharma RK. 2016a. Physical, chemical and phytoremediation technique for removal of heavy metals. *Journal of Heavy Metal Toxicity and Diseases* 1(2): 1–15.

Sharma SS, and Dietz KJ. 2009. The relationship between metal toxicity and cellular redox imbalance. *Trends in Plant Science* 14(1): 43–50.

Sharma SS, Dietz KJ, and Mimura T. 2016b. Vacuolar compartmentalization as indispensable component of heavy metal detoxification in plants. *Plant, Cell & Environment* 39(5): 1112–1126.

Shende P, Bhandarkar S, and Prabhakar B. 2019. Heat shock proteins and their protective roles in stem cell biology. *Stem Cell Reviews and Reports* 15(5): 637–651.

Shinozaki K, and Yamaguchi-Shinozaki K. 2000. Molecular responses to dehydration and low temperature: differences and cross-talk between two stress signaling pathways. *Current Opinion in Plant Biology* 3(3): 217–223.

Shukla D, Tiwari M, Tripathi RD, Nath P, and Trivedi PK. 2013. Synthetic phytochelatins complement a phytochelatin-deficient *Arabidopsis* mutant and enhance the accumulation of heavy metal (loid) s. *Biochemical and Biophysical Research Communications* 434(3): 664–669.

Siddiqui MH, Alamri S, Khan MN, Corpas FJ, Al-Amr, AA, Alsubaie QD, Ali HM, Kalaji HM, and Ahmad P. 2020. Melatonin and calcium function synergistically to promote the resilience through ROS metabolism under arsenic-induced stress. *Journal of Hazardous Materials* 398: 122882.

Silva R, Aguiar TQ, Oliveira R, and Domingues L. 2019. Light exposure during growth increases riboflavin production, reactive oxygen species accumulation and DNA damage in *Ashbya gossypii* riboflavin-overproducing strains. *FEMS Yeast Research* 19(1): 114.

Singh R, Jha AB, Misra AN, and Sharma P. 2019. Adaption mechanisms in plants under heavy metal stress conditions during phytoremediation. pp. 329–360. *In*: Vimal Chandra Pandey, and Kuldeep Bauddh (eds.). *Phytomanagement of Polluted Sites*. Elsevier.

Singh R, Singh S, Parihar P, Singh VP, and Prasad SM. 2015. Arsenic contamination, consequences and remediation techniques: a review. *Ecotoxicology and Environmental Safety* 112: 247–270.

Smirnova TA, Kolomiitseva GY, Prusov AN, and Vanyushin BF. 2006. Zinc and copper content in developing and aging coleoptiles of wheat seedlings. *Russian Journal of Plant Physiology* 53(4): 535–540.

Sorrentino MC, Capozzi F, Amitrano C, Giordano S, Arena C, and Spagnuolo V. 2018. Performance of three cardoon cultivars in an industrial heavy metal-contaminated soil: Effects on morphology, cytology and photosynthesis. *Journal of Hazardous Materials* 351: 131–137.

Srivastava V, Sarkar A, Singh S, Singh P, de Araujo AS, and Singh RP. 2017. Agroecological responses of heavy metal pollution with special emphasis on soil health and plant performances. *Frontiers in Environmental Science* 5: 64.

Sun K, Wang H, and Xia Z. 2019. The maize bHLH transcription factor bHLH105 confers manganese tolerance in transgenic tobacco. *Plant Science* 280: 97–109.

Talebi M, Tabatabaei BES, and Akbarzadeh H. 2019. Hyperaccumulation of Cu, Zn, Ni, and Cd in *Azolla* species inducing expression of methallothionein and phytochelatin synthase genes. *Chemosphere* 230: 488–497.

Tehseen M, Cairns N, Sherson S, and Cobbett CS. 2010. Metallochaperone-like genes in *Arabidopsis thaliana*. *Metallomics* 2(8): 556–564.

Terzi H, and Yıldız M. 2021. Proteomic analysis reveals the role of exogenous cysteine in alleviating chromium stress in maize seedlings. *Ecotoxicology and Environmental Safety* 209: 111784.

Tuomainen M, Tervahauta A, Hassinen V, Schat H, Koistinen KM, Lehesranta S, Rantalainen K, Häyrinen J, Auriola S, Anttonen M, and Kärenlampi S. 2010. Proteomics of *Thlaspi caerulescens* accessions and an inter-accession cross segregating for zinc accumulation. *Journal of Experimental Botany* 61(4): 1075–1087.

Ünlü M, Morgan ME, and Minden JS. 1997. Difference gel electrophoresis. A single gel method for detecting changes in protein extracts. *Electrophoresis* 18(11): 2071–2077.

Uraguchi S, Watanabe I, Yoshitomi A Kiyono M, and Kuno K. 2006. Characteristics of cadmium accumulation and tolerance in novel Cd-accumulating crops, *Avena strigosa,* and *Crotalaria juncea. Journal of Experimental Botany* 57(12): 2955–2965.

Uwah EI, Ndahi NP, Abdulrahman FI, and Ogugbuaja VO. 2011. Heavy metal levels in spinach (*Amaranthus caudatus*) and lettuce (*Lactuca sativa*) grown in Maiduguri, Nigeria. *Journal of Environmental Chemistry and Ecotoxicology* 3(10): 264–271.

Van Assche F, and Clijsters H. 1990. Effects of metals on enzyme activity in plants. *Plant, Cell & Environment* 13(3): 195–206.

van de Mortel JE, Villanueva LA, Schat H, Kwekkeboom J, Coughlan S, Moerland PD, van Themaat, EVL, Koornneef M, and Aarts MG. 2006. Large expression differences in genes for iron and zinc homeostasis, stress response, and lignin biosynthesis distinguish roots of *Arabidopsis thaliana* and the related metal hyperaccumulator *Thlaspi caerulescens. Plant Physiology* 142(3): 1127–1147.

van de Mortel JE, Schat H, Moerland PD, Van Themaat EVL, Van Der Ent, SJOERD, Blankestijn H, Ghandilyan A, Tsiatsiani S, and Aarts MG. 2008. Expression differences for genes involved in lignin, glutathione and sulphate metabolism in response to cadmium in *Arabidopsis thaliana* and the related Zn/Cd-hyperaccumulator *Thlaspi caerulescens. Plant, Cell & Environment* 31(3): 301–324.

Vasupalli N, Koramutla MK, Aminedi R, Kumar V, Borah P, Negi M, Ali A, Sonah H, and Deshmukh R. 2020. Omics approaches and biotechnological perspectives of arsenic stress and detoxification in plants. *Metalloids in Plants: Advances and Future Prospects*, pp. 249–273.

Verma PK, Verma S, Pande V, Mallick S, Deo Tripathi R, Dhankher OP, and Chakrabarty D. 2016. Overexpression of rice glutaredoxin OsGrx_C7 and OsGrx_C2. 1 reduces intracellular arsenic accumulation and increases tolerance in *Arabidopsis thaliana. Frontiers in Plant Science* 7: 740.

Vesterberg O, and Hansen L. 2019. Staining of proteins in polyacrylamide gels. pp. 123–134. *In*: David Garfin, and Satinder Ahuja (eds.). Handbook of Isoelectric Focusing and Proteomics *Electrofocusing and Isotachophoresis.* De Gruyter.

Wang C, Tan H, Li H, Xie Y, Liu H, Xu F, and Xu H. 2020. Mechanism study of Chromium influenced soil remediated by an uptake-detoxification system using hyperaccumulator, resistant microbe consortium, and nano iron complex. *Environmental Pollution* 257: 113558.

Weber M, Harada E, Vess C, Roepenack-Lahaye EV, and Clemens S. 2004. Comparative microarray analysis of *Arabidopsis thaliana* and *Arabidopsis halleri* roots identifies nicotianamine synthase, a ZIP transporter and other genes as potential metal hyperaccumulation factors. *The Plant Journal* 37(2): 269–281.

Wei L, Zhang J, Wang C, and Liao W. 2020a. Recent progress in the knowledge on the alleviating effect of nitric oxide on heavy metal stress in plants. *Plant Physiology and Biochemistry* 147: 161–171.

Wei L, Zhang M, Wei S, Zhang J, Wang C, and Liao W. 2020b. Roles of nitric oxide in heavy metal stress in plants: cross-talk with phytohormones and protein S-nitrosylation. *Environmental Pollution* 259: 113943.

Wei W, Chai T, Zhang Y, Han L, Xu J, and Guan Z. 2009. The Thlaspi caerulescens NRAMP homologue TcNRAMP3 is capable of divalent cation transport. *Molecular Biotechnology* 41(1): 15–21.

Wildes D, and Wells JA. 2010. Sampling the N-terminal proteome of human blood. *Proceedings of the National Academy of Sciences* 107(10): 4561–4566.

Wu F, Zhang G, and Dominy P. 2003. Four barley genotypes respond differently to cadmium: lipid peroxidation and activities of antioxidant capacity. *Environmental and Experimental Botany* 50(1): 67–78.

Xiong ZT. 1997. Bioaccumulation and physiological effects of excess lead in a roadside pioneer species *Sonchus oleraceus* L. *Environmental Pollution* 97(3): 275–279.

Xu B, Wang Y, Zhang S, Guo Q, Jin Y, Chen J, Gao Y, and Ma H. 2017. Transcriptomic and physiological analyses of *Medicago sativa* L. roots in response to lead stress. *PLoS One* 12(4): e0175307.

Xu J, and Thornton, I. 1985. Arsenic in garden soils and vegetable crops in Cornwall, England: implications for human health. *Environmental Geochemistry and Health* 7(4): 131–133.

Yadav SK. 2010. Heavy metals toxicity in plants: an overview on the role of glutathione and phytochelatins in heavy metal stress tolerance of plants. *South African Journal of Botany* 76(2): 167–179.

Yan JX, Devenish AT, Wait R, Stone TIM, Lewis S, and Fowler S. 2002. Fluorescence two-dimensional difference gel electrophoresis and mass spectrometry based proteomic analysis of *Escherichia coli. Proteomics* 2(12): 1682–1698.

Yang TPBW, and Poovaiah BW. 2002. Hydrogen peroxide homeostasis: activation of plant catalase by calcium/calmodulin. *Proceedings of the National Academy of Sciences* 99(6): 4097–4102.

Yang Z, Wu Y, Li Y, Ling HQ, and Chu C. 2009. OsMT1a, a type 1 metallothionein, plays the pivotal role in zinc homeostasis and drought tolerance in rice. *Plant Molecular Biology* 70(1): 219–229

Yang Z, and Chu C. 2011. Towards understanding plant response to heavy metal stress. *Abiotic Stress in Plants–Mechanisms and Adaptations* 10: 24204.

You J, and Chan Z. 2015. ROS regulation during abiotic stress responses in crop plants. *Frontiers in Plant Science* 6: 1092.

Zaid A, Mohammad F, Wani SH, and Siddique KM. 2019. Salicylic acid enhances nickel stress tolerance by up-regulating antioxidant defense and glyoxalase systems in mustard plants. *Ecotoxicology and Environmental Safety* 180: 575–587.

Zanganeh R, Jamei R, and Rahmani F. 2019. Role of salicylic acid and hydrogen sulfide in promoting lead stress tolerance and regulating free amino acid composition in *Zea mays* L. *Acta Physiologiae Plantarum* 41(6): 1–9.

Zeng XW, Qiu RL, Ying RR, Tang YT, Tang L, and Fang XH. 2011. The differentially-expressed proteome in Zn/Cd hyperaccumulator *Arabis paniculata* Franch. in response to Zn and Cd. *Chemosphere* 82(3): 321–328.

Zhai R, Su S, Lu X, Liao R, Ge X, He M, Huang Y, Mai S, Lu X, and Christiani D. 2005. Proteomic profiling in the sera of workers occupationally exposed to arsenic and lead: identification of potential biomarkers. *Biometals* 18(6): 603–613.

Zhang B, and Kuster B. 2019. Proteomics is not an island: multi-omics integration is the key to understanding biological systems. *Molecular & Cellular Proteomics* 18(8): S1–S4.

Zhang H, Xu W, Dai W, He Z, and Ma M. 2006. Functional characterization of cadmium-responsive garlic gene AsMT2b: A new member of metallothionein family. *Chinese Science Bulletin* 51(4): 409–416.

Zhang H, Xu Z, Guo K. Huo Y, He G, Sun H, Guan Y, Xu N, Yang W, and Sun G. 2020. Toxic effects of heavy metal Cd and Zn on chlorophyll, carotenoid metabolism, and photosynthetic function in tobacco leaves revealed by physiological and proteomic analysis. *Ecotoxicology and Environmental Safety* 202: 110856.

Zhang XD, Sun JY, You YY, Song JB, and Yang ZM. 2018. Identification of Cd-responsive RNA helicase genes and expression of a putative BnRH 24 mediated by miR158 in canola (*Brassica napus*). *Ecotoxicology and Environmental Safety* 157: 159–168.

Zhang Y, Xu YH, Yi HY, and Gong JM. 2012. Vacuolar membrane transporters OsVIT1 and OsVIT2 modulate iron translocation between flag leaves and seeds in rice. *The Plant Journal* 72(3): 400–410.

Zhao L, Sun YL, Cui SX, Chen M, Yang HM, Liu HM, Chai TY, and Huang F. 2011. Cd-induced changes in leaf proteome of the hyperaccumulator plant *Phytolacca americana*. *Chemosphere* 85(1): 56–66.

Zhi J, Liu X, Yin P, Yang R, Liu J, and Xu J. 2020. Overexpression of the metallothionein gene PaMT3-1 from *Phytolacca americana* enhances plant tolerance to cadmium. *Plant Cell, Tissue and Organ Culture (PCTOC)* 143(1): 211–218.

Zhou G, Li H, DeCamp D, Chen S, Shu H, Gong Y, Flaig M, Gillespie JW, Hu N, Taylor PR, and Emmert-Buck MR. 2002. 2D differential in-gel electrophoresis for the identification of esophageal scans cell cancer-specific protein markers. *Molecular & Cellular Proteomics* 1(2): 117–123.

Zhu JK. 2001. Plant salt tolerance. *Trends in Plant Science* 6(2): 66–71.

Zieske LR. 2006. A perspective on the use of iTRAQ™ reagent technology for protein complex and profiling studies. *Journal of Experimental Botany* 57(7): 1501–1508.

Zimeri AM, Dhankher OP, McCaig B, and Meagher RB. 2005. The plant MT1 metallothioneins are stabilized by binding cadmiums and are required for cadmium tolerance and accumulation. *Plant Molecular Biology* 58(6): 839–855.

Zuo S, Dai S, Li Y, Tang J, and Ren Y. 2018. Analysis of heavy metal sources in the soil of riverbanks across an urbanization gradient. *International Journal of Environmental Research and Public Health* 15(10): 2175.

16

Metallophytes

Ovaid Akhtar,[1,] Himanshu Sharma,[1] Ifra Zoomi,[2]
Kanhaiya Lal Chaudhary[2] and Manoj Kumar[1]*

ABSTRACT

Metallophytes are a group of specialized plants that can grow in heavy metal (HM) contaminated soils. They have one or more morphological, anatomical as well as biochemical features, which render them to grow in HM polluted sites. Broadly, metallophytes are categorized into two types; one which accumulates the HM in their different parts and others which avoid the high concentration of HM in soil. Former groups so-called, 'hyperaccumulators' are true metallophytes in the strict sense because they can extract and accumulate various HMs in their structure without any damage. To do so, they have specialized mechanisms evolved as a result of millions of years of co-evolution of ecotype in the HM polluted sites. In this chapter, the source of HM in the soils, their types along metallophytes of different HMs are listed. The underlying mechanisms of both the types of metallophytes and their bioprospection are discussed in detail.

1. Introduction

Metallophytes are unique plants with the capability to grow in substrates of moderate to high levels of heavy metals (HMs) concentration. These ecotypes are thought to be evolved as coevolution with change in the edaphic conditions of the soil. Their diversity and distribution are dependent upon the nature of metal(oid)s containing (metalliferous) sites. Metallophytes are distributed in various biomes throughout the globe and provide a specialized ecotype with the potential to restore degraded ecosystems.

As far as elemental distribution in plants is concerned, there are 17 elements, which are required by plants for their normal growth and development. These are

[1] Department of Botany, Kamla Nehru Institute of Physical and Social Sciences, Sultanpur-228118, India.
[2] Department of Botany, University of Allahabad, Prayagraj-211002, India.
* Corresponding author: ovaid.akhtar@hotamil.com

nitrogen, phosphorus, potassium, oxygen, hydrogen, carbon, calcium, magnesium, sulfur, iron, manganese, boron, zinc, molybdenum, copper, chlorine, and nickel (Clemens et al., 2021). Out of which, potassium (K), calcium (Ca), magnesium (Mg), iron (Fe), manganese (Mn), zinc (Zn), molybdenum (Mb), copper (Cu), and nickel (Ni) are HMs. These are essential for the plants but required in very minute quantities. Their exceeded concentration is, however, toxic to the plants. Other HMs, e.g., As, Cd, Pb, etc., have no use in the plants' metabolism and hence are always interfering with the cellular metabolic machinery of the plants. There are a few plants that are specialized as they have co-evolved along with the change in the physicochemical characteristics of the metalliferous site. These metallophytes can withstand the HM toxic conditions either by avoiding them or by accumulating them in a way discussed in the coming section. The former groups are categorized as 'metal tolerant', while later are categorized as 'hyperaccumulator'.

The roots of metal tolerant plants secrete various substances in the rhizosphere, thereby binding and immobilizing HM in the rhizospheric soil, thus, avoid the exposure of HM to the roots. Plants have unique properties of accumulating HM in a significant amount without being damaged. They do so in one or more ways including localization of HM into the vacuole, reducing HM into non-toxic forms (discussed in detail in coming section). For a plant to be categorized as a hyperaccumulator, it must be accumulating more than 0.01% for Cd, 0.1% for As, Co, Cu, Pb, Ni, Se, and 1.0% for Mn and Zn of shoot dry weight (Ginocchio and Baker, 2004). The efficiency of metallophytes is further enhanced in the association of soil microbes. There are several plant growth-promoting rhizobacteria (PGPR), fungi, actinomycetes, arbuscular Mycorrhizal (AM) fungi, which thrive in the rhizosphere or inside the roots of metallophytes. They can enhance, decrease or have no effect on HM accumulation by the metallophytes. The fate of their association depends upon the nature of metallophytes and the properties of the metalliferous site.

2. Metalliferous sites

Metalliferous sites are distributed throughout the globe. They are inhabited by metallophytes of various HMs. Metalliferous sites of recent origin may or may not be inhabited with plants. According to Baker et al. (2010), metalliferous sites are of three types; primary, secondary, and tertiary. In primary metalliferous sites, HMs are resultant of a natural process, i.e., mineralization or ore out croppings. Secondary metalliferous sites are produced by the mining activities on primary metalliferous sites. Tertiary metalliferous sites are created by the atmospheric deposition of HMs by the smelters and by alluvial deposition of metal-enriched substrates by sedimentation in riverbeds.

2.1 Properties of metalliferous sites

Properties of metalliferous sites depend upon the nature of origin and its location. The metalliferous sites may be containing HMs such as Cd, Cu, Pb, Ni, Zn, Hg, As, Cr, Fe, etc. The presence of HMs in the metalliferous site increases electrical conductivity, pH, and other physicochemical characteristics. Some of the metalliferous sites of the world are listed in Table 1.

Table 1: Some commonly studied metalliferous site of the world with their heavy metal contaminants.

Sr. No.	Nature of site	Country/region	Heavy metals	Reference
1.	Non-ferrous mine and smelter site	Southern Poland	Pb, Cd, Zn	Kucharski et al. (2005)
2.	Automobile waste dump	Port Horcourt	Pb, Cu, Zn, Cd, Cr, Ni	Iwegbue et al. (2006)
3.	Mining area	Northeastern China	Zn, Pb, Cu, Cd	Zhou et al. (2010)
4.	*Phyllostachys Praecox* forests	Linan	Hg, As, Cu, pb, Zn, Cd, Cr, Ni, Co, Mn	Yong et al. (2015)
5.	Mining area	China	As, Cu, Pb, Zn, Hg	Chen et al. (2018)
6.	Carlsberg ridge	Indian Ocean	Cu, Zn, Pb, Ag, Co, Cr, Ba, U, Th, V	Popoola et al. (2018)
7.	Tanneries waste dumping sites	India	Zn, Fe, Mn, Cr	Akhtar et al. (2019)

In a study of soil from a non-ferrous mine and smelter site in Southern Poland, Kucharski et al. (2005) reported that the site was heavily contaminated with HMs. It was laden with 9712 mg/kg Pb, 537 mg/kg Cd, and 11498 mg/kg Zn. Iwegbue et al. (2006) studied the HM pollution in soil of automobile waste dump in Port Harcourt and reported 510–570 mg/kg Pb, 343–397 mg/kg Cu, 423–435 mg/kg Zn, 32–38 mg/kg Cd, 11–13 mg/kg Cr, 16–17 mg/kg Ni. Zhou et al. (2010) studied HM pollution in the soil of lead-zinc mining area of Northeastern China and observed that the soil was contaminated with 0.56–11.81 mg/kg Zn, 9.68–122.93 mg/kg Pb, 3.78–47.40 mg/kg Cu, and 30–8010 mg/kg Cd. Yong et al. (2015) studied HM polluted areas in *Phyllostachys praecox* forests in Linan. They have reported 0.16 mg/kg Hg, 7.14 mg/kg As, 34.36 mg/kg Cu, 87.98 mg/kg Pb, 103.98 mg/kg Zn, 0.26 mg/kg Cd, 59.12 mg/kg Cr, 29.56 mg/kg Ni, 11.44 mg/kg Co, and 350.26 mg/kg Mn. Chen et al. (2018) studied the properties of a typical mining area in China that were contaminated with 78.70 mg/kg As, 37.96 mg/kg Cu, 160.19 mg/kg Pb, 202.79 mg/kg Zn, 0.31 mg/kg Hg. Popoola et al. (2018) studied the properties of metalliferous sedimentation in Wocan-1 and Wocan-2 on the Carlsberg Ridge, Indian Ocean, and reported 8510–51900 ppm Cu, 2890–8330 ppm Zn, 158–431 ppm Pb, 3.4–16.25 ppm Ag, 16.7–235 ppm Co, 17–32 ppm Cr, 30–190 ppm Ba, 4.7–19.6 ppm U, 0.2–1.1 ppm Th, and 150–266 ppm V. Akhtar et al. (2019) studied the characteristics of soil adjacent to clusters of tanneries at Jajmau, Kanpur Uttar Pradesh India and found that soil was contaminated with HMs. These were 1280 ppm Zn, 9745 ppm Fe, 181ppm Mn, and 32,562 ppm Cr.

The vegetation cover over the metalliferous sites constitutes entirely different types of plants, so-called metallophytes. These plants have novel physiology and metabolism that enable them to survive in soil which is unsupportive for other plants.

3. Metallophytes

Plants that are adapted to survive on metal(oid)s enriched soil and able to survive and reproduce there without suffering from phytotoxicity are termed metallophytes. Baker et al. (2010) classified metallophytes as 'obligate metallophytes' and 'facultative metallophytes'. Morrison et al. (1979) studied three metallophytes from Zaire and tested their heavy metal tolerant potential against Cu and Co. All the three plant species listed in Table 2 were moderate to a high level of tolerance to both the metals. Reeves and Brooks (1983) studied Pb and Zn accumulation in *Thlaspi rotundifolium* subsp. *cepaeifolium* and *Alyssum wulfenianum* growing on mine tailings and contaminated river gravels derived from lead-zinc mines in the Cave del Predil (Raibl) area of Northern Italy and reported that both the plants as excellent hyperaccumulators of the Pb as well Zn. They also have summarized other reported hyperaccumulators for various heavy metals from the other European region. These plants are listed in Table 2. Ginocchio and Baker (2004) summarized the metallophytes from various regions of Latin America and categorized the metallophytes into metal tolerant and metal hyperaccumulators. In their reports (listed in Table 2) 11 species of plants from Brazil were listed as hyperaccumulators of Ni. A total of six species from Chile were listed as metal tolerant of Cu. 25 species from Cuba were reported as hyperaccumulators of Ni and one species from the site was reported as hyperaccumulators of Cu. From Ecuador, three metal tolerant species of Zn, two of Pb, and As each were summarized. From Peru, three hyperaccumulator species of As and one metal tolerant species of Cu were listed in the commentary. They have further listed one species as Ni hyperaccumulator and one as metal tolerant from the Dominican Republic. From Venezuela, they have listed a single hyperaccumulator of Se, three each hyperaccumulator, and metal tolerant of Ni.

Schwartz et al. (2001) studied different types of metallophytes on the Zn smelter site of Mortagne-du-Nord France and measured the quantity of HMs extracted by them. These plants are listed in Table 2. Reeves et al. (2001) studied the metallophytes and accumulation of various HMs in them growing at metalliferous and serpentine soils of France. They have recorded a total of 16 species of plants belonging to the 10 genera as a hyperaccumulator of Zn and other HMs, which are listed in Table 2. Bothe and Słomka (2017) in their article cited the hyperaccumulator of HM heap at 'Schlangenberg' at Breinigerberg originally studied by Ernst (1982). The name and characteristics of these metallophytes are listed in Table 2. Claveria et al. (2019) studied Cu and As contents in the various plants belonging to the pteridophytes, gymnosperms, and angiosperm growing over Lepanto Cu-Au Mine, Luzon, Philippines. They have recorded a significant amount of HM accumulation in all the plants, which are listed in Table 2.

3.1 Types of metallophytes

Metallophytes are classified based on their distribution in metalliferous sites. Facultative metallophytes are found growing in both metalliferous as well as non-metalliferous sites, whereas, obligate metallophytes are found with restricted distribution over the metalliferous site of the world. Another division of

Table 2: Consolidated list of metallophytes of the worlds with their associated heavy metals.

Plant species	Heavy metal	Plant type	Country/region	References
Haumaniastrum katangense	Cu, Co	MT	Zaire	Morrison et al. (1979)
Haumaniastrum robertii	Cu, Co	MT		
Aeolanthus biformifolius	Cu, Co	MT		
Alyssum ovirense	Pb, Zn	H	Central Europe	Reeves and Brooks (1983)
Alyssum wulfenianum	Pb, Zn	H	Cave del Predil area	
Armeria maritima var. *haller*	Pb	H	Germany	
Polycarpaea synandra	Pb, Zn	H	Australia	
Thlaspi alpestre	Pb, Zn	H	Derbyshire	
Thlaspi calaminare	Zn	H	Western Europe	
Thlaspi rotundifolium	Zn	H	Central Europe	
Thlaspi rotundifolium sub sp. *cepaeifolium*	Pb, Zn	H	Cave del Predil area	
Armeria maritima	Zn	H	Zn smelter site of Mortagne-du-Nord France	Schwartz et al. (2001)
Arabidopsis halleri	Zn	H		
Arrhenatherum elatius	Zn	H		
Arabidopsis arenosa	Ni, Zn, Cd, Pb, Fe, Mn	H	Mine and smelter soil, France	Reeves et al. (2001)
Arabidopsis halleri	Ni, Zn, Cd, Pb, Fe, Mn	H		
Arabidopsis thaliana	Ni, Zn, Cd, Pb, Fe, K, Ca, Mg, Mn	H		
Armeria maritima sub sp. *halleri*	Ni, Zn, Cd, Pb, Fe, Mn	H		
Erophila verna	Ni, Zn, Cd, Pb, Fe, K, Ca, Mg, Mn	H		
Euphorbia cyparissias	Ni, Zn, Cd, Pb	H		
Luzula campestris	Ni, Zn, Cd, Pb, Fe, K, Ca, Mg, Mn	H		
Minuartia verna	Ni, Zn, Cd, Pb, Fe, K, Ca, Mg, Mn	H		
Pritzelago alpina	Ni, Zn, Cd, Pb, Fe, K, Ca, Mg, Mn	H		
Reseda glauca	Ni, Zn, Cd, Pb	H		
Rumex acetosella	Ni, Zn, Cd, Pb, Fe, K, Ca, Mg, Mn	H		
Stellaria holostea	Ni, Zn, Cd, Pb, Fe, K, Ca, Mg, Mn	H		
Thlaspi caerulescens	Ni, Zn, Cd, Pb, Fe, K, Ca, Mg, Mn	H		
Viola tricolor	Ni, Zn, Cd, Pb, Fe, K, Ca, Mg, Mn	H		

Table 2 contd. ...

...Table 2 contd.

Plant species	Heavy metal	Plant type	Country/region	References
Reseda lutea	Ni, Zn, Cd, Pb, Fe, Mn	H	Serpentine soil, France	
Thlaspi caerulescens	Ni, Zn, Cd, Pb, Fe, K, Ca, Mg, Mn	H		
Noccaea caerulescens	Zn, Pd, Cd	H	Heavy metal heap "Schlangenberg" at Breinigerberg	Bothe and Słomka (2017)
Minuartia verna	Zn, Pd, Cd	H		
Armeria maritima	Zn, Pd, Cd	H		
Silene vulgaris forma humilis	Zn, Pd, Cd	H		
Plantago lanceolata	Zn, Pd, Cd	H		
Lotus corniculatus	Zn, Pd, Cd	H		
Anthyllis vulneraria	Zn, Pd, Cd	H		
Festuca ovina	Zn, Pd, Cd	H		
Campanula rotundifolia	Zn, Pd, Cd	H		
Thymus serpyllum agg.	Zn, Pd, Cd	H		
Cladonia rangifera (podetium, lichen)	Zn, Pd, Cd	H		
Rumex acetosa	Zn, Pd, Cd	H		
Agrostis tenuis	Zn, Pd, Cd	H		
Achillea millefolium	Zn, Pd, Cd	H		
Euphrasia stricta	Zn, Pd, Cd	H		
Viola lutea sub sp. *calaminaria*	Zn, Pd, Cd	H		
Pimpinella saxifraga	Zn, Pd, Cd	H		
Adiantum sp.	Ni	H	Brazil	Ginocchio and Baker (2004)
Chromolaena sp.	Ni	H		
Cnidolcolus sp.	Ni	H		
Esterhazya sp.	Ni	H		
Heliotropium sp.	Ni	H		
Justicia lanstyakii	Ni	H		
Mitracarpus sp.	Ni	H		
Ruellia germiniflora	Ni	H		
Turnera subnuda	Ni	H		
Vellozia sp.	Ni	H		
Mimulus luteus sp.	Cu	MT	Chile	
Cenchrus echinatus	Cu	MT		
Erigeron barterianum	Cu	MT		

Table 2 contd. ...

...Table 2 contd.

Plant species	Heavy metal	Plant type	Country/region	References
Millinum spinosum	Cu	MT		
Nolana divaricate	Cu	MT		
Dactylium sp.	Cu	MT		
Ariadne sp.	Ni	H	Cuba	
Bonania sp.	Ni	H		
Buxus sp.	Ni	H		
Chionantus domingesis	Ni	H		
Euphoria sp.	Ni	H		
Garcinia sp.	Ni	H		
Gochnatia recurva	Ni	H		
Gochnatia crassifolia	Ni	H		
Gymnanthes recurva	Ni	H		
Koanophyllon grandiceps	Ni	H		
Koanophyllon prinoides	Ni	H		
Leucocroton	Ni	H		
Mosiera	Ni	H		
Ouratea	Ni	H		
Pentacalia	Ni	H		
Phidiasia lindavii	Ni	H		
Psidium	Ni	H		
Psychotria	Ni	H		
Rondeletia	Ni	H		
Sapium erythrosperum	Ni	H		
Savia	Ni	H		
Senecio	Ni	H		
Shafera platyphilla	Ni	H		
Tetralix	Ni	H		
Phyllanthus williamioides	Cu	H		
Baccharis amdatensis	Zn	MT	Ecuador	
Rumex crispus	Zn	MT		
Pennisetum clandestinum	Zn	MT		
Chenopodium ambrosioides	Pb	MT		
Pennisetum clandestium	Pb	MT		
Holcus lanatus	As	MT		
Pennisetum clandestinum	As	MT		

Table 2 contd. ...

...Table 2 contd.

Plant species	Heavy metal	Plant type	Country/region	References
Biden cinapiifolia	As	MT	Peru	
Senecia plumbens	Ni	H		
Leaythis ollaria	Se	H	Venezuela	
Waltheria americana	Ni	H		
Oyedea sp.	Ni	H		
Croton sp.	Ni	H		
Lepidaploa remotiflora	Ni	MT		
Borreria verticillate	Ni	MT		
Wedelia calycina	Ni	MT		
Prosopia alba	Zn	NA	Argentina	
Prosopis nigra	Zn	NA		
Prosopis alba	Cu, Zn, Ni, Sr, Cd, Bi	NA		
Larrea divaricata	Cu, Zn, Ni, Sr, Cd, Bi	NA		
Baccharis incarum	Au, As, Sb	NA		
Fabiana densa	Au, As, Sb	NA	Bolivia	
Dicranopteris linearis	Cu, As	H	Lepanto Cu-Au Mine, Luzon, Philippines	Claveria et al. (2019)
Histiopteris incisa	Cu, As	H		
Pityrogramma calomelanos	Cu, As	H		
Pteris vittate	Cu, As	H		
Nephrolepis hirsutula	Cu, As	H		
Pteris sp.	Cu, As	H		
Pinus sp.	Cu, As	H		
Thysanolaena latifolia	Cu, As	H		
Melastoma malabathricum	Cu, As	H		

H = (hyperaccumulator), **MT** = (metal tolerant), **NA** = (not available).

metallophytes is on basis of interaction with HMs. Consequently, these are divided into hyperaccumulator and metal tolerant groups.

3.1.1 Obligate metallophytes

Obligate metallophytes or 'strict metallophytes' are plants naturally occurring on metalliferous sites only. These plants show restricted distribution as they are unique in their genotype evolved along with the development of the metalliferous site. These have a tremendous power of tolerating HM pollution in the substrate. According to Pollard et al. (2014), 85–90% of the reported metallophytes are obligate. *Alyssum pintodasilvae* (Dudley, 1986), *Viola guestphalica*, and *V. lutea* subsp. *calaminaria* (Hildebrandt et al., 2007; Bizoux and Mahy, 2007) are some of them.

3.1.2 Facultative metallophytes

Facultative metallophytes have ecotypes growing both at the metalliferous as well as non-metalliferous sites of the same geographical area. According to Bothe and Słomka (2017), 10–15% of the reported metallophytes are facultative. *Arabidopsis helleri* (for Zn/Cd), *Pteris vittata* (for As), *Phytolacca americana* (for Mn/Cd), *Nyssa sylvatica* (for Co), *Biscutella laevigata* (for Tl), and *Alyssum bracteatum* (for Ni) are few among them (Pollard et al., 2014).

3.1.3 Hyperaccumulator

Hyperaccumulators are a class of metallophytes, which extract and accumulate HMs in their various parts (termed as phytoaccumulation or phytoextraction, Fig. 1) far beyond the concentration found in other plants. Table 2 lists the possible hyperaccumulator of various HMs. Although the concentration of HMs in the plants is very high, it doesn't harm the plants. The hyperaccumulators have developed novel mechanisms, through which they nullify the toxicity of HMs. These mechanisms are illustrated in Fig. 1. According to Rascio and Navari-Izzo (2011), there are about 450 hyperaccumulator plant species in the world. The number of these plant species reached around 500 (Pollard et al., 2014). They belong to diverse groups and evolved as a result of the selection of pressure from time to time. Hyperaccumulators are identified by at least three characteristics: (i) they have transporters in the root to extract HMs from the soil, (ii) they can translocate HMs from root to shoot very efficiently, and (iii) they have mechanisms to detoxify the HMs in various parts.

The uptake of HMs from the substrate into the roots is termed 'rhizofilteration' (Fig. 1). The process involves various transmembrane transporters, which are listed in Table 3. COPT (copper transporters) are the transmembrane proteins in the root cells, which uptake Cu from the substrate in the form of Cu^+ (Peñarrubia et al., 2015). ZIP (zinc-regulated, iron-regulated transporter-like proteins) family of transporters are involved in the cellular uptake of divalent cations like Zn, Cd, Fe, and Cu (Krishna et al., 2020). Another group of HM-transporters; HMA2 and HMA4 belonging to the HM–transporting subfamily of the P-type ATPases are involved in the uptake of Zn (Hussain et al., 2004). Mitra et al. (2014) summarized the NRAMP (Natural resistance-associated macrophage proteins), a highly conserved family of integral membrane proteins, which act as metal transporters for Mn, Zn, Cu, Fe, Cd, Ni, and Co. After uptake of HMs, hyperaccumulator may translocate HMs to shoots via vascular tissues, which again requires transmembrane loaders. *Arabidopsis* plasma membrane HM ATPase 4 (HMA4) transporter was recognized in loading Cd as well Zn to the xylem (Ceasar et al., 2020). Durrett et al. (2007) described another transporter of the MATE family, named FRD3 (Ferric Reductase Defective 3) is found in the plasma membrane of pericycle loads Fe and citrate into vascular tissues in the roots.Mitra et al. (2014) reviewed another transporter, YSL (Yellow strip 1 like) family protein responsible for nicotinamine-metal-chelators loading and unloading in and outside the xylem. Inside the plant cells of various tissues, HMs must be compartmentalized or reduced to a non-toxic form to avoid the phytotoxic effects. The ABC (ATP-binding cassette) transporter superfamily are involved in the localization of various HMs inside the vacuoles (Mitra et al., 2014).

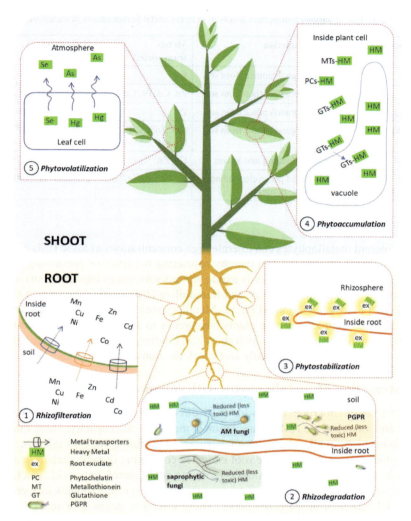

Figure 1: Figure showing various mechanisms of HM homoeostasis exhibited by metallophytes. ①**Rhizofilteration:** metallophytes uptake HMs by various transporters found in the root membrane, ②**Rhizodegradation:** root exudates influence the activities of soil microbes, which immobilize as well as detoxify HMs in the rhizosphere, ③**Phytostabilization:** HMs are immobilized in the rhizosphere by the root secretions, ④**Phytoaccumulation:** HMs are accumulated in the non-toxic forms or compartmentalized inside vacuole, ⑤**Phytovolatilization:** volatile HM contents are volatilized into the atmosphere through leaves of metallophytes.

The ABC transporters are driven by the hydrolysis of ATP. The HMs are also sequestered in the specialized epidermal cells of hyperaccumulator plants. In another way, hyperaccumulator plants bind HMs to various ligands inside the cell and detoxify them. Some volatile HM contaminants, e.g., Hg, As and Se are volatilized into the air from the leaves of these plants. This process is called 'phytovolatilization' (Fig. 1).

Table 3: List of various transporters involved in heavy metal homoeostasis in metallophytes.

Transporters	Location/function	Metals transported	References
COPT	Root membrane/uptake from soil	Cu	Peñarrubia et al. (2015)
ZIP	Root membrane/uptake from soil	Zn, Cd, Fe, Cu	Krishna et al. (2020)
HMA2, HMA4	Root membrane/uptake from soil	Zn	Hussain et al. (2004)
NRAMP	Root membrane/uptake from soil	Mn, Zn, Cu, Fe, Cd, Ni, Co	Mitra et al. (2014)
HMA4	Membrane/loading into xylem	Cd, Zn	Ceasar et al. (2020)
FRD3	Membrane of pericycle/loading into vascular tissues	Fe	Durrett et al. (2007)

3.1.4 Metal tolerant

Metal tolerant metallophytes can tolerate high concentrations of toxic HMs either by immobilizing them into the soil thereby restricting the entry or preventing the root to shoot translocation. These plants exude various substances from the roots into the rhizosphere, which bind, stabilize and immobilize HMs outside the roots a process so-called 'phytostabilization' (Fig. 1). Metallophytes with this capability are also termed 'metal excluders'. These plants have genes to synthesize phytochelatins and metallothionines in their cells, which show HM binding properties. These compounds sequester HMs inside the root only and check the translocation into the shoot.

4. Interaction of metallophytes with soil microbes

There are various microbes in the soil, which directly or indirectly affect the behavior of plants and modulate the HM accumulation or metal tolerant capacity. These microbes include bacteria, fungi, arbuscular mycorrhizal (AM) fungi, etc. Soil microbes in the rhizospheric zones thrive on the root exudates containing organic carbon in sugars, alcohols, and organic acids. Besides this, amino acids, sterol, fatty acids, nucleotide, flavanone, growth factors, and enzymes present in the root exudates influence the microbial activities in the rhizosphere. These microbes degrade or reduce various forms of HMs in the rhizosphere, a process termed as 'rhizodegradation' (Fig. 1). The term rhizodegradation is variously known as phytostimulation, rhizoremediation, microbe-assisted phytoremediation, plant-assisted bioremediation, plant-aided in-situ biodegradation, etc. Soil microbes do so by their dehydrogenase activity in the soil (Abou-Aly et al., 2021).

In an excellent review, Alford et al. (2010) summarized the role of various soil microbes on the potentiality of metallophytes, which is listed in Table 4.

4.1 Interaction of AM fungi and metallophytes

AM fungi are obligate mutualistic symbionts associated with the roots of higher plants (Kehri et al., 2018). These symbiotic soil fungi directly influence the HM homeostasis in metallophytes. AM fungi can reduce, enhance or have no significant effect on HM accumulation in the host plants. Their interaction is versatile and

Table 4: Showing effects of soil microbes on accumulation of various heavy metals in associated plant species.

Plant species	Soil microbes	Heavy metals	Effects ↑↓	References
Alyssum murale	*Microbacterium arabinogalactanolyticum*	Ni	32% ↑	Alford et al. (2010)
	Microbacterium arabinogalactanolyticum AY509225	Ni	46% ↑	
	Microbacterium liquefaciens	Ni	24% ↑	
	Microbacterium oxydans AY509222	Ni	41% ↑	
	Microbacterium oxydans AY509223	Ni	35% ↑	
	Sphingomonas macrogoltabidus	Ni	17% ↑	
Berkheya coddii	*Glomus intraradices*	Ni	167% ↑	
	Native AM fungi	Ni	45% ↑	
Pteris vittata	*Gigaspora margarita*	As	No effect	
	Glomus mosseae	As	33% ↓ to 31% ↑	
Sedum alfredii	*Burkholderia cepacia*	Cd	243% ↑	
	Burkholderia cepacia	Zn	96% ↑	
Solanum lycopersicum	Mixed inoculum of AM fungi	Cr	34.6% ↑	Akhtar et al. (2020)
Solanum lycopersicum	*Aspergillus terreus*	Cr	13.9% ↑	

↑ indicates the increased heavy metal accumulation, ↓ indicates the decreased heavy metal accumulation.

depends upon the nature of HM in the substrate, type of AM fungi, and plant species. In metallophytes, AM fungi usually enhance the HM accumulation, whereas, in non-metallophytes, it decreases. Whatever type the host pant is, most of the AM fungi reduce the translocation of HM from root to shoot (Akhtar et al., 2020; Aalipour et al., 2021; Adeyemi et al., 2021). A high root to shoot ratio of Cu, Pb and Zn was recorded in *Glycine max*, when inoculated with *Funneliformis mosseae* (Adeyemi et al., 2021). In metallophytes, e.g., tomato, the total accumulation of Cr was enhanced (Table 4) when inoculated with a mix-culture of AM fungi and *Aspergillus terreus* inoculation respectively (Akhtar et al., 2020).

4.2 Interaction of PGPR and metallophytes

In rhizospheric soils, there are various bacterial genera, e.g., *Azospirillum, Azotobacter, Bacillus, Burkholderia, Pseudomonas, Serratia*, etc., which can efficiently promote plant growths. These are termed plant growth-promoting rhizobacteria (PGPR). PGPR plays a very crucial role in the establishment of metallophytes on HM polluted sites. They provide an add-on to the metabolism of HM tolerance in metallophytes. The role of various PGPRs in the rehabilitation of metallophytes over HM contaminated sites is discussed by Zoomi et al. (2017) in detail. Accordingly, PGPR produces phytohormones, secrete siderophore, synthesize

ACC deaminase enzymes, fix nitrogen, and solubilize phosphate. Moreover, PGPR contains several functional negatively charged functional groups (e.g., sulfhydryl, carboxyl, hydroxyl, sulfonate, amine, amide, etc.), which can efficiently bind and immobilize various HMs. Recently, Abou-Aly et al. (2021) reported that *Bacillus cereus* (MG257494.1), *Alcaligenes faecalis* (MG966440.1), and *Alcaligenes faecalis* (MG257493.1) reduced the accumulation of Cu, Cd, Pb, and Zn in *Sorghum vulgare* concomitant to the enhanced growth performance.

5. Conclusions

The literature summarized in this book chapter provides a comprehensive overview of metalliferous sites, hyperaccumulator plants, metal tolerant plants, their diversity at a global level along with their interaction with soil microbes. Despite overwhelming reports worldwide, still there is a big gap between explored and unexplored metallophytes. Before mining or excavating any of the natural metalliferous sites, the vegetation cover over there must be studied regarding the HM accumulation and or tolerance. Otherwise, the result of millions of years of co-evolution in terms of metallophytes will have vanished. If properly explored, metallophytes could be utilized in post-mine rehabilitation and revegetation program. Hyperaccumulator plants can be utilized for the recovery of HMs from the polluted soil in an eco-friendly and sustainable manner, while metal tolerant or metal excluder metallophytes are best for revegetation of the metalliferous sites.

Acknowledgments

The authors are thankful to the Kamla Nehru Institute of Physical and Social Sciences, Sultanpur for providing Internet, library as well as laboratory facilities.

References

Aalipour H, Nikbakht A, and Etemadi N. 2021. Physiological response of Arizona cypress to Cd-contaminated soil inoculated with arbuscular mycorrhizal fungi and plant growth promoting rhizobacteria. *Rhizosphere*, p. 100354. https://doi.org/10.1016/j.rhisph.2021.100354.

Abou-Aly HE, Youssef AM, Tewfike TA, El-Alkshar EA, and El-Meihy RM. 2021. Reduction of heavy metals bioaccumulation in sorghum and its rhizosphere by heavy metals-tolerant bacterial consortium. *Biocatalysis and Agricultural Biotechnology* 31: 101911. https://doi.org/10.1016/j.bcab.2021.101911.

Adeyemi NO, Atayese MO, Sakariyawo OS, Azeez JO, Sobowale SPA, Olubode A, Mudathir R, Adebayo R, and Adeoye S. 2021. Alleviation of heavy metal stress by arbuscular mycorrhizal symbiosis in *Glycine max* (L.) grown in copper, lead and zinc contaminated soils. *Rhizosphere* 18: 100325. https://doi.org/10.1016/j.rhisph.2021.100325.

Akhtar O, Mishra R, and Kehri HK. 2019. Arbuscular mycorrhizal association contributes to Cr accumulation and tolerance in plants growing on Cr contaminated soils. *Proceedings of the National Academy of Sciences, India Section B: Biological Sciences* 89(1): 63–70. https://doi.org/10.1007/s40011-017-0914-4.

Akhtar O, Kehri HK, and Zoomi I. 2020. Arbuscular mycorrhiza and *Aspergillus terreus* inoculation along with compost amendment enhance the phytoremediation of Cr-rich technosol by *Solanum lycopersicum* under field conditions. *Ecotoxicology and Environmental Safety* 201: 110869. https://doi.org/10.1016/j.ecoenv.2020.110869.

Alford ÉR, Pilon-Smits EA, and Paschke MW. 2010. Metallophytes—A view from the rhizosphere. *Plant and Soil* 337(1): 33–50. https://doi.org/10.1007/s11104-010-0482-3.

Baker AJ, Ernst WH, van der Ent Antony, Malaisse François, and Ginocchio Rosanna. 2010. Metallophytes: The unique biological resource, its ecology and conservational status in Europe, central Africa and Latin America. *Ecology of Industrial Pollution* 18: 7–40.

Bizoux JP, and Mahy G. 2007. Within-population genetic structure and clonal diversity of a threatened endemic metallophyte, *Viola calaminaria* (Violaceae). *American Journal of Botany* 94(5): 887–895. https://doi.org/10.3732/ajb.94.5.887.

Bothe H, and Słomka A. 2017. Divergent biology of facultative heavy metal plants. *Journal of Plant Physiology* 219: 45–61. https://doi.org/10.1016/j.jplph.2017.08.014.

Ceasar SA, Lekeux G, Motte P, Xiao Z, Galleni M, and Hanikenne M. 2020. di-Cysteine residues of the *Arabidopsis thaliana* HMA4 C-terminus are only partially required for cadmium transport. *Frontiers in Plant Science* 11: 560. 10.3389/fpls.2020.00560.

Chen Y, Jiang X, Wang Y, and Zhuang D. 2018. Spatial characteristics of heavy metal pollution and the potential ecological risk of a typical mining area: A case study in China. *Process Safety and Environmental Protection* 113: 204–219. https://doi.org/10.1016/j.psep.2017.10.008.

Claveria RJR, Perez TR, Perez REC, Algo JLC, and Robles PQ. 2019. The identification of indigenous Cu and as metallophytes in the Lepanto Cu-Au Mine, Luzon, Philippines. *Environmental Monitoring and Assessment* 191(3): 1–15. https://doi.org/10.1007/s10661-019-7278-6.

Clemens S, Eroglu S, Grillet L, and Nozoye T. 2021. Metal transport in plants. *Frontiers in Plant Science* 12: 304. https://doi.org/10.3389/fpls.2021.644960.

Dudley TR. 1986. A new nickelophilous species *of Alyssum* (*Cruciferae*) from Portugal: *Alyssum pintodasilvae. Fedd Rep* 97: 135–138.

Durrett TP, Gassmann W, and Rogers EE. 2007. The FRD3-mediated efflux of citrate into the root vasculature is necessary for efficient iron translocation. *Plant Physiology* 144(1): 197–205. DOI: https://doi.org/10.1104/pp.107.097162.

Ernst WHO. 1982. Schwermetallpflanzen. pp. 472–506. *In*: Kinzel H. (ed.). Pflanzenökologie und Mineralstoffwechsel. Ulmer, Stuttgart.

Ginocchio Rosanna, and Baker AJ. 2004. Metallophytes in Latin America: A remarkable biological and genetic resource scarcely known and studied in the region. *Revista Chilena de Historia Natural* 77(1): 185–194.

Hildebrandt U, Regvar M, and Bothe H. 2007. Arbuscular mycorrhiza and heavy metal tolerance. *Phytochemistry* 68(1): 139–146. https://doi.org/10.1016/j.phytochem.2006.09.023.

Hussain D, Haydon MJ, Wang Y, Wong E, Sherson SM, Young J, Camakaris J, Harper JF, and Cobbett CS. 2004. P-type ATPase heavy metal transporters with roles in essential zinc homeostasis in *Arabidopsis. The Plant Cell* 16(5): 1327–1339. https://doi.org/10.1105/tpc.020487.

Iwegbue CA, Isirimah NO, Igwe C, and Williams ES. 2006. Characteristic levels of heavy metals in soil profiles of automobile mechanic waste dumps in Nigeria. *Environmentalist* 26(2): 123–128.

Kehri HK, Akhtar O, Zoomi I, and Pandey D. 2018. Arbuscular mycorrhizal fungi: Taxonomy and its systematics. *International Journal of Life Science Research* 6(4): 58–71.

Krishna TPA, Maharajan T, Victor Roch G, Ignacimuthu S, and Antony Ceasar S. 2020. Structure, function, regulation and phylogenetic relationship of ZIP family transporters of plants. *Frontiers in Plant Science* 11: 662. https://doi.org/10.3389/fpls.2020.00662.

Kucharski R, Sas-Nowosielska A, Małkowski E, Japenga J, Kuperberg JM, Pogrzeba M, and Krzyżak J. 2005. The use of indigenous plant species and calcium phosphate for the stabilization of highly metal-polluted sites in southern Poland. *Plant and Soil* 273(1): 291–305. https://doi.org/10.1007/s11104-004-8086-6.

Mitra A, Chatterjee S, Datta S, Sharma S, Veer V, Razafindrabe BH, Walther C, and Gupta DK. 2014. Mechanism of metal transporters in plants. *Heavy Metal Remediation. Nova Science, New York*, pp. 1–28.

Morrison RS, Brooks RR, Reeves RD, and Malaisse F. 1979. Copper and cobalt uptake by metallophytes from Zaïre. *Plant and Soil* 53(4): 535–539.

Peñarrubia L, Romero P, Carrió-Seguí A, Andrés-Bordería A, Moreno J, and Sanz A. 2015. Temporal aspects of copper homeostasis and its crosstalk with hormones. *Frontiers in Plant Science* 6: 255. https://doi.org/10.3389/fpls.2015.00255.

Pollard AJ, Reeves RD, and Baker AJ. 2014. Facultative hyperaccumulation of heavy metals and metalloids. *Plant Science* 217: 8–17. https://doi.org/10.1016/j.plantsci.2013.11.011.

Popoola LT, Adebanjo SA, and Adeoye BK. 2018. Assessment of atmospheric particulate matter and heavy metals: A critical review. *International Journal of Environmental Science and Technology* 15(5): 935–948. https://doi.org/10.1007/s13762-017-1454-4.

Rascio N, and Navari-Izzo F. 2011. Heavy metal hyperaccumulating plants: how and why do they do it? And what makes them so interesting? *Plant Science* 180(2): 169–181. https://doi.org/10.1016/j.plantsci.2010.08.016.

Reeves RD, and Brooks RR. 1983. Hyperaccumulation of lead and zinc by two metallophytes from mining areas of Central Europe. *Environmental Pollution Series A, Ecological and Biological* 31(4): 277–285. https://doi.org/10.1016/0143-1471(83)90064-8.

Reeves RD, Schwartz C, Morel JL, and Edmondson J. 2001. Distribution and metal-accumulating behavior of *Thlaspi caerulescens* and associated metallophytes in France. *International Journal of Phytoremediation* 3(2): 145–172. https://doi.org/10.1080/15226510108500054.

Schwartz C, Gérard E, Perronnet K, and Morel JL. 2001. Measurement of *in situ* phytoextraction of zinc by spontaneous metallophytes growing on a former smelter site. *Science of the Total Environment* 279(1-3): 215–221. https://doi.org/10.1016/S0048-9697(01)00784-7.

Yong LIU, Huifeng WANG, Xiaoting LI, and Jinchang LI. 2015. Heavy metal contamination of agricultural soils in Taiyuan, China. *Pedosphere* 25(6): 901–909. https://doi.org/10.1016/S1002-0160(15)30070-9.

Zhou X, Zhao Z, Zhang J, and Xue X. 2010, June. Characteristics of heavy metal pollution in the soil around lead-zinc mining area. pp. 1–4. *In: 4th International Conference on Bioinformatics and Biomedical Engineering*. IEEE. 10.1109/ICBBE.2010.5516817.

Zoomi I, Narayan RP, Akhtar O, and Srivastava P. 2017. Role of plant growth promoting rhizobacteria in reclamation of wasteland. pp. 61–80. *In: Microbial Biotechnology*. Springer, Singapore. https://doi.org/10.1007/978-981-10-6847-8_3.

17

Phytoremediation of Heavy Metals

Prasann Kumar,[1] *Bhupendra Koul*[2,]* and *Monika Sharma*[1]

ABSTRACT

Heavy metal pollution is one of the most serious concerns in the agricultural sector. It has emerged as a topic of discussion among various scientists as it has a serious impact on plants, humans, and the environment. Various remediation strategies have already been used to control the situation, but they have been very expensive, and even after effective results have not been found. One of the most important techniques for combating heavy metal pollution is the use of resistant plant varieties. Phytoremediation is a low-cost method of removing heavy metal contamination from soil. The primary goal of this chapter is to compile information about the sources of heavy metals and their effects on plant growth and the environment as well as to discuss the mechanism of heavy metal uptake by plants and consequent removal. Moreover, different types of phytoremediations in plants have been discussed.

1. Introduction

Heavy metals are a major environmental concern because they have a significant impact on the quality of the land, water, and environment. With today's growing population comes an increase in industrialization, which results in the release of various heavy metals into the environment. The release of these pollutants has toxic effects on plants, animals, and the environment (Koul and Taak, 2018). These contaminants are transported to non-contaminated areas as dust or leachates via soil, sewage, and sludge (Gaur and Adholeya, 2004). Various methods like chemical,

[1] Department of Agronomy, School of Agriculture, Lovely Professional University, Jalandhar, Punjab, India-144411.
[2] Department of Biotechnology, School of Bioengineering and Biosciences, Lovely Professional University, Jalandhar, Punjab, India-144411.
* Corresponding author: bhupendra.18673@lpu.co.in

thermal, etc., were previously used to control or remove pollution caused by heavy metals, but they were costly and ineffective. Chemical methods for heavy metal pollution removal are risky and expensive, resulting in a large amount of sludge (Rakhshaee et al., 2009). Thermal and chemical processes are extremely difficult to carry out because they degrade soil quality and inhibit plant growth (Hinchman et al., 1995). Heavy metal contaminated soil must be managed properly and disposed of in a specific location so that it does not degrade soil quality further. Another method for removing heavy metal pollution from the soil is soil washing, which can help remove some contaminants from contaminated soil and allow for eco-friendly disposal of the contaminated soil. These methods are extremely expensive and result in a large amount of sludge. Overall, safe disposal of pollutants is not possible, which may have further consequences for human and animal health. Heavy metal pollutants must be treated further before they can be disposed of. These physical and chemical methods for heavy metal remediation are effective in creating a good environment for plant growth and removing the contaminants' biological activities (Gaur and Adholeya, 2004). Various environmental methods are being used to determine the presence and dispersal of heavy metal ions in the land, wastewater, and environment (Shtangeeva et al., 2004).

Phytoremediation has recently emerged as an effective method for removing heavy metal pollutants from the land, water, and environment. It is a popular and low-cost method of removing contaminants from the environment. The plants that are used to accumulate heavy metal pollutants are referred to as hyperaccumulators (Cho-Ruk et al., 2006). These plants can remove heavy metal pollutants from the soil. These hyperaccumulators undergo various processes of heavy metal contaminant translocation, bioaccumulation, and degradation to completely remove them from the soil (Hinchman et al., 1995). Some plant species are very effective at absorbing highly toxic heavy metals such as lead, cadmium, chromium, arsenic, and so on. Another method of phytoremediation is phytoextraction, which can be used to extract pollutants by utilizing the ions of beneficial elements. Nickel, manganese, zinc, iron, copper, and other beneficial elements are required for proper plant growth and development throughout its life cycle. Different metals have unknown biological functions, and hyperaccumulator plants can accumulate them as well (Cho-Ruk et al., 2006). The overarching goal of this chapter is to discuss heavy metal remediation techniques and methods using hyperaccumulator plants to understand the causes of heavy metal pollutants within the plant material. This research is necessary to learn more about the causes of heavy metal pollutants and the phytoremediation of heavy metals, which is related to the degradation techniques used to remove heavy metal pollutants (Yan et al., 2020).

1.1 Heavy metals sources and effects

Heavy metals are present in the soil due to a variety of natural and anthropogenic factors. Heavy metal accumulation in the soil is primarily caused by waste disposals by various mining industries and smelting processes, volatilization of various fertilizers, insecticides, pesticides, and herbicides (Fig. 1). After that, they get to mix with the water is transported from one location to another and pollute rivers and

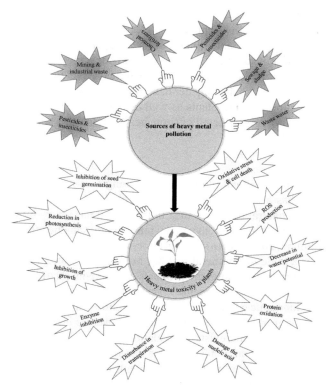

Figure 1: Sources of heavy metal pollutants and toxicity caused to plants.

lakes, as well as infecting our groundwater. After they enter the water body form their the entry into the plant system via water becomes easy and thereafter disrupts the normal functioning of the plants. In urban areas, sewage and sludge are disposed near water bodies, polluting the soil and transporting it from one location to another. This polluted water is sometimes used for irrigation by farmers and introduced into the plant system. As a result, various heavy metals are introduced into the plant system, affecting product quality and, ultimately, population health. These metals enter plants via root transport membranes.

Heavy metals are elements with a density of > 4 g cm^{-3} that have a toxic impact. As, Cu, Cd, Pb, Ni, Zn, and other heavy metals are common examples. These are found naturally in the soil and earth crust (Lasat, 2000). Heavy metals are extremely difficult to control or remove because they do not degrade through biodegradation. Some heavy metals are distinguished from others because they can accumulate in plant cells and cause harm even at low concentrations. The toxicity mechanism is determined by the heavy metal's type and form, which can cause serious problems for vegetation and living organisms. Heavy metal pollution is caused by industrial and agricultural sources. Chemical fertilizers, pesticides, insecticides, and other chemical sprays are examples of agricultural resources (Resaee et al., 2005). They can also be found in various forms and from various sources, such as acids, bleach, sodium hydroxide, medical instruments, household materials, paints and fillings, and textiles, and are also present in fluorescent lamps, electrical appliances, pharmaceuticals, cosmetics, oils, detergents, and other products. Even though heavy metals are used in a variety of industries for a variety of purposes, they pose several hazards. Because of the rise in pollution, the use of such heavy metals is now restricted in some areas (Musselman, 2004). In general, heavy metals with a higher concentration in the soil can be more toxic to plants as well as the human and animal populations (Pehlivan et al., 2005). A brief detail of some heavy metals has been discussed in the forementioned headings.

1.1.1 Arsenic

Arsenic (As) is a heavy metal with an atomic weight of nearly 75 and an atomic number of 33. It has a very high melting point of 817°C. As a result, it cannot be easily degraded by natural processes in soil. Because it has a very high boiling point and a higher vapor pressure, the heavy metal ions become attached to the soil particles (Mohan and Pittman, 2007). It is a silver-grey crystalline solid that exists in the earth's crust. Arsenic is a toxic heavy metal that has no odour or taste. It can bind to other particles to form organic or inorganic products or elements (National Ground Water Association, Copyright, 2001). Arsenic reacts with other gases in the environment, forming various inorganic compounds that are commonly used in wood preservation. Organic arsenic compounds, on the other hand, are used as insecticides or pesticides in a variety of crops. Arsenic exists in oxidation states with valencies of -1, 0, $+3$, and $+5$ (Mohan and Pittman, 2007). Arsenic can be found in a variety of chemical and physical states in bodies of water and sediments (Hasegawa et al., 2009). Arsenic exists in various forms in the earth's crust, and when it reacts with environmental gases, it changes states and forms organic and inorganic elements. Arsenic compounds found in the environment include arsenous acids, arsenates, and arsine. Arsenite and arsenate are the two most important forms of arsenic, both of which are inorganic and found in water bodies (Mohan and Pittman, 2007). Arsenic compounds are classified into three major groups. Inorganic arsenic compounds, organic arsenic compounds, and arsine gas. Arsine is extremely toxic and may endanger human and animal populations. Arsenic oxides such as trioxide arsenic compound, trichloride, and sodium arsenate are toxic and difficult to degrade. Arsenite creates highly reduced redox conditions in the environment, polluting the land and water (Hasegawa et al., 2009). Arsenic is one of the most dangerous heavy metals because it is extremely toxic and can cause a variety of problems for both nature and humans (Chutia, 2009). The toxicity of arsenic varies according to plant species, pH, redox reactions, and microorganism activity,and oxidation state. Furthermore, inorganic forms of arsenic are far more toxic than organic forms (Ampiah-Bonney et al., 2007; Vaclavikova et al., 2008). Similarly, trivalent compounds are more harmful and have a higher solubility in water than other compounds. DNA and other protein molecules are broken down by trivalent compounds (Vaclavikova et al., 2008). Arsenic pollutants are persistent and contaminate groundwater quality (Chutia, 2009). Arsenic is a toxic heavy metal that causes cancer in humans and animals (Yusof and Malek, 2009).

1.1.2 Lead

Lead is a silver-grey heavy metal with the atomic number 82. Because of its high boiling and melting points, it cannot be easily degraded. There are four isotopes of lead metal that are abundant. It has a valency of four electrons and is poorly soluble in water. It has two oxidation states and only two electrons are ionized. It is less water-soluble than other compounds such as nitrate, chloride, and many salts. Lead is the most abundant heavy metal, and it can be found in a variety of forms in nature. The main sources of lead are industrial activities, vehicle exhaust, dust particles, and so on. Because it is present in higher concentrations in the land and water, the 2+ oxidation state of lead is extremely toxic to plants as well as humans. This type of lead is nonbiodegradable, making it difficult to reduce its toxicity in plants. Because of its toxicity, it can persist in the soil for an extended period. Heavy metal pollution from lead has an impact on both biological and physiological processes in plants (Pehlivan et al., 2009). Other sources of lead pollution in soil and water include industrial releases, fuels and oils, metal invasion, and other pollutants that severely pollute the environment and biological system. Lead metal is found in the upper layers of soil. Because it is mobile, it accumulates at a site, making the remediation process difficult to carry out. Pollution cannot be controlled unless the soil is remedied (Traunfeld and Clement, 2001). Lead heavy metal pollution is harmful to plants and animals.

1.1.3 Mercury

The Hg exists in various forms and states in nature. It is a silver-white liquid that has no odour. Mercuric compounds can be found in powder or salt form. In the environment, mercury combines with other elements such as sulphur, chlorine, oxygen, and so on to form inorganic mercury compounds. Mercury also combines with carbon compounds to form various organic mercury compounds (Musselman, 2004). Mercury is a liquid in nature, with a low melting point, but it is extremely toxic. Because of its low melting and boiling points, it is extremely useful in industry (Chang et al., 2009). It is found in soil in various forms to react with other compounds. Mercury pollutants can solubilize free ions that are held together by static forces. These pollutants react with sulphide, carbonates, chlorides, hydroxides, and other pollutants to form inorganic and organic mercury compounds. Mercury has an oxidation state of 2+, which is commonly used to perform various functions. Mercury, a heavy metal, is converted from one form to another by the activity of bacteria (Rodriguez et al., 2005). Mercury is a major pollutant that pollutes the environment. It accumulates in plants, humans, and other organisms, causing severe toxicity. In the case of water sources, mercury contaminants can accumulate in fish and other organisms, posing health risks. Organomercury salts and compounds are more toxic in the environment and can cause a variety of negative effects. Heavy metal mercury interferes with photosynthesis and the oxidation-reduction process of plants, as well as reduces the activity of various plant functions. It can also reduce the uptake of minerals and nutrients from the soil, causing significant harm to the ecosystem (Sas-Nowosielska et al., 2008).

1.1.4 Cadmium

Cadmium is a very dangerous heavy metal in the environment that has a significant impact on the ecosystem. It is found in very low concentrations in the earth's crust but is extremely toxic. Cadmium concentrations are higher in phosphates and sedimentary rocks. Cadmium is a heavy metal that is primarily found in alloys, automobile batteries, and industrial processes. The release of cadmium into the environment is increasing day by day as the population's access to resources expands. The use of cadmium in car batteries is on the rise these days, and it has become a potentially hazardous source of pollution to the environment. Cadmium (Cd) toxicity also affects plant growth and results in a significant decrease in the plant's yield potential (Wu et al., 2007) by causing chlorosis and necrosis in the leaves and disrupting various physiological and biochemical functions. Cadmium's toxicity reduces the quality of the product, which ultimately affects human health. Cadmium pollutants enter the body through cigarette smoking or food consumption. The entry of these pollutants directly through the skin is not as likely or as unlikely. There are numerous ways for entry of Cd in humans like drinking polluted water, eating contaminated food, smoking cigarettes, and so on (Paschal et al., 2000).

Other cadmium sources include industrial exposure, ore smelting, mining, and battery manufacturing. It is also found as a trace element in various vegetables (Satarug et al., 2003). It usually enters the circulatory system and travels to other parts of the body with the blood. It causes a variety of harm to both humans and animals. This metal is carcinogenic and has the potential to cause chronic anemia. It disrupts calcium regulation in biological systems and causes blood vessel coagulation. Various studies have been conducted in the past to remove cadmium pollutants from the land, water, and environment, but none of them has yielded effective results. However, these methods are not only more expensive, but they also do not produce efficient results. Cadmium interferes with photosynthesis and thus slows plant growth and development. It further reduces the quality as well as the quantity of yield produced. Cadmium is a toxic heavy metal that harms plant growth and yield. It also disrupts the biological processes of the plant and is transported from one part to another. The phytoremediation process is effective in controlling Cd metal pollution, but higher concentrations of cadmium cannot be easily absorbed or degraded by plant roots.

1.1.5 Chromium

Chromium is found in low concentrations in the environment. It occurs naturally in various oxidation forms in the earth's crust. The oxidation states found in the earth's crust range from 2 to 6 (Jacobs and Testa, 2005). The chromium (III) oxidation state is the most common in nature, followed by chromium (IV) and chromium (VI). Chromium in the environment is released by various industries as well as man-made factors or activities. Industrialization is a major source of heavy metal chromium release into the environment. Chromium (VI) is released into the environment by a variety of anthropogenic factors that are carcinogenic and cause a variety of harmful effects in plants, animals, and humans. Chromium in its second oxidation state is less toxic than chromium in its sixth oxidation state. Chromium (III) is abundant and toxic, causing a variety of negative effects in both humans and plants. Because chromium is used in a variety of industrial processes, it is a major environmental contaminant (Cohen et al., 1993). Inhaling chromium pollutants can lead to cancerous diseases in humans and animals. Generally, seafood has a higher concentration of chromium. When it enters the human body, it attacks the lungs, causing infection. Asthma, cancer of the respiratory system and throat, and other diseases are caused by heavy metal chromium. Inhaling a higher concentration of chromium can cause nose infections and ulcers. Chromium (VI) causes more harm to animals than any other form. A high concentration of chromium in agricultural soil and water can inhibit plant growth and development. The toxicity of chromium (Cr) causes a decrease in photosynthetic activity and a decrease in plant growth. Inhibiting photosynthetic activity reduces the rate of enzymatic processes, gas exchange, chlorophyll content, and N and CO_2 assimilation in the plant system. Chromium (Cr) toxicity causes a large structure to form on the leaves of the barley plant, resulting in a decrease in quality yield by slowing growth and development. These structures formed on the leaves include chloroplast thickening and swelling, as well as membrane cell disruption. Chromium (Cr) toxicity reduces photosynthetic rate, chlorophyll content, transpiration rate, CO_2 accumulation, and stomatal conductance regulation (Ali et al., 2013). Overall, Cr reduces the quality of the product as well as the yield.

1.1.6 Nickel

Ni is a micronutrient that aids in the conversion of urea to ammonia via the enzyme urease (Gajewska and Skodowska, 2009). Nickel deficiency can cause urea accumulation in the leaf and shoot tips, causing toxicity. Nickel is an important nutrient that plants require for growth and development. The concentration or level of Ni that a plant uses is determined by the plant's need and growth factor. It is required at all stages of plant life, from germination to maturation (Sreekanth et al., 2013). A higher concentration of nickel, on the other hand, can cause toxicity in both plants and animals. The major sources of nickel pollution are sewage, sludge, industrial activities, and so on, which can pollute agricultural land, water, and the environment. With the increase in industrial activities, the concentration of nickel is increasing day by day. Nickel toxicity symptoms can appear between 0.15 and 0.80 mM/kg plant dry matter (de Queiroz Barcelos et al., 2017). The higher the level of nickel in the plant, the more toxic it is to the plant and inhibits germination and enzymatic activities. Nickel toxicity interferes with plant physiological and biochemical processes by affecting chlorophyll synthesis and photosynthesis efficiency, plant's relative water content, and causes oxidative damage (Gajewska et al., 2012). Moreover, Nickel toxicity causes deficiency of iron, zinc, and other essential heavy metals (Gautam and Pandey, 2008). Ni toxicity reduces nitrogen availability in the plant and also inhibits its further translocation from root to shoot part (Yusuf et al., 2011). Several studies have been conducted in this field to remove heavy metal toxicity so that the plant can function properly. Nickel toxicity reduces the uptake of various macro and micronutrients from the soil, causing significant damage to the plant. It is observed in a study that the toxicity of nickel inhibits the translocation of nitrogen in the plant material (Ameen et al., 2019). It also causes negative effects on the growth and development of the plants so the quality of the product decreases.

1.1.7 Heavy metals toxicity in plants

Heavy metals are toxic compounds that have an impact on plant and population growth. When heavy metals are transported from one location to another, they cause toxicity in all parts of the plant. It has negative effects as it travels from the root to the shoot. Heavy metals primarily affect the cellular activities of the plant system, resulting in a decrease in all plant life processes, including various physiological and biochemical processes, potentially reducing plant development and yield (Singh and Prasad, 2011). The toxicity of heavy metals causes a decrease in photosynthesis in plants, as well as a decrease in plant growth. Inhibiting photosynthetic activity reduces the rate of enzymatic processes, gas exchange, chlorophyll content, and N and CO_2 assimilation in the plant system. Heavy metal toxicity reduces all plant life processes that are important for plant development and damages plant cells and other organs. Moreover, these toxicants enhanced the formation of reactive oxygen species (ROS) in plants. The buildup of ROS can harm plant cell metabolism and almost all other physiological components, lowering the quality and quantity of the plants.

2. Mechanism of heavy metal uptake by plants

Several studies have been conducted to describe the uptake and transport pathway of heavy metal pollutants within the plant system. All of the plant's activities, from absorption to the removal of heavy metal contaminants, are influenced by a variety of factors. Hyperaccumulator plants are used as both accumulators and excluders to remove heavy metal ions from soil and water (Sinha et al., 2004). Heavy metal contaminants accumulate in the aerial tissues of these plants, where they are removed to safe sites. These are the plants that aid in the biodegradation of heavy metal contaminants and the biological transformation of these particles from toxic to nontoxic forms within plant tissues. The excluders in the plant's biomass prevent the uptake of contaminants. Micronutrients and macronutrients can be absorbed by plants from the soil. They are capable of absorbing nutrients in both low and high concentrations. Similarly, hyperaccumulator plants can absorb or accumulate toxic heavy metal ions. Plant roots can produce a variety of metabolic and catabolic reactions, as well as chelating agents, to aid in the degradation and removal of heavy metal ions from soil and water bodies. The increase and decrease in soil pH, solubility, and different chemical reactions caused by root exudates will aid in further modifying the functions of heavy metal pollution degradation or extraction. The redox reaction occurring within plant tissues can solubilize contaminants in both low and high concentrations. Hyperaccumulators have a unique mechanism for micro and macronutrient translocation and storage. These plant pathways can store toxic elements and convert them into essential nutrients for plant use. As a result, these nutrients are thought to be more effective in heavy metal phytoremediation. The plant system's transport mechanism contains an optimal number of protein molecules. These are found in plant cells and tissues to transport various elements such as enzyme ATPase, energy molecules, and proton molecules, allowing for the effective uptake of elements at higher and lower concentrations. To carry out phytoremediation of heavy metal pollutants, various physiological and biochemical processes occur within the plant material. ATPase enzymes generate electrochemical energy to consume it for heavy metal ion uptake. A variety of heavy metal ions are taken up by the transport mechanism and detoxified to a certain level so that they do not harm the environment. Interaction between metal ions can cause problems during the translocation of heavy metal contaminants. As heavy metal ions are absorbed by the root zone, they are transported to the shooting part via xylem activities, along with minerals and nutrients, for degradation and harvesting to extract the metal ions from land and water. Because root harvesting is not possible, metal ion translocation to shoot parts is required. The transport mechanism is regulated by various enzymes and protein molecules, which also play roles in the plant's physiological and biochemical activities. Plants, in general, absorb trace elements based on their needs. They accumulate elements in the 10–12 ppm range that are normally required to carry out all of the processes of plant life. Hyperaccumulator plants, on the other hand, can absorb higher concentrations of heavy metal ions. Hyperaccumulator plants, which are effective in the remediation of heavy metal contaminants from land and water sources, can take up to thousand ppm concentrations. It is also necessary to understand the form and state of toxicity of metal ions in which they are stored in plant cells to determine their level of toxicity and avoid their negative effects. To determine the level of toxicity of heavy metals inside plant cells, various techniques are used. The toxicity of metal ions stored in vacuoles is easily determined. The evapotranspiration phenomenon from the leaf surface is also used in the phytoremediation of heavy metal contaminants. Heavy metal ions that are absorbed by the root zone are also transpired by the shoot portion as water evaporates from the leaf surface. The metal ions are transported within the shoots by the process of evapotranspiration. Ions are transferred from the root to the shoot of hyperaccumulator plants, where they are degraded, modified, and detoxified into safer elements. Pollutants that have degraded accumulate in plants, which are then dried and burned to remove the toxic compounds. Hyper accumulators are plants that transport and accumulate heavy metal ions in them and then remediate those pollutants. These contaminants or toxic compounds are moved from the root to the shoot portion of the plant and degraded in the plant cells or tissues. Hyperaccumulators are plants that can survive in any environment and do not require a lot of care. It can generate enough biomass to completely degrade heavy metal pollutants (Salido et al., 2003). These plants can accumulate or absorb heavy metal ions hundreds to thousands of times more than non-accumulator plants. Non-accumulator plants, also known as excluders, are plants that cannot absorb highly toxic compounds. Some microorganisms that live in the root zone of plants, such as bacteria and fungi, can mobilise metal ions and aid in their absorption by the root tips. Organic compounds are more easily removed by these microorganisms than nonorganic compounds (Erdei et al., 2005).

3. Methods of phytoremediation

Heavy metal pollution remediation has become a common occurrence in recent years. It is the process of improving the soil to increase production in terms of both quality and quantity. It generally refers to the use of green plants to remove heavy metal contaminants from soil and water (Raskin et al., 1997). Phytoextraction,

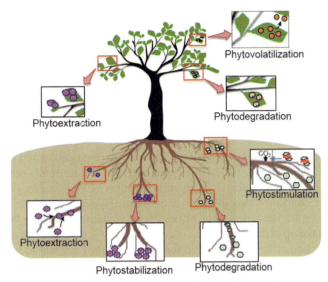

Figure 2: Phytoremediation an integrated approach.

phytodegradation, hemofiltration, phytostabilization, phytovolatilization, and photodegradation are all types of phytoremediation. Phytoremediation is a highly effective process because it is both cost-effective and risk-free. Phytoremediation of heavy metals entails the use of plants or hyperaccumulator plants to absorb a large number of heavy metal pollutants from the soil and water, thereby reducing environmental toxicity (Fig. 2).

3.1 *Phytoextraction*

The pollutants of heavy metals, either organic or inorganic, are absorbed by the roots of the plant in this type of phytoremediation, and the removal of these contaminants is possible with the removal of the plant from the soil surface. This method, also known as vegetal assimilation, involves plants absorbing heavy metal ions from the soil via their roots. This technique is very effective in the remediation of contaminated areas because hyperaccumulator plants can extract heavy metal ions from the subsoil and remove them by removing the plant itself or harvesting the plant. This method is very effective, but it is time-consuming because it takes a long time from seed sowing to harvesting. As a result, it cannot be used in highly contaminated areas. It is also difficult to select hyperaccumulator plants in heavily contaminated areas. Another requirement is that hyperaccumulator plants not be seasonal because they must be harvested later. After harvesting, the plants used in the phytoremediation process are safely burned in an incinerator. This technique is also known as phytomining because it allows access to mineral ore cultivation. Some metals are gaining access as a result of these methods, such as gold and nickel, which are forming again as a result of these processes (Pivetz, 2001; Gaur and Adholeya, 2004). When compared to other plants, hyperaccumulator plants can absorb 100 times more heavy metal pollutants.

Plants from the Brassicaceae, Euphorbiaceae, Asteraceae, Lamiaceae, and other families can be used as hyperaccumulators, absorbing a large number of heavy metal ions inside them. Following that, the plant debris is dried, burned, and recycled by making compost, among other things (Memon et al., 2000).

3.2 *Phytodegradation*

Degradation refers to the breaking down of something into smaller pieces. Phytodegradation is the process of degrading heavy metal ions using hyperaccumulator plants to recover contaminated soil or water bodies. The degradation of heavy metals by plants or vegetation is also known as vegetal degradation. Different metabolic and catabolic processes occur within plant material, which can aid in the degradation of heavy metal pollutants. Phytodegradation degrades a wide range of organic and inorganic contaminants. This type of phytoremediation can be applied to various soil and subsurface water bodies. The main advantage of this process is that the degradation occurs within the plant material itself, as physiological processes conduct the absorption and degradation of heavy metal ions. It does not rely on any external input or microorganisms for the degradation process, as well as the stimulation and removal of various products carried out within the plant to restore soil health (Pivetz, 2001). The degradation of heavy metal ions is determined by the hyperaccumulator plants' absorption capacity, type, and life cycle. It can also be influenced by the structure and texture of soil particles. When compared to insoluble particles, soluble particles are more difficult to degrade. The particles of insecticides, herbicides, and chemical fertilizers are degraded by the plants' various enzymatic activities. The enzymes stimulated in the plant system during its life cycle also degrade sewage water and sludge wastewater (Memon et al., 2000). Heavy metal pollutants can be degraded by various physiological and biochemical activities that occur within the plant system (Fig. 2).

Table 1: Harmful effects of heavy metals.

Metal	Effects
Arsenic	Arsenic pollutants enter the cellular mechanism of the plants and destroy the physiological functioning of the plant, which ultimately decreases the quality of the produce. Interferes with animal cells and decreases the synthesis of ATP and other biological processes.
Lead	It is a toxic metal that harms human health. It leads to short term memory, impaired development and learning disabilities in children.
Mercury	It can lead to anxiety, depression, hair loss, insomnia, restlessness, memory loss, nephrotoxic and cancer of the throat and stomach etc. to the population.
Cadmium	This metal is carcinogenic in nature and can cause chronic anaemia. It interferes with calcium regulation in biological systems.
Nickel	It shows allergenic effects. Inhalation can cause cancer of the nose, lungs, stomach and throat. It can leads to hair loss also.

3.3 Phytostabilization

This phytoremediation process is also carried out through the use of roots. This is also known as the root stabilization process because it results in soil stabilization. Heavy metal accumulators tolerate the level of toxicity of heavy metals and reduce the cause through the processes of metal ion absorption, degradation, sedimentation, and reduction. Heavy metal pollutants become immobilized around the roots of heavy metal accumulators and undergo a variety of processes such as sedimentation, absorption, degradation, reduction, and so on (Mirsal, 2004). A study was conducted on maize plants to investigate copper absorption and mitigation using arbuscular Mycorrhiza fungi. To reach his conclusions, the researcher employs a variety of copper doses (Turkoglu et al., 2006). It was a lab study, and an experiment was carried out in pots to estimate the study's results. Pollutant absorption is observed to be very low in pots with a higher concentration of copper. The result is caused by a change in soil pH. A significant decrease in the pH of the soil results in a lower absorption ratio of the plant. Fungus application alters the behavior of various organic acids such as malic acid, citric acid, oxalate acid, and others, as well as several enzymatic activities within the plant system. The overall conclusion of this experiment was that arbuscular fungi can be used to mitigate or phytoextract copper from the maize plant. The use of these fungi aids in the phytoextraction of copper from contaminated soil through the root zone of plants. This method prevents heavy metal pollutants from being transported by wind, water, or any other means, and it allows for efficient heavy metal remediation. The interaction between the root zone, soil, and environment, which plays a major role in the phenomenon of phytoextraction, can allow for the effective remediation of heavy metal pollutants.

3.4 Phytovolatilization

The process of evaporation is the same as the process of volatilization. We must choose accumulators based on the depth of the water level for phytovolatilization. It is also known as vegetal evaporation, and it refers to the volatilization of pollutants from plant roots. It is extremely difficult to estimate root depth during the phytoremediation process. The roots of the accumulator plants should be deep for lower water levels, and the water should be pumped for shallow-rooted plants to adjust the absorption rate of the pollutants. The primary goal of this process is to convert high toxicity compounds into less toxic forms. Heavy metals, for example, mercury, arsenic, cadmium, and others are highly toxic compounds or pollutants that are converted to lower toxicity through the phytovolatilization process. The release of these chemicals or heavy metal pollutants into the atmosphere is hazardous to the environment as well as the human and animal populations (Gaur and Adholeya, 2004). As a result, it is critical to mitigate the effects of heavy metal toxicity on the land, water, and environment. These contaminants can be removed from plants through the evaporation and transpiration processes. This is a very simple method of heavy metal phytoremediation because, as we know, water is absorbed by the plant roots and transported to the shoot parts via the vascular system. It is then removed

Table 2: Phytoremediation mechanism and pollutants (Source: Gupta et al., 2000).

Technique	Mechanism	Pollutants
Phytoextraction	Accumulation of heavy metal pollutants in shoot parts to be harvested.	Pollutants of Copper, Nickel, Molybdenum, Zinc, Lead, Cadmium, etc., can be extracted.
Phytodegradation	Eradication of pollutants	Organic compounds, phenols, herbicides, etc.
Phytostabilization	Absorption, complexation and precipitation.	Inorganic compounds of As, Cd, Pb, Zn, Ni, Hg, etc.
Phytovolatilization	Transpiration by leaves and shoot portion	Organic and inorganic compounds of As, Hg and Cd.
Rhizofiltration	Accumulation and absorption by roots and precipitation occur.	Organic and inorganic compounds of heavy metals Cd, Zn, Ni, Cu, Cr, etc.

from the soil by the process of evaporation and transpiration from the leaves. Similarly, heavy metal ions or pollutants are absorbed by the roots along with water molecules and are extracted from the shoot portion via transpiration and evaporation. The volatilization of heavy metal contaminants that occurs here is referred to as phytovolatilization. The activities of a poplar tree provide an example of this type of phytoremediation. Heavy metal pollutants are volatilized in some plants, such as oilseed crops, and are removed from the plants in gaseous form (Wang et al., 2007). Different tree species are also used in the phytoremediation of heavy metal pollutants because their deeper root systems can absorb pollutant ions from the deeper level (Ghosh and Singh, 2005) (Table 2).

3.5 *Rhizodegradation*

It refers to the decomposition of heavy metal contaminants in the root zone as well as the activities of microorganisms. Various amino acids, organic acids, minerals, enzymes, and metabolic processes are taking place within the plant system or in the root zone of hyperaccumulators to remediate or decompose heavy metal ions from the soil. The main advantage of this method is that the decomposition of

Table 3: Species of hyper-accumulating plant (Source: Chaney et al., 2000).

Heavy metal	Plant species
Cadmium	*Thlaspi caerulescens*
Zinc	*Thlaspi caerulescens*
Copper	*Aeolanthus biformifolius*
Arsenic	*Pteris vitata*
Tin	*Biscutella laevigata*
Copper	*Emblica officinalis*
Lead	*Eclipta alba*
Zinc	*Tribulus terrestris*
Chromium	*Phyllanthus amarus*
Mercury	*Curcuma longa*

pollutants takes place in the kit's natural environment, causing no further harm to the environment (Gaur and Adholeya, 2004; Rakhshaee et al., 2009). Contaminants found in various insecticides, pesticides, chemical fertilizers, and hydrocarbons are degraded by photodegradation, which includes species such as mint, red berry, and Lucerne (Gaur and Adholeya, 2004; Vanli, 2007; Pulford, and Watson, 2003; Aybar et al., 2015). Plant roots are used in this remediation process to degrade heavy metal contaminants.

3.6 Rhizofiltration

The use of plant roots for the filtration of toxic and nontoxic compounds from the plant system is referred to as rhizofiltration. In this process, contaminants are absorbed by the roots via various biotic or abiotic processes and then transported by the plant from one location to another. Physiological functions within the plant material develop their activities during the process to sustain the extraction of pollutants from the plant itself. This method aids in the removal of heavy metals from wastewater, sewage sludge, underground soil, and water, among other things (Shtangeeva et al., 2004; Sogut et al., 2002) (Fig. 2). This phytoremediation method is effective in removing highly toxic compounds from wastewater, and the plants used in this method should be well adapted to their surroundings. If the plants have a larger root system, they are grown directly in water using the hydroponics technique. As a result, the plants with roots are transferred to the contaminated water body, where the plants can adapt and carry out further processes. The hyperaccumulator plants are grown directly in wastewater or polluted water, allowing them to saturate the pollutants or molecules in the wastewater and remove them through the hemofiltration process. After heavy metal ion saturation, the plant is harvested and dried or burned for safer pollutant disposal. This method is commonly used against aquatic and terrestrial plants, as well as from various water bodies to the environment (Salt et al., 1998; Mirsal, 2004).

4. Limitations of phytoremediation

Along with so many benefits, there are always drawbacks that change the role of any process or activity that occurs. Phytoremediation, on the other hand, is a highly effective technique for removing heavy metal contaminants from the land, water, and environment (Clemens, 2001; Tong et al., 2004; Le Duc and Terry, 2005; Karami and Shamsuddin, 2010; Mukhopadhyay and Maiti, 2010; Naees et al., 2011; Ramamurthy and Memarian, 2012). It is a novel approach to heavy metal remediation by plants. Phytoremediation is a time-consuming process that necessitates a specific harvesting period for the plants. The accumulation and degradation of heavy metal pollutants can have an impact on plant growth rate and biomass production. The removal efficiency of heavy metal ions is also influenced by the hyperaccumulator plants' growth and physiological functions. Metal ions that are tightly bound in the soil cannot be easily absorbed by hyperaccumulator plants. The availability of beneficial metal ions in the soil has also decreased. It is commonly used or applied in high metal contaminated areas where other crop plant growth cannot be sustained. Heavy metal pollutants must be properly maintained and disposed of because metal ions can enter our food chain and harm human, animal, and plant health.

5. Future prospects

A phytoremediation is a new approach to pollution remediation that is yielding better results. Heavy metal pollution is a growing problem nowadays, and remediation is required to remove pollutants from the land, water, and environment. The technique of phytoremediation is also a new technique that scientists are interested in because it uses the plant system to remove heavy metal pollutants from the land and water. It is a relatively new method that is critical for understanding the movement of heavy metal ions within the plant system. It is entirely a field study. Until now, most scientists have conducted lab and field studies on heavy metal contamination remediation processes. It is a field study that is very interested in removing metal ions from the ecosystem. Because there is an open environment in field studies, and experiments are conducted in the real world and under climatic conditions, the results of lab studies differ from those of field studies (Ji et al., 2011). Climate factors such as rainfall, humidity, temperature, nutrients, soil health, and soil qualities can all have an impact on the phytoremediation process (Vangronsveld et al., 2009). The efficiency of hyperaccumulator plants to absorb or degrade heavy metal ions can also be observed or determined in the field. It is critical to grow plants in highly contaminated areas to find better plant products. There is a need to increase the diversity of hyperaccumulator plants so that heavy metal pollution can be reduced or the quality of plant products can be improved for the benefit of the population. For the production of hyperaccumulator plants, breeding and transgenic techniques can be used (Pollard et al., 2002). There is an urgent need to pique scientists' interest in this site so that the production of hyperaccumulator plants can be increased to reduce metal pollution in the environment. More research in this area is needed to achieve better results. Gene identification and modification can aid in the development of high-efficiency hyperaccumulator plants. Transgenic plants play an important role in this process as well, so the development of transgenic plants could be a new factor in this field (Thakur, 2006). On this basis, an experiment is also carried out to investigate the effect of combining different traits in a single plant species. The ability of a plant species with multiple characteristics to remediate metal ions or not is of great interest in this field. Gene modifications can be used to learn more about hyperaccumulator plants and their ability to carry out phytoremediation processes. Knowledge of biochemistry also assists researchers in learning more about phytoremediation (Saraswat and Rai, 2011). This is a great approach to heavy metal ion removal because it has many advantages as well as challenges.

5.1 Phytoremediation an integrated approach

A phytoremediation process is an integrated approach that necessitates extensive knowledge in a variety of fields, including soil health, soil chemistry, soil chemical, and physical properties, plant metabolism, various catabolic and metabolic processes, and plant physiological functions (Fig. 2). This process is also intertwined with other fields because knowledge of microbiology, biochemistry, and biotechnology can aid in the development of more characters or hyperaccumulator plants to study more about this process. Various studies will be conducted in this area in the future to remediate heavy metal pollutants in the ecosystem.

6. Conclusions

Heavy metal pollution is a major source of concern for everyone. It is increasing day by day as industrialization and other daily activities increase. With today's growing population, it is becoming increasingly difficult to produce and feed quality products to the populace. Heavy metal pollution contaminates the land and water and may pose serious health risks to the population. It is critical to address this issue to maintain the quality of food products. To remove heavy metal pollutants, various chemical and physical processes are used, but these are extremely difficult to remediate. These methods are very expensive, and there are no effective results. As a result, a process known as phytoremediation is used to remove heavy metal pollutants from soil and wastewater bodies. The positive aspects of this method are: (a) low operational cost, (b) plants can be grown directly on contaminated water or soil, (c) it is aesthetically pleasing, (d) after harvesting, plants can be used for biofuel production, (e) and metals can be recovered from the plants (phytomining). Phytoremediation is linked to other fields of study such as soil microbiology, biotechnology, soil health and quality, plant metabolic and catabolic processes, and plant physiology. It is critical to grow plants in highly contaminated areas to find better plant products. There is a need to increase the diversity of hyperaccumulator plants so that heavy metal pollution can be reduced or the quality of plant products can be improved for the benefit of the population. For the production of hyperaccumulator plants, breeding and transgenic techniques can be used. There is an urgent need to pique scientists' interest in this site so that the production of hyperaccumulator plants can be increased to reduce metal pollution in the environment. More research in this area is needed to achieve better results. More changes in this research can help to understand and improve the mechanism of phytoremediation by hyperaccumulator plants. Only knowledge of soil physical and chemical processes can explain the uptake of heavy metal pollutants. Understanding the translocation of heavy metal pollutants within the plant system also necessitates an understanding of plant physiology. It is an effective technique that requires a lot of attention to become even more beneficial in the future. This technique is expected to become more efficient and commercially viable in the future.

Acknowledgements

All the authors are thankful to Lovely Professional University, Punjab for the consistent support.

Authors' contributions

The chapter has been written by Prasann Kumar, Bhupendra Koul, and Monika Sharma. The collection and study of different reviews are also done by the authors.

Conflict of interest statement

The authors state that they have no interest in conflicts.

References

Ali S, Farooq MA, Yasmeen T, Hussain S, Arif MS, Abbas F, and Zhang G. 2013. The influence of silicon on barley growth, photosynthesis and ultra-structure under chromium stress. *Ecotoxicology and Environmental Safety* 89: 66–72.

Ameen N, Amjad M, Murtaza B, Abbas G, Shahid M, Imran M, and Niazi NK. 2019. Biogeochemical behavior of nickel under different abiotic stresses: toxicity and detoxification mechanisms in plants. *Environmental Science and Pollution Research* 26(11): 10496–10514.

Ampiah-Bonney RJ, Tyson JF, and Lanza GR. 2007. Phytoextraction of arsenic from the soil by *Leersia oryzoides*. *International Journal of Phytoremediation* 9(1): 31–40.

Aybar M, Bilgin A, and Sağlam B. 2015. Removing heavy metals from the soil with phytoremediation. *Artvin Çoruh University Journal of Natural Hazards and Environment* 1: 59–65.

Chaney RL, Broadhurst CL, and Centofanti T. 2010. Phytoremediation of soil trace elements. *Trace Elements in Soils*, 311–352.

Chang TC, You SJ, Yu BS, Chen CM, and Chiu YC. 2009. Treating high-mercury-containing lamps using full-scale thermal desorption technology. *Journal of Hazardous Materials* 162(2-3): 967–972.

Cho-Ruk K, Kurukote J, Supprung P, and Vetayasuporn S. 2006. Perennial plants in the phytoremediation of lead-contaminated soils. *Biotechnology* 5(1): 1–4.

Chutia P, Kato S, Kojima T, and Satokawa S. 2009. Arsenic adsorption from aqueous solution on synthetic zeolites. *Journal of Hazardous Materials* 162(1): 440–447.

Clemens, STEPHAN. 2001. Developing tools for phytoremediation: towards a molecular understanding of plant metal tolerance and accumulation. *International Journal of Occupational Medicine and Environmental Health* 14(3): 235–239.

Cohen MD, Kargacin B, Klein CB, and Costa M. 1993. Mechanisms of chromium carcinogenicity and toxicity. *Critical Reviews in Toxicology* 23(3): 255–281.

de Queiroz Barcelos JP, de Souza Osorio CRW, Leal AJF, Alves CZ, Santos EF, Reis HPG, and dos Reis AR. 2017. Effects of foliar nickel (Ni) application on mineral nutrition status, urease activity and physiological quality of soybean seeds. *Australian Journal of Crop Science* 11(2): 184–192.

Erdei L. 2005. Phytoremediation as a program for decontamination of heavy-metal polluted environment. *Acta Biologica Szegediensis* 49(1-2): 75–76.

Gajewska E, and Skłodowska M. 2009. Nickel-induced changes in nitrogen metabolism in wheat shoots. *Journal of Plant Physiology* 166(10): 1034–1044.

Gajewska E, Bernat P, Długoński J, and Skłodowska M. 2012. Effect of nickel on membrane integrity, lipid peroxidation and fatty acid composition in wheat seedlings. *Journal of Agronomy and Crop Science* 198(4): 286–294.

Gaur A, and Adholeya A. 2004. Prospects of arbuscular mycorrhizal fungi in phytoremediation of heavy metal contaminated soils. *Current Science*, 528–534.

Gautam S, and Pandey SN. 2008. Growth and biochemical responses of nickel toxicity on leguminous crop (Lens esculentum) grown in alluvial soil. *Res Environ Life Sci* 1: 25–28.

Ghosh M, and Singh SP. 2005. A review on phytoremediation of heavy metals and utilization of its by-products. *Asian J Energy Environ* 6(4): 18.

Gupta D, Singh LK, Gupta AD, and Babu V. 2012. Phytoremediation: An efficient approach for bioremediation of organic and metallic ions pollutants. pp. 213–240. *In*: Mohee R, and Mudhoo A (eds.). *Bioremediation and Sustainability: Research and Applications*. Hoboken, NJ: John Wiley & Sons, Inc.

Gusman GS, Oliveira JA, Farnese FS, and Cambraia J. 2013. Arsenate and arsenite: the toxic effects on photosynthesis and growth of lettuce plants. *Acta Physiologiae Plantarum* 35(4): 1201–1209.

Hasegawa H, Rahman MA, Matsuda T, Kitahara T, Maki T, and Ueda K. 2009. Effect of eutrophication on the distribution of arsenic species in eutrophic and mesotrophic lakes. *Science of the Total Environment* 407(4): 1418–1425.

Jacobs JA, and Testa SM. 2005. Overview of chromium (VI) in the environment: background and history. Chromium (VI) handbook, 1–21.

Ji P, Sun T, Song Y, Ackland ML, and Liu Y. 2011. Strategies for enhancing the phytoremediation of cadmium-contaminated agricultural soils by *Solanum nigrum* L. *Environmental Pollution* 159(3): 762–768.

Karami A, and Shamsuddin ZH. 2010. Phytoremediation of heavy metals with several efficiency enhancer methods. *Afr J Biotechnol* 9: 3689–3698.

Koul B, and Taak P. 2018. Biotechnological Strategies for Effective Remediation of Polluted Soils. Springer. pp. 1–240.

Lasat MM. 1999. Phytoextraction of metals from contaminated soil: a review of plant/soil/metal interaction and assessment of pertinent agronomic issues. *Journal of Hazardous Substance Research* 2(1): 5.

LeDuc DL, and Terry N. 2005. Phytoremediation of toxic trace elements in soil and water. *J Ind Microbiol Biotechnol* 32: 514–520.

Memon AR, Aktoprakligil D, Özdemir A, and Vertii A. 2001. Heavy metal accumulation and detoxification mechanisms in plants. *Turkish Journal of Botany* 25(3): 111–121.

Mirsal Ibrahim A. 2004. Soil pollution: origin, monitoring, and remediation, IstEd. Germany, Springer 1: 5–11.

Mohan D, and Pittman Jr. CU. 2007. Arsenic removal from water/wastewater using adsorbents—a critical review. *Journal of Hazardous Materials* 142(1-2): 1–53.

Mukhopadhyay S, and Maiti SK. 2010. Phytoremediation of metal-enriched mine waste: a review. *Global J Environ Res* 4: 135–150.

Musselman JF. 2004. Sources of Mercury in Wastewater, Pretreatment corner. http://www.cet-inc.com/cmsdocuments//7%20-%20Sources%20of%20in%20Wastewater% 20.pdf.

Naees M, Ali Q, Shahbaz M, and Ali F. 2011. Role of rhizobacteria in phytoremediation of heavy metals: an overview. *Int Res J Plant Sci* 2: 220–232.

National Ground Water Association. 2001. Copyright 2001. Arsenic. What you need to know http://www.ngwa. org/ASSETS/A0DD107452D74B33AE9D5114EE6647ED/Arsenic. pdf.

National Risk Management Research Laboratory (US). 2000. Introduction to phytoremediation. National Risk Management Research Laboratory, Office of Research and Development, US Environmental Protection Agency.

Negri MC, Hinchman RR, and Gatliff EG. 1996. Phytoremediation: using green plants to clean up contaminated soil, groundwater, and wastewater (No. ANL/ES/CP-89941; CONF-960804-38). Argonne National Lab., IL (United States).

Paschal DC, Burt V, Caudill SP, Gunter EW, Pirkle JL, Sampson, EJ, Miller DT, and Jackson RJ. 2000. Exposure of the US population aged 6 years and older to cadmium: 1988–1994. *Archives of Environmental Contamination and Toxicology* 38(3): 377–383.

Pehlivan E, Özkan AM, Dinç S, and Parlayici Ş. 2009. Adsorption of Cu2+ and Pb2+ ion on dolomite powder. *Journal of Hazardous Materials* 167(1-3): 1044–1049.

Pivetz BE. 2001. Ground Water Issue: Phytoremediation of Contaminated Soil and Ground Water at Hazardous Waste Sites pp. 1–36.

Pollard AJ, Powell KD, Harper FA, and Smith JAC. 2002. The genetic basis of metal hyperaccumulation in plants. *Critical Reviews in Plant Sciences* 21(6): 539–566.

Porter JR, and Sheridan RP. 1981. Inhibition of nitrogen fixation in alfalfa by arsenate, heavy metals, fluoride, and simulated acid rain. *Plant Physiology* 68(1): 143–148.

Pulford ID, and Watson C. 2003. Phytoremediation of heavy metal-contaminated land by trees—a review. *Environment International* 29(4): 529–540.

Rakhshaee R, Giahi M, and Pourahmad A. 2009. Studying the effect of cell wall's carboxyl–carboxylate ratio change of Lemna minor to remove heavy metals from aqueous solution. *Journal of Hazardous Materials* 163(1): 165–173.

Ramamurthy AS, and Memarian R. 2012. Phytoremediation of mixed soil contaminants. *Water Air Soil Pollut* 223: 511–518.

Raskin I, Smith RD, and Salt DE. 1997. Phytoremediation of metals: using plants to remove pollutants from the environment. *Current Opinion in Biotechnology* 8(2): 221–226.

Rodriguez L, Lopez-Bellido FJ, Carnicer A, Recreo F, Tallos A, and Monteagudo JM. 2005. Mercury recovery from soils by phytoremediation. *In:* Lichtfouse E, Schwarzbauer J, and Robert D (eds.). *Environmental Chemistry.* Springer, Berlin, Heidelberg.

Roy S, Labelle S, Mehta P, Mihoc A, Fortin N, Masson C, Leblanc R, Châteauneuf G, Sura C, Gallipeau C, Olsen C, Delisle S, Labrecque M, and Greer CW. 2005. Phytoremediation of heavy metal and PAH-contaminated brownfield sites. *Plant and Soil* 272(1/2): 277–290.

Royer MD, and Smith LA. 1995. Contaminants and remedial options at selected metals contaminated sites-a technical resource document (No. CONF-9504110-). *Environmental Protection Agency.* Cincinnati, OH (United States).

Salido A L, Hasty KL, Lim JM, and Butcher DJ. 2003. Phytoremediation of arsenic and lead in contaminated soil using Chinese brake ferns (*Pteris vittata*) and Indian mustard (*Brassica juncea*). *International Journal of Phytoremediation* 5(2): 89–103.

Salt DE, Smith RD, and Raskin I. 1998. Phytoremediation. *Annual Review of Plant Biology* 49(1): 643–668.

Saraswat S, and Rai JPN. 2011. Complexation and detoxification of Zn and Cd in metal accumulating plants. *Reviews in Environmental Science and Bio/Technology* 10(4): 327–339.

Sas-Nowosielska A, Galimska-Stypa R, Kucharski R, Zielonka U, Małkowski E, and Gray L. 2008. Remediation aspect of microbial changes of plant rhizosphere in mercury-contaminated soil. *Environmental Monitoring and Assessment* 137(1): 101–109.

Satarug S, Baker JR, Urbenjapol S, Haswell-Elkins M, Reilly PE, Williams DJ, and Moore MR. 2003. A global perspective on cadmium pollution and toxicity in non-occupationally exposed population. Toxicology Letters 137(1-2): 65–83.

Sharma P, and Dubey RS. 2005. Lead toxicity in plants. *Brazilian Journal of Plant Physiology* 17(1): 35–52.

Shtangeeva I, Laiho JVP, Kahelin H, and Gobran GR. 2004. Phytoremediation of metal-contaminated soils. Symposia papers presented before the division of environmental chemistry. In American Chemical Society, Anaheim, Calif, USA, http://ersdprojects. science. doe. gov/workshop% 20pdfs/ California (Vol. 202004, p. p050).

Singh A, and Prasad SM. 2011. Reduction of heavy metal load in the food chain: a technology assessment. *Reviews in Environmental Science and Bio/Technology* 10(3): 199–214.

Singh N, Ma LQ, Vu JC, and Raj A. 2009. Effects of arsenic on nitrate metabolism in arsenic hyperaccumulating and non-hyperaccumulating ferns. *Environmental Pollution* 157(8-9): 2300–2305.

Sinha RK, Herat S, and Tandon PK. 2007. Phytoremediation: role of plants in contaminated site management. In Environmental bioremediation technologies (pp. 315–330). Springer, Berlin, Heidelberg.

Söğüt Z, Zaimoğlu BZ, Erdoğan R, and Doğan S. 2002. Su Kalitesinin Arttırılmasında Bitki Kullanımı (Yeşil Islah-Phytoremediation), Türkiye'nin Kıyı ve Deniz alanları IV. Ulusal Konferansı, 5–8.

Sreekanth TVM, Nagajyothi PC, Lee KD, and Prasad TNVKV. 2013. Occurrence, physiological responses and toxicity of nickel in plants. *International Journal of Environmental Science and Technology* 10(5): 1129–1140.

Thakur IS. 2006. Environmental Biotechnology. IK International, New Delhi.

Tong YP, Kneer R, and Zhu YG. 2004. Vacuolar compartmentalization: a second-generation approach to engineering plants for phytoremediation. *Trends Plant Sci* 9: 7–9.

US Department of Health and Human Services. 2005. Public health service, agency for toxic substances and disease registry. *Toxicological Profile for Tin and Tin Compounds* pp. 1–376.

Vaclavikova M, Gallios GP, Hredzak S, and Jakabsky S. 2008. Removal of arsenic from water streams: an overview of available techniques. *Clean Technologies and Environmental Policy* 10(1): 89–95.

Vangronsveld J, Herzig R, Weyens N, Boulet J, Adriaensen K, Ruttens A,Thewys T, Vassilev A, Meers E, Nehnevajova E, van der Lelie D, and Mench M. 2009. Phytoremediation of contaminated soils and groundwater: lessons from the field. *Environmental Science and Pollution Research* 16(7): 765–794.

Vanlı Ö. 2007. Pb, Cd, B Elementlerinin Topraklardan Şelat Destekli Fitoremediasyon Yöntemiyle Giderilmesi (Doctoral dissertation, Fen Bilimleri Enstitüsü).

Wang FY, Lin XG, and Yin R. 2007. Inoculation with arbuscular mycorrhizal fungus *Acaulospora mellea* decreases Cu phytoextraction by maize from Cu-contaminated soil. *Pedobiologia* 51(2): 99–109.

Wu F, Zhang G, Dominy P, Wu H, and Bachir DM. 2007. Differences in yield components and kernel Cd accumulation in response to Cd toxicity in four barley genotypes. *Chemosphere* 70(1): 83–92.

Yan A, Wang Y, Tan SN, Yusof MLM, Ghosh S, and Chen Z. 2020. Phytoremediation: a promising approach for revegetation of heavy metal-polluted land. *Frontiers in Plant Science*, 11.

Yusof AM, and Malek NANN. 2009. Removal of Cr (VI) and As (V) from aqueous solutions by HDTMA-modified zeolite Y. *Journal of Hazardous Materials* 162(2-3): 1019–1024.

Yusuf M, Fariduddin Q, Hayat S, and Ahmad A. 2011. Nickel: an overview of uptake, essentiality and toxicity in plants. *Bulletin of Environmental Contamination and Toxicology* 86(1): 1–17.

Index